INFRASTRUCTURE DESIGN, SIGNALLING AND SECURITY IN RAILWAY

Edited by **Xavier Perpinya**

Infrastructure Design, Signalling and Security in Railway
Edited by Xavier Perpinya

Published by InTech
Janeza Trdine 9, 51000 Rijeka, Croatia

Publishing Process Manager Sasa Leporic

Technical Editor Teodora Smiljanic

Cover Designer InTech Design Team

Additional hard copies can be obtained from orders@intechopen.com

Infrastructure Design, Signalling and Security in Railway, Edited by Xavier Perpinya
p. cm.
ISBN 978-953-51-0448-3

For EU product safety concerns:
IN TECH d.o.o., Prolaz Marije Krucifikse Kozulić 3, 51000 Rijeka, Croatia,
info@intechopen.com or visit our website at intechopen.com.

Meet the editor

Dr Xavier Perpiñà was born in Almenar, Spain, in 1976. He received the B.S. degree in physics, the M.Phil. degree in electronic engineering, and the Ph.D. degree from the Universitat Autònoma de Barcelona, Bellaterra, Spain, in 1999, 2002, and 2005, respectively. In 1999, he was with the Institut de Microelectrònica de Barcelona-Centre Nacional de Microelectrònica (IMB-CNM), Spanish Research Council (CSIC), Bellaterra, Spain, where he worked in the clean room and, then until 2005, he began his research activity with the Power Devices and Systems Group, IMB-CNM. From 2005 to 2007, he was with Alstom Transport, where he developed studies on thermal management and power-converters reliability. He is currently a Contracted Researcher with IMB-CNM and his research deals with thermal investigations and reliability studies in power devices and packaging. He has authored and coauthored more than 60 research papers in international conferences and refereed journals. He also belongs to the scientific committee of EUROSIME conference and THERMINIC workshop.

Contents

Preface

Railway transportation has become one of the main technological advances of our society. Since the first railway system used to carry coal from a mine in Shropshire (England, 1600), a lot of efforts have been made to improve this transportation concept. One of its milestones was the invention and development of the steam locomotive, but commercial rail travels became practical two hundred years later. Currently; electric railway traction chains have become a better solution than the traction systems with generating power on board (e.g., diesel or steam-based systems). This could not be possible without the advances experienced throughout the years in power electronics, mechanics and materials engineering. In terms of performances, ERTCs show the highest power-to-weight ratio, fastest acceleration and highest traction effort on steep gradients of the railway traction scenario. Other of their advantages includes less noise, lower maintenance requirements of the traction units, and a higher rational use of energy respecting and preserving the environment (e.g., energy harvesting systems as regenerative brakes or no greenhouse gasses' emissions). Obviously, their main disadvantages are the capital cost of the electrification line, depending on a trade-off between the distance and traffic volume of the service line.

In fact, the evolution of railway transportation could not be possible without the simultaneous growth of railway infrastructures, signalling and security. They are responsible for supporting, controlling and coordinating railway traffic. The high number of railway commercial lines around the most important cities in the world, as well as the requirements of the current business market, has made of them a key factor for the development of many commercial activities. Obviously, their design and characterization is not easy at all, becoming a much more complex procedure than those performed in the railway earlier stages.

This book provides the reader an overview of railway systems from several countries, some details on modelling for railway infrastructure design and characterisation, and finally the implementation of signalling procedures, security protocols and infrastructures. Besides, it reports on research progress on these issues. During the preparation of this book, I asked the authors to add recent research findings and future works in this area and cite latest references in the chapter. For this reason, a variety of novel approaches in the covered topic are detailed in this book. Insightful and reader-friendly descriptions are presented to nourish readers of any level, from practicing and knowledgeable electrical engineers to beginning or professional researchers. All interested readers can easily find noteworthy materials in much greater detail than in previous publications and in the references cited in these chapters. This book includes twenty chapters that were authored by world-wide well-known researchers. Each chapter was written in an introductory style beginning with the fundamentals, describing approaches to the hottest issues and concluding with a comprehensive discussion. The content in each chapter is taken from many publications in prestigious journals and conferences and followed by fruitful insights. The chapters in this book also provide many recent references for relevant topics, and interested readers will find these references helpful when exploring these topics in further detail.

This book is divided into three parts. Part 1 consists of six chapters devoted to describe how the railway systems have been developed in several countries and their socio-economical impact. Part 2 consists of seven chapters devoted to providing some ideas on safety and reliability issues. Finally, part 3 consists of seven chapters devoted to parameters monitoring in railway scenario for safety and reliability purposes.

We hope that this book will fulfill the need for publication on infrastructure design, signalling, and security in railway, as well as being useful for engineers and scientists interested in learning about or developing any system related to this topic. Furthermore, this can be used as a text book for engineering advanced undergraduate and graduate students interested in learning about the topics raised in this book.

Xavier Perpinya
Institut de Microelectronica de Barcelona,
Campus Universitat Autónoma de Barcelona, Barcelona,
Spain

Part 1

Railway Systems in the World

1

The Role of Light Railway in Sugarcane Transport in Egypt

Hassan A. Abdel-Mawla
Department of Ag. Engineering, Al-Azhar University, Assiut, Egypt

1. Introduction

The first section of the Egyptian standard railway for public transport service started at 1854. Fifteen years later, the first light railway network established to serve sugar industry southern Egypt. A light railway network initiated through the area considered for sugarcane production whenever a modern sugar mill established. The light railway represented the mechanism that continuously convey and feed each sugar factory with raw material of sugarcane produced in wide farm areas around the mill.

As a principle transport system, the light railway networks started transport service simultaneously with the beginning operation of each sugar mill. Whenever a modern sugar mill constructed, a light railway net established for its own cane transport service. The first light railway network started service at the west bank of Nile at 1869 when the first modern sugar mill started operation at Armant (Ar. 691 *km* south Cairo). At 1896 the second oldest light railway was initiates at the west bank of Nile to serve cane transport to Nagaa-Hamadi factory (N. H. 553 *km* south Cairo). At the early stage of the 20th century, two light railway networks started cane transport service in Abo-Qurkas (AQ. 267 *km* south Cairo) in 1904 and in Kom-Ombo (KO. 834 *km* south Cairo) in 1912 when two sugar factories begin operation at these two locations. Other four light railway networks were established within the period from 1963 to 1987 in Edfo (Ed. 776 *km* south Cairo), Quse (Qu. 573 *km* south Cairo), Dishna, (Di. 573 *km* south Cairo) and Gerga (Ge. 502 *km* south Cairo) when the sugar mills started there **(Afifi 1988).**

Based on the data of the annual report of the Sugar Counsel 2010 and former reports, continuous change of the role of narrow railway system has been recorded over the last two decades. Figure 1 shows the development of the light railway system contribution to the transport of vegetative cane delivered as row material to sugar industry. Road transport strongly competes as cane transport mean due to constant improvement of infield roads and the availability of road vehicles. On the other hand, the decline of the narrow railway system contribution may partially refer to the expansion of cane plantations outside the light railway net. The chapter discusses the existing conditions and the expected future of the role of light railway initiated for cane transport in Egypt. Alternative road transport vehicles may replace the narrow railway because of availability in addition to transport cost. It seems like the conditions of narrow railway system of cane transport in Egypt has some similar aspects of that of South Africa as reported by **Abdel-Mawla (2001)**. **Malelane (2000)** concluded that the economics of each cane transport system establish the optimum mix of transport mode in South Africa. The availability of road transport given the limitations of fixed rail siding placement and infield haulage distances.

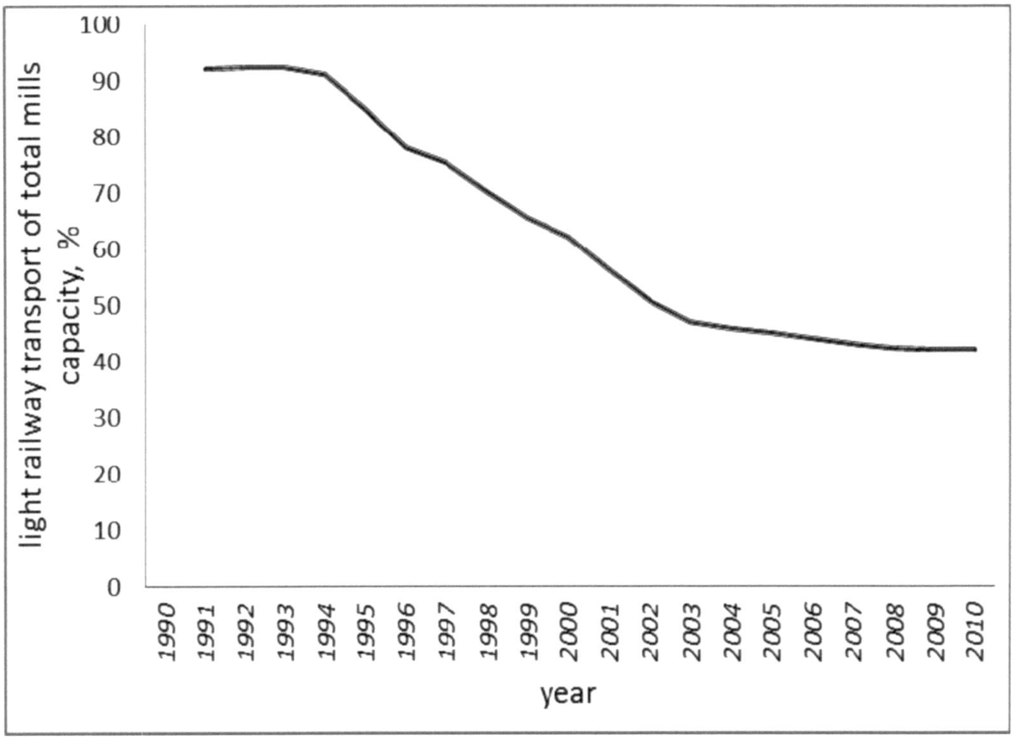

Fig. 1. Development of the role of light railway system for feeding row materials to sugar industry.

2. Light railways line expansion

The narrow railway network and whole stalk wagons represented the principle cane delivery system especially for the old constructed sugar industry. The regions at which the narrow railway expanded for sugarcane transport occupy continuous areas along both sides of the Nile. Sugar factories located at the both Nile banks where narrow railway and whole stalk wagons receive the cane transported cross Nile by the help of a crane at certain ports. The railway lines started at the back and side gates of the sugar mill and branched along the infield roads through the cane production area. The main narrow railway lines near the mill gates include several grand unions and large number of switches.

Over 2200 km of the narrow gauge railways expanded to maintain feeding sugar industry with the raw materials that represented in sugarcane produced from the adjacent areas on both sides of the Nile as shown on the map Figure (2).

Fig. 2. Light railway expansion areas in the Nile Valley.

Infield roads on which the narrow railway lines constructed may be expanded on a side of an irrigation or drainage channel may cross several bridges and may cross the main railway line of Upper Egypt. Double light railway lines may be expanded on the main roads to maintain easy motion of cane trains travel to or coming from several infield lines connected to the main lines by unions. The sub branches of the narrow railways may be double lines that include a main rail line on which the loaded train move and an auxiliary line for the travel of empty train coming from the mill. This arraignment of auxiliary rail line for the travel of empty train may be limited to certain locations to maintain smooth motion on the light railway lines. Infield railway lines are single lines on which a train moves either empty or loaded.

Figure 3 shows a map of the second oldest narrow railway network (1896) that established to feed Naga-Hammady mill with row cane. The 115 years old cane transport narrow railway network of 420 km long still efficiently working by the help of seasonal maintenance. In this particular region, the contribution of narrow railway transport system may currently exceed 60% of the daily mill capacity.

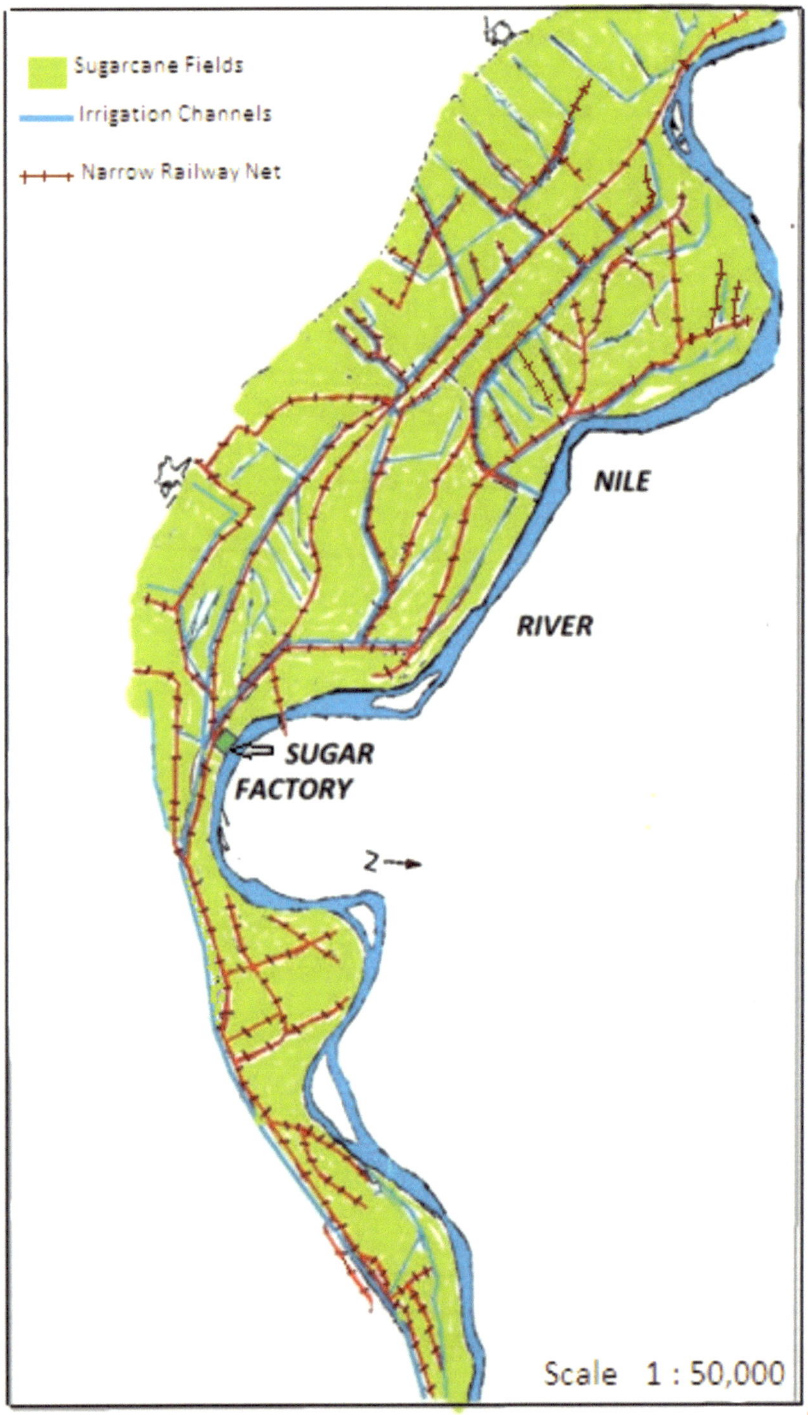

Fig. 3. A map of light railway network of NH. sugar factory established 1896 (Courtsy, Sugar & Int. Idustry Company)

3. Light railway system transport elements

3.1 Light railway lines

The narrow railway lines established for cane transport initiated with similar gauge of 2 feet (61 cm) that represent the inside width between the rails (Figure 4). The narrow track sleepers are fabricated from cold formed steel plates of 2 m width. The ballast-less narrow track constructed by arranging the steel sleepers 0.75 to 1 m apart directly on the road soil (Figure 5). The two feet spaced rails are fixed to the sleepers with bolts and clamps.

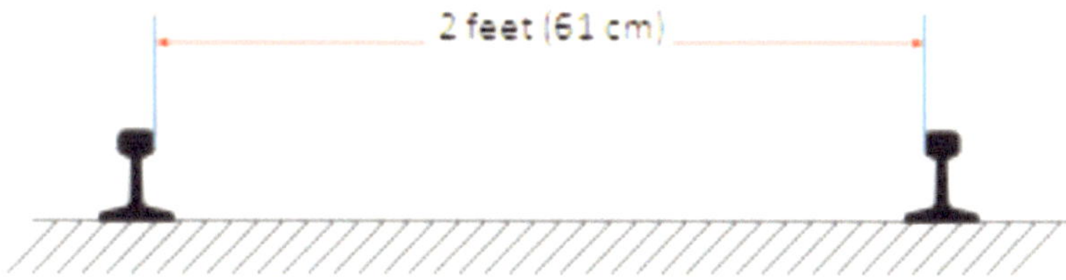

Fig. 4. Size (gauge) of the Egyptian light railway for cane transport

Fig. 5. A balastless narrow railway expanded on a bank of an irrigation channel

Fishplates are used to connect the ends of rails along the track. A short space is left between the ends of the rails for thermal expansion. Since this sort of rail lines are ballast-less expanded on dirt roads with considerably wide interval between sleepers, the alignment of the rail ends at the point of fishplate connection is not always secured. To overcome the probable vertical misalignment at the expansion gap, a short single bolt rail plate is used. Figure 6 show the single bolt alignment short rail piece. Whenever the train is coming from any direction, the near end of the plate is aligned to the end of the rail, carrying the train wheel and turn around the pin to be aligned to the front rail. This simple arrangement largely reduces hard sudden impact, reduces rapid wear and breakdowns of the rail wagons undercarriage.

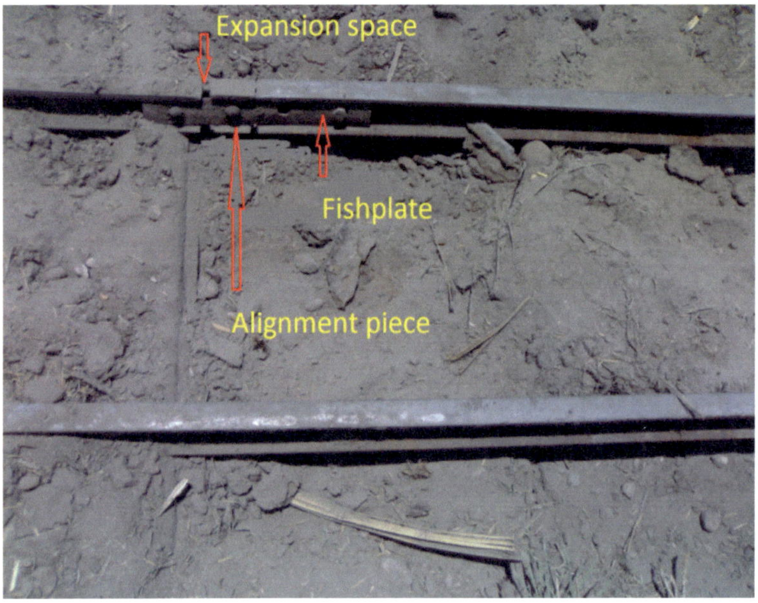

Fig. 6. Rail ends connection

3.2 Light railway locomotives

Variable sizes of locomotive are available to pull the light railway sugarcane train. Locomotives of variable types have been imported mainly from Germany, Romania, Japan, and Slovakia. Based on the statistics, old and new German and Romanian types represent the major numbers of locomotives belong to the sugarcane transport system. The sizes and function of locomotives of the narrow railway cane transport system may be:

- Locomotives of 250 hp and more have been used to work on the main narrow railway lines. Most of these locomotives are of 350 hp operated to pull the empty wagons to the field and pull back the loaded train to the mill. Experienced operators have been employed to drive such locomotives to ensure the train travel safety. The locomotive driver should be memorizing the location of large number of switches and be sure each is switched to the proper direction on his way either to the field or back to the mill. The driver should also be aware about the location of infield roads which the train crosses. A person is assigned to help the driver during the trip. Figure 7 shows one of the narrow gauge railway locomotives of 350 hp that used to pull the sugarcane train.

- Locomotives of power from range from 150 to 250 hp are used to pull the loaded rail wagons inside the mill yard. Such locomotives are used for pulling the wagons for weighing and for unloading. Intensive maneuvering operations may be required inside the mill yard to move the loaded wagons toward the unloading line. The small locomotives also used to clear the discharged wagons from the yard to the departure lines to save more room for the trains coming from the fields.
- Locomotives of power less than 150 hp are used to move the unloaded wagons away from the unloading line. These types of locomotives perform a lot of maneuvering operations inside the mill yard to collect the empty wagons and to move the empty train to the departure line.

Fig. 7. A narrow gauge locomotive on the way to the field

3.3 Light railway whole-stalk cane wagons

Since all the sugar factories followed one company, the light railway wagons fabricated for cane transport size variation is very limited. The wagons designed to be whole stalk loaded parallel to the longitudinal axel of the wagon. Unlike the Australian cane bins described by **Lynn (2008)** show large variation of wagons size that carry chopped cane.

The wagon has two bogies each of four steel wheels on which a rectangular steel flat surface is fixed. Steel columns are bolted vertically to the outer side of the rectangular flat surface that form a basket that hold cane parallel to the longitudinal axle of the wagon. The ground clearance to the bottom surface of the wagon around 60 cm. The wheel diameter may be 32 cm from the flange side and 24 cm from the wheel trade side. The light railway wagon flat load surface may be 6 to 7 m in length and 1.5 to 1.8 m in width and the side columns are 1.4 to 1.6 m in height. Wheel base from the center of the rear wheel of the rear bogie to the centre of front wheel of the front bogie ranged from 5 to 5.7 m.

The loading volume inside the wagon may be ranged from 14 to 18 m^3. The cane is loaded parallel to the longitudinal axle of the wagon. The load may be expanded up to 1 m over the wagon side columns to permit higher capacity of the wagon.

Transverse steel channels welded to the loading surface of the wagon to permit passing the chains under the load while unloading the wagon in the mill. Figures (8) and (9) shows isometric and projection drawings of the light railway cane transport wagon.

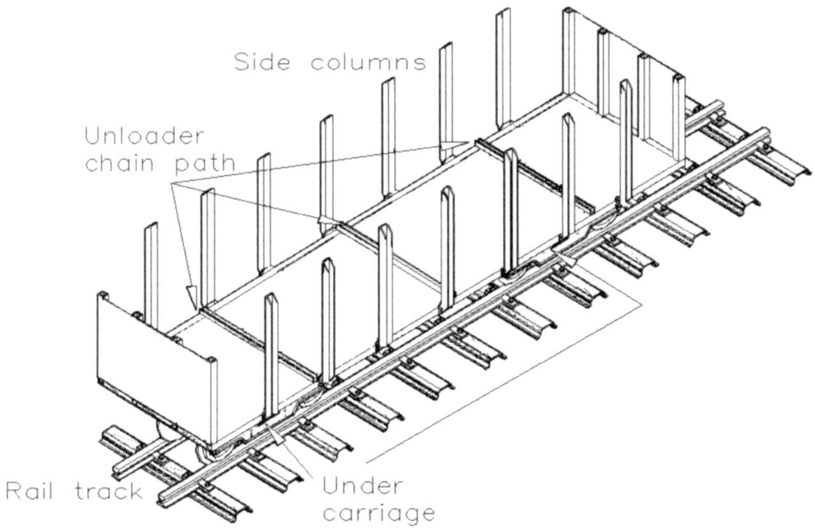

Fig. 8. Isometric of the cane transport light rail-wagon

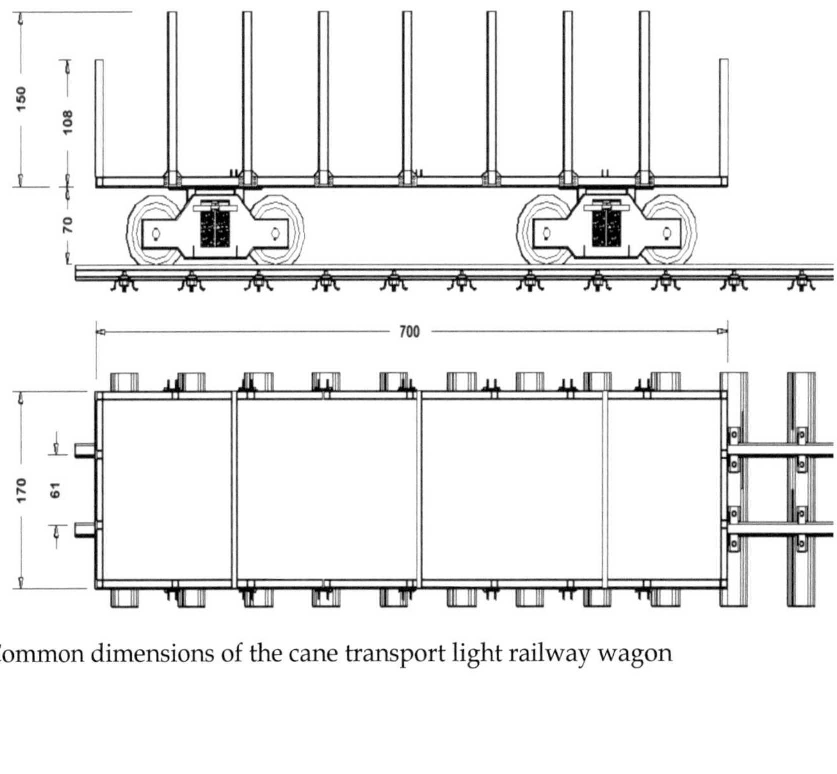

Fig. 9. Common dimensions of the cane transport light railway wagon

3.4 Light railways system operation schedule

The principle objectives of railway wagons operation schedule may include:

1. To secure uniform diurnal arrival of the current of railway wagons to the mill.
2. To reduce the probability of loaded wagons delivery delay.
3. To face the overload transport due to accidental conditions.
4. To secure overnight operation of the mill.

Figures 10 and 11 show the trains while transporting sugarcane. The operation of the light railway system for cane supply to the mill has to be performed according to a pre-defined schedule. The mill seasonal operation period should be approximately estimated based on the daily capacity of the mill, cane production area and average production of the unit area. The average data of the recent juicing seasons would be helpful in that concern.

The size of the railway wagons fleet required for a sugar mill may be determined according to variable conditions. The mill daily capacity represents the total mass of raw materials has to be supplied to the mill around 24 hours. Row cane delivery Schedule plan should determine the quantities of sugarcane to be transported by road vehicles. General estimation of the average rail wagon capacity should be estimated based on the past season data. Cycle time of the rail wagon transport trip should also be clear and specified. In addition to several other factors related to harvesting, infield transport and loading, the rate of the rail wagons breakdowns occurred during the season should be considered.

The rate of row materials delivered by the light railway wagons around the day should be managed by the mill administration to reduce the waiting time at the unloading queue. The mill administration may have to consider the following steps to estimate the numbers of the rail wagons, pull locomotives and operation team around the day:

- The labor operation is arranged into three shifts which are; morning shift that last from 7 am to 3 pm, evening shift from 3 pm to 11 pm and night shift from 11 pm to 7 am.
- Road transport is limited to the diurnal period and vehicles may continue arrive to the unloading queue till the evening. Therefore the supply of road transport may be limited to the morning and evening shifts.
- Supply of sugarcane row materials to the mill during the night shift depend mainly on the light railway system.
- Diurnal operation of the light railway wagons should be considered to secure the shortage of road transport supply to maintain continuous operation of the mill.
- The operation of the light railway wagons is arranged as diurnal and night fleets. The number of railway wagons required for diurnal operation and those required for overnight operation should be estimated. Wagon/s with certain card number/s assigned to transport the cane of certain farmer. A locomotive pulls the empty train to certain region and then pulls the loaded wagons back at specific time.
- Specific time duration is determined for mill yard departure and arrival of each of the diurnal operated trains and the overnight operated trains.

Androw and Ian (2005) reported that several mill regions within the Australian sugar industry are currently exploring long-term scenarios to reduce costs in the harvesting and rail transport of sugarcane. These efficiencies can be achieved through extending the time window of harvesting, reducing the number of harvesters, and investing in new or upgraded infrastructures. As part of a series of integrated models to conduct the analysis, we developed a capacity planning model for transport to estimate the (1) number of locomotives and shifts required; (2) the number of bins required; and (3) the delays to harvesting operations resulting from harvesters waiting for bin deliveries. The schedule developed to operate the Egyptian system may have similar objectives **(Abdel-Mawla 2011)**. For example, the second oldest sugar mill (N. H.) started operation in 1896, the light railway system used to transport almost 100% of the cane delivered to the mill. At present, the light railway wagons deliver only 50% of the mill daily capacity. The mill holds the most long light railway network (410 km) expanded through the cane fields. The mill also has 1700 light railway wagons ready for operation. Large amount of field data concerning crop, field, environment and labors required for the proper design of the light railway operation schedule. Concerning the determination of the rail wagon numbers, the basic data presented in Table 1 may be necessary.

Item	Value
Mill capacity	= 1.7 million ton/season
Estimated season duration	= 140 days
Daily supply	= 12000 tons/day, approximately
Required hourly supply	= 500 tons/h
Average rail wagon load	= 9 tons

Table 1. Basic data required to estimate the number of light rail wagons.

Table 2 presents estimation of the narrow railway wagons fleet size required to secure adequate supply of the mill daily capacity of cane row materials.

Shift		Shift duration	Required cane supply ton	Light railway contribution		Required wagons	Departure time		Return time		Mill yard waiting
				%	ton		From	To	From	To	h
Diurnal	Morning	7 am 15 pm	4000	50	2000	223	7 am	16 pm	12 am	19 pm	6-10
	Evening	15 pm 23 pm	4000	10	400	45					
Night		23 pm 7 am	4000	90	3600	400	19 pm	00 am	23 pm	6 am	10

Table 2. Estimation of the railway wagon fleet size

The efficiency of the narrow railway cane transport system may be largely improved by reducing transport cycle time as follow:

- Reducing the time of the loaded wagon waiting in the mill yard.
- Mechanize cane loading operation.
- Improve the rail line management related to switches and signalling system.

Fig. 10. A narrow rail train is loaded with cane and ready for pull

Fig. 11. Train loaded with on the way back to the mill

4. Light railway wagons loading and unloading

4.1 Loading

The cane transport administration of the mill distributes the empty light railway wagons according to the schedule. The driver of the locomotive leaves the wagons in the trans-loading site scheduled for cane delivery. Farmers bring the cane from inside fields to the location at which the wagons loaded. The common activity is to start loading the wagons in the morning. Loading may be done manually or mechanically according to the availability of mechanical loaders.

4.2 Manual loading

The light railway of loading surface 60 to 70 cm high from the ground surface may be loaded manually (Figure 12). Two labors start carrying cane bundles, climb a ladder and place them inside the wagon. Even though the manual loading is considered adverse operation, it may permit some important advantages to obtain a higher density load such as:

- The labor loaders may fit the cane bundles tightly to ensure efficient use the whole volume of the wagon.
- The labor loading may permit employing a knifeman who is working over the wagon to cut the curved parts of the cane Figure 12. After the labor place the cane bundle in the loading area, the knife man cut the uneven parts of the cane stalks to facilitate higher density load. This activity may be important specially if the cane is taken from a lodged field.
- The labor loading may also permit a better chance to expand the load by force fit vertical columns of cane stalks when the load level become over the steel side columns of the wagon.

Fig. 12. Labor loading of rail-wagons with sugarcane

4.3 Mechanical loading

Few mechanical cane loaders were available till the Aswan Mechanization Company established at 1980. At that time large number of Bell type cane loaders imported and operated. Even though, the company stops purchasing new loaders and the majority of their loaders become old, the farmers bought those old loaders, rebuild them and bring them to operation again **(Abdel-Mawla 2010)**. Recently, other tractor mounted loaders may be locally developed and operated for cane loading. Figure (13) shows mechanical loading of the light rail-wagons using a tractor mounted loader developed by the author 2011.

Light railway system also designed to handle the cross Nile transported cane. The system depends on the similarity in design and size of the cane holding bins fixed on the ship to that of the light railway wagons. Actually, light rail wagon frames fixed on the ship each of them have certain code number. Farmers load their cane each in certain frames on the ship. After the ship load complete, it travels across Nile to the unloading crane. A light railway line passes opposite to the crane. The crane lift the load conserving its dimension and structure and place it into a wagon (taking the same code number) waiting on the rail line. As soon as the rail wagon receives the load it pulled away waiting for pull to the mill. Another light rail wagon pulled to the crane loading area as indicated in Figure (14).

The mechanism of unloading the light rail-wagons to the mill conveyor may vary from mill to another. A crane that carry the loaded wagon up then inverse the wagon to discharge the load over the conveyor may be found in Kom-Ombo mill. The empty wagon then returned back to the rail line and pulled away to give the chance to another wagon to be unloaded. The other common unloading mechanism may include a crane that left the wagons load with help of chains and place it over the conveyor. The unloaded wagon then moved and another one advanced toward the crane. Figure 15 show the chain un-loader which commonly used in sugar mills.

Fig. 13. Mechanical loading of light railway wagons using a tractor mounted loader (developed by the author).

Fig. 14. Light railway system handle cane transported cross Nile

Fig. 15. Unloading light rail wagons.

5. Light railway problems

5.1 Problems related to rail track wear out

According to **Abdel-Mawla (2000)**, the narrow railway network faces breakdown problems due to the wear of long parts of the rail track. Currently, the light railway transport around 40% of the total cane delivered to sugar mills as general average for the eight sugar mills.

As previously explained some of the light railway systems started about 140 years ago. The old narrow railway expanded on the infield roads have been facing problems of steel components worn out. In spite of continuous seaseonal maintenance, the railway network have several corroded parts. Some of the narrow railway tracks constructed on the clay soil of the infield roads which is in the same level of the neighbor fields. The sleepers, bolts, fishplats and other parts of the rail track gradually covered by the road dirt. Moisture of underground water as well as moisture infeltrated from irrigation water may reach the rail track. The clay soil preservs moisture around the buried track causing intensive rust of th steel parts.

In the routin maintenance, the labors uncover the rail and change the wear-out parts that are easily to descover. Figure 16 show the rusted steel sleepers of the narrow railroad track.

Some parts of the old light railway network may become out of service because of the intensive breakdowns due to wear out. In most cases the track should be completely replaced otherwice several accedents expected due to loaded wagons turn a side or track climb where intensive losses may be occurred . Whenever such accedents repeated, the farmers abstain from transporting by the light railway and go for road transport even though it is more costly.

Fig. 16. Balastless narrow rail track showing intensive ruste of sleepers buried in the clay soil

5.2 Problems related to system operation

The light railway system employed for cane transport may be considered a slow system where the loaded vehicle wait for long time to be pulled back to the mill. The empty rail wagons distributed to several fields by a distribution locomotive. After these wagons been loaded with cane, the distribution locomotive move them to certain location where the stuff responsible for the train operation attach the loaded rail wagons together. The train loaded with can attached to the pull locomotive and start move back to the mill. Therefore it may last for long time before the train reach the mill. Actually, the train has to travel at limited speed (10-15 km/h) to avoid the accedents may occure at the intersections of the railroad and infield roads. Also train has to stop at the railroad switches where the locomotive driver or his helper has to swich it himself. The railroad swiches may be abused by young farmers, so that the locomotive driver himself should be sure about its position before cross.

The longer duration from the time of loading to the time of weighing the wagon load in the mill is critical for the farmer. The moisture losses from the vegitative load may be of high rate specially in such hot dry weather. Science the mony value of the load will be determined according to its weight, farmers may prefer to go for faster transport system to avoid vehicle load weight losses.

5.3 Problems related to farmers behaviour and road conditions

Some other railroad tracks may be constructed on the irrigation channel banks. In such cases, water pipes passes under the railroad track to convey irrigation water from the channel to the field. Intensive soil erosion may occurred under the rail track because of the repeated activities while opening and closing the irrigation pipes as shown in Figure17. Some other farmers may park their animals on the railroad whenever they are out of the season which may be a reason of soil erosion under the railroad and/or the loosen of the track and sleepers. In several cases, the narrow railroad trak expanded on the same infield road on which farmers, animals and equipment move. Therefore, some parts of the track may be covered with dirt Figure 18. Also, some equipment drivers do not maintain the safety of the narrow railroad track while moving.

Fig. 17. Soil Erosion under the railroad.

The narrow railroad has to be doublicated at several locations. The main track is for the loaded train travel from the field to the mill. The auxilary track established at certain locations for the travel of the empty train coming from the factory to the field. The additional track also maintain the manuver of the locomotives while collecting the loaded wagons together and the manuver of the pull locomotive to turn in front of the loaded train before pulling it back to the mill. It has been observed that intensive herbs may grow on the auxiliary patrs of the railroad Figure 19. Intinsive herbs may cause wagon wheels climb off the track.

Fig. 18. The narrow railroad constructed on the middle of an infield road with parts covered with dirt

Fig. 19. Herbs intensively grow and harm the auxiliary railroad

6. Light railways transport system maintenance

6.1 Equipment maintenance

Routine maintenance of Locomotives has been continuously done during the operation season. After the operation season end (in June), seasonal inspection of the locomotives started at the mill workshop. Important repair should be accomplished to make the total locomotive power ready before the next operation season start at the end of December. Some locomotives purchased during 1960's still working by the help of continuous maintenance and repair. The rail wagons maintenance also take place at the end of the season. Replacing old were up or broken bearings, gracing, replacing the twisted columns, and welding broken parts may be the major activities done to rail wagons. Replacing wear out wheels, broken springs and repair damaged bogies are also common activities of the rail wagons maintenance. Some old wagons may become out of service, the staff may decide to consider them salvage and forward a report to replace them. The new rail wagons for cane transport fabricated in the heavy equipment assembly factory belongs to the sugar company in Cairo to replace the salvage wagons.

6.2 Rail track maintenance

Maintenance of the rail track start after the operation season end in June and should be finished at December before the new season start. Technicians walk over the rail track inspecting the type and location of the breakdowns (Figure 20). After localizing breakdowns, technicians uncover the wear out parts of the track to perform maintenance and repair activities. The operation of railway network maintenance may include clear dirt or weeds that cover the track, tightening loose bolts and nuts and replace wear parts. Replace wear up sleepers may be the most common activity during the maintenance season Figure 21. The rails parallelism and rail gage should be also inspected and adjusted. The final step of the narrow railroad maintenance is to test and adjust the rail level Figure 22. A monthly report has to be forward to the narrow railway engineering administration showing the completed job.

Fig. 20. Two technicians inspect the probable breakdowns of the rail track

Fig. 21. Replacing the corroded steel sleepers of the rail track

Fig. 22. Balancing the level of the rail track

7. Light railways future

As previously discussed, each narrow railway network constructed and started operation simultaneously with the sugar factory initiation. The light railway networks belong to the old sugar factories initiated during the 19th century and those initiated at the early period of the 20th century used to supply 100% of the mill daily capacity. The narrow railway network expansion has been very limited compared to the expansion of sugar cane production area. According to **Soltan and Mohammed (2008)**, the area of sugarcane has been increased from less than 200,000 acres at 1980's to about 350,000 acres at 2008. Therefore the light railway of sugarcane transport stay constant and the cane area expanded more than 40 % outside the network. Since the role of the narrow railway sugarcane transport system declined from about 90% to about 40%. Considering the 40% decline because of cane area expansion outside the network, therefore declined of the light railway system transport may only be about 10%. Actually the percent contribution of the cane transported may be decreased but the total tonnage transported by the narrow railway system may be increased because the average unit area production increased and the mills now working at their full capacities.

The change of some farms to road vehicle may be because of the availability of their own vehicles or the advantages offered by road transport. The most important advantage offered by road transport is the short duration of transport cycle that save the excessive moisture losses that reduces the total weight and money value of the wagon load. In contrast, applications have been forward from several farmer groups to expand the light railway sugarcane transport network to their plantations. The light railway network of sugarcane transport may grow parallel to the cane production area whenever narrow rail tracks expanded according to the applications forward to the sugar company from farmers.

The sugar company has been developing experiences of light railway track maintenance, wagon fabrication and locomotive repair to maintain long life and efficient operation of the system. Constant efforts have been exerted by the company to replace locomotives and rail wagons which become out of service. The sugar mills may have hundreds of locomotives most of them compatible to the 2 feet rail gauge and more than 10,000 railway wagon for whole stalk loading. The company has been improving the level of locomotive maintenance and the design of the railway wagons to facilitate better role of the system.

The light railway sugarcane transport system was always able to transport cane with lower cost as indicated in Figure 23. Finally it may be concluded that the role of the light railway sugarcane transport system did not actually declined but remain constant while the mill capacity and the cane production increased. Since the alternative transport represented in road transport operated diurnal and it is difficult to use the road vehicles as storage bin, a minimum contribution of the narrow railway transport have to be conserved. The minimum role of the light railway system transport may be equivalent to the percent of daily capacity of the mill required for night shift. Reference to Figure 1 it could be observed that the role of railway transport system is not expected to show more decline. The sugar company organized special administration for narrow railway engineering that construct the rail track, fabricate wagons and been responsible for the system maintenance.

Australia may be considered as one of the countries achieved the most important development in the field of light railway transport of sugarcane. In his comment to the future of the sugarcane light railway of Australia, **John Browning (2007)** stated that "Cane railways will continue to surprise and to interest, and they will remain "special" to the men who operate them, to the many visitors to the areas in which they run, and to those who simply love railways". It has been recommended that, some of the modern techniques developed in countries such as Australia to control the light railway sugarcane transport cycle time should be considered.

Finally, the light railway for sugarcane transport represents the backbone of the raw material feeding system for sugar industry in Egypt. The system has several advantages compared to road transport such as lower transport cost, higher reliability, higher stability and minimum accidents occurred. Application of the advanced techniques for minimizing transport cycle duration expected to help for regaining the pioneer role of the light railway transport system. Practical ideas to increase wagons capacity and to improve mill yard management have been currently developed to speed up the system. Light railway transport system will continue being the familiar lovely transport system for sugarcane farmers.

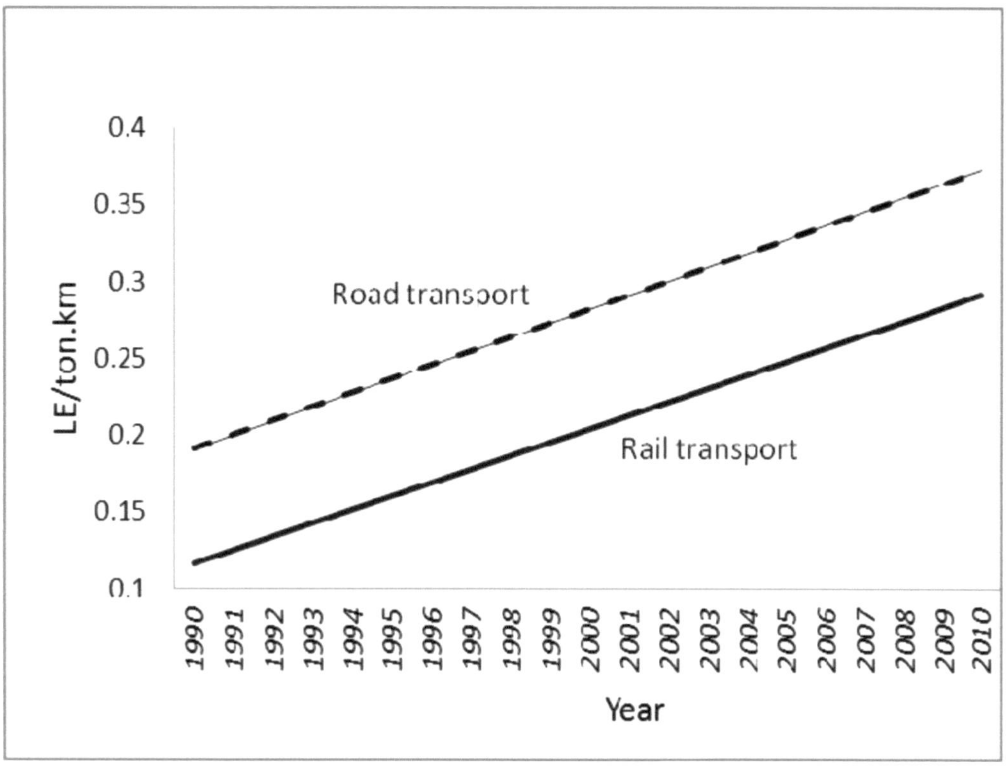

Fig. 23. Cost of light railway transport compared to road transport

8. Acknowledgements

The author wishes to announce that part of the data was collected through a project financed by the Egyptian Science and Technology Development Fund. The help of the members of the Sugar and Integrated Industry Company is also acknowledged.

9. References

Abdel-Mawla (2011) Expert system for selecting cane transport system. *Egyptian Sugar Journal. Vol. 4, June 2011: 161-178*

Abdel-Mawla (2010) Efficiency of mechanical cane loading in Egypt. *Sugar Tech. 2010, vol. 12, n°2, pp. 108-114 [7 page(s) (article)]*

Abdel-Mawla H A. (2000) Analysis of cane delay of traditional delivery systems: *Paper presented to the MSAE, Menofia Univ.:25-26 October 2000.*

Abdel-Mawla H A. (2001) Alternative cane to mill delivery systems. *MJAE 18 (3): 647-662.*

Affifi, F. (1988)) Sugar production in Egypt. Central Council for Sugar Crops. Ministry of Agriculture and Land Reclamation.

Andrew H. and Ian D. (2005) A simulation model for capacity planning in sugarcane transport. Computers and Electronics in Agriculture. Elsevier Science Publishers B. *V. Amsterdam, The Netherlands Volume 47 Issue 2, May, 2005*

John Browning (2007) Queensland sugar cane railways today. Light Railway Research Society of Australia. *http://www.lrrsa.org.au/LRR_SGRb.htm*

Malelane, M. (2000) Evaluation of Cane Transport Modes From Loading Zone to Mill to Minimize Transport Costs. ISSCT Agricultural Engineering Workshop. South Africa 23-28 July, 2000. *http://issct.intnet.mu/past-workshops/agriabs3.html#i*

Lynn Z. (2008): An Introduction to Modeling Queensland's Sugar Cane Railways . *1 www.zelmeroz.com/canesig*

Sugar Crops Council (1990-2010) Annual reports: The percent cane transported by light rail wagons compared road transport: 64-74.

Soltan F. H. and Mohammed I. N. (2008) Sugar industry in Egypt. *Sugar Tech. 10 (3): 204-209*

2

Topological Analysis of Tokyo Metropolitan Railway System

Takeshi Ozeki
Faculty of Science and Technology, Sophia University
Japan

1. Introduction

1.1 Railway system reflects the real world

Leading concept of the topological analysis of railway network systems is based on the fact that the topology of railway networks reflects the real world. It is believed because strong mutual interactions between railway systems and real worlds continue through longer periods of their growth: An eventual growth in a regional economy due to opening such a new shopping plaza may require extension of a railway system, verves, a scheduled extension of a railway may result in a growth in regional economy due to rapid increase in town population, for instance. In this way, the growth of railway system and regional activity affects their growth mutually. In context, the railway system topology reflects the real world: In other words, they "entangle" each other.

This leading concept agrees with that of Brin and Page, co-founders of Google: they reported, in their first paper on "Google"(Page and et al, 1990), that it was a great surprise the PageRank is obtained purely mechanically from the topology of Web page links. Their surprise is the discovery of the fact that the network is entangled with real world. The "Google" approximates a Web surfer as a random walker in Markov process and combines the dominant eigenvector of Markov process with a list of coincidence for a inquiry as the PageRank (Langville-Mayer, 2006)

This leading concept grows up as a mathematical platform using multimodal non-linear Markov process approximation so that it is applied to analyse Tokyo Metropolitan Railway System.

It is no doubt that there have been established platforms to analyse the dynamics of railway network systems based on growing supercomputer power. On contrary, our platform can be said as providing abstractive viewpoint based only on network topology so that it is expected to illustrate different new worlds for the railway system engineers.

1.2 Family network approximation: Rosary network

Network topologies have been discussed as scale free networks mainly in a field of complex systems from the end of the previous century. The scale free network science is expected to provide potential methods to analyse various network characteristics of complex systems.

However, there is no network model suitable for analysing railway systems. Then, the rosary network in series of family the network was proposed as suitable one for railway system networks as shown in Fig.1.1 (Ozeki, 2006).

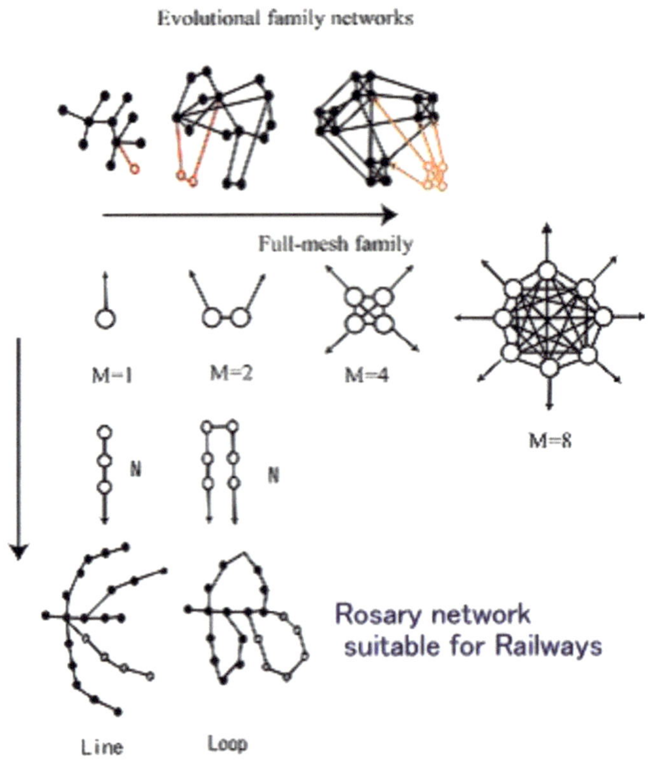

Fig. 1.1. Family network Series including Rosary Networks

Historical flows of complex systems are very interesting competitions between abstraction and computation: Origin of complex systems was introduced by Prigogine based on coupled nonlinear differential equations and sophisticated chemical experiments (Prigogine, 1981). It was followed by distributed agent model supported by rapidly growing computational power. However, for analysis of huge network systems the distributed agent model was suffered by computational complexity explosion in 1990'. Then, abstractive approaches such as scale free networks become to share exploring complex network systems. Topological analysis of railway network is backed by these historical flows.

The Watts-Strogatz's small world evolves from regular lattice networks to the Erdos-Renyi's random networks(Erdos, 1960) by random rewiring links with a given probability (Watts, 1998). The Watts-Strogatz's small world having fixed number of nodes is discussed as a static network. On the other hand, the scale-free network of Barabasi-Albert (BA model) introduces the concept of growing networks with preferential attachment (Barabashi, 1999). One of characterizations of networks is given by the connectivity distribution of $P(k)$, which is the probability that a node has k degrees (or, number of links). In the scale free networks based on BA model, the connectivity distribution follows the power law, in which $P(k)$ is approximated to $k^{-\gamma}$, having the exponent $\gamma=3$. The real world complex networks are analysed to find various scale free networks having various exponents, which are covered in references (Newman, 2006). For an example, it is well known that social infrastructure networks, such as power grids, as egalitarian networks, follow the power law with exponent 4 (Barabasi, 2002). There were many trials reported to generate models with larger exponents for fitting these real-world networks (Newman, 2006): Dorogovtsev *et al* (Dorogovtsev, 2000) modified the preferential attachment probability and derived the exact asymptotic solution of the connectivity distribution showing the wide range of exponents $\gamma = a+2$, where a is the attractiveness. However, there was no network generation model suitable for analysing railway systems.

In context, "the evolutional family networks" generated by "a group entry growth mechanism" with the preferential attachment was proposed in ICCS2006 (Ozeki, 2006): growth mechanism employed is group entry having constituent family connected in full-mesh, line and loop. This is suitable to simulate the railway system: as shown in Fig.1.1, a graph in the bottom looks like a railway system; We call it "Rosary network approximation" that will be discussed in the case of Tokyo metropolitan railway system in section 2. Various characteristics will be analysed based on the Multi-modal Markov transition approximation in section 3.

1.3 Birds with a feather flock together

We point out that nonlinear effects are inevitable in the passenger flow analysis. Since the Google is an infrastructure in daily life same as railway system, we refer the Google: the Google is characterized by a single dominant mode: In linear Markov transition, the asymptotic state is always the dominant mode. However, a Japanese adage: "people wish to get together to the place where people get together" or "Birds with a feather flock together" is important in real world to determine such PageRanking. The Google assumes such tendency is reflected in the page link network. Here, we point out it is not always sufficient, and demonstrate a Markov engine with the third-order nonlinear interaction reflecting such tendency to retrieve a real world, correctly.

We demonstrated the new engine to retrieve *the largest three stations* in respect of number of passengers in Tokyo Metropolitan Railway Network, in section 4.

1.4 TSUBO: Impulse response of network

We discuss "key stations of railway network dynamics" by analogy with "Tsubo in Shiatsu".

In Japan, "Shiatsu" is a popular therapy by pressing "shiatsu point" to enhance the body's natural healing ability and prevent the progression of disease. Shiatsu points are called "Tsubo", in Japanese. Their locations and effects are based on understanding of modern anatomy and physiology. The concept of "Tsubo" has been used as a strategy in re-activation of an old city, such as Padova, Italy (Horiike, 2000). He calls it "the Point Stimulus". The "Point Stimulus Response" corresponds to the impulse response of the network system, that is, the temporal state variation in the Markov transition to the delta-function with negative sign of initial state. We can evaluate the node activity by its response to the point stimulus.

We will discuss "Tsubo" of Tokyo metropolitan railway system in session 5.

2. Scale free characteristics of railway network

We show here a large railway system, such as Tokyo metropolitan railway system, that indicates characteristics of scale free networks: "station" corresponds to "node", and "track" to "link". This section is based on our paper presented in ICCS 2006. (Ozeki 2006)

2.1 Growth mechanism of Rosary

A growth step of a railway network is modelled as illustrated in Fig.2.1 (a): a rosary that consists of M stations connected in a shape like a rosary is added to an old railway network. There are two cases of its constituent: one is like a rosary having two jointing links as shown in Fig.2.1 (a) left, the other is like a snake having one jointing link as shown in Fig.2-1right. Fig.2.1 (b) is a rosary network generated this growth mechanism: assuming the fraction of snakes in constituent groups to be 10% and growth step 11 for convenience to grapes its perspective. This topology is drawn by a free-software: Cytoscape (http://www.cytoscape.org/download.html). The initial constituent is a group #0~#8 and the total number of stations is 65. The degree distribution is illustrated in Fig.2.1(c) (the "degree" denotes the number of links of a node). The degree distribution follows the power law with exponent of -4 as shown in Fig.2.1 (c).

2.2 Multimodal analysis of Rosary network

Before analysing Tokyo Metropolitan Railway System, it seems better to analyse this small rosary network. We assume a passenger in the rosary railway network as "a random walker", that is equivalent to multimodal Markov transition approximation (refer Appendix 1). The dominant mode of the multimodal Markov transition corresponds to the stationary state of passenger distribution that is illustrated in Fig.2.2 (c). The eigenvector of dominant mode has a peak at station #2, and mountains in the dominant eigenvector are illustrated in Fig.4.3 (a): the original station group #0~#8 corresponds to the first mountain in the figure, and the followings are illustrated in blue rosaries. The eigenvalue of the rosary network is shown in Fig.2.2 (a): The #64 eigenvalue of 2.773 corresponds to the dominant mode. The 2nd mode has negative largest eigenvalue. The mode competition among these modes in nonlinear multimodal Markov transition is discussed in section 4.

This rosary network has no real world so that it is difficult to show the substructure analysis. Next we discuss a actual rosary network.

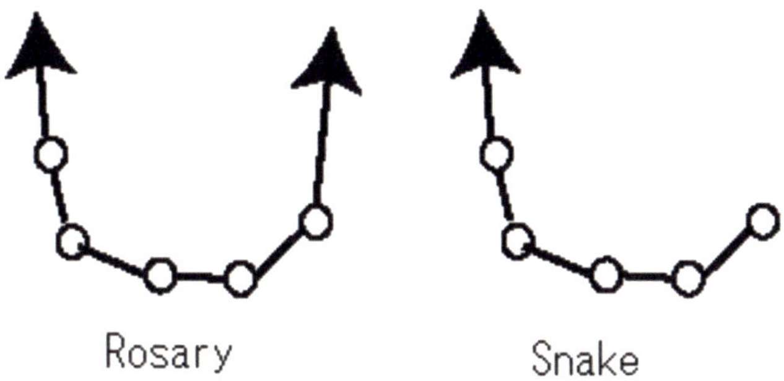

(a) Constituents of Rosary network

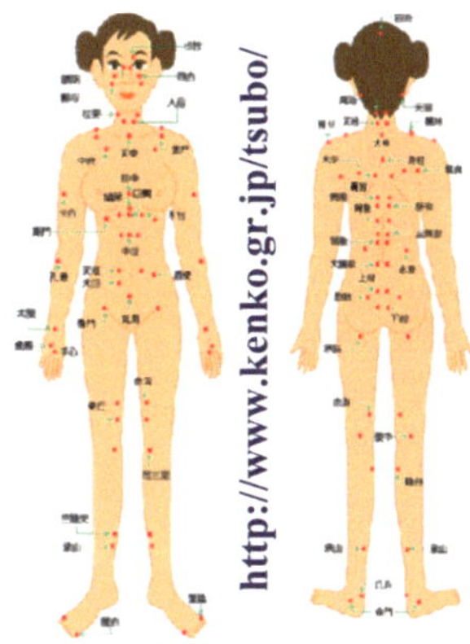

"tsubo"map

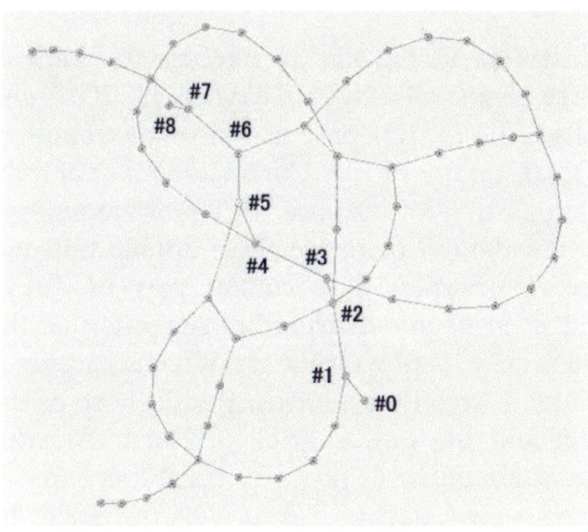

(b) A small rosary network generated.

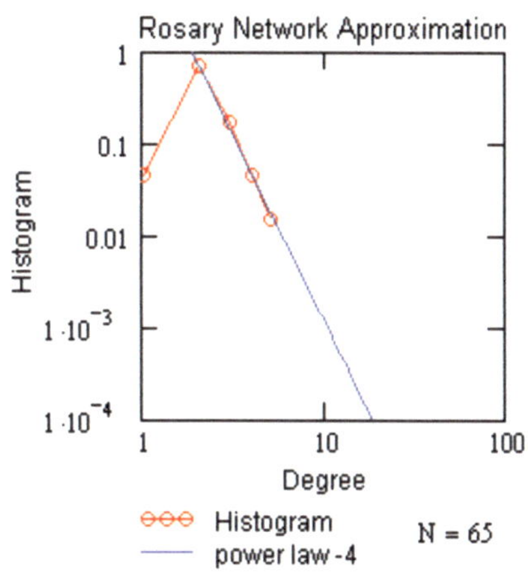

(c) Power law

Fig. 2.1. Rosary network model suitable for analyzing railway systems.

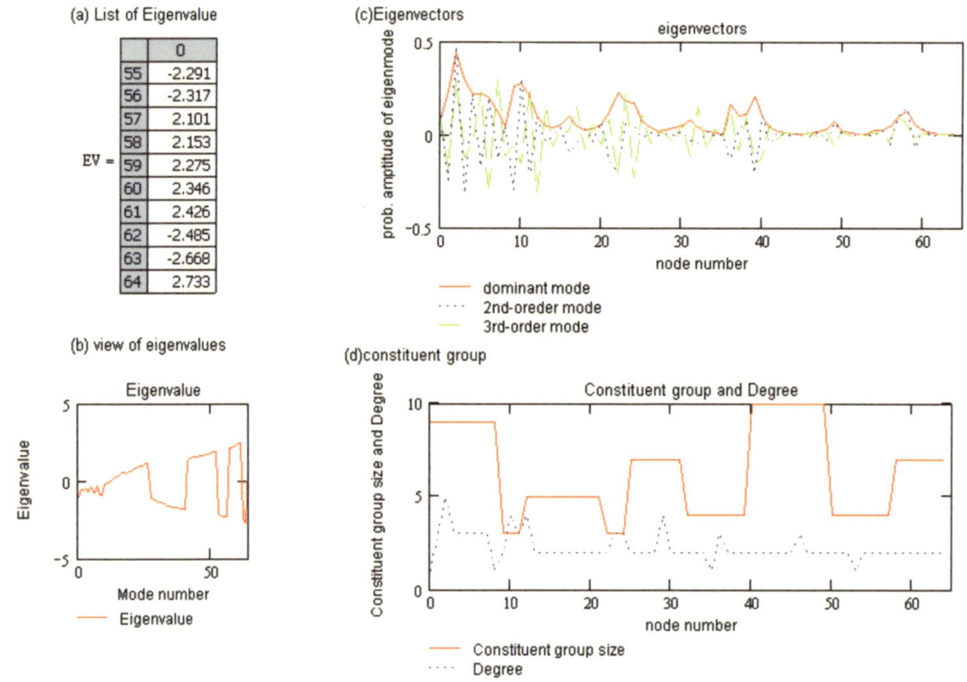

Fig. 2.2. Mode Structure of Rosary Network

3. Analysis of Tokyo metropolitan railway system

Tokyo metropolitan railway system is illustrated in Fig.3.1: (a) denotes the Map of contemporary Tokyo metropolitan railway system (Rail Map of Tokyo Area, 2004) and (b) denotes the map of Edo in 18th century. The central part of Tokyo metropolitan railway is truncated to have the number of total stations of 736. The total number of links is 1762. The number of links is counted topologically: for instance, we count the number of links between Tokyo and Kanda as 1, even though there are three double railways between them. Fig.3.2 (a) depicts degree distribution of a central part of Tokyo Metropolitan Railway System. The excellent fit in degree distribution suggests that the growth mechanism of Tokyo railway system is coincident with the growth mechanism of rosary network. The exponent measured to be 4, which is coincident with those of the small rosary network shown in Fig.2.1 (c) and the power grids of North America (Barabasi, 1999). It is surprising to find that the number of nodes in constituent rosary networks is M=3, which is reasonable in the central part of Tokyo with respect to its complexity. A real world railway network is well approximated by our rosary network model.

3.1 Substructures of Tokyo

The main issue is the extraction of an authentic centre (Tokyo, Shinbashi, Shinagawa) and a new metropolitan centre (Shinjuku, Ikebukuro, Shibuya). The later corresponds to the centre and the outskirts of Edo as shown Fig.3.1 (b).

(a) Map of Tokyo Metropolitan Railway System

(b) Map of Edo in Tokugawa Era of 18th Century

(http://onjweb.com/netbakumaz/edomap/edomap.html)

Fig. 3.1. Tokyo Metropolitan Railway Network System

A distorted hexagonal in Fig.3.1 (a) is "*Yamanote* Circular Line" which includes several well-known stations such as Tokyo, Shinbashi, Shinagawa, Shibuya, Shinjuku and Ikebukuro etc.

Fig.3.2 summarizes the mode structures of the network. In a list of eigenvalues illustrated in the right, we focus on the following two modes; the dominant mode #733 with eigenvalue +4.738 has larger probability at *Shinagawa* (station number #8), *Shinbashi* (#11) and *Tokyo* (#13) as shown in middle left panel of Fig.3.2. The constituent stations of the dominant mode are illustrated in Fig.3.1 (a) on the *Yamanote* circular Line.

The second mode #735, having negative largest eigenvalue of -4.271, has larger probability at *Shinjuku* (#0), *Shibuya* (#3) and *Ikebukuro* (#25) as illustrated in middle right panel of Fig.3.2. The constituent stations of mode #735 are also illustrated by *italic character* on the *Yamanote* circular line, in Fig.3.1 (a).

3.2 Orthogonal features of substructures

It is interesting that the dominant mode #733 extracts the central structure of business and government of Metropolitan Tokyo. This area also corresponds to the main structure in Edo metropolitan area. (Tokyo was called *Edo* in 18th century.)

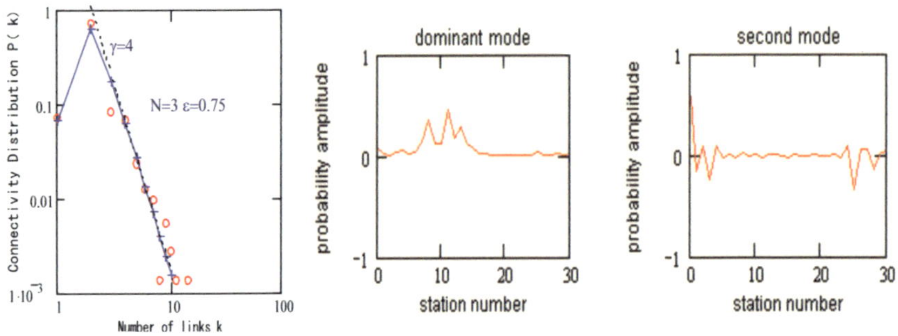

	0
726	-3.306
727	3.821
728	-3.378
729	-3.391
730	-3.445
731	4.081
732	4.423
733	4.738
734	-3.947
735	-4.271

Left: Degree distribution, Middle left: dominant mode; Shinagawa (8), Shinbashi (11), Tokyo (13); Middle right: second mode: Shinjuku (0), Shibuya (3),Ikebukuro(25) Right: List of eigenvalues

Fig. 3.2. Multimodal Analysis of Tokyo Metropolitan Railway System

The bridge of *Nihonbashi* is the original point of national roads including the *Tokaido* (presently *root 1*) in Edo era as shown in Fig.3.1 (b). The main business was blooming along the *Tokaido*, and the political organization was concentrated between the *Edo castle* and the *root 1*. It can be said that the central structure of contemporary Tokyo succeeds the main structure of the *Edo* metropolitan of which population was exceed one million in 18th century.

On the other hand, the second mode #735 is successor of the *Edo* outskirt villages located in the lower part of Fig.3.1 (b). The eigenmode effectively extracts orthogonal substructures in variety of viewpoints: The dominant mode retrieves the dominant political and business area of present Tokyo metropolitan. The second mode retrieves its most growing area that was the outskirt of *Edo*.

It is suggestive that the probability amplitude of the second mode, illustrated in the middle right panel of Fig.3.2, is positive at *Shinjuku* (station number #0) and negative at *Shibuya* (#3) and *Ikebukuro* (#25). Historically, *Shinjuku*, as the fourth hosting station of *Edo*, leads the others in this outskirt area. It is interesting because this mode profile has strong relations with mode competition in nonlinear Markov transition, as will be discussed in section4.

The further interpretation of probability amplitude remains in being unexplored. The data mining technology may be useful to reveal it.

3.3 An interesting eigenmode extracts the *Kohoku* new town project

The study of the Kohoku Newtown project is an old graduation thesis of our laboratory, when a different definition of transition matrix was used in Markov transition (Ozeki, 2009). Fig.3.3 (a) denotes the eigenvector of the 200th eigenmode of Tokyo Metropolitan Railway system. It consists of three station groups, named *Azamino / Nagtsuda* group (*Denentoshi* line), *Shin-yokohama* group (*Yokohama* line) and *Kikuna / Ohkurayama* group (*Toyokyu* line). Those are excited coherently and simultaneously in the same phase, as shown in Fig.3.3 (a) A speculation suggests that a zone, encircled by three lines, might be successfully developed as a triangle business park. It is our great surprise to find that "the *Kouhoku* Newtown project" was promoted from 1965 to 1996, exactly in this zone. Fig.3.3 (b) denotes the railway network map of the project, which shows *Az* for *Azamino* and *Sy* for *Shinyokohama*. *Nagatusuda* marked by *Na* was not recognized as key stations of the Newtown, but presently "*Yokohama* city plans" includes it as the *Yokohama Silicon valley*: It is well known that *Nagatsuda* includes the campus of Tokyo Institute of Technology. This network analysing engine points out the importance of *Nagatsuda* to provide TIT as Stanford University of Silicon valley.

Unfortunately, the network graph, used hitherto, does not include the *Blueline* subway that is one of the main constructions in the *Kouhoku* Newtown project. In next, we will discuss the evaluation of *Blueline* project.

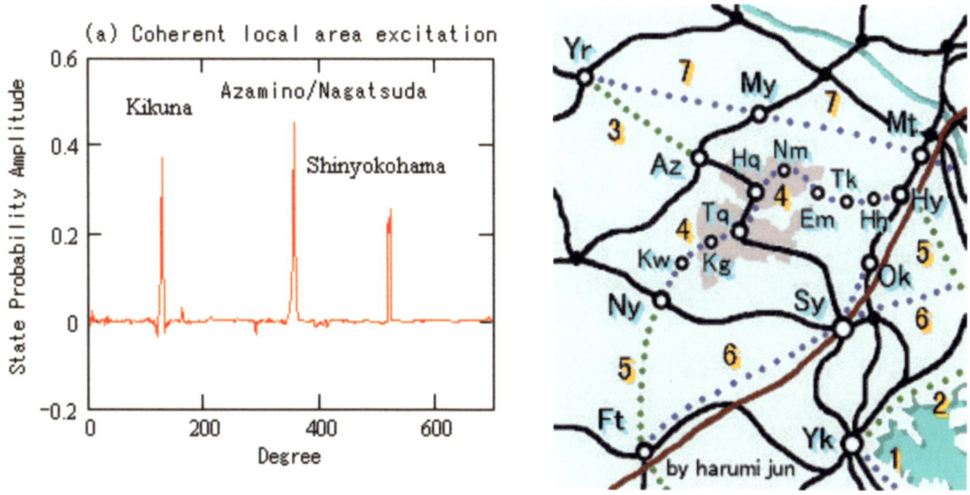

(a) The eigenvector #200 coherently excited (b) Kohoku Newtown Project:Kohoku Newtown encircled by the #200 eigen-mode:Sy=Shin-Yokohama, Ok=Ohkura yam, Az=Azamino (http://www.yk.rim.or.jp/~harujun/ntown/ntftr.html.)

Fig. 3.3. Eigenvector of *Kohoku* Newtown.

3.4 Evaluation of a new subway project

Here we would like to introduce an interesting application of our network analysis platform: it is a blind evaluation method of network modification project. As introduced in the previous subsection, we try to evaluate the project of "Blueline". The method is the variation of node entropy before and after *Blueline* inauguration, as illustrated in Fig.3.4 (a). The station group with increasing in their node entropy includes *Totsuka* (#94), *Sakuragicho*(#84) and *Kannai* (#85). On the contrary, the station group with decreasing node entropy includes *Hodogaya*(#104) and *Kita-Kamakura*(#106).

We can show a supporting data for this evaluation in Fig.3.4 (b). The number of annual passengers of *Hodogaya* station shows abrupt drop in 1999 when the *Blueline* service was started. This blind-evaluation method presently only provides the variation of passenger flow of modified network, but it seems a powerful tool for network system design in future.

4. Nonlinear phenomena in passenger flow in Tokyo metropolitan railway system

Here we would like to point out that nonlinear phenomena are important in passenger flow analysis. First of all, it should be noted that there are two types of nonlinear phenomena in the third order nonlinear interaction (Agrawal, 1989): one is SFM (Self Phase Modulation) that is equivalent to "Like Button" in Facebook, that is, transition probability to a node having the same opinion increases. (Please refer Appendix to find details including notations.)

This nonlinear Markov transition process is expressed mathematically by the following;

$$(q_i)_{n+1} = \sum_j A_{i,j}(q_j)_n(1+\gamma(q_i)_n(q_j)_n) \qquad (1)$$

In this expression, the nonlinear term $\gamma(q_i)_n(q_j)_n$ might be recognized as "like button": in case of the state $(q_i)_n$ of node $\#i$ having the same sign with $(q_i)_n$, the transition probability from node j to i increases when the nonlinear coefficient γ is positive. In other words, "like button" is a tool to express our personal opinion that controls routing of information in Facebook. It might be reasonable that nonlinear phenomenon in rush hours is recognized as SPM because most of passengers in rush hours have more sharp intention to reach their destinations.

The other is called XPM (Cross Phase Modulation) that is equivalent to "curious bystanders", that is, the transition probability to a node having many "curious bystanders" increases. It is shown mathematically as following:

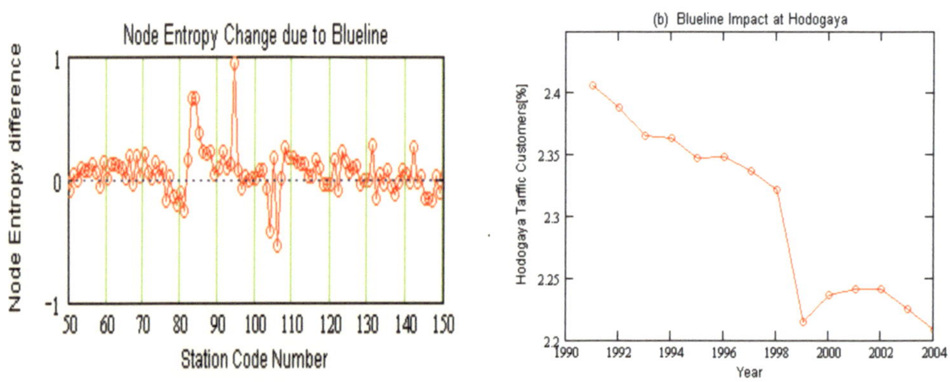

Fig. 3.4. Eigenvector Variation due to the Blueline.:(a) Node Entropy Change due to Blueline. (d) Evolution of Hodogaya Traffic Customers. Operation of Blueline was 1999.

$$(q_i)_{n+1} = \sum_j A_{i,j}(q_j)_n(1+\gamma\sum_{k,m} A_{i,k}A_{i,m}(q_k)_n(q_m)_n) \qquad (2)$$

In this expression, the nonlinear term $\sum_{k,m} A_{i,k}A_{i,m}(q_k)_n(q_m)_n$ might be recognized as effect of "curious bystander" because the transition probability from node j to i increases when $(q_k)_n(q_m)_n$ is positive. In other words, XPM expresses "Birds with a feather flock together".

To make our intention of introduction of SPM and XPM clear, we show their import differences in network dynamics:

Final target is discussion of mode competition between the authentic centre and the new growing metropolitan centre. And the third-order non-linear interaction is inevitable to show that the largest three stations are those in the new growing metropolitan region in Metropolitan Tokyo.

4.1 Emergence of instability

It seems better to introduce "Emergence" in the small rosary network discussed in session 2, before we discuss more complicated Metropolitan Tokyo railway system. Fig.4.1 illustrates the temporal variation of mode probability (the squared mode amplitude) of the nonlinear Markov transition based on Eq.4.1 applied to the small rosary network shown in Fig.2.2. The initial condition is a random distribution of node probability amplitude. SPM with medium $\nu = 0.1$ leads to the dominant mode as the stationary state of the rosary network. It should be noted that the right panel illustrates the temporal variation of the mode amplitudes: The 2nd and 3rd order modes have negative eigenvalues so that the mode amplitudes change their sign at each Markov transition. The dominant mode having positive eigenvalue does not oscillate.

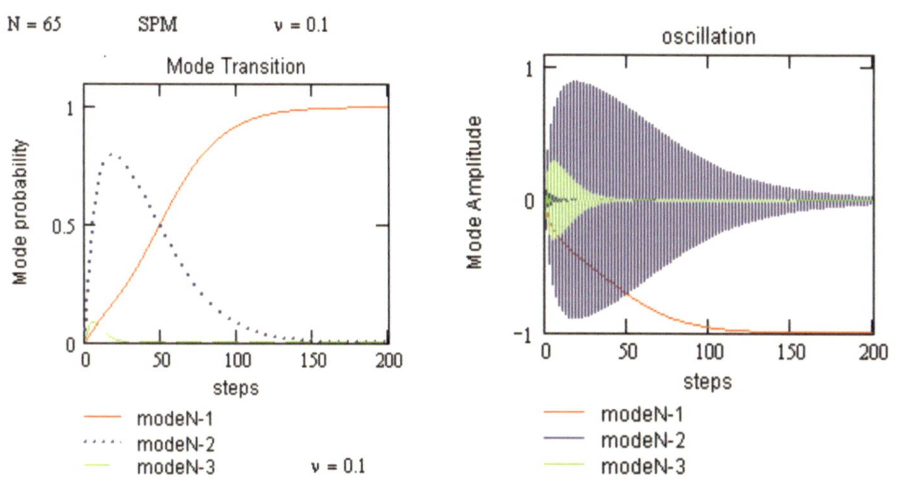

Left diagram indicates mode probability (squared mode amplitudes) and the right diagram denotes the temporal variation of mode amplitudes.

Fig. 4.1. Temporal variation of modes in SPM nonlinear Markov transition from random mode amplitude distribution as initial condition.

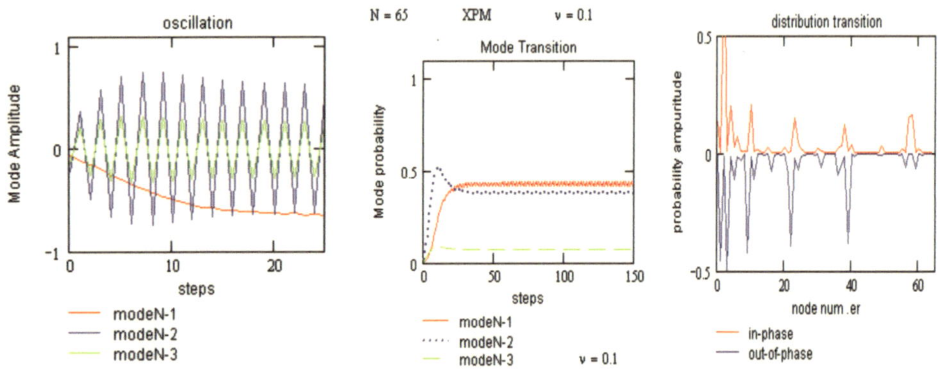

The light panel denotes the temporal variation of mode amplitudes. The middle panel denotes the temporal variation of mode probabilities. The right panel denotes two phase of probability amplitude.(the red race inverted in sign for clearness.)

Fig. 4.2. XPM Markov transition of the Rosary network:

On the other hand, Fig.4.2 illustrates the temporal variation of mode amplitudes in the case of XPM nonlinear Markov transition based on Eq.4.2. In this case the rosary network shows "emergence" of a kind of instability: The temporal variation of mode amplitude oscillates as shown in left panel of Fig.4.2. At stationary state, two phases are shown in the right panel of Fig.4.2. At "in-phase" denoted by red, the passengers gather on node #2 and #4 . In "out-of-phase" denoted by blue, the passengers shift to node #1,#3,#9,#22 and #39 coloured by yellow in the map of Fig.4.3 left. Those nodes can be reached from node #2 or #4 within one step. The oscillation is recognized to be sustainable transition between two groups of nodes.

This kind of oscillation has not been reported in real world, yet. One of convenient interpretation is that average of two states is assumed to be observable; that is, we assume the average state corresponds to observable phenomenon in the real world. Fig.4.4 (b) illustrates the average state corresponds to the instability in the XPM Markov transition. Markov transition approximation of a large-scale network has a limitation due to delay time to obtain information of nodes connected to a node, at each transition, so that it might be reasonable to take average of oscillating states, just mentioned above.

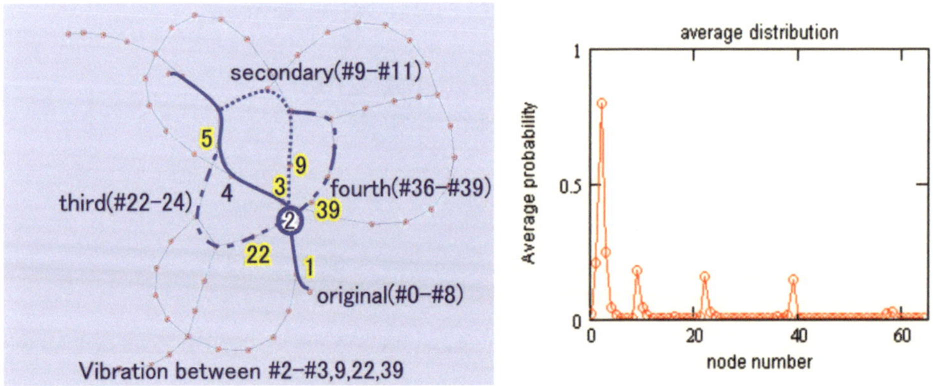

Fig. 4.3. (a) Sustainable oscillation between two groups of nodes (b) Average probability distribution

However, it should be noted that there are many intuitive samples of oscillations in the real world. This oscillation has strong relation with the network controllability and stability . These issues are discussed in Appendix C.

Since available data of the passenger flow analysis in Tokyo railway system are dairy data average over a year, it is reasonable to use the average of probability distribution of Markov transition approximation.

4.2 The largest three stations of Tokyo metropolitan railway

One of the targets is to extract outstanding patterns from huge network system: In linear systems, the dominant mode corresponds to such an outstanding pattern. This understanding coincides with that of the basic Google in which one assumes that passengers in Tokyo railway system can be approximated as random walkers in the linear Markov process. Its stationary state is the dominant mode.

The real world data, however, tell us that the largest three stations, in respect of passenger number, are *Shinjuku, Ikebukuro* and *Shibuya*: *Shinjuku* had 3.2 millions per a day as its number of passengers, *Ikebukuro* 2.6 millions, and *Shibuya* 2.3 millions, in 2006. This pattern does not coincide with the dominant mode.

We should overcome this discrepancy

First we introduce SPM Markov transition of Eq.4.1 to analyse the passenger distribution pattern. The initial condition of probability amplitude is a uniformly random distribution normalized by Euclidean norm. Fig.4.5 (a) depicts temporal variation of mode probability to reach dominant mode. The passenger distribution obtained is shown in Fig.4-5 (b), that corresponds to the authentic (political and business) centre of Tokyo: *Shinbashi, Shinagawa* and *Tokyo* are the dominant stations.

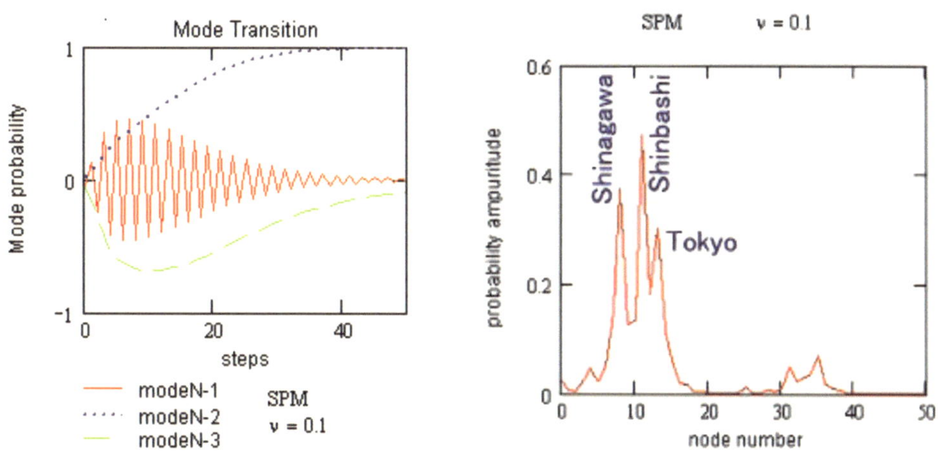

a) Temporal variation of mode amplitude

(b) The stationary state corresponds to the dominant mode.

Fig. 4.5. SPM Markov Transition of Tokyo Metropolitan Railway System.

Secondary, we introduce XPM Markov transition of Eq.4.2 to analyse the passenger distribution pattern. Fig.4.6 (a) depicts temporal variation of mode probability obtained for XPM. Mode #735 (N-1) having negative eigenvalue of -4.271 oscillates continuously and Mode #732 (N-4) having positive eigenvalue of +4.423 reaches to stationary state of -0.612. This instability corresponds to sustainable commuting of random walkers between the two phases as shown in the middle panel of Fig.4.6.

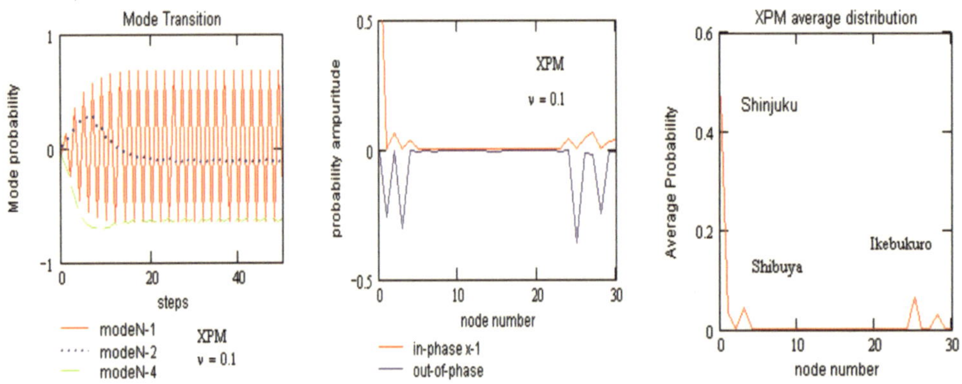

Left: Temporal variation of mode amplitude, Middle: Two phase of oscillation. Right: Average Probability Distribution.

Fig. 4.6. XPM Markov transition of Tokyo Metropolitan Railway System

According to the discussion on the rosary network in subsection 4.1, the average observable state corresponds to three largest stations of Tokyo Metropolitan Railway system, as shown in the right panel of Fig.4.6. This pattern of passenger distribution might be recognized as a central zone of entertainments or young people's activity, comparing to the authentic centre of Tokyo discussed in SPM Markov transition.

These analyses suggest the importance of nonlinear interaction in passenger distribution of railway network systems. This non-linearity of passenger flow reflects a human nature such as "Birds with a feather flog together".

4.3 Network dynamics and Markov process

The most basic assumption of the nonlinear Markov transition is the synchronous transition among all of the nodes in the network. The probability amplitude of higher-order mode varies in sign at nodes so that the superposition in transition causes complicated interferences among various routes of transition.

These multiple path interference may cause oscillation and dominates dynamics of network system. The multiple path interference may have relation with chromatic number in local structure.

The possibility of sustainable oscillations, including relation of chromatic number, was reported in the ICCS (International Conference of Complex System) in Boston, July 2011. However, no experimental evidence is reported yet. (Ozeki, 2011)

5. "Tsubo" of network system

"Tsubo therapy" inspired us to study "Impulse response" in network dynamics: The impulse response in electrical circuit theory provides "frequency response function", through the Fourier Transform, that makes it possible to analyse various dynamics of the circuit system. We named "an impulse applied at a node" as "point stimulus", after professor H. Horiike, architect: winner of Grand Prix of the Dedalo-Minosse International Award'02, Italy. (Horiike, 2000); The point stimulus is expressed as the initial condition $-\delta(i,p)$ of the Markov transition, where $\delta(..)$ is the kronecker delta, and p denotes the node where the impulse is applied. So far, this study is at dawn and a lot of unexplored remains. This section is based on our paper presented at KDIR2010 (International conference on Knowledge Discovery and Information Retrieval 2010).

5.1 Point stimulus response of Tokyo metropolitan railway system

Here we would like to show examples of point stimulus responses in Tokyo Metropolitan Railway System as illustrared in Fig.5.1: the upper row denotes those of "*Shinjuku*", "*Harajuku*" and "*Shibuya*". These point stimulus responses are dumping oscillations having fairly large amplitude of #735 mode. As discussed in subsection 4.2, since *Shinjuku* and *Shibuya* are the two of largest stations in Tokyo Metropolitan Railway System, and their degrees are sufficiently large, it is reasonable that they have larger point stimulus responses. On the contrary, "Harajuku" shows fairly large point stimulus response, but is a small station from viewpoint of its degree. The degree of *Harajuku* is only two compared to 11 of *Shinjuku*. In real world, "*Harajuku*" is a small station, but a famous down of youths and fashions. It can be said that the larger point stimulus response of "Harajiku" suggests that the point stimulus response is a reasonable tool to evaluate station activity.

The lower row denotes point stimulus responses of "*Shinagawa*","*Yurakucho*" and "*Tokyo*". As discussed in subsection 4.2, those stations belong to the group of "authentic centre of Metropolitan Tokyo". The dumping oscillations occur in eigenmode #734 that is the mode having negative second largest eigenvalue of -3.947. "Shinagawa" and "Tokyo" are fairly large station but the point stimulus responses are not so large that may reflect declining of those areas in 2006. Recently it can be said there are many successful projects to refresh these areas, such as *Shinagawa* intercity project. On the contrary, *Yurakucho* shows a relatively larger point stimulus response compared to small degree of 4. It can be said that the point stimulus response well reflects the town activity of *Yurakucho*.

Fig.5.2 denotes the cases that point stimulus responses reflect the miscellaneous station activities. The cases of "*Akihabara*" with degree of 7, "*Megro*" with degree of 4 and "*Otsuka*" with degree of 2 are illustrated.: The point stimulus response well reflects the declining activity of "*Akihabara*" in spite of various projects for actiovation. "Meguro" and "*Otsuka*" seem to have larger point stimulus responses than their actual activities. It requires further studies whether these discrepancies suggest the chance of investments for town activation or not. We demonstrate the point stimulus response as one of interesting tool of checking the activity of node. It is expected that the point stimulus response is a clue to find the real affects of networking on a node.

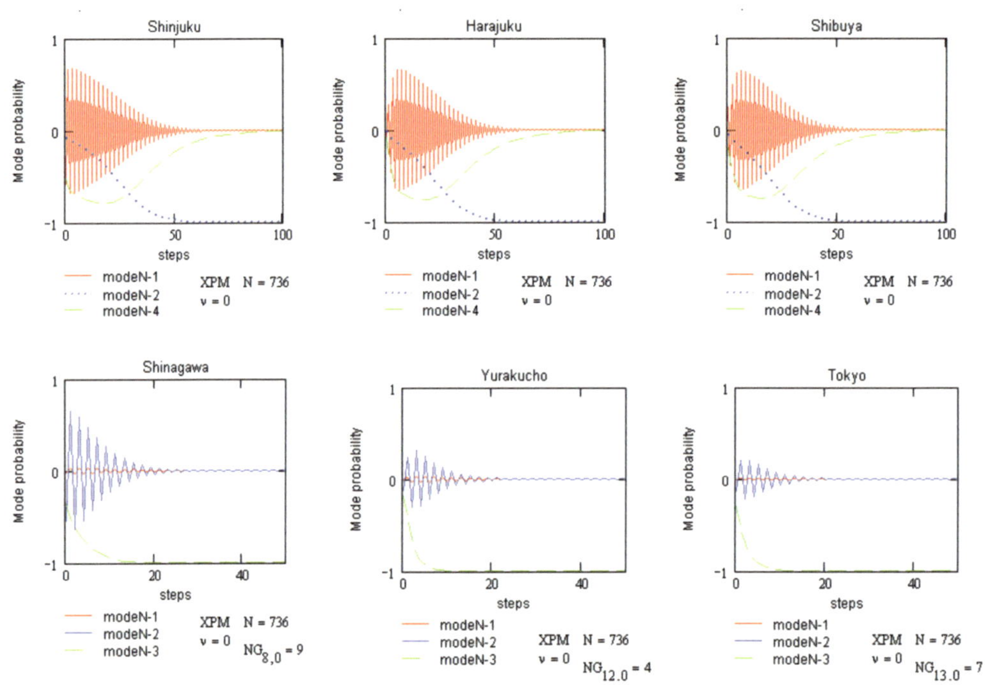

Fig. 5.1. Reasonable correlations of "point stimulus responses" with station activities.

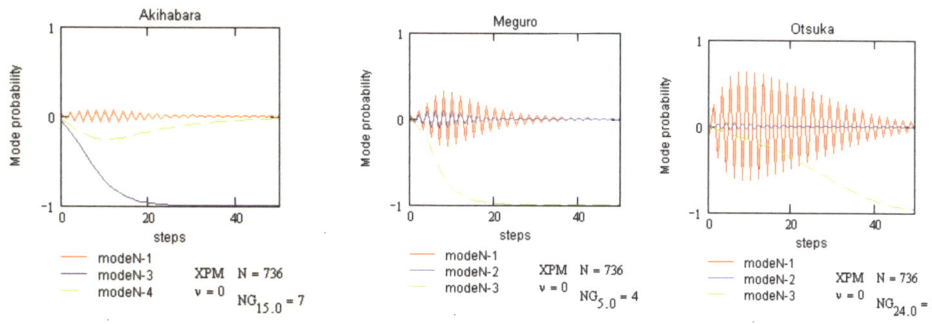

Fig. 5.2. Miscellaneous "point stimulus response"

6. Conclusion

We demonstrate a new network analysis based only on network topology of Tokyo Metropolitan Railway System. It is in highly abstractive and seems like a metaphor without any rigorous physical approval. However, many of analysis seem to illustrate the truth of railway system from abstractive viewpoint.

Rene Descartes wished all of the world could be described mathematically, then, as his first step, the analytical geometry was innovated. Prigogine, the originator complex systems, declared a "new alliance" between natural sciences and human sciences to solve global issues of human beings.

Here, we report a tiny effort of topological analysis of railway systems in this context. It is our wish to explore the horizon of our new mathematical platform as a tool for supporting intuitive power of railway system designers.

The multimodal analysis based non-linear Markov transition approximation is still in its dawn. There are the vast amounts of works unexplored for the future.

7. Appendix1 – Mathematical platform

This section is based on our paper presented in ICCS (International Conference on Complex Systems) 2011.

7.1 A new rule introduced in mathematical platform

After the complex systems was originated by Ilya Prigogine from various foundations including irreversibility and self-organization in nonlinear dynamics (Prigogine, 1996), Barabasi introduced scale free networks for describing interaction between structures or constituents of complex network systems (Barabasi2002). On the other hand, Brin and Page simulated a web surfer by Markov transition through network linking web pages (Brin, 1998). Then they found with their surprise, in spite of personal inherence of the Page Rank, that the web-network graph can rank the page importance mechanically by its dominant eigenvector (Brin, 1998, Langville 2006). We were inspired from theses historical flows to construct a multi-modal platform of Markov transition with nonlinear interaction for analysing complex networks.

A new rule of Euclidian norm, introduced in normalization in Markov transition approximation, provides us a new multi-modal description of complex systems: The Markov transition approximation of Page-Ranking in Google uses only the dominant mode, because their elements of the eigenvector are positive semi-definitive, and so, can be recognized as the probability finding a web surfer on a web page.

However, the elements of the eigenvectors of higher order modes are not positive semi-definitive, so that a new rule is necessary for defining the probability to find the web surfer on the page in higher order mode. It is well known that, in the power method for analysing eigenmode of such an adjacency matrix, the state vector at each transition is normalized by its Euclidian norm to prevent divergence. Since the sum of squared elements of the state vector is normalized to unity, the squared elements of the state vector can be recognized as the probability finding the web surfer on each node, just analogous to the quantum physics. In this way, we can describe the multi-modal behaviours of complex network systems employing the nonlinear Markov transition approximation (Ozeki, 2009).

7.2 Topology dependent characteristics of various networks

For analysing topology dependent characteristics, it is necessary to generate scale free networks having different topological characteristics, as references. Here we employ the Granvetter's family network series [Ozeki2006:]. Using family network series, various topology dependent characteristics of scale free networks can be reviewed, in terms of chromatic number, degree correlation, clustering coefficient, and network entropy. Among these reviews, we report a sustainable oscillation caused by the unique eigenmode structure of BA-network. This topology dependent instability, which arises from mode competition in a special mode structure, named "skew degenerate modes", is observed in the most popular BA networks (Barabasi, 2002). The skew mode is discussed more in A.3.

7.3 A mathematical platform of network multimodal analysis

Now we summarize a mathematical platform for network analysis in multi-modal scheme. The platform is based on the Markov transition to approximate the variation of network state: A symmetric adjacency matrix, providing an orthogonal eigenvector set, is suitable for multi-modal analysis of a network system. However, divergence in the Markov transition using the adjacency matrix as the transition matrix is a serious problem. Here, we apply a mathematical procedure, being used in "the power method" (Langville, 2006), for preventing the divergence: In Markov process approximation, the variation of network state given by

$$\hat{q}_{n+1} = A \cdot \hat{q}_n \tag{A1}$$

is described explicitly in the power method by

$$\hat{q}_{n+1} = A \cdot \hat{q}_n / || A \cdot \hat{q}_n || \qquad \text{(A2),}$$

where $\hat{q}_n$ is the state vector at the nth transition step. The state vector is nomalized with respect to the Euclidean norm$||A\cdot\hat{q}_n||$ after each transition step. This mathmatical idea used in the power method assures the stablity and also assures the linear properties of the Markov transition.

Furthermore, this idea lead us to read the state vector $(q_i)_n$ as a probability amplitude. The probability is defined by $(p_i)_n = |(q_i)_n|^2$ for finding a random walker at the node "i ", because the sum of probability $(p_i)_n$ over all nodes is normalized to unity as shown in Eq.A2.

The eigen-equation is $A\cdot\phi_i^{(m)} = \lambda_m \cdot \phi_i^{(m)}$ where λ_m is the eigenvalue of mode " m ", and $\phi_i^{(m)}$ is its eigenvector. The eigenvectors can be coincident with the asymptotic solution of Eq.A2 in the power method.

7.4 Markov transition with weak non-linearity

It is essential for the network analysis platform to be capable to analyse nonlinear phenomena. We introduce a non-linear Markov transition as follows: the nonlinear interaction in Markov transition means that transition from node "j" to node" i" is *affected* by the probability amplitude $(q_k)_n$ at node" k" linked to the node i , that is

$$(q_i)_{n+1} = \sum_j A_{i,j}\cdot(q_j)_n + \sum_{j,k} \nu\cdot A_{j,i}\cdot A_{k,i}(q_j)_n\cdot(q_k)_n \tag{A3}$$

where ν is a measure of nonlinear interaction. Eq.A.3 includes implicitly the normalization process as shown in EqA.2. This expression agrees with the definition of the Markov process, that is, the transition is determined only by the states at the present step n. The 3rd order nonlinear Markov transition is introduced in section 4.

8. Appendix 2 – Variety of topological parameters in family networks

The family network series, visualized in Table A1, provides variety of topological parameters of networks, such as degree correlations, clustering coefficients, and network entropies. These parameters are plotted in Fig.A.1 to understand details. The red line in the figure denotes a typical variation of the degree Pearson correlations depending on the constituent family size M of family network series. A" typical variation" means that the degree correlations shown in Fig.A.1 is the mean value of those calculated for 10 samples of networks, having about 100 total nodes, for each. The BA network is known as a disassortative network, that is, nodes with low degrees are more likely to be connected to the nodes with high degrees, and vice versa. The family networks with larger size M of constituent family become to be assortative, that is, nodes with a given degree are more likely to have links with nodes of similar degree. These Pearson coefficients of degree correlation (Soramaki, 2007) are illustrated in Fig.A.1 as a red line.

	BA Network (M=1)	Pair Network (M=2)	Trio Network(M=3)	Quartet Net (M=4)
Family Network Topology (ref. ICCS2006,id405)	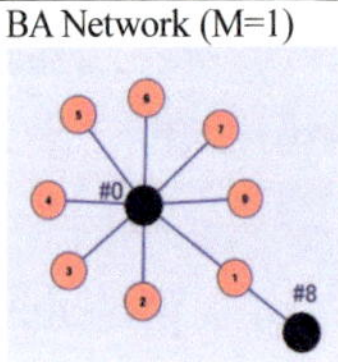	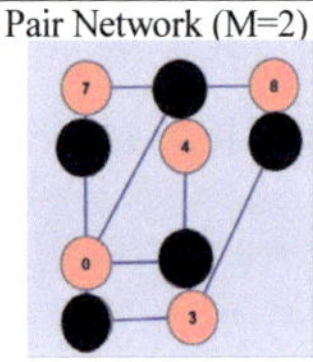	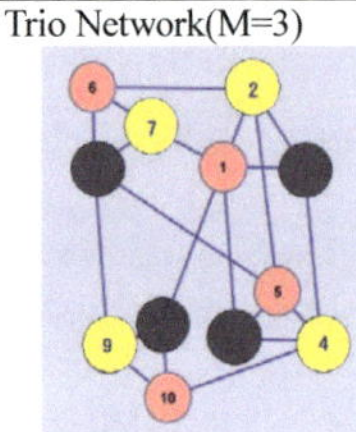	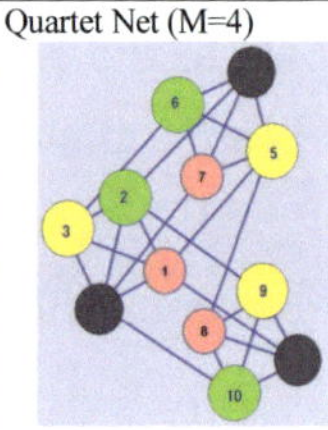
Temporal Response (Non-linear Markov)	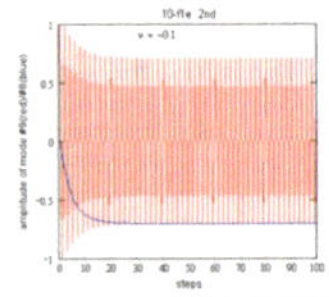	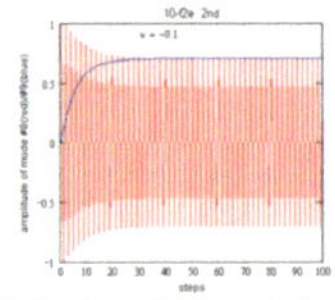	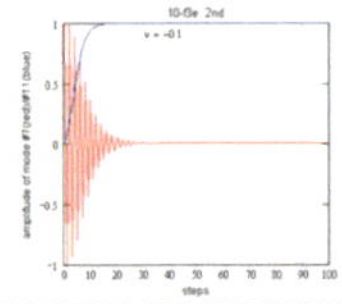	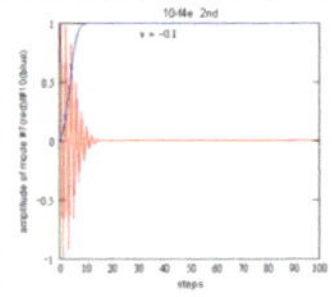
Chromatic Number	2	2 or 3	More than 3	Morethan4
Entropy	1.0	1.9	2.0	2.0
Degree Correlation	-0.30	-0.10	0.11	0.15
Clustering Coefficient	0	0	0.24	0.37
Asymptotic Exponent	3	4	5	6

Table A.1. Topology Dependent Network Dynamics

A clustering coefficient of a node is defined by the ratio of the actual number of links among neighbours of the node over the number of potential links among them. The clustering coefficient of the network is the mean clustering coefficient over all of nodes. The blue line in Fig.A.1 denotes the clustering coefficients of family networks. The family network with larger *M* has higher clustering coefficient.

Three kinds of entropy can be defined in multimodal description:

The first one is the node entropy NE_i that is defined by the sum of Shannon entropies over all of modes, that is,

$$NE_i = -\sum_m p_i^{(m)} \ln p_i^{(m)} . \tag{A4}$$

The second is the mode entropy ME_m that is defined by the sum of Shannon entropies over all of nodes, that is,

$$ME_m = -\sum_i p_i^{(m)} \ln p_i^{(m)} . \tag{A5}$$

The third is the network entropy that is defined by the sum of node (or mode) entropies over all of nodes (or modes), that is,

$$NetE = \sum_i NE_i = \sum_m ME_m . \tag{A6}$$

The network entropy is plotted by black line with diamonds in Fig.A.1, corresponding the family network shown in Table A1.

The variety of topological parameters of the family network provides a possibility of better approximation for a given network topology: We can approximate a network topology generated by the family network growth mechanism, with selecting the size M of constituent family randomly at the entry to meet its statistics, such as size of household (Ozeki, 2009).

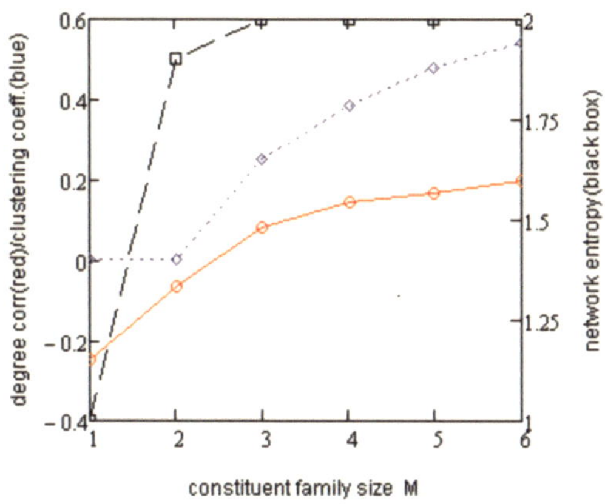

Red line denotes the degree correlation using Pearson's formula and Blue dotted line denotes the clustering coefficient. Black line with diamonds denotes the network entropy with the right-hand scale.

Fig. A.1.1. Topological parameter variation of family network series

9. Appendix 3 – Skew degeneracy

In Table A1, family network series are illustrated with the smallest number of colours: the chromatic number of a graph, in the third row in Table A1, is the smallest number of colours such that no two adjacent nodes share the same colour (http://mathworld.wolfram.com/chromaticNumber.html). This chromatic number is strongly related to the symmetry of the graph.

The BA network has the same chromatic number as the coupled harmonic oscillators, which consist of a long chain of masses and springs. Dyson analysed the coupled oscillators in 1953 to find mode pairs having the same eigenvalue in absolute value but different in sign (Dyson, 1953).

Degeneracy generally refers to objects having the same eigenvalue but different in eigenvectors, whereas the skew degeneracy, we named, refers to objects having different eigenvalues with respect to the sign of the eigenvalue but having the same probability distribution, that is the square of the eigenvectors normalized with respect to the Euclidean norm.

For an example, Fig.A.3.1 (1) shows the eigenvalue of the adjacency matrix of the BA network M=1 illustrated in Table A1, that includes two pairs of skew degeneracy modes. Fig.A.3.1 (2) illustrates the eigenvectors corresponding to a skew degenerate mode pair: Blue line denotes the dominant mode #8 having eigenvalue of +2.853, and Red line with circles denotes the mode amplitude of mode #9 having eigenvalue of -2.853. Fig.A.3.1. (3) denotes their probability distributions of mode #8 and #9 that coincides with each other.

EV =

	0
0	0
1	0
2	0
3	0
4	0
5	0.927
6	-0.927
7	0
8	2.853
9	-2.853

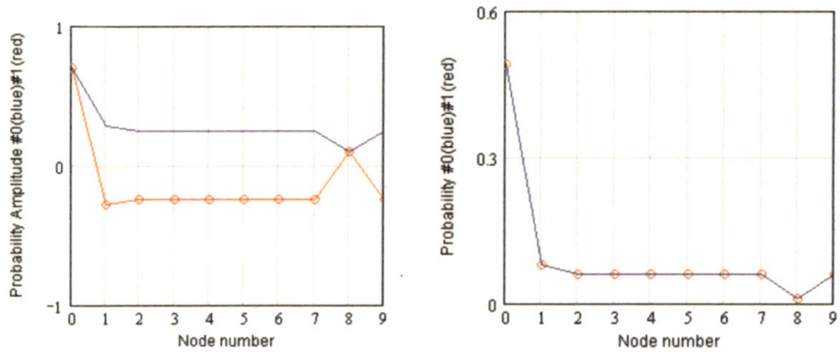

(1) (2) (3)

(1)Eigenvalue (2) Mode amplitudes: Blue line denotes the dominant mode (#8) having eigenvalue of 2.853 Red line with circles denotes the mode amplitude of mode #9 having eigenvalue of -2.853 (3)Probability Distribution :Both of skew modes coincide.

Fig. A.3.1. Skew Degenerate Mode Pair of BA network (M=1)

The family network with M=2 has a possibility having mode pairs of skew degeneracy. However, the other family networks with larger M than 3 do not show the skew degeneracy.

9.1 Temporal response of skew degenerate modes in nonlinear Markov transition

The nonlinear Markov transition of Eq.A.3 can be converted to the description of the nonlinear interaction of mode amplitudes for getting clearer image, as the following:

$$(a_m)_{n+1} = \lambda_m \cdot (a_m)_n + \sum_{i,m',m''} \nu \cdot \lambda_{m'} \cdot \lambda_{m''} \cdot \phi_i^{m'} \cdot \phi_i^{m''} \cdot \phi_i^{m} \cdot (a_{m'})_n \cdot (a_{m''})_n \quad \text{(A7)},$$

where the modes are defined by the linear adjacency matrix. It should be noted that the equivalency of Eq.A.3 and Eq.A.4 is limited only for the case of very weak non-linearity considered.

The transient response of the network having the skew degeneracy shows the sustainable oscillation in the nonlinear Markov transition as shown in the second row of Table A.1. The initial conditions are the modes with the negative largest eigenvalue. The skew mode pair survives in mode competition so that the random walker continues to commutate between two states that are the superposed states of the skew degenerate modes with in-phase and out-of-phase, respectively. The two states correspond to the group of the black and the red nodes in the BA -network, so that the random walker commutes between the node groups of red and black.

Fig.A.3.2 illustrates these situations clearly; the probability amplitude distribution is given by $(q_i)_n = \sum_m (a_m)_n \cdot \phi_i^{(m)}$, that corresponds to the superposition of two competing modes #8 and #9 illustrated in Fig.A.3.2 (2): The mode amplitude $(a_9)_n$ in red of Fig.A.3.2 (1) continues to oscillate between $+1/\sqrt{2}$ and $-1/\sqrt{2}$ whereas $(a_8)_n$ in blue grows up to $-1/\sqrt{2}$ so that the superposition of two competing modes corresponds to red line of in-phase and blue line of out-of-phase as shown in Fig.A.3.2 (2).

The node patterns illustrated by red and blue lines coincide with the chromatic groups shown in the first row of Table A.1.

On the other hand, the family networks with $M \geq 3$ show quicker transition to the stationary state corresponding to the dominant mode, as shown in the second row of Table A.1.

It is shown the following; the topology dependent instability dominates the temporal response in controlling the network system so that the network topology determines the dependability of the system, in a sense.

10. Appendix 4 – Family network series as reference

10.1 Network growth mechanism of family network

The session 2 of reference(Ozeki, 2006) should be read as follows: The asymptotic connectivity distributions of the full-mesh family networks are derived by the method reported by Dorogovtsev et al. At initial time t=1, the number of constituent family is one so that the number of nodes is *M*. We assume the node attractiveness is given by A+*M*-1 where the number of links is *M*-1, so that the total attractiveness is M(A+M-1). At time t=t, the total attractiveness of the network is M(A+M-1)t+M(t-1), where the last term M(t-1) is the contribution of the weak ties. By replacing these in equations $p(k,s,t+1)$, then the connectivity distribution p(k) is given by

$$p(k) = \frac{1}{2} \frac{\Gamma(2M+2A+1)}{\Gamma(M+A)} \frac{\Gamma(k+A)}{\Gamma(k+M+2A+1)} \tag{A10}$$

We obtain the asymptotic exponent $\gamma = M+1+A$.

10.2 Network stabilization by topological improvement

The network dynamics such as stability of the network system depends on the topology of the network system. Family network series gives us typical dynamics variations, as shown in TableA.1, as a reference. These understanding seem helpful to design network such as railway system.

So far there is no experimental evidence showing these transient behaviours of networks yet, but we can imagine several examples intuitively as follow:

a BA network with 100 nodes is illustrated in Fig.A.4.1 (1). The node # 0, and #1 and #2 are larger hubs. We might assume it as an ancestry of a family struggle, or an organization map just after consolidation of three small consanguineous companies. This topology consists of 26 pairs of skew degenerate modes and shows sustainable oscillation from an initial condition of random probability amplitude distribution as shown in Fig.A.4.1 (2). This might correspond to longer periods of struggles or troubles in this network system.

An intuitive method to prevent these troubles is to span a new link between two hubs, node #1 and #2 as shown in Fig.A.4.1 (3). This method is confirmed to be effective to convert the sustainable oscillation to quicker transition to stable state, by the non-linear Markov transition simulation, as shown in Fig.A.4.1 (4).

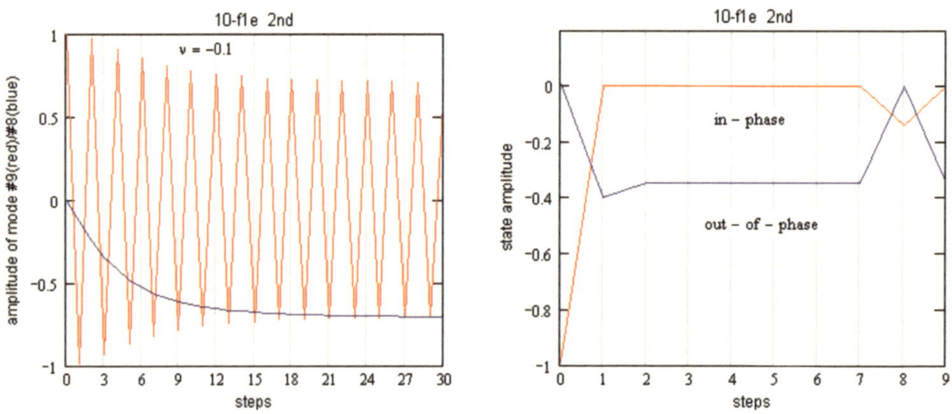

(1) (2)

(1) mode amplitude $(a_9)_n$ in red and $(a_8)_n$ in blue, (2) the state amplitude of superposition.

Fig. A.3.3. In-Phase and Out-of-Phase Superposition of Skew Degenerate Mode Pair

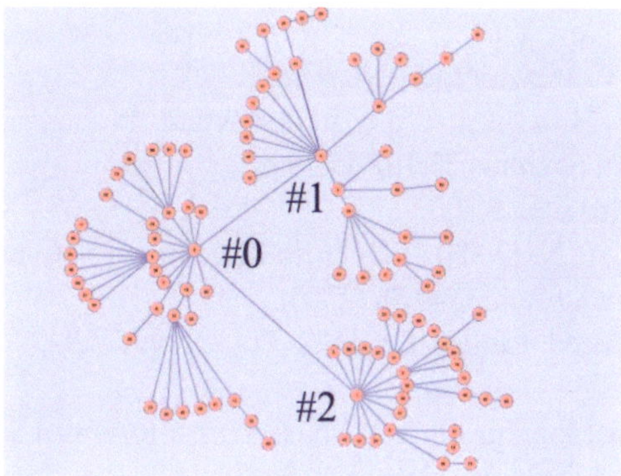

(1)

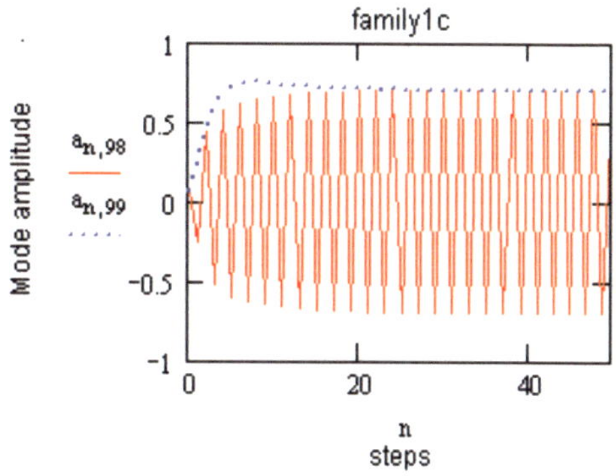

(2)

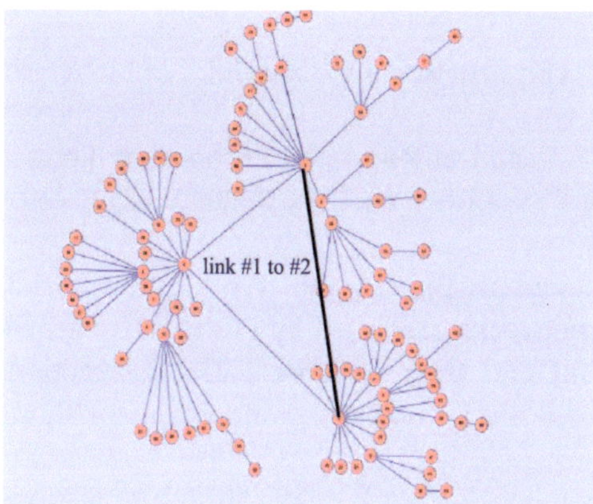

(3)

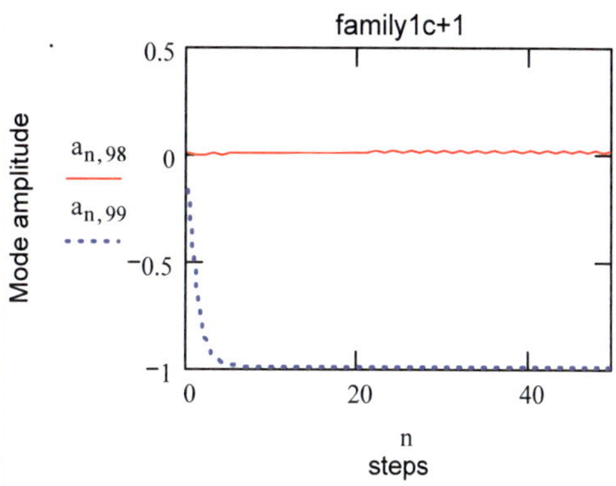

(4)

(1) BA network with 100 nodes, (2) Sustainable oscillation of skew degenerate mode pair, (3) Topological improvement by connecting #1 and #2, (4) The topological improvement can convert the sustainable oscillation to quicker transition to stable state.

Fig. A.4.1. Topological Improvement of Network System Stability

11. References

Agrawal ,G.P,(1989)"Nonlinear Fiber Optics", Academic Press, San Diego.

Barabasi,A. L.,Albert, R. and Jeong, H.(1999) " Mean field theory of scale free random networks", Physica A.272,173.

Barabasi, A. L, and Albert, R. (2002) "Emeregence of scaling in random networks", Science286, 509.

Barabasi, A. L, (2002)"*Linked*", Penguin Group, New York

Buchanan, M. (2002) "Nexus", W.W. Norton & Company Ltd., New York.

Brin, S. and Page, L. (1998)" The anatomy of a large-scale hypertextual Web search engine",Computer Networks and ISDN Systems,33:107-17,1998

Erdos, P. and Renyi, A.(1960) Publ.Math. Inst.Acad.Sci.5, 17.

Dorogovtsev, S. N, Mendes J F and Samukhin, A. N. (2000) "Structure of growing networks with preferential linking". *Phys. Rev. Lett.* 85, 4633-4636 (2000).

Dyson, F.J.(1953)"The Dynamics of a Disordered Linear Chain.", Phys.Rev.Vol.92,No.6, 1331.

Haken, H. (1978) "Synergetics, An Introduction. Nonequilibrium Phase Transitions and Self-Organization in Physics, Chemistry and Biology", Springer, Berlin.

Horiike, H.(2000), private communication.

Granovetter, M.(1973) "The strength of weak ties.", *American Journal of Sociology*, 78,1360-1380.

Langville, A and Mayer, C, (2006) "*Google's PageRank and Beyond*". Princeton University Press, Princeton and Oxford.

Newman, M., Barabasi, A.L. and Watts, J. (2006)"The structure and Dynamics of Networks" Princeton Univ.Press.

Ozeki,T.,and Kudo,T.,(2009) Invited paper,IEICE Technical Report,"A Proposal of Network Evaluation Method and Its Applications" vol.109 , no.220 , IN2009 -65, PN2009-24,pp13-21.

Ozeki,T.(2006) http://necsi.org/events/iccs6/viewpaper.php?id=405 We found a mistake in session 2 of our "Evolutional family networks generated by group-entry growth mechanism with preferential attachment and their features.", The International Conference on Complex Systems, id 405,Boston. The asymptotic exponent should be read as M+A+1

Ozeki,T.(2010), "Multimodal Analysis of Complex Network-Point stimulus response depending on its location in the network-" Knowledge Discovery and Information Retrieval 2010, Valencia, Spain,2010.

Prigogine,I .(1996)" The end of certainty", The Free Press, London

Soramaki, K, Bech, M L, Arnold, J, Glass, R J and Beyeler, W E, (2007)"The topology of interbank payment flows", *Physica A: Statistical Mechanics and its Applications* Vol 379, Issue 1, June 2007, pp 317-333.

Tsujii,S.(1983), "Transmission Circuits", Colona Publication Inc. Tokyo.

Watts, D. J. and Strogatz, S. H.(1998),"Collective dynamics of "small-world" Networks", Nature 393,440-442.

3

Privatization Versus Public Funding on the Atacama Desert Railway – An Interpretation

Jose Antonio Gonzalez-Pizarro
Universidad Católica del Norte, Antofagasta, Chile

1. Introduction[1]

Before the arrival of railways at the Atacama desert, there were only paths and tracks inherited by caravans exchanging products - the complimentarity of John Murra's ecological ground - between the Cordillera puna and the Pacific coast. The cart roads of mining prospectors from the time of the manure and copper cycles corresponding to the republican beginnings in the plateau would be added to this pre-Hispanic sequence. But, with the discovery of potassium and silver nitrate in the middle plateau during the 1860s-1870s, the multi-ethnic population settlement of Europeans and Latin Americans could be possible, the Chilean one being the most numerous. Human settlement started on the coast; Tocopilla and Mejillones in the 1840s, preceded by Cobija in 1825; then, Antofagasta in 1866. These were seaports prepared for merchandise shipping and traffic to Bolivia. Further south, that is, from parallel 24, settlement occurred in Taltal in 1858 and Paposo small port in 1865. It was the need of connectivity among the productive mining sites of the desert and communication with Potosi from the coast what made it possible to visualize a more modern and efficient means of transport in terms of capacity, speed, and safety: the railway.

As any history of techniques and means of transport, railways fall within a space and time framework connected to the well-known history of the desert geography: a process marked by landmarks, opportunities, and a long-lasting process, to use French historian Fernand Braudel's categories (Braudel, 1970). In this historic context, the Atacama desert railway is the reflection of various aspects of the social, economic, political, and natural resource history of this territory. But, at the same time, the railway together with the telegraph symbolized the keen desire for progress, science potencial, and Comte's positivistic philosophy for contemporary people, as expressed by engineer Matias Rojas-Delgado, the first Antofagasta Mayor in 1872, that is, that science should lead to the rationality of mining work and railways would open the frontiers of the unknown. "Order and Progress" was a motto that triggered private ventures and set the basis for organizing a new society in this place (Gonzalez-Pizarro, 2002). This changed the image that had been inherited from Spanish chroniclers and the first republican travellers. Science and technology that poured on the desert owing to the changes made to the nitrate leaching system thought of in

[1]This paper is part of Fondecyt Project 1100074 and Nucleo Milenio "Ciencia Regional y Políticas Públicas", 2011.

England by James Shanks and introduced by Santiago Humberstone in the sodium nitrate industry, along with the demand for locomotives and wagons made in Europe, opened a new positivistic reading of the desert (Gonzalez-Pizarro, 2009).

Geography and scientific knowledge influenced the design of railway communication. We must remember that this geographic knowledge, including its maps, has its landmark in the explorations of Rodolfo A. Philippi, a German scientist hired by the Chilean government to travel in the desert from 1853 to 1854. His conclusions contained a certain geographic determinism: "it is extremely difficult, not to say impossible, to build railways or electric telegraphs in the desert... the many ravines, 150 to 200 m deep, that constantly cut through the current road... these ravines would be avoided by a line located further west, but no water would be found in it" (Philippi, 2008: 132). Philippi added that important findings related to mining wealth could change his assumptions. This geographic conditioning of the desert considered the world's most arid remained as a barrier that would demand more accurate topographic studies since it would be affected by the presence of phenomena resulting from the climate, such as *camanchaca,* the *optical illusion* that made man be lost in this huge space. The Chilean government sponsored other studies of the desert, in both the middle plateau, as entrusted to the French geographer Pedro A. Pissis, and along the coast to its Navy Captain, Francisco Vidal-Gormaz in the 1870s, before the war. The increase of geographic knowledge became outstanding in Alejandro Bertrand's (Bertrand, 1885) and Francisco San Román's work in the Atacama Desert Exploration Commission (San Román, 1896). The physical description, mapping, mining potentiality studies, and fundamental toponymy of the desert were Bertrand's and San Roman's main contributions.

Thanks to San Roman's effort, scientific work could show that the Atacama desert nomenclature had narrowed regarding the colonial territory. In his work **Desierto y Cordilleras de Atacama** (1896), he established a close relation between what was understood as the Atacama desert and what was known from the towns and industries established in the plateau:

"The long stretch of Chilean territory running from the wild valley of Huayco to the nitrate pampas where the Loa river runs, encompassing between both extreme borders in all Chile's width, which extends from the Pacific coast to the Andes crest, is what was properly taken as the *Atacama Desert* until the beginning of this century. This denomination has been more and more restricted to the north as general progress and mining findings were populated or made exploration in this territory accesible, founding towns and creating industries in them. However, as a mere geographic title and, above all, as significant of an arid zone and production exclusively due to the mineral kingdom, traditions and customs still keep that denomination for all the territory that includes two Chilean provinces today, Atacama and Antofagasta" (San Román, 1896: III-IV).

The geographic knowledge and naming of ravines, mountains, and other territory accidents made it possible to conquer the land. Let's say that the Antofagasta-Bolivia Railway Company (A.B.R.C or F.C.A.B, for its acronym in Spanish) was a reference for Santiago Astronomy Observatory until 1910 to localize parallels 23° and 24° south latitude in the Antofagasta province *hinterland.*

The expansion of the Atacama desert railways shows that they were connected to nitrate findings in 1866 and silver in 1870, when the territory was divided in terms of sovereigneity.

The Ceasefire Agreement between Chile and Bolivia in 1866 and 1874 moved the border to parallel 24 S latitude: Chile owned the land to the south and Bolivia to the north, as long as we are interested in noting, without putting emphasis on articles regarding custom's rights or the absence of taxes to Chilean nitrate activity during 25 years. This situation would change as a result of the Pacific War, 1879-1883, in which Chile took over the territory in dispute (Lagos-Carmona, 1981; Gonzalez-Pizarro, 2005).

The geographic sector for the main railway activity was the one connecting Mejillones-Antofagasta axis to the hinterland, where the main nitrate mines are located (the central or Bolivian canton) with a northeast orientation, that is, the interconnection among Carmen Alto nitrate mine, Salinas station, Caracoles mine - currently located in Sierra Gorda county - and the extension to Calama, where the network was expanded to connect with Bolivia.

The beginning of the railway in this area mixes unsuccesful projects, proposals accepted but not implemented, and outlines that became real. In 1872, the French explorer André Bresson (Bresson, 1997: 180) stated that the railway would change a five-day long and deadly journey from the coast to Potosi by a comfortable five-hour one. Exactly so, the Atacama desert, in being dominated by mining techniques and the introduction of the railway became, from a "naturalistic-deterministic" nature under a colonial perspective, a nature viewed from a "pragmatic-utilitarian" optics (Gonzalez-Pizarro, 2008).

The exploitation of Mejillones manure and Cobija commercial activity stimulated Bolivia to do a public bid to build a railway connecting Cobija and Calama during Jorge Cordoba's administration in 1856. This, along with other projects such as Gibbson's and Arrieta's in September 1863 to build a railway to Mejillones, or Roberto Brown's in August 1868 to build a railway from Cobija to Potosi, were projects discussed in La Paz government, but did not turn out successful. It is important to highlight Avelino Aramayo's project presented in 1861: "Connection of the Pacific Ocean and some point in Bolivia" with Mejillones railway, which would be supported by the government in 1863. Aramayo would get a loan in London by being a partner of Mr. Samuel Morton-Petto-Barnett and Mr. Edward Ladol-Betts, based on a study of the German engineer Hugo Reck, giving Mejillones manure companies as a guarantee. Avelino Aramayo's partners broke in 1871, the Bolivian government keeping its interest in the route starting in Mejillones. So, they made a bid for this purpose in the same year. Twenty-seven projects were put forward, but none of them was successful, including Gustavo Bordes's and partners' and that of José Manuel Brown, Marcial Martinez, and Enrique Meiggs (Bravo, 2000:53-58). According to Gomez-Zubiela, another Aramayo's project had been passed by the government through a resolution passed on May 22, 1872. This fact would be the obstacle for the "famous railway constructors" appearing during Melgarejo's government, Milbourne Clark and Carlos Watson, among them. The suspension of the railway construction in 1874 affected the whole mining and commercial area (Gomez-Zubiela, 1999; Perez, 1997). It was the time of the great Caracoles silver ore deposit discovery in 1870, so that Mejillones railway could connect this small port to this small place and its mines. However, Aramayo's project sponsored by La Paz government became an obstacle for other projects such as the concession requested by Milbourne Clark & Co. on January 14, 1873, for a Mejillones-Caracoles railway, or by Felipe Iturriche to connect Cobija and Calama in March 1873 (Gonzalez-Pizarro, 2010: 919).

The big British investments in the nitrate industry, along with the Chilean ones, would influence the *de facto* orientation of railway policies. Milbourne Clark & Co., created on March 19, 1869, was funded by Chilean capitals - Jose Santos Ossa, Francisco Puelma,

Agustin Edwards - and British investments - Milbourne Clark, George Laborer, and Casa Gibbs from Valparaíso. The official negation to Milbourne Clark & Co. did not prevent it from continuing with its railway project, changing its beginning from Antofagasta to its hinterland. Under the name "Nitrate Company and Antofagasta Railway" established in October 1872, its railway arrives at Salar del Carmen on December 1, 1873; Salinas in 1877; Pampa Central in 1881; and Pampa Alta in 1883. At this time, it was estimated that the railway construction from Antofagasta, the establishment of the big nitrate treatment plant in this port, and manure exhaustion would result in the abandonment of the Mejillones railway (Mandiola & Castillo, 1894: 3vta). The consequences of the Mejillones-Caracoles railway failure affected its workers negatively due to the high transport costs resulting from operations with the Antofagasta railway company (Bravo, 2000). For Bowman, the future of the Antofagasta port was determined by the railway construction (Bowman, 1924: 81).

2. The appearance of the Atacama desert railway giant: The Antofagasta and Bolivia Railway Company Ltd.

The thriving nitrate activity in the Antofagasta province was slowed down at the south of parallel 24° after the war because of the government tax legislation that favored Tarapaca and affected the economic activities of the zone (Gonzalez-Pizarro, 2009a). At that time, the ideology of political liberalism prevailed in Chile, thus being reflected in economic policies, which would influence concessions to foreign companies.

At the beginning of the 1870s, Huanchaca Company had been established in Bolivia with the participation of Chilean investors, Melchor Concha y Toro, among others and Bolivians such as Aniceto Arce, all of them interested in the exploitation of silver from Pulacayo and Huanchaca. The ceasefire between Chile and Bolivia in 1884 again brought under discussion the mining production export from Huanchaca via Calama to Antofagasta port. Economic interests, mainly in mining, became powerful in outlining new railways.

Blakemore suggests that at the beginning of 1887, the Nitrate Co. and Antofagasta Railway had sold its railway and other rights to Huanchaca Company (Blakemore, 1996: 49). Nevertheless, the first Antofagasta Railway Co. Report shows 1886 as the exact date and gives details about the contribution to the merger between Nitrate Co. and Huanchaca Co., more particularly regarding the amount corresponding to the railway: 2,600,000 pesos distributed in: Permanent Rails: 1,600,000 pesos; Rolling Stock: 740,000 pesos, Antofagasta Machine Shop: 150,000 pesos; Dock: 50,000 pesos; and Antofagasta Real Estate: 60,000 pesos, according to Gomez-Zubiela.

The merger between Antofagasta Nitrate Co. and Huanchaca Co. gave rise to the most important and trascendental railway company of the Atacama desert on November 28, 1888: *The Antofagasta and Bolivia Railway Company Ltd.* (A.B.R.C), legalized by the Chilean government on April 2, 1889 and the Bolivian Congress on December 8, 1888. The demands of the preandean topography made engineering work essential in order to overcome natural barriers. This was done by the English engineer Josiah Harding who, between 1883 and 1887, lay the rails from Pampa Alta to Ascotan lake, the bridge reaching 3,956 m.a.s.l. It was November 1883. The famous Conchi duct was in operation until 1918, when a similar one was built (Greve, 1944: 507). The newly established town of Uyuni in Bolivia received the first engine from the brand new company on October 30, 1889. The railway extension from Antofagasta to Oruro on May 15, 1892, under Arce's government, also stimulated the modernizing boom process of the

Bolivian city, together with tin exploitation. It was even stated that civilization was measured in railway kilometers (Mendieta-Parada, 2006: 211). Liberalism years in Bolivia had led to a huge influence of British capitals on the railway network and the connection of the railway with the Pacific Ocean, where "Playa Blanca Metallurgical Installations" was located, connected to Huanchaca Co. built between 1888 and 1892 and in operation until 1902 (Ahumada, 1999; Mitre, 1981; Calderon, 2003). The disappearance of Huanchaca Co. let Antofagasta-Oruro railway under the administration of only the A.B.R.C. in 1903. It is important to note that the arrival of the railway at Calama influenced work usually done by mule packs and alfalfa pasture grounds dedicated to cart transport in the nitrate pampa (Nuñez, 1992).

The prowess of extending the railway from Antofagasta to Bolivia is also shown in the various railway stations that had to be built along its way:

Km from Antofagasta	Altitude	RailwayBranches Stations
0	34	Antofagasta
4	60	Playa Blanca
14	295	Sgto.Aldea
21	408	La NegraArranque hacia cantón A. Blancas
29	554	Portezuelo
35	561	O'HigginsOf.Savona,Pissis y Domeyko Arranque hacia Boquete.
48	573	Uribe
59	682	Prat
70	783	Latorre
83	893	Cuevitas
96	1,014	BaquedanoOf.Ercilla,Astoreca,J.F.Vergara
109	1,164	Cerrillos
117	1,231	El BuitreOf.Sgto.Aldea.
120	1,279	Santa Rosa
122	1,285	Carmen Alto Of.F.Puelma,Condell, Celia
128	1,341	Salinas Of.Lastenia,Aurelia,Carmela Blanco Encalada.
133	1,369	PeinetaOf.Ausonia, Cecilia
136	1,383	CentralOf.A.Edwards.
144	1,414	UniónOf.Anita,Candelaria,Luisis Angamos, Araucana.
148	1,431	PlacillaOf.María, Curicó
154	1,470	SolitarioOf.Filomena, Perseverancia
162	1,534	La NoriaOf.Aconcagua
170	1,623	Sierra Gorda Arranque para Caracoles
179	1,727	Cochrane
205	2,142	Cerritos Bayos
238	2,265	CalamaA Chuquicamata y El Abra
269	2,641	Cere
299	3,015	Conchi
312	3,223	San Pedro
340	3,772	Polapi
360	3,955	Ascotán
387	3,729	Cebollar
412	3,692	San Martín
435	3,696	OllagüeA Collahuasi

Altitudes referring to railway stations close to the Chile-Bolivia border demanded hiring qualified manpower that attracted Bolivian immigration (Gonzalez-Pizarro, 2008a).

The railway hired many technical and non-technical workers and was not far from the problems between capital and work, proper of political liberalism and the so-called local "social matter". As Blakemore states, A.B.R.C. hired manpower, clearly distinguishing the laborer segment (heads of gang, boatmen, port loaders, firemen, trackwalkers, and service personnel) from employees, who enjoyed a higher social status (railway station chiefs, foremen, inspectors and office personnel) (Blakemore, 1996: 155).

Some specific facts hurt the company's reputation in the regional community. Indeed, there was criticism regarding its operations because wagons used to run off the rails and put workers' safety in risk due to lack of personnel or victimization affecting men who built Pampa Alta railway station telegraph line, as dennounced by the local newspaper *La Mañana* on October 1, 1902. Although the railway company contributed to the city with the creation of a Fire Department, "Bomba Ferrocarril", in September 1902, its expansion in the province, such as the petition of land in Mejillones at the end of 1903, was criticized in the National Congress. However, when the government asked engineer Emilio de Vidts to study Mejillones opening in 1905, A.B.R.C. started its greatest investment: The construction of Mejillones machine shop, considered South America's greatest one, with a 300-house camp (Panades, 1990)

The most complex social situation affecting *The Antofagasta and Bolivia Railway Co. Ltd* was the rejection to the petitionary on January 29, 1906, through which the so-called "Mancomunal Obrera de Antofagasta" representing the laborers of the main companies in the city, including the railway, requested one hour and a half for lunch. This request was accepted by all companies, except the English A.B.C.R. administration. The boatmen and Orchard Industry and Smelter workers supported the railway laborers. The strike committee called for a meeting on Febrary 6; while the government sent naval forces and the A.B.R.C administration provided foreigners with guns. Bishop Luis Silva-Lezaeta's eclesiastic mediation was not accepted by the company. The strike ended in a laborers' massacre in Colon Square on February 6, 1906. Two days after, under the pressure of the news, the National Congress passed the first social law: the law of housing for laborers (Gonzalez-Pizarro, 2009b)

The importance of A.B.R.C. in the zone was enormous, particularly when the Peace and Friendship Treaty between Bolivia and Chile was signed in 1904. On the one hand, the treaty stated that the Atacama desert, which was part of the Antofagasta province, would belong to Chile and, on the other hand, Chile would have to build a railway to connect Arica and La Paz, along with allowing free transport from the altiplanic country to Pacific ports. The Antofagasta-Oruro railway perfectly met the requirements stated in the treaty. If, for Bolivia, *The Antofagasta and Bolivia Railway Co. Ltd.* was so important due to railway policy control, obtaining advantages for mineral exports and mechandise imports (Aramayo, 1959; Gómez Zubiela, 1999, Informe, 1959), what was observed in Antofagasta regional economy was as important. Three issues were controversial for Antofagasta and Calama City Halls and mining companies: water supply and its cost for consumers, real estate valuation, and the price for transporting cargo and passengers.

Water supply in the desert was essential for productive activities and also for the feasibility of the railway company since it used steam engines. The problem was the monopoly concerning water rights and its distribution to the urban cities and insdustries in the nitrate pampa.

Huanchaca Co. had purchased the water rights owned by Enrique Villegas, an entrepreneur and regional politician, in 1887. For this reason, when Huanchaca Co. transferred its rights to *The Antofagasta and Bolivia Railway Co. Ltd.*, it was not only connected to railway materials, but also to its water rights. In addition, a law passed on January 21, 1888, allowed Huanchaca Co. to "provide the city of Antofagasta and other territories that can be supplied by the Loa River with tap water" (Anguita, 1912: 64-65). Also, Huanchaca could extend the time to do the necessary work for supplying Antofagasta with water. By a law passed on September 23, 1890, the deadline was extended until October 1, 1892. But it was not known that Huanchaca had become *The Antofagasta and Bolivia Railway Co. Ltd* in 1889. In this way, a situation similar to Tarapaca, concerning John T. North's activities, arose during Balmaceda's government, that is, the direct or indirect monopoly of natural resources, nitrate and water, railway control, or what historian Hernan Ramírez-Necochea would call, the danger of "northification" (Ramirez, 2007: 81). When travelling to Antofagasta in March 1889, Balmaceda visited some railway stations of the central network and promised "the expropriation of all private railways in the whole Republic" (Sagredo, 2001: 144), something probably expressed at the warmth of welcome in every northern City Hall, but these ideas, as many others, did not come true (Blakemore, 1991).

A.B.R.C. tap water supply included industries, nitrate mines at special prices, and the cities of Antofagasta and Calama, among others. The water price and mainly free supply to public services were the focus of a permanent struggle between the company and the Antofagasta City Hall. By a decree on July 30, 1904, the company could "use and enjoy Palpana waters and then obtained the concession of Ujuna Grande, Puquios , and Siloli waters" (Arce: 1930: 263). In addition, it established a policy to fight for free water resources in favor of Antofagasta public services and City Hall (Blakemore, 1996: 104). Antofagasta water supply was criticized because of the tariff charged for houses and the government negation for allowing the company to increase tariffs (Recabarren, 2002). This was a recurrent issue until the late 1960s.

The basic issue for Calama City Hall was the discrepancy between real estate and the valuation established by A.B.R.C and the amount of tax to pay, an issue analyzed from 1915 to 1936 since it strongly affected the City Hall budget (Mondaca, Segovia & Sanchez, 2011).

For mining activity, the basic issues were the railway breach with the liabilities agreed on with nitrate companies, particularly Antofagasta Nitrate Co., the most important Chilean company in this line of business until 1907. Among complaints detailed by Isaac Arce, Pampa Alta administrator, in 1906 are: the towing machine service because its itinerary affected work; the negation to transport workers; the demand for loading wagons in the least possible time; the prohibition to use own cars in the railway; the transport of rails and crossbeams by the nitrate company; placing switches and habilitating by-ways; storage and wagon charges for transporting forage; and excess charges for transporting merchandise, materials, coal, and gunpowder (Archivo, 1906).

The placement of nitrate company rails in the central canton, which finally reached the railway, was hindered by the broad concession given to Milbourne Clark. Nevertheless, since 1887 attempts had been made to build railways independent of A.B.R.C , such as the one between Pampa Alta and San Pedro de Atacama in December 1887. In October 1893, Carlos A. Watters is allowed to build a railway, taking kilometer 20 of the current A.B.R.C. as a starting point. In August 1899, Enrique Barra made a request to build a steam engine railway from Chuquicamata copper mine to Antofagasta-Bolivia railway, being approved in December (Gonzalez-Pizarro, 2008).

In general, the nitrate companies of the central canton had to operate with A.B.R.C. to export their production through Antofagasta port. Nitrate companies usually had private trains inside their premises, between nitrate concessions and supply yards, using locomotives and wagons. One of the most important companies, the Antofagasta Nitrate Co., was equipped with 41 80-250 h.p. steam engines; 8 90-150 h.p. electric engines; 250 6m^3 nitrate wagons; and 540 1-2 m^3 nitrate wagons, apart form 470 other wagons, all of which shows a panorama of the thriving activity in the desert (Gonzalez-Pizarro, 2003: 135).

The expansion of American capital, mainly Guggenheim Brothers', not only in Chuquicamata - connected to Huanchaca Co. from 1899 until its shut-down -, but also in the powerful nitrate industry, when purchasing *Anglo-Chilean Nitrate Company* and *Lautaro Nitrate Company* in the 1920s and 1930s, led to a better planning between A.B.R.C. and the railway stations in the pampa and nitrate mines. Schedules allowed identifying railway stations and the type of railway machinery in use. In 1929, A.B.R.C. took charge of all the nitrate railways, committing to the conservation and repair of the existing ones. The railways became A.B.R.C. property and nitrate companies agreed on a monthly pay for each crossing and meters run.

A.B.R:C. and Chile North Railway established a new train service in 1929 (Thompson, 2003). Two mixed trains weekly covered the distance between Baquedano and El Toco and viceversa so as to give better service to nitrate mine inhabitants. The trains stopped in Baquedano to make connections with passenger trains travelling between Mejillones and Calama. In this way, passengers could arrive at Antofagasta at 3 p.m. In addition, passengers travelling to Baquedano sorroundings could make different combinations to move throughout the pampa and railway stations until arriving at Calama since a direct train to this city was available.

In the mid-1931, A.B.R.C changed the train schedule to Calama, affecting the itinerary of the train running downward, which was scheduled for Wednesday, Thursday, and Saturday. This made *Lautaro Nitrate* organize its own transport service for mail and workers' transport.

To have an impression of the railway service and how it connected different places in the nitrate pampa, let's take a look at the schedule in 1929:

UPWARD TRIP				DOWNWARD TRIP	
Railway stations	Train 83		Railway Stations	Train 84	
	Sunday & Thursday			Monday & Friday	
	Arrival	Departure		Arrival	Departure
Antofagasta	----	9:35			
Baquedano	----	13:45	Toco	----	7:00
La Rioja	15:01	15:05	Chacance	8:02	8:07
Deseada	16:01	16:20	Miraje	8:26	8:31
Los Dones	16:43	17:00	B. Astoreca y Los Dones	8:59	9:01
Lynch	17:18	17:32	Lynch	9:35	9:36
B. Astoreca y Los Dones	18:02	18:05	Los Dones	9:48	9:56
Miraje	18:29	18:40	Deseada	10:15	10:20
Chacance	18:57	19:10	La Rioja	11:10	
Toco	20:12	---	Baquedano	12:05	----
			Antofagasta	15:04	----

Changes made in May 1931 did not greatly affect schedules in the nitrate mines, but they highlighted the importance of Salinas railway station for mixed trains, both regular and special, and also the internacional train.

So, on Monday, the regular mixed train went up to Salinas at 1:15 p.m. and down to the same railway station at 11:15 a.m., following the usual itinerary; on Tuesday, the regular mixed train did not go up, but went down to the station at 11: 15 a.m.; while the international train went up at 1: 15 a.m., making connections between Salinas and Union stations; on Wednesday, the regular mixed train went up at 1:15 p.m. and did not go down, following the usual itinerary; the mixed train from Calama went down on this day and the international train went down, arriving at 6:20 p.m., making the connection above. On Thursday, there were no trains going up and down. On Friday, the regular mixed train did not go up, but 11:15-a.m. train went down, following the usual itinerary; the international train made a stop at Salinas at 11:15 a.m., making the usual connection. On Saturday, the regular mixed train went up at 1:15 p.m. and did not go down; the international train passed by Salinas at 6:20 p.m. On Sunday, only the regular mixed train ran, making a stop at Salinas at 1:15 p.m., when going up, and at 11:15 a.m., when going down (Archivo Historico, 1929).

International train trips from Antofagasta to Bolivia underwent difficulties when making a stop at Salinas station. Complaints included passengers' delay and change of second-class wagons from the mixed train arriving from Calama to add them to the international train. An service of "auto-gondolas" (small old buses) with a capacity of about 25 passengers from "María Elena" and "Chacabuco" nitrate mines was the only means for arriving at Salinas station.

The American administrators of Chuquicamata copper ore deposit signed an agreement with A.B.R.C. to use the railway for exporting the metal resource to international markets. The Chuquicamata train inaugurated in 1914 was connected to the main A.B.R.C. branch. The so-called Chuquicamata branch started from San Salvador station located to the north of Calama and arrived at Punta de Rieles, in a 10-km trip. In addition, A.B.R.C. built Conchi Viejo- El Abra branch in 1906 with a 19-km length to give service to the copper and silver exploitations existing to the north of Calama (Thomson-Angerstein, 1997; Castro 1984). The company profits depended on production levels which, in turn, depended on the copper pound quotation in the London stockmarket. However, *Chile Exploration Company,* the American company exploiting Chuquicamata mine, established *The Chile Exploration Co. Railway* to connect Chuquicamata and San Salvador, providing electricity in 1925. The first Chuquicamata general manager, Fred Hellman, built a railway inside the mine to transport materials and workers. It remained active until the appearance of big trucks in the early 1950s (Monterrey, 2009).

In the 1970s, *The Antofagasta and Bolivia Railway Co.Ltd.* was involved in the difficulties affecting the Chilean politics and economics, until purchased by the Chilean entrepreneur with Croatian ancestors, Andronico Luksic, at the end of 1979 (Blakemore, 1996). In the 1980s, the frequency of the international train was once a week. At present, its acivities focus on commerce to and from Bolivia, along with CODELCO copper shipping.

3. Nitrate railways of El Toco, Aguas Blancas, and Taltal cantons: Private interests

El Toco canton was located between parallels 21° and 23° and meridians 70° and 69° , including 14 nitrate mines. It was the only canton using Shanks and Guggenheim systems since "Pedro de Valdivia" and "María Elena" nitrate mines were located in it, the latter being the last nitrate mine in operation.

Aguas Blancas canton was located between parallels 23° and 24° and meridians 70° and 69°, including 22 nitrate mines.

Taltal canton, below parallel 25° and between meridians 70° and 69°, included 26 nitrate mines.

In the mid-1833, the English man, Edward Squire, built a railway in El Toco canton to connect El Toco nitrate mines with a port between Loa river and Cobija, as established by a law passed on January 23, 1888, which legalized this branch. The railway purchased by *Anglo-Chilian Nitrate and Railway Co. Ltd.* was inaugurated in 1890. *Anglo-Chilian* connected Jose Francisco Vergara nitrate mine in 1910. In 1927, this company was purchased by Guggenheim who could build branches to Pedro de Valdivia and María Elena nitrate mines. This led to the appearance of other stations between Maria Elena and Tocopilla: Tupiza, Cerrillos, Colupito, Central, Barriles, and Tigres. The Central station railway branched off to provide service to nitrate mines located to the SE: Maria Elena, J.F. Vergara, Coya up to Miraje station, and also to the NE to arrive at Ojeda, Puntillas, and El Toco stations, where it branched off again to include other nitrate mines.

Steam engines operated until 1958, being replaced by diesel engines. Their itinerary in the 1950s was scheduled on a weekly basis for passengers and cargo. This itinerary has the most curves and gradients, operating until today and owned by the Chile Chemical and Mining Society which, established in 1968 as a mixed company and then belonging to the State, is now in private hands.

The history of Taltal canton nitrate railway is related to the government decision in 1878 to make prospections for a railway connecting nitrate productive activity with the port. In 1880, the proposal presented by Alfredo Quaet- Faslem was accepted. He transferred his rights to Jorge Stevenson, who established *Taltal Railway Company Ltd.*,. Supported by John Meiggs, Stevenson could build the railway to Refresco in a short time. Between 1887 and 1928, the network enlarged, its shareholders making big profits. Canchas station railway branched off to the NW, in the direction of Santa Luisa and Alemania stations up to Lautaro. To the E, there was a group of five nitrate mines, Flor de Chile being the most notable, whose branch located in Ovalo station branched off again to the NW to connect Caupolican and Bascunan nitrate mines. After the 1930-1931 crisis, the company determined the destiny of the Shanks system nitrate mines, getting rid of, as Ian Thomson states, rolling stock in 1940, *Taltal Railway Company Ltd.* being sold to the Chilean company "Rumie & Sons" in 1960. This company provided service to the last three nitrate mines. Alemania nitrate mine ended operations and the nitrate train was dismantled between 1977 and 1979 (Thomson, 2003).

Coloso-Aguas Blancas railway history is one of the most intricate of the type. Since 1880, nitrate people had been asking the government the construction of a railway for Aguas Blancas nitrate mines and were fighting for an extension of what had been accepted for Juan Besterrica, Juan Vera, and Francisco Mirada to build the rails for a steam train between Antofagasta and Aguas Blancas (Rojas, 1883). In January 1884, the Congress discussed a railway from Antofagasta to Aguas Blancas since the previous one had been rejected. In 1896, Rafael Barazarte was given permission to build a railway between Paposo and Desierto ore deposit. In 1886, Arturo Prat Mining Co. and Taltal Railway Co. received the approval to build a railway between the port and the company's mining installations. On the next year, they were allowed to extend the railway to Cachinal. This issue was again dealt with in August 1889 by J. Phillips in order to build a steam railway between these two places. On September 1, 1897, approval was given to Jose Antonio Moreno to build and exploit a railway between Paposo and Desierto ore deposit (Gonzalez-Pizarro, 2008: 37-38). But the most relevant railway connecting Aguas Blancas nitrate mines was the one requested on November 28, 1898 by the firm "Granja & Domínguez", which would build a railway between Antofagasta and Aguas Blancas. Permission was given a month later, the construction beginning in March 1899. Work done in 1900 showed that the firm was not using Antofagasta piers, but the habilitation of a site to the south of Antofagasta, a fact that revealed the firm strategy to avoid the opposition of the City Hall, boatmen unions, and A.B.R.C. On January 1, 1902, Coloso was ranked as a minor port. In March 1902, the railway connected "Pepita" nitrate mine with Coloso. Carrizo, La Negra, and Varillas stations were built between Coloso and Pepita. At Barazarte station, the branch to the east led to Yungay station with two branches including 90% of the canton nitrate mines. To the SE, it led to Rosario nitrate mine. The railway was open to the public for passenger transport. When Baltazar Dominguez died in 1902, his heirs sold the railway and nitrate belongings to Matias Granja. When this one died, Coloso-Aguas Blancas railway became involved in one of the most commented scandals of the time, mixing business and politics, which in turn involved the government at that time - 1907. This is what some authors have called "the famous affair of Granja house" (Recabarren, Obilinovic, Panades, 1989: 61-67). Finally, the transfer of the railway from Granja to W.R. Grace in 1908 ended in the railway being in the hands of *The Antofagasta (Chili) and Bolivia Railway Company Ltd.* It was 1909. In this way, all the private railways in the province were in the hands of English A.B.R.C. capitals. The train continued operations as long as nitrate mines working with the Shanks system could be profitable. In 1932, some branches were dismantled and it definitely disappeared in 1961.

In the nitrate pampa, the mines operating with private trains ruled railway jobs. Each nitrate mine had a Traffic Chief in charge of keeping the rails in good state and do necessary repairs. He was the direct boss of the engine driver, whose main job was to keep the boiler water at the right level and check the good state of all the engine keys and valves; firemen, dedicated to keep the engine throroughly clean, manage fire, and burn the coal or oil; and lastly, brakemen, in charge of taking care of the brakes of the train in motion. Railroad workers worked in the rails. There were also the so-called engine starters in charge of the lamps of trains in motion and lighting the engine fire at dawn. (Macuer, 1930: 170-172; Gonzalez-Pizarro, 2003:312-314).

4. The Noth longitudinal railway: The state intervention

The construction of the North Longitudinal Railway started when the government decided to extend the fiscal railway from Pueblo Hundido, in Copiapo province, to Pintados, in Tarapaca. Reasons of national safety and territory integration lay behind this venture. A public bid, after some failures, was awarded to *Chilian Northern Railway Co. Ltd* in 1910.

The well-known North Longitudinal, the famous *Longino,* was finally inaugurated on January 10, 1919 (Thomson, 2003), although the definite exploitation of de journey between Iquique and Calera started on March 19, 1930 (En Viaje, 1960: N° 325).

The North Longitudinal was connected to Santiago and Valparaíso trains. It included several stations, starting in Pueblo Hundido, followed by Altamira, Catalina, Balmaceda, Los Vientos, Lacalle, Agua Buena, Aguas Blancas, Oriente, Palestina, Desierto, Baquedano, (to Antofagasta and Calama), Rioja, Deseada, Los Dones, Lynch, Miraje, Chacabuco, El Toco, Santa Fe, and Quillagua. In 1966, the trip from Calera to Iquique was scheduled for Sunday (with a connection to and from Antofagasta) using first-class, second-class, and buffet wagons, the latter being the most complete; on Thursday, there was a trip from Baquedano to Calama, with second-class and buffet wagons; on Saturday, the train arrived at Antofagasta, with the same wagons as on Thursday; on Tuesday, there was a train with second-class and buffet wagons arriving at El Toco. The trip from Iquique to Calera was scheduled for Monday and Thursday; the trip from Antofagasta to El Toco, Tuesday and Saturday (En Viaje, 1966: N° 388)

Unlike what happened to A.B.R.C., which would later add Chuquicamata copper mining production to its load transported to Antofagasta, the North Longitudinal had to overcome economic difficulties in time. Ian Thomson, the English railway specialist, only slightly states the adverse picture of the Longitudinal crossing the desert: "A longitudinal railway to travel along one of the world's most arid zones had been built; (i) where agricultural production is practically none; (ii) where population is non-existing, except for a very limited number of small cities; (iii) where mining production is also scarce, and; (iv) which would have very limited expectations to make long journeys to and from the country's central and south zones" (Thomson, 2003:48). In the 1950s, its material had not been replaced and maintainance costs were high. The accelerated disappearance of Shanks system nitrate mines involved passenger and cargo loss. Only in the mid-1950s, the State Railway Co. considered connecting appealing desert locations to its tourist agenda. It was the beginning of a tourist massification boom in the Chilean north, supported by the railway as a non-elitist popular means of transport, as experimented in the U.S.A. (Sheffer, 2001). This happening helped discovering the northern geography as a tourist landscape, together with a State policy to support northern cities. In 1934, the State Railway Co. started publishing a *Tourist Guide* which did not strongly stimulated visits to northern "tourist" attractions, but those in the south. Nevertheless, another State Railway Co. publication, *En Viaje,* included places located in the preandean locations and the city of Antofagasta in its pages in the late 1940s. Paradoxically, the North Logitudinal did not arrive at these places (except the city of Antofagasta, where A.B.R.C. had built a railway branch between Baquedano and the port, based on an agreement with *The Chilian Northern* in 1921). In this way, the main cities could be known and a "leisure and winter tourism" could be institutionalized in the northern zone, where the railway and roads made recreation possible (Gonzalez-Pizarro, Ms).

The North Longitudinal Railway administration was transferred to A.B.R.C. in 1919 under the name *Chilian Northern*, a situation that remained until October 1957, when the government decided to transfer *Chilian Northern* to the State Railway Co., authorizing A.B.R.C. to administrate it until May 1961. The most popular Atacama dersert train operated until June 9, 1975.

The northern novelist Hernan Rivera-Letelier would strongly evoque the famous *Longino* in his work **Trains go to Purgatory.**

5. The Antofagasta-salta railway: From citizen initiative to bi-state concretion

One of the railways having the greatest support by citizens, after Antofagasta-Salinas and Caracoles, was the Antofagasta-Salta railway.

In November 1966, *En Viaje* magazine director, Manuel Jofre, wrote that this venture had started in Argentina, by naming a study commission, but it soon found opposition on both sides because "some sectors considered that it was against the interest of farmers in the south of both Chile and Argentina. For this reason, the project was delayed" (Jofre, 1966:17).

It was precisely the A.B.R.C. - *The Antofagasta and Bolivia Railway Co. Ltd* - which made the first studies in 1888 to connect the Argentinian northeast with the Chilean north, through a group of engineers. One of them, Luis Abd- El Kader, with Italian-Arab ancestors, was greatly influencial in the urban planning of Antofagasta, where, as Thomson & Angerstein state, a connection with *Argentina Grand Central Railway* would be looked for with a design starting from Sierra Gorda station, going through Caracoles mine, San Pedro de Atacama, and Aguas Calientes and then arriving at the Argentinian territory through Huaytiquina. These authors conclude that this study had great advantages, a steady income from local transport because the railway crossed an area very rich in minerals, among others (Thomson- Angerstein, 1997: 172-173). The project did not succeed probably owing to the territory dispute, solved in 1899, of the Atacama puna between Chile and Argentina.

A new impulse to this venture came from the coincidence between Mejillones re-foundation efforts made by the government and A.B.R.C. request for land to install its machine shop bewteen 1904 and 1906, on the one hand, and the Argentinian renewed effort made by engineer Manuel Sola who, in 1905, called the government attention to the huge advantages Mejillones offered to export the agricultural and cattle production from Salta and the new Andes territory through its port. According to Sola, "All the input for men and animal survival easily find a market in this province. Cattle is imported from Salta (Argentinian Republic) and the south of Chile; flower, from California; rice, sugar, and fruits, from Peru; tobacco, from La Habana and Bolivia; wine, beer, cereals, beans, vegetables, barley, dry grass, and another hundred products from the south of Chile" (Solá, 1906:19). Sola's ideas were supported by other Argentinian reports such as Dr. Arturo S. Torino's in 1906 and exposed to the Argentinian Congress by the Minister of Foreign Affairs at that time, Estanislao Zeballos. On the following year, July 1907, Horacio Fabres, Manuel Maira, and Santiago Zanelli requested the government authorization to build a transandean railway to connect Mejillones and Salta. After putting it off several times, the railway was inaugurated in June 1911(Sociedad Nacional de Agricultura, 1922: 5-6).

But there was also another issue in this connectivity: the Chilean government authorization for the construction of a new Antofagasta port based on Law N° 2390, passed on September 7, 1910, which was fruitful in 1913 when the Port Commission reported the connectivity - Antofagasta natural attraction zone - of the Mejillones-Salta projected railway, which should take advantage of the new installations in the future (Gonzalez-Pizarro, 2010[a]).

This proposal was supported by citizens - laborers' unions, political parties, and commercial and industrial associations –; disseminated in Open City Hall Meetings; and also supported by the establishment of the Salta Pro-Railway Executive Commission. Since April 5, 1920, multitudinal meetings in the form of Open City Hall Meetings were organized in favor of the proposal. For mayor Maximiliano Poblete-Cortes, the railway was of "national conveniente because it will help in the development of one of the country's most important regions; it will attract a big part of Argentinian commerce to the Pacific; and, therefore, there will be an increase in freight and cargo for our merchant marine; some of our industries, such as the nitrate one, will increase their production to fertilize land producing sugar cane, cotton, etc. Other industries such as shoe-making, canned food, and maybe other ones will have a safe market. Concerning regional coexistence, we believe no one can deny it. At present, the life and progress of Antofagasta and the whole region are closely related to the development and prosperity of nitrate and copper industries (Gonzalez-Pizarro, 1994, 1995, 1999, 2002).

The same response was given by Argentina, where a Pro-Pacific Railway Commission was established in 1921, its director being Luis de los Rios. This commission organized "various acts that gave prestige and widely disseminated Salta population's mood". The railway construction also lead to a geopolitical mistrust view from the military prism, while in Antofagasta, civilians reaffirmed their conviction of the integration with the Argentinian northeast" (Benedeti, 2005).

Government actions from both sides led to the investment budget agenda agreed on by the Chilean Minister of Foreign Affairs, Ernesto Barros Jarpa, and the Argentinian Envoy Extraordinary and Minister Plenipotentiary in Chile, Carlos M. Noel, on April 25, 1922. The project, however, could only be improved in 1928. In 1930, Argentina had already built the railway from Salta to San Antonio de los Cobres. Finally, the railway was constructed in Augusta Victoria station sector in 1937, after recovering from the 1930-1932 world crisis (Thomson, 2003, 2006).

During the early 1940s, the government increased its contribution to speed up the railway construction. Curiously enough, as stated by Alejandro Benedetti, the railway that had been thought of by Argentina to improve the Andes Territory would arrive late, when the territory had disappeared in 1943 to favor Salta, Jujuy, and Tucuman provinces. This was a bad sign. The Chilean north still cherished hopes for the railway, months before being inaugurated because, apart from the work of many people, "the cost of living in the northern provinces will be cheaper… it will avoid the current supply difficulties and complications due to the scarcity of cargo ships… The railway does not only have an economic mission, but it also takes the torch of progress, culture, mutual knowledge and, therefore, people's physical and spiritual welfare everywhere" (Szigethy, 1948: 67-68).

On February 20, 1948, the President of Argentina, Juan D. Peron, inaugurated the railway in the Argentinian sector, with the presence of Antofagasta Mayor, Juan de Dios Carmona.

In 1949, the Antofagasta-Salta railway was scheduled on a weekly basis, the journey taking two days: the train included regular and buffet wagons. It started from Antofagasta on Sunday and arrived at Salta on Tuesday.

Nevertheless, high transport costs, exceeded by truck competente in time, did not meet the expectations of both regions. A.B.R.C. operated the railway, as contrated with the government in the 1920s, and as stated by Ian Thomson, assigned "old engines left over from other operations to cargo trains (and) in the mid-1950s, assigned relatively modern steam engines for passengers' service" (Thomson, 2006: 145) until 1964. In the 1960s, the State Railway Co. started operations on the rails. At the end of 1970, passengers' trips were cancelled. In 1990, it was transferred to Ferronor S.A., a Production-Fostering Corporation (CORFO, for its acronym in Spanish) company, privatized in 1996. At present, the railway operates only sporadically.

6. References

Ahumada, María T. 1999. El Establecimiento Industrial de Playa Blanca en Antofagasta. Antofagasta: Ediciones Santos Ossa.

Anguita, Ricardo. 1912. Leyes promulgadas en Chile desde 1810 hasta el 1° de junio de 1913. Santiago de Chile: Imprenta, Litografía i Encuadernación Barcelona, Tomo III.

Aramayo, Cesáreo. 1959. Ferrocarriles bolivianos. Pasado, presente y futuro. La Paz: Imprenta Nacional.

Arce, Isaac. 1930. Narraciones Históricas de Antofagasta. Antofagasta: Imprenta Moderna.

Archivo Escuela de Derecho, Universidad Católica del Norte. 1906. Archivo de Isaac Arce. Carpeta varia "Personal y Salitrera".

Archivo historico, Universidad Católica del Norte. 1929. Archivo Salitrero Oficina Chacabuco: Caja "Medios de Transporte. Años 1920-1939.Transporte y Comunicaciones". Circular de la Administración del F.C.A.B, Antofagasta, 29 de mayo de 1929.

Benedetti, Alejandro. 2005. "El ferrocarril Huaytiquina, entre el progreso y el fracaso. Aproximaciones desde la geografía histórica del territorio de los Andes", Revista Escuela de Historia, Salta, enero-diciembre, N° 4, 123-165. Disponible en http://wwwscielo.org.ar/scielo.php?script=sci_arttext&pid=S1669-90412005000100007&ing=es&nrm=iso. Consulta el 14 de agosto de 2011.

Bertrand, Alejandro. 1885. Memoria sobre las cordilleras del desierto de Atacama i rejiones limítrofes. Santiago de Chile: Imprenta Nacional.

Blakemore, Harold. 1991. "¿Nacionalismo frustrado? Chile y el salitre, 1870-1895" en Harold Blakemore, Dos estudios sobre salitre y política en Chile (1870-1895). Editado por Luis Ortega. Santiago de Chile. Departamento de Historia. Universidad de Santiago de Chile.

Blakemore, Harold.1996. Historia del Ferrocarril de Antofagasta a Bolivia 1888-1988. Traducción de Juan Ricardo Couyoumdjian y Beatríz Kase. Santiago de Chile: Impresos Universitarios S.A.

Bowman, Isaiah. 1924. Desert Trails of Atacama. New York: American Geographical Society.

Braudel, Fernando, 1970. La historia y las ciencias sociales. Madrid: Alianza Editorial.

Bravo Quezada, Carmen G. 2000. La Flor del Desierto. El mineral de Caracoles y su impacto en la economía chilena. Santiago de Chile: Dibam, Lom Ediciones, Centro de Investigaciones Diego Barros Arana.

Bresson, André. 1997 [1886]. Una visión francesa del Litoral Boliviano (1886). La Paz: Stampa Grafica Digital.

Calderón G. Fernando, Coordinador. 2003. Formación y evolución del espacio nacional. La Paz: Ed.Ceres.

Castro, Marina T. 1984. F.C.A.B. Una ruta de Nostalgias. Antofagasta: s.p.i.

En Viaje, 1960. "Ferrocarriles vencedores del desierto y la montaña". Santiago: noviembre, Número 325

En Viaje, 1966. "Valores de pasajes sencillos en trenes expresos, ordinarios y mixtos, entre las principales estaciones de Santiago a Calera, Iquique y ramales". Santiago: febrero, Número 388.

Jofré N. Manuel. 1966. "Antofagasta: terminal FF.CC. Internacionales", Revista En Viaje, noviembre, Número 397.

Gómez Zubiela, Luis R. 1999. Ferrocarriles en Bolivia. Del anhelo a la frustración 1860-1925. Tesis de Licenciatura en Historia. Universidad Mayor de San Andrés, La Paz. Una versión en Políticas de Transporte Ferroviario en Bolivia 1860-1940. Disponible en www.boliviaenlared.gm/.../politica-transporte-ferroviario-bolivia.pdf. Consulta 19 de agosto 2011.

Gonzalez Pizarro, José A. 1994. "El FF.CC. de Antofagasta a Salta: Regionalización e Integración", Actas del II Seminario Internacional de Integración Subregional, Universidad Nacional de Jujuy-Universidad Católica de Salta.

Gonzalez Pizarro, José A. 1995. "La expresión regionalista en Antofagasta: base social, demanda comercial y canalización política. El FF.CC. de Antofagasta a Salta en 1920-1930" Primer Encuentro de Historia Económica y Social, Universidad de Santiago de Chile.

Gonzalez Pizarro, José A. 1999. "El ferrocarril que dio vida a la región" y "El desarrollo de la complementación económica" en Noa-Norte Grande. Crónica de dos regiones integradas. Buenos Aires: Embajada de Chile en Argentina, pp. 87-93- 150-154.

Gonzalez Pizarro, José A. 2002. "Espacio y política en Antofagasta en el ciclo salitrero. La percepción del desierto y el sentimiento regionalista, 1880-1930" en Viviana Conti-Marcelo Lagos, Compiladores, Una Tierra y tres naciones. El litoral salitrero entre 1830 y 1930. Jujuy: Universidad Nacional. Unidad de Investigación en Historia Regional, pp. 251-290.

Gonzalez Pizarro, José A. 2003. La pampa salitrera en Antofagasta. La vida cotidiana durante los ciclos Shanks y Guggenheim en el desierto de Atacama. Antofagasta: Corporación Pro Antofagasta.

Gonzalez Pizarro, José A. 2005. "Chile y Bolivia (1810-2000)" en LACOSTE, P. (Compilador) Argentina - Chile y sus vecinos. Mendoza: Editorial Caviar Bleu. Colección Cono Sur, Tomo I, pp. 335-392.

Gonzalez Pizarro, José A. 2008. "Conquering a natural boundary. Mentalities and technologies in the communication paths of the Atacama desert", Revista de Geografía Norte Grande, Pontificia Universidad Católica de Chile, Número 40, septiembre, pp. 23-46.

Gonzalez Pizarro, José A. 2008[a]. "La emigración boliviana en la precordillera de la región de Antofagasta: 1910-1930. Redes sociales y estudio de casos", Revista de Ciencias Sociales, Iquique, Número 21, pp. 61-85.

Gonzalez Pizarro, José A. 2009. "Contrasting Imaginaries: The Atacama Desert perceived from the Region and Observed from the Nation", Revista de Dialectología y Tradiciones Populares. Antropología. Etnografía. Folklore. Madrid, Vol. LXIV, N° 2, Julio-diciembre, pp. 91-116.

Gonzalez Pizarro, José A. 2009a. "The province of Antofagasta. Creation and consolidation of a new territory in Chile: 1888-1933", Revista de Indias. Madrid, Vol. LXX, N° 249, mayo-agosto, pp. 345-380.

Gonzalez Pizarro, José A. 2009b. "La huelga/masacre de la Plaza Colón: 6 de febrero de 1906 en Antofagasta. Las lecciones para la historia" en Pablo Artaza, Sergio González Miranda, Susana Jiles Castillo (Editores), A cien años de la masacre de Santa María de Iquique. Santiago: Lom Ediciones, pp. 211-239.

Gonzalez Pizarro, José A. 2010. "La influencia de la legislación municipal boliviana en Antofagasta, 1879- 1888. Un capítulo desconocido en la historia del derecho público chileno", en Estudios en Honor de Bernardino Bravo Lira, Premio Nacional de Historia. 2010. Revista Chilena de Historia del Derecho, Facultad de Derecho de la Universidad de Chile, Número 22, Tomo II, 913-937.

Gonzalez Pizarro, José A. 2010ª. "El Puerto Fiscal y la ciudad de Antofagasta" en Liliana Cordero Vitaglic, Coordinación y Edición, Historia Gráfica del Puerto de Antofagasta. Abril 1919-Agosto 1929. Antofagasta: Universidad Católica del Norte-Gobierno de Chile, Gobierno Regional de Antofagasta- Consejo Regional- Consejo Nacional de la Cultura y las Artes, Impresión Graficandes, pp. 9-27

Gonzalez Pizarro, José A. Ms. Geografía del desierto y turismo de la naturaleza. La revista En Viaje y la mirada sobre el paisaje nortino: 1945-1966. En Revista de Geografía Norte Grande, Pontificia Universidad Católica de Chile, aceptado y a publicarse en el número 52, septiembre de 2012.

Greve, Ernesto. 1944. Historia de la Ingeniería en Chile. Santiago de Chile: Imprenta Universitaria, Tomo III.

Informe Económico Ferrocarril de Antofagasta a Bolivia. 1959. The Bolivia Railway Company. Anexo. Un comentario del Dr. Humberto Fossati. Oruro: Universidad Técnica de Oruro. Departamento de Extensión Cultural.

Lagos C. Guillermo, 1981. Historia de la frontera de Chile. Los tratados de límites con Bolivia. Santiago de Chile: Editorial Andrés Bello.

Macuer Llaña, Horacio. 1930. Manual Práctico de los trabajos en la Pampa Salitrera. Valparaíso: Talleres Gráficos Salesianos.

Mandiola, Juan-Castillo, Pedro. 1894. Guía de Antofagasta. Antofagasta: Imprenta El Industrial.

Mendieta Parada, Pilar. 2006. "Oruro: ciudad moderna y cosmopolita 1892-1930" en Ximena Medinacelli, Coordinación, Ensayos Históricos sobre Oruro. La Paz: Sierpe Publicaciones.

Mitre, Antonio. 1981. Los Patriarcas de la Plata. Estructura socioeconómica de la minería boliviana en el siglo XIX. Lima: Instituto de Estudios Peruanos.

Mondaca R. Carlos- Segovia B. Wilson-Sánchez G. Elizabeth. 2011. Historia y Sociedad del Departamento del Loa. Calama, una mirada desde los archivos. El municipio y la construcción social del espacio 1879-1950. Calama: Orizonta Producciones Digitales.

Monterrey C. Nancy. 2009. Chuquicamata. Otras voces te recuerdan. Antofagasta: Sergraf Ltda.

Nuñez Atencio, Lautaro.1992. Cultura y conflicto en los oasis de San Pedro de Atacama. Santiago de Chile: Editorial Universitaria.

Panades, Juan. 1990. "La maestranza del Ferrocarril Antofagasta a Bolivia en Mejillones". Anexo I. En Julio Pinto Vallejos- Luis Ortega Martínez, Expansión minera y desarrollo industrial: un caso de crecimiento asociado (Chile 1850-1914). Santiago de Chile: Universidad de Santiago de Chile, pp. 113-136.

Pérez, Alexis. 1997. "El intercambio comercial Bolivia-Chile y el tratado de límites de 1874" en Rossana Barragán, Dora Cajías, Seemin Qayam, Compiladoras, El Siglo XIX.

Bolivia y América Latina. La Paz: IFEA-Embajada de Francia- Coordinadora de Historia, Muela del Diablo Editores.

Philippi, Rodulfo .A. 2008 [1860]. Viaje al desierto de Atacama. Estudio Preliminar de Augusto Bruna-Andrea Larroucau. Santiago de Chile: Biblioteca Fundamentos de la Construcción de Chile (Facultad de Historia, Geografía y Ciencia Política, Pontificia Universidad Católica de Chile-Fundación RA Philippi de Estudios Naturales-Biblioteca Nacional de Chile- Cámara Chilena de la Construcción), Tomo 39.

Ramirez N. Hernán. 2007. Balmaceda y la contrarrevolución de 1891. En Hernán Ramírez Necochea, Obras Escogidas. Selección, edición y estudio preliminar Julio Pinto. Santiago de Chile: Lom Ediciones- Consejo Nacional de la Cultura y las Artes.

Recabarren R. Floreal. 2002. Episodios de la vida regional. Antofagasta: Corporación Pro Antofagasta- Universidad Católica del Norte.

Recabarren R. Floreal-Obilinovic A. Antonio-Panades V. Juan. 1989 [1983 1° Ed).Coloso. Una aventura histórica. Antofagasta: Imprenta Atelier.

Rojas D. Matías. 1883. El Desierto de Atacama i el Territorio Reivindicado. Antofagasta: Imprenta de El Industrial.

Sagredo Baeza, Rafael. 2001. La gira del presidente Balmaceda al norte. El inicio del "crudo y riguroso invierno de su quinquenio", (verano de 1889). Santiago de Chile: Lom Ediciones-Universidad Arturo Prat- Centro de Investigaciones Diego Barros Arana.

San Roman, Francisco. J. 1896. Desierto y Cordilleras de Atacama. Santiago de Chile: Imprenta Nacional. 2 tomos.

Sheffer, M. 2001. See America First. Tourism and National Identity, 1880-1940. Washington: Laborersonian Institute Press.

Sociedad Nacional De Agricultura, 1922. Ferrocarril trasandino de Antofagasta a Salta. Santiago de Chile: s.p.i.

Solá, Manuel. 1906. Ferrocarril Trasandino de Salta a Mejillones o Antofagasta. Salta: Imprenta y Tipografía El Cívico.

Szigethy, Teodoro de. 1948. "El Ferrocarril de Salta a Antofagasta y su importancia", Revista En Viaje, febrero, Número 172, pp. 66-68.

Thompson, Ian. 2003. Red Norte: La historia de los ferrocarriles del norte chileno. Santiago de Chile: Publicación patrocinada por el Instituto de Ingenieros de Chile, Imprenta Silva.

Thomson, Ian Thomson. 2006. "Los ferrocarriles del Capricornio Andino", en Angel Cabezas, María Isabel Hernández, Lautaro Núñez, Mario Vásquez, Comité Editor, Las Rutas del Capricornio Andino. Huellas milenarias de Antofagasta, San Pedro de Atacama, Jujuy y Salta. Santiago de Chile: Consejo de Monumentos Nacionales, Santiago de Chile, pp. 137-149.

Thomson, Ian- Angerstein, Dietrich. 1997. Historia del ferrocarril en Chile. Santiago: Dibam-Centro de Investigaciones Diego Barros Arana.

4

Competitiveness and Sustainability of Railways

Dave van der Meulen and Fienie Möller
Railway Corporate Strategy CC
South Africa

1. Introduction

1.1 Failure and success: Competitiveness and sustainability

The world's railway population spans many outcome variations between failure and success. The study of differences is of course the foundation of scientific research: The ability to understand what drives such differences facilitates cognitive positioning of railways for success, or more specifically, competitiveness and sustainability. The authors have pioneered research that contributed some understanding, on a journey that commenced with research to describe the global railway setting, and ultimately applied multivariate statistical analysis to discover how railways adapted to their particular settings. The findings were published piecemeal as they emerged (e.g. International Heavy Haul Association 1997, 2007, and 2009; Railway Gazette International, 2006; Transport Research Arena 2012; and World Congress on Railway Research, 2003, 2006, and 2008), and have been integrated for the first time in this chapter to present a global overview of the railway industry. As put forward here, the research foundation is still evident, but only sufficient detail to support the storyline has been retained. The present objective is to emphasize interpretation and significance of the findings for future railway positioning, within the available page limit. Reference will of course be made to the underlying research where appropriate.

1.2 Building a foundation for research into railway positioning

The research stream originated in a need to redress South Africa's colonial railway heritage of narrow track gauge, light axle load, low speed, small vehicle profile, steep gradients, and monolithic state ownership, which attributes posed many challenges. Some were amenable to technological solution—for example it developed successful heavy haul operations—but others remained endemic. Solutions to similar challenges emerged around the world during railway renaissance and ensuing railway reform. However, the need to control for many differences among such solutions and their settings deterred research to understand which interventions work and which do not. It will become evident, for example, that debating the merits of vertical integration and vertical separation can miss the point that either can work if a railway is inherently competitive, while neither will work if it is inherently uncompetitive. Unlike other more or less homogeneous modes such as airlines, which are challenged to differentiate their offerings, rail's attributes are so heterogeneous that their variance boggles the unaided human mind. Gradually awareness dawned that only high-level numerate research would make progress.

The field of corporate strategy is a well established major in business administration and business leadership. It addresses the total enterprise and how various functions interact to achieve objectives. Thus while all enterprises encompass several common functions, e.g. human capital, information technology, and so on, they must also manage their distinct core business, whether that be banking, mining, whatever, or in the case of railways, moving goods and people to support their logistics and mobility needs. However, googling *railway corporate strategy* returned items dominated by the authors' enterprise and their publications. The unique contribution of the present research stream is therefore a grounded understanding of corporate strategy with respect to the core business of railways.

2. Railway positioning

2.1 Competitiveness fundamentals

To comprehend railway positioning, it is helpful to examine railway competitiveness vis-à-vis that of other transport modes by considering their degrees-of-freedom-of-movability. Three degrees-of-freedom-of-movability (e.g. aerial- and submarine transport) offer high, spatial movability, but at relatively high cost. Next, two degrees-of-freedom-of-movability (e.g. unguided surface transport) offer lower, surface movability at lower cost. Last, one degree-of-freedom-of-movability (e.g. guided surface transport) offers only limited, linear movability, back and forth on its guideway. To the extent that limited movability reduces the value of their offering, guided surface transport modes such as railways must offer compensating advantages to compete effectively against other modes that offer higher movability, including door-to-door transport.

Axiomatically, such compensating advantages should inhere in technologies that differentiate guided surface transport from other transport modes. A vehicle-guideway pair ensures precise application of vertical loads, and safe application of lateral loads: Wheel-rail contact mechanics develop vertical- and lateral force components, technologies named *Supporting* and *Guiding* by Vuchic (2007: 449), which can sustain respectively heavy axle load and high speed. One may leverage Supporting and Guiding by combining two or many vehicles, to scale capacity as required, a technology the authors named *Coupling*. Supporting, Guiding, and Coupling are the three *genetic technologies* that distinguish guided surface transport from all other transport modes: *Inherent competitiveness* is defined here, and is measurable as, the extent to which such modes exploit their genetic technologies.

Note that the preceding two paragraphs have been generalized to guided surface transport, of which railways is a subset. Other guided surface transport modes also exist, which do not use steel-wheel-on-steel rail. We shall return to them in the context of urban guided transit in §3.3 and §3.5. Until then, this chapter addresses railways.

Cross-breaking Bearing and Guiding, in Figure 1, yields four railway market spaces. For this purpose, speed in tens of km/h is low speed, speed in hundreds of km/h is high speed; axle load above 25 tonnes is heavy. Three market spaces feature high competitiveness, by exploiting two or more of rail's genetic technologies—namely Heavy Haul (Supporting and Coupling), High-speed Intercity (Guiding and Coupling), and Heavy Intermodal[1] or Double

[1]Heavy intermodal traffic has heavy axle load: It is distinct from light traffic that simply transfers from one mode to another.

Stack (Supporting, Guiding, and Coupling). All three have demonstrated robust sustainability in competition with other transport modes.

Axle Load	Low Speed	High Speed
Heavy	Heavy Haul	Heavy Intermodal (Double Stack)
Light	Classic Passenger, General Freight, Urban Rail	High-speed Intercity, Regional Rail

Fig. 1. Railway market spaces by Axle Load and Speed

The fourth market space is potentially weak—light axle load combined with low speed exploits neither Supporting nor Guiding genetic technology. Failure to exploit the remaining genetic technology, Coupling, in general freight- and classic long-distance passenger rail applications exacerbates their weakness, hence competitors erode their markets. Depending on whether economic-, political-, or social objectives determine their destiny, such railways are respectively eliminated, protected, or subsidized.

2.2 The special case of urban rail

Human passengers as payload do not achieve high axle load by railway standards, even in double deck vehicles. Furthermore, the comfort criteria and physical laws pertaining to acceleration, coasting, retardation, and station dwell time, maximize the capacity function at around 80km/h. Therefore urban rail cannot maximally exploit either the Supporting- or the Guiding genetic technology: It is confined to a potentially weak market space, in which it can exploit only the Coupling genetic technology. By forming vehicles into trains, it can achieve shorter mean headways than would be attainable by the same number of autonomous vehicles, thereby maximizing passenger capacity per direction per unit time. However, where urban rail does not exploit that one and only genetic technology to realize its capacity potential, it is vulnerable to competition from the rubber-tyred modes Automated Guided Transit, Bus Rapid Transit, and Monorail. The maximum speed of all urban guided transit modes is similar, so winners must leverage headway to maximize capacity. With no pretence at high axle load, rubber tyred modes exploit their consistent high adhesion to encroach on rail's eminent domain.

2.3 The railway renaissance

From bleak prospects in the first decade after World War II, railways in many countries have learned to exploit the competitive strengths that inhere in rail's genetic technologies. The following events mark the course of their learning:

In 1964, Japan introduced the world's first commercial high-speed intercity trains (Japanese Railway, 1964); they exploited the Guiding genetic technology to reach speeds of 210km/h, and leveraged high capacity by exploiting the Coupling genetic technology. Today, high-speed trains attain average speeds in excess of 300km/h, and move 20 000 passengers per hour per direction.

In 1972, a landmark article (Tracks to, 1972) recognized heavy haul as a distinct market space. By then, Supporting and Coupling technologies and equipment to Association of American Railroads specifications had spread abroad to dedicated railways conveying bulk-commodities. Today, heavy haul lines can move 400 million tonnes per year in trains of 300 cars or more.

In 1980, the United States' Staggers Act deregulated its railways: The ensuing wave of innovation among other triggered introduction of double stack container trains (Levinson, 2006). They enhanced inherent competitiveness through increasing axle load despite conveying low-density high-value freight in containers, and leveraged it further with the Guiding and Coupling genetic technologies.

In 1989, the fall of the Berlin Wall tipped the balance of power across the world toward those advocating democratic, consensual, free-market-oriented governance (Friedman, 2006), an ongoing process that stimulated economic globalization. In the railway supply industry, the resultant increased competition and trade rationalized many nationally-fragmented system integrators into fewer strong global brands. Concurrently, accelerating agglomeration in developing economies has expanded the urban rail market. The number of cities has proliferated by some seventy in the last decade.

Economic globalization has of course been transforming all four abovementioned railway market spaces: As examples, the *Global Rail Freight Conference 2007* in New Delhi reflected that transformation in its title, and World Congress on Railway Research 2008 reflected it in its theme *Towards a global railway.*

The foregoing four events revitalized railways in those countries that appreciated the imperative to enter as many of rail's inherently competitive market spaces as applied to them. Their accumulation across all railway countries has become known as the *railway renaissance*. It has precipitated a substantial body of data, able to support research into the modalities. However, even as the railway mode enters its third century as a strong competitor, many railways still have not integrated seamlessly into global logistics and intelligent mobility; they look different from one another, and even from many other global service industries. So how does one undertake research that will lead to some understanding of the differences among them?

3. Railway adaptation: A research paradigm

3.1 Background

When informally comparing railways, which had not joined the renaissance, to those that had, the latter seemed to have acquired a modicum of consistent identity, or *corporate citizenship*. The latter concerns an enterprise's profitability and sustainability; balancing stakeholder expectations, including those of customers, suppliers, and communities in which it operates; maintaining sustainable partnerships with all levels of government; and accepting its role in developing countries (World Economic, n.d.). Railways attain this standing when their corporate citizenship resembles that of other global service industries, such as airlines and logistics service providers. Corporate citizenship therefore provided a sensible perspective on which to found the research reported here. It supports a social sciences behavioural approach, because human behaviour drives enterprises.

3.2 Research in a dynamic, global setting

Railway countries have adapted themselves for competitiveness and sustainability to varying degrees in a globalized industry. Hence, in addition to rail's genetic technologies, which address their inherent competitiveness, it is necessary to control for setting variables, which influence their positioning. The research design must seamlessly compare railways in command economies with those in free economies, open access with vertical integration, heavy haul with transnational operators, and so on. It must also compare monolithic national railways, which may publish comprehensive statistics, with entities whose data are consolidated at a higher level, and with small operators whose data are confidential.

Corporate citizenship is by definition an ongoing process that requires observations over time. Behaviour implicitly includes a time scale—snapshot data cannot observe it. A behavioural approach can naturally support the foregoing requirements. One of the authors developed a methodology for longitudinal railway corporate strategy research using large samples in a doctoral dissertation (Van der Meulen, 1994), which methodology underlies the research reported here.

Fortuitously, the global population of cities and countries with railways is sufficiently small to avoid sampling, yet, using longitudinal research, at the same time sufficiently large to support multivariate statistical analysis. It has been mentioned that line haul- and urban rail are positioned differently, i.e. in respectively inherently competitive- and inherently weak market spaces. They were therefore researched separately, first line haul and thereafter urban rail. The necessary methodological distinctions start immediately below, and have been maintained throughout the rest of this chapter as appropriate.

3.3 The research questions

The authors formulated their research questions within the context of an enterprise's corporate citizenship, as represented by its Contribution to Society, Core Business, Social Investment, and Engagement in Public Policy, as well as resources deployed to set about its task. In respect of the three market spaces that demarcate line-haul railways, they hypothesized the existence of some number of underlying longitudinal, or time-dependent, relations among variables associated with positioning line haul railways. The research question was therefore: *Can one identify archetypal railway corporate citizenships within the global setting*?

In respect of urban rail, the market space is somewhat different. A subsidy is generally present, so the responsible authority tends to deal directly with public policy aspects. Furthermore, multiple guided transit modes in a city are not unusual, so urban transit solutions tend to be more complex. The authors therefore hypothesized that positioning the various urban guided transit modes in particular cities reflected attributes of their ever changing economic- and social setting vis-à-vis attributes of the various transit modes. Their research question was therefore *Which country- and city green- and socio-economic attributes and relations fit guided transit solutions to particular cities?*

The two different research questions reflect the essential difference between positioning line haul rail in market spaces where rail can be inherently competitive, versus positioning urban rail in a market space where it may be inherently uncompetitive. Nevertheless, the research design was set up to examine positioning, the action, and fit, the outcome, over time in both situations.

3.4 Line haul railways

3.4.1 Variables and their definitions

For the purpose of this chapter, line haul railways transport goods or persons over long distances or between cities. The authors measured the interaction between them and their settings by the following variables that reflected rail's corporate citizenship as well as its genetic technologies and their naturally competitive market spaces. Pending the outcome of statistical analysis, they were placed in the following groups for convenience:

Business Group represents the way in which railways deal with their task (Variables Infrastructure Operator Diversity, Train Operator Diversity, Information Technology Leverage, Total Road Network-, Motorways- and Paved Roads Percentage).

Competitiveness Group represents the way in which railways position themselves to compete in their chosen or allotted market spaces (Variables Research & Development Level, Relative Maximum Axle Load, Relative Maximum Speed, Distributed Power Presence, Heavy Haul Presence, High-speed Intercity Presence, Heavy Intermodal Presence, Motive Power Type, and Attitude to Competition).

Contribution Group describes the railways' contribution to their society (Variables Network Coverage, Transport Task—Freight- and Passenger Traffic Volume, Employment Created, and Initiative Source).

Networkability Group describes the extent and gauge of track, and the contiguous network beyond a country's borders (Variables *Narrow-, Standard-, and Broad Gauge; Networkability;* and *Strategic Horizon*).

Ownership Group describes industry structure (Infrastructure-operations Separation, Infrastructure- and Rolling Stock Ownership Locus, and Infrastructure- and Rolling Stock Commitment Horizon).

Society Group describes the railway setting (Variables Country (Name), Economic Freedom, Population, Gross National Income, Physical Size, Determinism, and Climate-change Position).

Sustainability Group describes adaptation and fit (Variables Infrastructure- and Rolling Stock Investment Capacity, Stakeholder Satisfaction Level, Service Reputation, Safety Reputation, Subsidy Influence).

Time Group represents passage of time, a prerequisite for longitudinal research (Variable *Calendar Year*).

The operational definitions of the foregoing forty-four variables, plus their measurement scales, exceed the space available in this chapter: Full details may be found at www.railcorpstrat.com/Downloads/feb2008/WCR2008%20Line%20Haul%20Operational%20Definitions.pdf.

3.4.2 Identification and selection of cases

Whatever the detail institutional arrangements, national governments typically either own railways, or regulate to varying degrees railways that they do not own, within their jurisdictions. Exceptions do of course exist where railway operations crisscross national boundaries by agreement or directive, as in the North American Free Trade Agreement and the European Union respectively. The authors therefore elected to examine railways by country.

Some railway attributes are independent of track gauge, but the latter does drive inherent competitiveness. There is no evidence that railways on track gauge of less than yard/meter/3′-6″ are sustainable: The authors therefore excluded data for narrower track gauges, irrespective of the gauge mix in a country. They used the Railways/Train Operators section of Railway Directory to define the set of line haul railways. The above criteria yielded 113 countries. Some of them included suburban and regional passenger operations, which are strictly not line-haul. However, the complementary set of global railway data is the City Railways section of Railway Directory, which the authors used for urban railways: Together these two sections represent the entire global population of railways. On that scale, they were disinclined to niggle about classification of boundary cases.

3.4.3 Construction of a database

Observations were predicated on the natural affinity between corporate citizenship and public domain data. Metric data was extracted from Railway Directory (2002-2007), Jane's World Railways (2005-2006, 2007-2008), or the Internet, and non-metric data was extracted by content analysis from International Railway Journal and Railway Gazette International. The detail measurement methodology has been reported by Van der Meulen & Möller (2006, 2008b). The Internet was used liberally to verify data to ensure internal consistency. The longitudinal database, containing one hundred and thirteen line-haul railways by country, populated with data for the six years 2002-2007, for each railway, gave a population (and sample) size of 113 x 6 = 678 cases, and is available at www.railcorpstrat.com/Downloads/WCRR2008%20Line%20Haul%20Database.xls.

3.4.4 Statistical analysis

The authors applied multivariate statistical analysis to the database to examine simultaneously relations among multiple variables, and multiple cases. They selected Factor Analysis, to analyze relations among a large number of variables and then to explain them in terms of a smaller number of latent variables, and Cluster Analysis, to reduce a large number of cases to a smaller number of clusters. Statgraphics Centurion XV was used to analyze the data. They culled variables with low communalities that contributed noise rather than insight (i.e. those that appeared in the Operational Definitions file, but which are absent from Table 1), after which the data set arrayed thirty-seven variables and 678 cases, for a total of 25 086 observations. Statistical analysis stops at the Factor Loading Matrix in Table 1, and at the Icicle Plot available at www.railcorpstrat.com/Downloads/WCRR2008%

20Line%20Haul%20Icicle%20plot.xls . Deeper discussion on the statistical intervention is available in Van der Meulen & Möller (2008b). Latent variable- and cluster names, and the following discussion, reflect the authors' interpretation of their knowledge of the variables in the research setting.

3.5 Urban guided transit

3.5.1 Enlarging the scope

The authors next applied broadly the same research methodology to urban rail. Although a previous paper (Van der Meulen & Möller, 2008a) passed peer review and revealed a constructive distinction between positioning urban rail in cities in developed countries and in developing countries, they were less than satisfied with its overall predictive validity. Importantly, the research did not address and therefore could not explain the ascent of the rubber-tyred competitors Automated Guided Transit (Vuchic, 2007, p.455), Bus Rapid Transit, and Monorail, against which heavy- and light rail must compete for investment funding. The authors therefore enlarged the scope of their research in the light-axle-load, low-speed market space from urban rail to urban guided transit, by including the modes mentioned below (Van der Meulen & Möller, 2012):

Heavy Metro maximally exploits rail's genetic technologies in urban settings. Included are the rubber-tyred systems found on some Paris Métro lines, and similar systems elsewhere: Despite rubber-tyred Supporting and Guiding, their gleaming running rails and wheel flanges indicate that these steel components are not redundant.

Light Rail and trams were merged, as neither attains fully controlled right of way. By definition, at 10-11 tonnes/axle, exploitation of rail's Supporting genetic technology is weak. Likewise the built environment constrains Guiding, and typically only a small number of vehicles are coupled: Technically, their inherent competitiveness is marginal.

Light Metro takes Light Rail to the next level with fully segregated right-of-way. Light axle load minimizes the cost of elevated structures, while small vehicle profiles minimize the cost of underground works. Driverless operation offers consistent performance and operational flexibility free from the labour issues that disturb manned systems.

Automated Guided Transit e.g. VAL and similar, offers consistently higher acceleration and higher retardation than steel-on-steel, although rubber tyres constrain axle load. As for automated Light Metro, light axle load and small vehicle profile minimize the cost of civil works. Automated operation offers consistent, precise high performance.

Monorail excels where pre-existing built environment admits only elevated structures with small physical footprint. Transit-grade monorails have converged on rubber-tyred straddle systems. Capacity and performance is comparable to Automated Guided Transit: Once again, automated operation offers consistent, precise high performance.

Bus Rapid Transit reputedly rolls out faster at lower cost than comparable rail systems. Its inclusion in guided transit is justified by its narrow concrete runway to support relatively heavy 12-13 tonne axle load, plus emerging virtual guidance by lane tracking systems. Bi-articulated buses even emulate rail's Coupling genetic technology.

Variable	Factor 1	Factor 2	Factor 3	Factor 4	Factor 5	Factor 6	Factor 7	No factor	Factor 8
Relative Maximum Speed	***0.78***	0.34	-0.02	0.26	0.13	-0.01	0.02	-0.03	0.21
Gross National Income	***0.76***	0.03	0.22	0.22	0.36	-0.01	0.15	-0.01	0.03
Motorways	***0.76***	0.10	0.14	0.15	0.12	0.01	0.01	-0.26	0.11
Information Technology Leverage	***0.70***	0.18	0.24	0.07	0.24	-0.01	0.20	0.06	0.10
High-speed Intercity Presence	***0.66***	0.28	0.03	-0.02	0.08	-0.10	0.04	-0.14	0.35
Country Economic Freedom	***0.64***	-0.22	0.31	-0.15	0.30	-0.11	0.18	0.21	-0.12
Paved Roads	***0.63***	0.13	-0.17	0.42	-0.04	-0.02	-0.01	0.14	0.01
Research and Development Level	***0.56***	0.46	0.37	-0.08	0.11	-0.01	0.06	-0.08	0.32
Electric Traction	***0.47***	0.42	-0.19	0.33	0.24	0.03	-0.02	0.27	-0.04
Network Coverage	0.23	***0.85***	0.27	0.03	0.20	0.02	0.02	0.08	0.11
Country Population	-0.05	***0.84***	0.12	-0.31	-0.10	0.00	0.04	-0.14	0.03
Employee Count	0.31	***0.81***	-0.02	0.28	-0.02	0.04	-0.02	0.18	0.08
Total Road Network	0.21	***0.80***	0.32	-0.13	0.19	0.04	0.06	0.04	0.02
Passenger Traffic Volume	0.60	***0.69***	0.00	0.05	0.16	-0.03	0.04	0.15	0.07
Country Physical Size	-0.35	***0.62***	0.40	-0.32	-0.01	0.01	-0.01	0.03	0.12
Freight Traffic Volume	0.39	***0.62***	0.35	0.27	0.16	0.01	0.02	0.24	0.15
Heavy Intermodal Presence	0.03	0.09	***0.82***	0.08	-0.02	0.09	0.06	-0.09	0.08
Distributed Power Presence	0.04	0.25	***0.76***	-0.01	0.00	0.04	0.05	0.04	0.16
Heavy Haul Presence	0.03	0.36	***0.73***	-0.03	-0.04	0.12	0.03	0.07	0.22
Infrastructure Ownership Locus	0.04	0.05	***0.67***	-0.29	0.31	-0.13	-0.03	0.01	-0.16
Relative Maximum Axle Load	0.15	0.09	***0.65***	0.47	0.13	0.01	-0.17	0.22	0.18
Infrastructure Operator Diversity	0.22	0.05	***0.62***	0.12	-0.11	0.01	0.03	-0.23	-0.13
Narrow Gauge	-0.09	0.20	-0.04	***-0.84***	0.05	-0.04	0.00	-0.12	0.01
Networkability	0.29	0.04	0.00	***0.76***	0.22	0.04	0.00	-0.07	-0.03
Standard Gauge	0.33	0.30	0.24	***0.49***	0.27	0.01	-0.01	-0.47	0.08
Infrastructure-operations Separation	0.29	0.12	-0.11	0.18	***0.81***	-0.05	0.07	0.04	0.18
Train Operator Diversity	0.31	0.12	-0.05	0.16	***0.80***	-0.04	0.12	-0.01	0.16
Rolling Stock Ownership Locus	0.17	0.09	0.47	-0.16	***0.68***	-0.16	0.12	-0.06	-0.03
Rolling Stock Commitment Horizon	0.00	0.00	0.09	0.01	-0.08	***0.90***	-0.01	0.02	-0.02
Infrastructure Commitment Horizon	-0.07	0.03	0.03	0.06	-0.05	***0.90***	0.01	0.06	-0.05
Calendar Year	-0.03	-0.04	-0.02	-0.01	0.07	0.05	***0.81***	-0.03	0.05
Climate-change Position	0.26	-0.04	-0.03	-0.20	0.17	0.03	***0.59***	0.23	-0.08
Rolling Stock Investment Capacity	0.18	0.41	0.17	0.21	-0.01	-0.04	***0.48***	-0.21	0.20
Infrastructure Investment Capacity	0.15	0.41	0.14	0.12	0.05	-0.18	***0.46***	-0.02	0.01
Broad Gauge	-0.02	0.23	0.00	0.13	0.02	0.09	0.04	***0.88***	0.04
Attitude to Competition	0.16	0.13	0.05	0.13	0.07	-0.17	0.14	0.03	***0.72***
Subsidy Influence	0.17	0.07	0.14	-0.12	0.13	0.08	-0.06	0.00	***0.67***

Table 1. The line haul factor loading matrix

3.5.2 Variables and their definitions

For the purpose of this chapter, urban guided transit offers mobility to persons within cities. The authors measured the fit between guided transit modes and their settings by the following variables that reflected their corporate citizenship. Pending the outcome of statistical analysis, they were placed in the following groups for convenience:

Business Group represents the amount of competition or support that urban guided transit faces in performing its task (variables Bus-, Car-, and Motorcycle Populations; Fuel Price; Motorways, Highways; and Secondary plus Other Roads Distance).

City Group describes the close urban setting (variables City Name; Surface Area; Metropolitan Population; Population Growth Rate; World Cities Score; Green Cities Score; and Smart Card Application).

Contribution Group describes guided transit's contribution to its society (variables Inaugural Year, Number of Operators, Status of Project, Network Coverage, Rolling Stock Fleet, Passenger Journeys, Number of Routes, Number of Stations, and Employee Count). They were measured separately for each of the urban guided transit modes, namely Heavy Metro, Light Rail and Trams, Light Metro, Automated Guided Transit, Monorail, and Bus Rapid Transit.

Country group describes the broad national setting (variables Country Name; Agricultural Land; Agriculture, Value Added; Alternative and Nuclear Energy; CO_2 Emissions; Electric Power Consumption; Energy Use; Exports of Goods and Services; Foreign Direct Investment; Forest Area; GDP; GNI per Capita; Gross Capital Formation; High-technology Exports; Imports of Goods and Services; Improved Sanitation Facilities, Urban; Improved Water Source, Urban; Industry, Value Added; Inflation, GDP Deflator; Internet Users; Life Expectancy at Birth; Merchandise Trade; Mobile Cellular Subscriptions; Out-of-pocket Health Expenditure; Population Growth; Population, Total; Public Spending on Education; Services, Value Added; and Surface Area). These variables were selected from World Bank Development Indicators: Themes identified by content analysis of Time magazine for the period July 2009 to June 2010 suggested the twenty-eight indicators actually used out of 298 available.

Society Group describes governance and societal attributes of the setting (variables *Economic Freedom Index* and *Income Inequality*).

Time Group represents passage of time, a prerequisite for longitudinal research (variable *Calendar Year*).

Operational definitions, measurement scales, and source references, either documentary or uniform resource locator, for each of the abovementioned variables, are available at www.railcorpstrat.com/Downloads/Sep2011/TRA%202012%20Operational%20Definitions.pdf.

To emphasize the difference between the line-haul rail and urban guided transit datasets, note that the latter does not include the Competitiveness-, Networkability-, and Ownership groups of the former. Urban guided transit was researched separately because it occupies a potentially low competitiveness market space; it does not naturally network with other railways and frequently cannot; and it is generally vertically integrated under a public authority, so ownership aspects recede into the background. Furthermore, urban rail responds to authority initiative rather than market initiative as more generally applies to line haul rail: Subsidies are generally present so sustainability is inherently secure. It was therefore not considered necessary to describe and measure subsidy and sustainability.

3.5.3 Identification and selection of cases

The research included the entire population of cities for which sufficient data could be found to populate the database in respect of the transit modes that served them. The City Railways section of Railway Directory (2009-2011) defined a minimum set of urban railways. Cities with one or more of automated guided transit, bus rapid transit, and monorail were added from websites listed under the applicable operational definitions at www.railcorpstrat.com/Downloads/Sep2011/TRA%202012%20Operational%20Definitions.pdf.

The longitudinal research design captured the adaptation dynamics of the global urban transit industry for the three consecutive years 2009-2011. To add a fourth, projected year, 2012, greenfields- and brownfields projects were also included, their various stages of progress measured on a five-point scale (Proposed 1, Feasibility Study 2, In Design 3, Under Construction 4, and Operational 5). The latter value of course also applied to all existing systems for the years 2009-2011.

Where necessary, raw data for agglomerations with more than one guided transit system were adjusted to match them to the population and area that they served. Details of the affected agglomerations accompany the applicable operational definitions.

3.5.4 Construction of a database

The authors constructed a new, dedicated, urban guided transit database using the variables and cases mentioned above. The Microsoft Excel file comprises two complementary data subsets, namely Countries and Cities, and is available at www.railcorpstrat.com/Downloads/Sep2011/TRA%202012%20Database%20and%20Factor%20Loading%20Matrices.xls. It gathered 330 cities with guided transit in sixty-eight countries, each with four years' data for the period 2009-2012, for a total of 1320 cases. The database thus contains (1320 cases) x (98 variables) = 129 360 observations.

3.5.5 Statistical analysis

In previous research, Van der Meulen & Möller (2008a) had used factor analysis to reduce the initial variables to a smaller set of latent variables. However, the many variables required to describe country settings in sufficient detail tended to unduly dominate some of the latent variables. Therefore, reflecting the research question, exploratory factor analysis was first undertaken separately for Country- and for City descriptive variables, using Statgraphics Centurion XV software. From the initial 36 country variables, it found seven latent variables, namely *Country Stature, Economic Development Level, Energy Demand Level* and alter ego *Alternative Energy Acceptance, Services Contribution to GDP; Trade Contribution to GDP,* and *Societal Development Level.* From the initial 60 city variables, it also found seven latent variables, namely *Heavy Metro Position, Automated Guided Transit Position, Monorail Position, Light Metro Position, Light Rail Position,* and *Green City Impediments.* The authors named the latent variables in the light of the variables that loaded onto them, within the context of the urban rail industry setting: The separate factor loading matrices are available at www.railcorpstrat.com/Downloads/Sep2011/TRA%202012%20Database%20and%20Factor%20Loading%20Matrices.xls, while a diagram showing which variables by name loaded onto each Country- and City latent variable is available at www.railcorpstrat.com/Downloads/Sep2011/TRA%202012%20Latent%20Variables%20Diagram.pdf. Thereafter, structural equation modeling using EQS 6.1 software found relations among these latent variables. The path diagram in Figure 2 shows the significant standardized regression coefficients as arrows pointing to the dependent latent variables. Positive correlations indicate support, negative correlations indicate opposition. Interpretation follows in §4.2.1. A detailed report on the structural equation modeling intervention is available on the authors' website at www.railcorpstrat.com/Downloads/Sep2011/TRA%202012%20SEM% 20Report.pdf .

4. Findings

4.1 Line haul railways

4.1.1 The factor loading matrix

Exploratory factor analysis extracted seven latent variables plus one single variable, shown in boldface italics in Table 1: They represent activities by which railways position their corporate citizenship in respect of their core business. Interpretation of the latent variables follows, with a reminder that §4.1 does not address urban rail: The latter will be addressed in §4.2.

4.1.2 Positioning passenger rail

The variables Relative Maximum Speed, Gross National Income, Motorways Percentage, Information Technology Leverage, High-speed Intercity Presence, Economic Freedom, Paved Roads Percentage, R&D Level, and Electric Traction, all loaded positively onto the latent variable Positioning Passenger Rail. Their effects are therefore mutually supportive. Relative Maximum Speed anchors Positioning Passenger Rail. Based on rail's Guiding genetic technology, it enabled the railway renaissance to meet passengers' high-speed expectations on dedicated high speed lines. It even created new markets, such as China's high-speed overnight electric multiple unit services, which extend their reach beyond the constraints of a working day. Such innovations facilitate rail's contribution beyond peak oil, when high fuel prices could curb air travel.

Significantly, high national income and economic freedom associate concurrently with motorways and paved roads, and with high-technology passenger railway attributes, i.e. high relative maximum speed, information technology leverage, high-speed intercity presence, electric traction, and high R&D level. Evidently road competition stimulates high-speed railways, which require high technology to remain competitive. It is therefore noteworthy that the R&D function has migrated from railway operators to industry: Emerging brand- and model competition among system integrators is comparable to that between Airbus and Boeing in the aircraft industry.

4.1.3 Exploiting opportunities

The variables Network Coverage, Country Population, Employment Creation, Total Road Network, Passenger Traffic Volume, Country Physical Size, and Freight Traffic Volume all loaded positively onto the latent variable Exploiting Opportunities. It suggested competitive and cooperative symbiotic relations among a country's transport infrastructure (Network Coverage and Total Road Network), its stature (Population and Physical Size), and rail's contribution to the economy (Employment Created, Passenger Traffic Volume, and Freight Traffic Volume). It demarcated the space in which Enlightened-, Progressive-, and Assertive Railways actualize their corporate citizenship as discussed in §4.1.10.

Large countries, or smaller countries with large contiguous networks beyond their borders, are prime railway locations. Notwithstanding that, the inherently competitive applications Heavy Haul, High-speed Intercity, and Heavy Intermodal, do not load on this latent variable: *Positioning Passenger Rail, Exploiting Opportunities,* and *Positioning Freight Rail,* therefore present mutually exclusive corporate citizenship positioning opportunities for railways.

Real world examples reflect both actualization and absence thereof. Large developing countries with high rail traffic volumes, such as Brazil, China, India and Russia are substantially redeveloping their railways to increase their contributions to their respective national transport tasks. Europe's high-speed railways, and North America's heavy freight railways, illustrate strong performance in particular market spaces. China's Freight- and Passenger Dedicated Lines, as well as India's Freight Dedicated Corridors and its emerging interest in high speed, illustrate ability to position railways in separate market spaces where opportunities are sufficient. By contrast, Europe has the population, area, and traffic to support substantial rail freight presence, yet substantial freight volume continues to move by road, for reasons that will become clear in the next section.

4.1.4 Positioning freight rail

The variables Heavy Intermodal Presence, Distributed Power Presence, Heavy Haul Presence, Infrastructure Ownership Locus, Relative Maximum Axle Load, and Infrastructure Operator Diversity all loaded positively onto the latent variable Positioning Freight Rail. It suggested that competitive freight railways, manifested by heavy intermodal-, heavy haul-, and distributed power presence, associated with high relative maximum axle load, privately owned infrastructure, and competing infrastructure operators. Examples are preservation of competition among railways in the North American Free Trade Agreement (Canada, Mexico, and the United States), and competition among parallel iron ore railways in Australia's Pilbara and Québec's North Shore.

Highly competitive and sustainable positioning of freight railways is evident in the member countries of the International Heavy Haul Association (Australia, Brazil, Canada, China, India, Russia, South Africa, Sweden-Norway, and the United States); the double stack container trains of the North American Free Trade Agreement and Australia, China, India, and Saudi Arabia; and the emerging dedicated rail freight corridors in China and India (Dedicated Freight, 2010). By contrast, the constituents of *Positioning Freight Rail* are absent in Europe: Indeed the notion of a rail freight dedicated network has already been rejected (European Freight, 2008). It is therefore unsurprising that European rail freight struggles to compete with road freight (Heydenreich & Lehrmann, 2010), and it will be interesting to observe whether the evolving rail freight network (Jackson, 2011a) will turn the tide.

Interestingly, while both freight- and passenger railways use information technology, the latent variable *Information Technology Leverage* loaded only onto *Positioning Passenger Rail*, but is absent from *Positioning Freight Rail*. This suggested that freight rail's ideal corporate citizenship is that of competent carrier, and that logistics management belongs elsewhere. It supports the assertion that few railways have had the management capability to integrate acquired logistics companies efficiently and effectively (Reinhold & Gasparic, 2009).

4.1.5 Exploring horizons

The variables *Narrow Gauge* (negative), *Networkability*, and *Standard Gauge* loaded onto *Exploring Horizons*. The signs indicated that Narrow Gauge opposed networkability, while Standard Gauge reinforced it. Standard gauge track allows network- and train operators to explore ever-wider horizons. This is evident in several initiatives to connect the standard gauge networks of China, Europe and the Middle East. Note from Table 1 that Broad Gauge did not load onto any latent variable: From a networkability perspective it is an independent variable, like the real world examples.

By contrast, many narrow gauge railways must forego participation in long-haul business. While they have achieved modest success in heavy haul, arguably the only viable post-renaissance narrow gauge application, heavy haul railways are usually short, and do not naturally network with one another. Queensland's are interesting—approximately parallel systems from multiple coalmines to several ports. South Africa's are on opposite sides of the continent. Brazil's Estrada de Ferro Vitória a Minas is essentially a single purpose operation. None of them establish a basis for continental scale networkability.

4.1.6 Pursuing competition

The variables *Infrastructure-operations Separation*, *Train Operator Diversity*, and *Rolling Stock Ownership Locus* all loaded positively onto the latent variable *Pursuing Competition*. Vertical separation, multiple train operators, and private rolling stock ownership constitute the basis of liberal on-rail competition with open access to infrastructure, which has emerged notably in the European Union and Australia. Latent variables are mutually exclusive: *Pursuing Competition* introduces competition in the market in settings that are physically unable or politically unwilling to support competition for the market among multiple infrastructure operators as in *Positioning Freight Rail*. Whether vertical separation benefits the railway industry and its stakeholders has been debated since Sweden first implemented it in 1987: There are many arguments for and against (Jackson, 2011b). However, *Pursuing Competition* is only one of a suite of applicable corporate citizenship latent variables. It is evident that in instances where vertical separation has not worked as expected, that inherent competitiveness has also fallen short. Consider, for example, that open access has met EU expectations for passenger operators, but missed them for freight, while the freight-oriented Australian Rail Track Corporation network has met expectations: In Europe, passenger rail positioning is inherently competitive, but not so freight, while in Australia freight rail positioning is inherently competitive.

4.1.7 Aligning assets

The variables *Rolling Stock Commitment Horizon* and *Infrastructure Commitment Horizon* both loaded positively onto the latent variable *Aligning Assets*. It suggested aligning infrastructure and rolling stock investment for appropriate periods, to avoid competitiveness being eroded by obsolescence. Without competition to demand ever-increasing performance, railways are not incentivized to replace existing assets with higher performing assets. If they do not routinely raise the bar of their genetic technologies by increasing axle load, speed, and/or train length, it becomes difficult to justify new- or upgraded assets. Railways then contemplate refurbishment and rehabilitation, often leading indicators of unsustainability. Sometimes they deploy new rolling stock on existing infrastructure—a palliative that may fail to realize the new trains' full performance potential. The countervailing value of private ownership emerged in *Positioning Freight Rail* and *Pursuing Competition*. Sustainable private enterprise works assets hard or works them out.

4.1.8 Greening the image

The variables Calendar Year, Climate-change Position, Rolling Stock Investment Capacity, and Infrastructure Investment Capacity all loaded positively onto the latent variable Greening the Image. The anchor roles of Calendar Year and Climate-change Position suggested that actors outside rather than inside the railway industry were actually setting the pace of greening.

Greening the Image reinforces *Positioning Passenger Rail* and *Positioning Freight Rail*, but the benefits of a greener but uncompetitive mode are insignificant: The real challenge is to increase rail's competitiveness and thereby shift traffic to a greener mode by ecological adaptation. Rail's green credentials are undisputed: High speed railways accept steeper gradients that minimize environmental impact; heavy axle load attracts traffic from road to rail. State-of-the-art high-speed trains, and hybrid diesel locomotives with intelligent driving aids, reduce energy consumption per passenger journey and per ton-km. However, high speed and heavy axle load are uncomfortable bedfellows: Hence, in principle and to the extent that it is viable, physically separate dedicated freight- and passenger infrastructure promotes greening.

Passenger trains tend to be lighter and more frequent, so recovery of their regenerated braking energy poses the lesser challenge. However, *Positioning Freight Rail* promotes longer and heavier trains, so recovery of their regenerated braking energy poses the greater challenge. In particular, many heavy haul railways descend from mine to port, and several are potentially net energy generators over the empty-loaded round trip (Van der Meulen, 2010). Maximum regenerative braking should be the point of departure. However, while on-board battery storage on hybrid diesel locomotives might be worthwhile, such systems cannot deal with a net surplus. Furthermore, regenerating all instantaneously surplus energy requires locomotives to control the same load on downgrades that they haul on upgrades. While this requires symmetrical up- and downgrades, many heavy haul routes have asymmetrical grades that oblige loaded trains to dissipate potential energy through dynamic or friction braking on descending grades, which reduces sustainability. Even if gradients supported full regenerative braking, matching a three phase supply grid to single phase overhead traction supply is the next challenge. This introduces the concept of smart grids and open systems, which one hears about, but not yet in railway traction.

4.1.9 Constraining downside

The variables *Attitude to Competition* and *Subsidy Influence* both loaded positively onto the latent variable *Constraining Downside*. It suggested that encouraging competition, while applying subsidy to influence the beneficiary, could constrain downside in adverse situations. A country's railway industry is only as competitive as government will allow or encourage. Where appropriate, governments traditionally subsidized railways directly, but their role is changing. Instead of simply assuming responsibility for runaway expenses, they now tend to recognize railways as worthy corporate citizens that they influence through instruments such as investing to raise competitiveness, public-private partnerships, tax incentives, and so on. Two examples are the United States' Passenger Rail Improvement and Investment Act of 2008 and its American Recovery and Reinvestment Act of 2009, which provided seed investment for high speed rail, to be matched by state funding for operational support (Boardman, 2010).

4.1.10 Cluster analysis

Whereas factor analysis finds relations among the variables in a database, the multivariate procedure cluster analysis finds relations among the cases in a database, countries in this instance. Applying cluster analysis, to the 2007 data only, reduced the 113 countries in the population to five clusters, which the authors named Fortuitous Railways, Insecure Railways, Enlightened Railways, Progressive Railways, and the quasi-cluster Assertive Railways. The procedure maximizes within-cluster homogeneity and maximizes between-cluster heterogeneity. Mentioning cluster members by name is restricted here, because some ups and downs have occurred over time. Full details are nevertheless available in Van der Meulen & Möller (2008a), while discussion of key issues follows.

Fortuitous Railways clustered twenty medium-sized countries. Their *Relative Maximum Axle Load* was the only high attribute, the rest rating either moderate or low. They were standard- or broad gauge state railways, redeemed by axle load that happened to be sufficiently heavy to support basic competitiveness. The authors named them *Fortuitous Railways* because they lacked attributes with which to project a distinctive corporate citizenship. Demonstrating that railway renaissance is advancing, several Middle Eastern countries, which are making substantial railway investment, have probably moved from the Fortuitous cluster to one of the Enlightened-, Progressive-, or Assertive clusters.

Insecure Railways clustered fifty-four medium-sized countries. It had no high attributes, had generally moderate attributes, and had low competitiveness, i.e. low maximum axle load and -speed; no distributed power-, heavy haul-, or heavy intermodal presence; predominantly narrow track gauge and low networkability. The authors named them *Insecure Railways* because they failed to exploit any of rail's strengths, and hence could be vulnerable to external threats or withdrawal of political support. Many have colonial origins, which possibly denied them wherewithal to actualize the positioning latent variables detailed in §4.1.2 to §4.1.9: In countries where line haul railways are justified at all, rebalancing of global power from developed- to developing countries could well redress this legacy, one way or another.

New Zealand, which returned full circle during the present research, characterizes the Insecure Railways cluster. Privatized in 1995, it soon fell short of expectations, running down assets along the way (New Zealand, 2008). The government repurchased the infrastructure in 2004, and the operations in 2008, thereby re-nationalizing railways. The events are unsurprising: Narrow track gauge on an island, and other handicaps, precluded it from *Positioning Passenger Rail, Exploiting Opportunities, Positioning Freight Rail*, or *Exploring Horizons*, not to mention the other positioning latent variables. Government's skepticism regarding the role of rail (KiwiRail debates, 2009) appeared justified.

Enlightened Railways clustered twenty small countries, mainly European Union members, -candidates, or -applicants, plus South Korea. They rated high on relative maximum axle load and -maximum speed, electric traction, networkability, information technology leverage; paved roads, economic freedom, and gross national income. All other variables were moderate, while freight technology was low—no distributed power-, heavy haul-, or heavy intermodal presence. The name reflected their enlightened approach to rail reform by encouraging competitive railway positioning per the latent variable *Pursuing Competition*.

Progressive Railways clustered France, Italy, Spain, Japan, Germany, and the United Kingdom. They rated high on R&D level, relative maximum speed, high-speed intercity presence, electric traction, attitude to competition, standard gauge, train operator diversity, information technology leverage, total road network, motorways, network coverage, freight traffic volume, passenger traffic volume, employee count, economic freedom, population, gross national income, infrastructure investment capacity, and rolling stock investment capacity. These attributes supported competitive high-speed passenger services in developed economies, while freight technology rated low—distributed power, heavy haul, and heavy intermodal were absent. All other attributes rated moderate. However, state involvement was still present, and infrastructure operator diversity was essentially absent, hence actualization of the full spectrum of positioning latent variables was circumscribed. The name speaks for itself.

The Enlightened- and Progressive clusters needed to position their rail freight for competitiveness. Setting aside Japan and South Korea, where respectively geography and present politics constrain networkability, this reduces to a European matter. The uncertain outlook of Europe's rail freight (Reinhold & Gasparic, 2009) is not unexpected. The latent variables *Positioning Passenger Rail* and *Positioning Freight Rail* indicate that positioning freight- and passenger rail are distinct corporate citizenship activities: To date, these clusters have hardly actualized *Positioning Freight Rail*. The prospect of heavy freight that will predominantly run on a dedicated Trans-European Freight Network (European Rail, 2007) is therefore encouraging. If successful, the notion of general freight transport increasingly being undertaken by relatively light containerized trains that resemble passenger trains in terms of loads exerted on the infrastructure, average speed, reliability and performance, should be expected to rearrange the latent variable *Positioning Freight Rail*, and possibly other latent variables too.

Assertive Railways formed a quasi-cluster, i.e. statistically independent railways that are icicle plot neighbours but did not actually cluster. They are not insignificant railways, as the following two selected above-the-median examples illustrate.

The United States rated high on research and development level, relative maximum axle load and -speed; distributed power-, high-speed intercity-, and heavy intermodal presence; attitude to competition; standard gauge; infrastructure operator diversity; information technology leverage; total road network; infrastructure- and rolling stock ownership locus; network coverage; freight traffic volume; economic freedom; gross national income; physical size; infrastructure investment capacity; and subsidy influence. The US' competitive private enterprise and technology leadership have established a formidable freight railway corporate citizenship. Trucking is both a tough competitor and a symbiotic supporter, mainly in the intermodal market space. However, its comparatively high rail freight market share, one of the highest in the world, has moved shippers to seek increased competition and strengthened federal oversight (Kimes, 2011). Comparing European and US outcomes to freight and passenger separation challenges in terms of *Positioning Passenger Rail* and *Positioning Freight Rail*, note that Europe's is generally a *minus freight* outcome; the US' is generally a *minus passenger* outcome. Both actively seek to introduce the missing positioning latent variable. However, as mentioned under *Progressive Railways* above, Europe's rail freight still needs to demonstrate turnaround. Noting that the US is the only country with strong underpinning for the latent variable *Constraining Downside*, its recent modest stimulus funding (High speed, 2009) was on cue, but the quantum seemed unlikely to create dedicated passenger corridors, while admission of 145km/h intercity trains on conventional mixed-traffic routes might have diluted its potential impact. Subsequent progress has been ambivalent (Six-year high, 2011; Governor halts, 2011), with only California committing to commence construction of a 350km/h system in 2012 (California High, 2011). Evidently the latent variables *Positioning Passenger Rail* and *Positioning Freight Rail* are so robustly rooted in rail's genetic technologies that they do not readily yield to political expediency.

China rated high on R&D level, relative maximum speed; distributed power-, high-speed intercity-, and heavy intermodal presence; electric traction; attitude to competition; motorways; paved roads; freight traffic volume; employee count; population; physical size; and infrastructure- and rolling stock investment capacity. Its rapid network growth and technological development, and its immense railway corporate citizenship, have drawn global admiration. Initially, it mixed high-speed passenger and freight on an upgraded, network: For a country eagerly actualizing the latent variable *Exploiting Opportunities*, that outcome aligned uneasily with the latent variables *Positioning Passenger Rail* and *Positioning Freight Rail*. However, in recent times it has vindicated the research findings by rapid expansion actualizing *Positioning Passenger Rail* through the emerging Passenger Dedicated Line (PDL) network (Li-ren & Li, 2010), and *Positioning Freight Rail* through a second heavy haul line to augment the 400 million-tonnes-per-year Daqin line (Second heavy, 2009) with an ultimate objective of a 10000km heavy haul network. Heavy axle load is absent from the variables listed above, but is set to increase to 30 tonnes on heavy haul lines (Seizing the, 2009). China's actualization of *Positioning Passenger Rail* and *Positioning Freight Rail* has positioned it as world's busiest heavy haul railway, and operates the first trains in the world timetabled to run at an average of over 300km/h (China's star, 2010).

To do justice to members of the Assertive Railways cluster requires more space than the page limit of this chapter. Because they are a quasi-cluster and not a true cluster, it is not possible to discuss them in generic terms as has been done for the Fortuitous-, Insecure-, Enlightened-, and Progressive Railways clusters. Instead each one requires discussion of its individual attributes. Further examples may be therefore be found in Van der Meulen & Möller (2008b).

4.2 Urban guided transit

4.2.1 The path diagram

Section 4.1 has interpreted the statistical findings of rail's three inherently competitive market spaces. Next, in urban transit language, §4.2.2 to §4.2.10 will interpret the path diagram in Fig. 2 with respect to rail's inherently weak market space. More detail is available in Van der Meulen & Möller (2012).

4.2.2 Green city impediments

The latent variable *Green City Impediments* mediated between the country setting and urban rail solutions in particular cities. Noting carefully the relative directions of their signs, and that double negative is positive, the latent variables *Societal Development Level* (-0.389), *Alternative Energy Acceptance* (-0.256), *Economic Development Level* (-0.150), and *Services Contribution to GDP* (-0.149), opposed it, while *Country Stature* (0.075) supported it. From the perspective of populous, large countries that feature urban guided transit, larger is evidently not greener: Rather, positive societal development, minus alternative energy acceptance, minus economic development, and minus services contribution, associate with green cities.

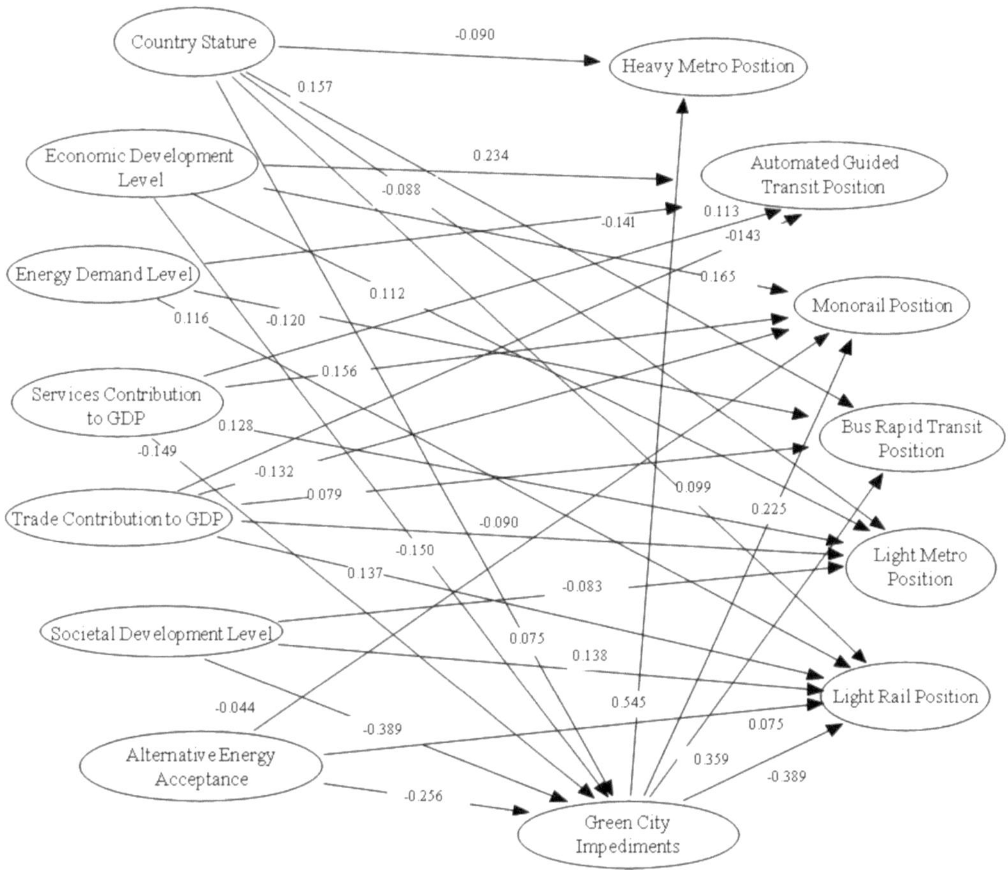

Fig. 2. Path Diagram showing standardized regression coefficients as arrows pointing to dependent latent variables

4.2.3 Positioning heavy metro

Green City Impediments (0.545) supported the latent variable *Heavy Metro Positioning*; *Country Stature* (-0.090) opposed it. The authors interpreted these relations to mean that the variables that impede green cities, namely *Area*, *Population*, and *Population Growth*, also drive the Heavy Metro mode, rather than that Heavy Metro is not green. The road network- and vehicle population variables loaded onto *Country Stature*: Its negative sign thus suggests that countries that enjoy advanced development might ultimately come to oppose Heavy Metro. Alternatively, one could interpret the relation to mean that large cities with large populations and high population growth might find Heavy Metro out of reach, and opt for Bus Rapid Transit instead. To emphasize its pre-eminence among other urban guided transit modes, note that *Heavy Metro Position* attracted loading by the variables *World Cities Score* and *Smart Card Application*. It appears that, until the present, leading cities considered Heavy Metro to be a must have feature.

4.2.4 Positioning automated guided transit

Economic Development Level (0.234) and Services Contribution to GDP (0.113) supported the latent variable Automated Guided Transit Positioning; Trade Contribution to GDP (-0.143) and Energy Demand Level (-0.141) opposed it. The authors interpreted this to mean that Automated Guided Transit fitted into developed, service economies, together with Monorail and Light Metro. See also §4.2.9 and §4.2.10 for additional interpretation.

4.2.5 Positioning monorail

Green City Impediments (0.225), Economic Development Level (0.165), and Services Contribution to GDP (0.156) supported the latent variable Monorail Positioning; Trade Contribution to GDP (-0.132) and Alternative Energy Acceptance (-0.044) opposed it. As for the Heavy Metro mode, the variables that impede green cities also drive the Monorail mode. Like Automated Guided Transit and Light Metro, Monorail fitted into developed, service economies. See also §4.2.9 and §4.2.10 for additional interpretation.

4.2.6 Positioning bus rapid transit

Green City Impediments (0.359), *Country Stature* (0.157), and *Trade Contribution to GDP* (0.079) supported the latent variable *Bus Rapid Transit Positioning*; *Energy Demand Level* (-0.120) opposed it. As for the Heavy Metro- and Monorail modes, the variables that impede green cities also drive Bus Rapid Transit. However, noting that *Country Stature* supports Bus Rapid Transit, one would naturally expect to find Bus Rapid Transit in large, rapidly growing cities, such as Jinan in China. See also §4.2.9 and §4.2.10 for additional interpretation.

4.2.7 Positioning light metro

Services Contribution to GDP (0.128) and Economic Development Level (0.112) supported the latent variable Light Metro Positioning; Trade Contribution to GDP (-0.090), Country Stature (-0.088), and Societal Development Level (-0.083) opposed it. Like Automated Guided Transit and Monorail, the Light Metro mode fits into developed, service economies, although not yet assertively due to its comparatively recent emergence. Opposition by Country Stature suggested that it fits well into smallish, more intimate cities. Opposition by Societal Developmental Level appeared counter intuitive, but it might indicate that high-capacity public transport is unwanted in such cities. See also §4.2.10 for additional interpretation.

4.2.8 Positioning light rail

Societal Development Level (0.138), Trade Contribution to GDP (0.137), Energy Demand Level (0.116), Country Stature (0.099), and Alternative Energy Acceptance (0.075) supported the latent variable Light Rail Positioning; Green City Impediments (-0.389) opposed it. The double negative (minus impediments) indicated that Light Rail actually supported green cities: It was the only urban guided transit mode for which all correlation coefficients that pointed to it were found to mutually reinforce its fit in a city.

4.2.9 Energy awareness

Either *Energy Demand Level* or *Alternative Energy Acceptance* opposed each of the three rubber-tyred modes, Automated Guided Transit, Monorail, and Bus Rapid Transit. They have higher rolling resistance than steel tyred modes, and all things being equal, also higher energy consumption. Evidently, countries that already had high energy demand, and those that accepted alternative energy, were sensitive to unduly increasing their energy demand. By contrast, *Energy Demand Level* and *Alternative Energy Acceptance* both supported Light Rail: Its greenness therefore offsets its marginal inherent competitiveness.

4.2.10 Trade awareness

The latent variable *Trade Contribution to GDP* opposed the higher technology automated modes, i.e. Automated Guided Transit, Monorail, and Light Metro; whereas it supported the lower-technology Bus Rapid Transit and Light Rail. Evidently there was an inverse relationship between technology and trade, which would manifest itself as the conservative tactic of initially deploying advanced technology close to its origin or support base.

5. Discussion

5.1 Positioning passenger rail

The latent variable *Positioning Passenger Rail* has become a prominent function of railway corporate citizenship. High speed rail has established itself as a formidable competitor against airlines in most developed countries. It is also considered to stimulate developing countries, and has become an aspirational objective in newly industrialized countries. Countries such as Brazil, India, Iran, Morocco, Russia, Thailand, and Turkey, already have, or are committed to acquiring high speed rail systems. China sees it as an environmentally responsible mode for journeys up to six hours. Dedicated passenger corridors have become the norm. China's 2010 High Speed Rail Conference sealed high speed as the way to do long-haul passenger rail for the future. *Positioning Passenger Rail* is a useful indicator of which variables should be within the high speed rail frame of reference.

5.2 Positioning freight rail

The latent variable *Positioning Freight Rail* has also become a prominent function of railway corporate citizenship. Heavy Haul is the prime solution for moving high volumes of bulk commodities, as is Heavy Intermodal for moving high volumes of containers long distances overland. Dedicated freight corridors have emerged in countries such as China and India, which have the traffic volumes to justify them. Aside from the essential technical attributes of heavy freight railways, the latent variable has highlighted the fundamental role of competition in freight transport, getting right down to private infrastructure ownership and then pitting competing railways against each other. Freight transport is generally a ruthless, low margin, and very competitive market, not well suited to a government player. It is therefore difficult to see why some governments continue to see freight transport as a core government function (Amos, 2006). Looking ahead to §5.4, it is evident that not all rail reforms have faced this issue.

5.3 Positioning urban rail

The findings on urban rail positioning have provided interesting insights into ecological adaptation in the light-axle-load low-speed market space. While Heavy Metro remains unchallenged at the highest capacity level, it appears that rubber-tyred guided transit modes have breached rail's pre-eminent status. One would therefore expect Light Rail, and by extension Light Metro, to be more vulnerable to competition from rubber-tyred systems by virtue of their low axle load. Notwithstanding that potential weakness, their good green credentials are attractive to smaller cities that value inherent environmental friendliness over the expediency of simply moving people under the weight of popular demand. Automated Guided Transit, Monorail, and Bus Rapid Transit do however seem well positioned to drive a wedge between the heavy- and light poles of urban rail. From there, one should expect them to win in medium cities such as Automated Guided Transit in Lille in France, and Bus Rapid Transit in Jinan in China.

5.4 The role of rail reform

This chapter commenced with observing differences among railways. In the context of this Section, variables in the Ownership Group and the Society Group were particularly relevant. Having researched and explained the modalities of railway positioning, and finding that renaissance is an achievable aspiration, it is now apposite to return to the original differences. Why do the Fortuitous and Insecure clusters still exist? Note that countries that have advanced have in many instances liberalized their institutional arrangements, or are in process of doing so. There are of course exceptions, such as China and India, but the weight of evidence from the Fortuitous and Insecure clusters, the majority of whose members are state-owned, suggests that unless state ownership results in operational efficiency and capital investment that positions railways for competitiveness and sustainability, it is an impediment rather than a facilitator.

Vertical integration versus vertical separation and open access has been another persistent reform issue. It is helpful to benchmark positions against the United States industry model. Interoperability is near seamless, and continental-scale haul distances transcend geographically-defined railroad franchises. Extensive leasing separates rolling stock ownership from liveries and reporting marks. Symbiotic trackage rights give access to infrastructure of others. Unrestricted interchange of 32.4 tonne-per-axle vehicles give the lie to assertions that heavy haul and vertical separation cannot co-exist. All told, US railroads are effectively infrastructure managers, not unlike those in Europe, and for purpose of this argument its actual vertical integration is less than what the name suggests. Note nevertheless critical differences regarding private infrastructure ownership, and extensive though not ubiquitous competition between parallel railroads. While shippers will attest that it is not a perfect market, it does suggest that anything less free is more constraining than it need be.

Note that this Section applies to line haul railways only. Urban rail is generally a local government responsibility, which institutional arrangements are appropriate.

6. Future research

The research journey has revealed complex relations that underlie railway positioning. As the renaissance progresses, railways migrate to clusters that exhibit more advanced corporate citizenship. However, the adaptation process has fragmented the industry into more competing entities, whose data become less accessible due to commercial sensitivities, in many instances providing no more than minimum legal requirements. The present research was fortunately conducted in a window of opportunity that has all but closed in countries whose line-haul railways have advanced the furthest. Due to the public nature of urban rail, that has not yet happened to its data. It would therefore be difficult to replicate the line haul research with current data. It might even be pointless because, having identified the Fortuitous and Insecure clusters and the determinants of their position, there is arguably no more insight to extract. It would be more productive to learn from the Enlightened, Progressive, and Assertive railways. However, that would reduce the population size by 65%, which would require a new research design.

7. Conclusions

The research stream described in this chapter has developed a statistical research approach to global railway positioning, both line haul- and urban, from a corporate citizenship perspective.

In supporting the hypothesized existence of some number of underlying longitudinal, or time-dependent, relations among variables associated with positioning line haul railways, the research has found the eight latent variables *Positioning Passenger Rail, Exploiting Opportunities, Positioning Freight Rail, Exploring Horizons, Pursuing Competition, Aligning Assets, Greening the Image,* and *Constraining Downside.* These latent variables represent actualization of a railway's corporate citizenship with respect to its core business.

In supporting the hypothesis that positioning the various urban guided transit modes in particular cities reflected attributes of their ever changing socio-economic setting vis-à-vis attributes of the various transit modes, the research found the seven country-related latent variables *Country Stature, Economic Development Level, Energy Demand Level, Services Contribution to GDP; Trade Contribution to GDP, Societal Development Level,* and *Alternative Energy Acceptance,* and the seven city-related latent variables *Heavy Metro Position, Automated Guided Transit Position, Monorail Position, Light Metro Position, Light Rail Position,* and *Green City Impediments.* The relations found between country- and city latent variables were presented in a path diagram that shows regression coefficients from the socio-economic setting to particular guided transit solutions. It represents the positioning of those solutions with respect to their country and city settings.

The foregoing outcome has developed grounded understanding of railway positioning in all four of rail's market spaces, the three in which it is inherently competitive, namely

Heavy Haul, High-speed Intercity, and Heavy Intermodal, as well as the market space in which it is potentially weak, namely Urban Rail. This has made it possible to understand and to predict with reasonable certainty what will be the outcome of a particular railway positioning intervention, or to analyse a situation and design appropriate remedial intervention.

Like the top-down research design, its application is primarily top down. As examples, the authors have used it as a framework for national rail development strategy, national railway economic regulation, national railway policy, rail's contribution to national transport planning, national passenger commuter rail technology, development of strategic rail plans at provincial- and regional levels, high level positioning of major rail corridors, and conceptual design of a regional rail corridor. In short, it provides high-level insight with which to assess the viability of policy options.

8. Acknowledgements

The authors express their appreciation for constructive comments to peer reviewers at many conferences, colleagues in the global railway industry, and colleagues in the South African railway industry. The work reported here is much richer for it.

9. References

Amos, P. (2006). *Railway reform: Vertical integration and separation*, World Bank, Accessed 2011-10-16, Available from: www.euromedtransport.org/En/image.php?id=1129

Boardman, J. (2010). Fast-tracking the future, *Railway Gazette International*, Vol.166, No.7, pp.44-47

California High-speed Rail Authority (2011). *Chairman issues statement*, Accessed 2011-10-16, Available from: www.cahighspeedrail.ca.gov/10102011-leg.aspx

Dedicated Freight Corridor loan (2010). *Railway Gazette International*, Vol.166, No.9, p.7

European Freight and Logistics Leaders Forum (2008). *NEWOPERA The rail freight dedicated lines concept final report*, Author, ISBN 978-3-00-0275700-1, Brussels, Belgium, p.157

European Rail Research Advisory Council (2007). *Strategic Rail Research Agenda 2020*, Brussels, Belgium, p.13

Friedman, T.L. (2006). *The world is flat*, Penguin, ISBN 978-0-141-03489-8, London, England

Governor halts Florida high speed project (2011). *Railway Gazette International*, Vol.167, No.3, p.10

Heydenreich, T. & Lehrmann, M. (2010). How to save wagonload freight, *Railway Gazette International*, Vol.166, No.9. pp.126, 128, 130

High speed scramble (2009). *Railway Gazette International*, Vol.165, No.8, p.19

Jackson, C. (2010). China's star blazes a high speed trail, *Railway Gazette International*, Vol.166, No.2, p.3

Jackson, C. (2011a), Rail freight network starts to evolve, *Railway Gazette International*, Vol.167, No.3, pp.48-52

Jackson, C. (2011b). Does separation really work? *Railway Gazette International*, Vol.167, No.6, p.3

Jane's World Railways (2005-2006, 2007-2008). Englewood, Colorado, United States of America, IHS

Japanese Railway Engineering (1964). *Special Issue – New Tokaido Line*, Vol.5, No.4, pp.1-56

Kimes, M (2011). Showdown on the railroad, *Fortune*, Vol.164, No. 5. pp.81-88

KiwiRail debates the way forward (2009). *Railway Gazette International*, Vol.165, No.11, pp39-42

Levinson, M. (2006). *The box*, pp. 261-262, Princeton University, ISBN 978-0-691-12324-0, Princeton, New Jersey, United States of America

Li-ren, D. & Li, D. (2010). Planning the world's biggest high speed network, *Railway Gazette International*, Vol.166, No.12, pp.41-46

New Zealand nationalizes (2008). *Railway Gazette International*, Vol.164, No.6, p.344

Railway Directory (2002-2007, 2009-2012). Hamburg, Germany, DVV Media Group

Reinhold, T. & Gasparic, C. (2009). Facing the moment of truth, *Railway Gazette International*, Vol.165, No.12, pp. 24-27

Second heavy haul line gears up to handle more coal (2009). *Railway Gazette International*, Vol.165, No.9, p.40

Seizing the opportunity (2009). *Railway Gazette International*, Vol.165, No.9, pp.31-36

Six-year high speed plan announced (2011). *Railway Gazette International*, Vol.167, No.3, p.10

Tracks to carry the big mineral hauls (1972). *Railway Gazette International*, Vol.128, No.2, pp.49-53

Van der Meulen, R.D. (1994). Some relations of corporate strategy content to organizational environment by comparing a capital-intensive service industry across selected societies, Doctoral dissertation, University of Pretoria, *Dissertation Abstracts International*, Vol.55, No.5, p.1336-A.

Van der Meulen, R.D. (2010). Heavy haul railway electrification—experiences and prospects, *Proceedings of the Joint Rail Conference*, JRC2010-36151 [CD-ROM]. Urbana, Illinois, United States of America, ASME, IEEE, ASCE, TRB, AREMA, and University of Illinois.

Van der Meulen, R.D. & Möller, L.C. (2006). Railway globalization: Leveraging insight from developed- into developing regions. *Proceedings of the 7th World Congress on Railway Research* [CD-ROM], Montréal, Québec, Canada

Van der Meulen, R.D. & Möller, L.C. (2008a). Strategies for sustainable mobility: Urban railways as global corporate citizens, *Proceedings of the 8th World Congress on Railway Research*, G.2.2.2.1 [CD-ROM], Seoul, Korea

Van der Meulen, R.D. & Möller, L.C. (2008b). Ultimate interoperability: Line-haul railways as global corporate citizens, *Proceedings of the 8th World Congress on Railway Research*, PN.1.2 [CD-ROM], Seoul, Korea

Van der Meulen, R.D. & Möller, L.C. (2012). European- and global urban guided transit: Green- and socio-economic fit, *Transport Research Arena*, Athens, Greece

Vuchic, V.R. (2007). *Urban transit systems and technology*, Wiley, ISBN 978-0-471-75823-5, Hoboken, New Jersey, United States of America

World Economic Forum (n.d). *Corporate global citizenship,* Accessed 2011-10-05, Available from: www.weforum.org/issues/corporate-global-citizenship

5

Structural and Kinematic Analysis of EMS Maglev Trains

Zhao Zhisu
National University of Defense Technology
China

1. Introduction

Maglev train is a new means of transport and an integration of the latest high-techs in the field of track-bound transportation system. Over the past half century, the research on vehicle structure has been always a very active area. The researchers realized there is great difference between the movements of the maglev train with that of the conventional rail vehicles. For designing maglev vehicle, creation of a new mechanism is necessary, and then the mechanism is converted to a specific machine to compose vehicles. In this process, machine and mechanism kinematics analysis are indispensable prerequisites. Study of kinematics analysis method and theoretical is the forefront of researching for the structure of maglev train. This chapter aims to introduce author's the latest research outcome.

2. Structure of maglev trains

EMS maglev trains have two basic structures which are represented by German Transrapid and Japanese HSST. Chinese and Korean mid-low speed maglev trains are in these two basic structures now.

2.1 Outline of structure development of ems maglev trains

The structure of EMS maglev trains has changed through a rigid aircraft - flexible coupling - modularization structure process. Based on the vehicles levitation running in the air, naturally a structure type of rigid spacecraft has been designed by researchers, namely the whole vehicle in rigid structure. It takes Japanese HSST-01(Yoshio Hikasa & Yutaka Takeuchi, 1980) (Fig.1) and German Transrapid 02(J.L.He et al., 1992) (Fig. 2) as the representatives of this vehicle.

The vehicle shakes violently when they are experimentally running at a high speed. Both kinds of vehicles are non-manned and the researchers design a new kind of maglev vehicle structure for solving the manned riding comfort. This structure separates the car body and running gear first and a secondary suspension system is sets up with buffer spring between them. It takes Japanese HSST-02(Yoshio Hikasa & Yutaka Takeuchi, 1980) (Fig.3) as the representative.

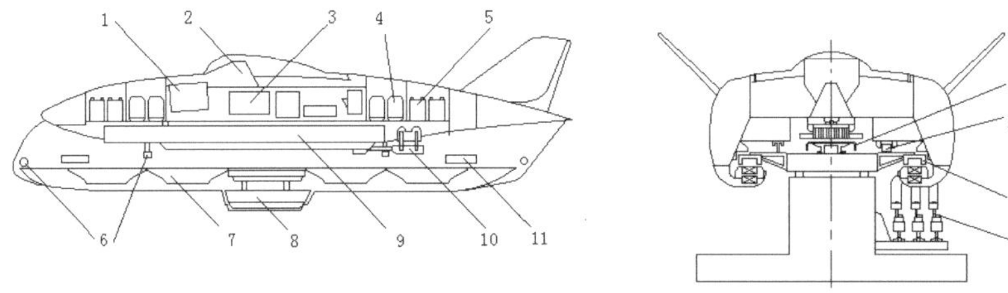

1.Electricity box, 2.Instument panel, 3.Automate control unit, 4.Thyrist chopper, 5.Battery, 6.Gas sensor, 7.Levitation magnet, 8.Power collector, 9.Linear induction motor, 10.Hydraulic brake, 11.Saving skid, 12.Reaction plate, 13.Brakage, 14.Anchor rail, 15.Power rail

Fig. 1. HSST-01 Maglev Vehicle

Fig. 2. Transrapid-02

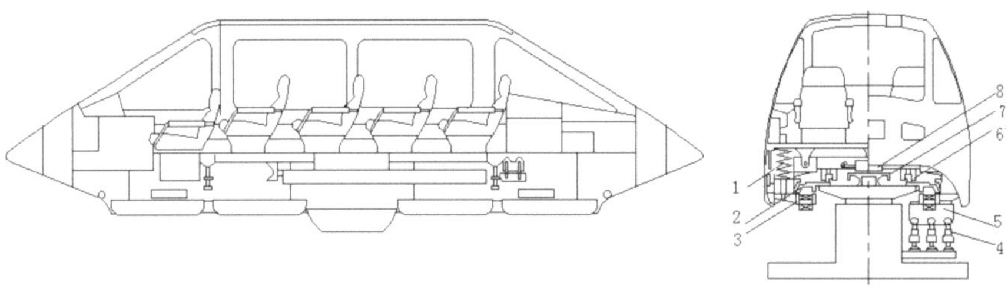

1. Secondary suspend, 2.Anchor rail, 3.Levitation magnet, 4.Power rail, 5.Power collector, 6.Hyraulic brake, 7.Reaction plate, 8.Linear induction motor.

Fig. 3. HSST-02

However, Vibration problem is still unresolved by use of this structure when the train are running at a high speed, because the gap size between magnetic track and suspension electromagnet is acquired by gap sensors which are generally laid for four. The four points should be controlled independently and may not in the same plane (for example, track error, vehicle passing transition curve, asynchronous dynamic adjustment of all points, etc.), but for the rigid or elastic support system in which the bogies are still rigid, the four sensors are installed in a comparatively rigid plane, so this is a conflict. After a long period of experiments and researches, a new kind of modularized vehicle structure (TEJIMA Yuichi, et al., 2004; Seki & Tomohiro, 1995; Maglev Technical Committee, 2007) (Fig.4, 5) is invented. The car body and running gear are separated and jointed by the secondary suspension system in which the four control points of bogies are decoupled, so the vibration problem of vehicles are solved perfectly.

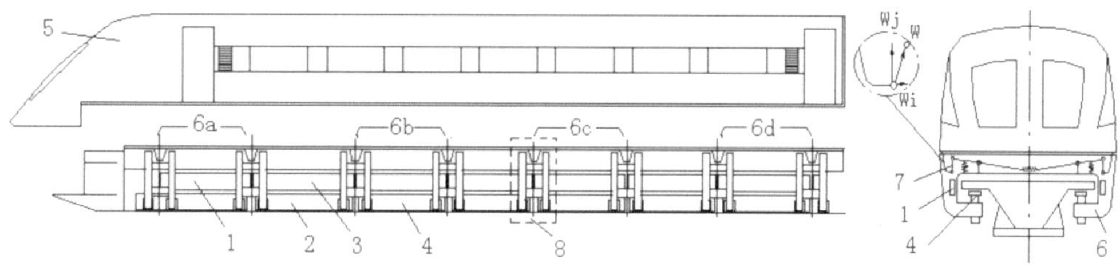

1. Guidance magnet, 2.Overlap magnet, 3.Brake magnet, 4.Levitation magnet, 5.Car body, 6.Maglev bogie, 7.Secondary suspension system, 8.Levitation frame

Fig. 4. Transrapid 08

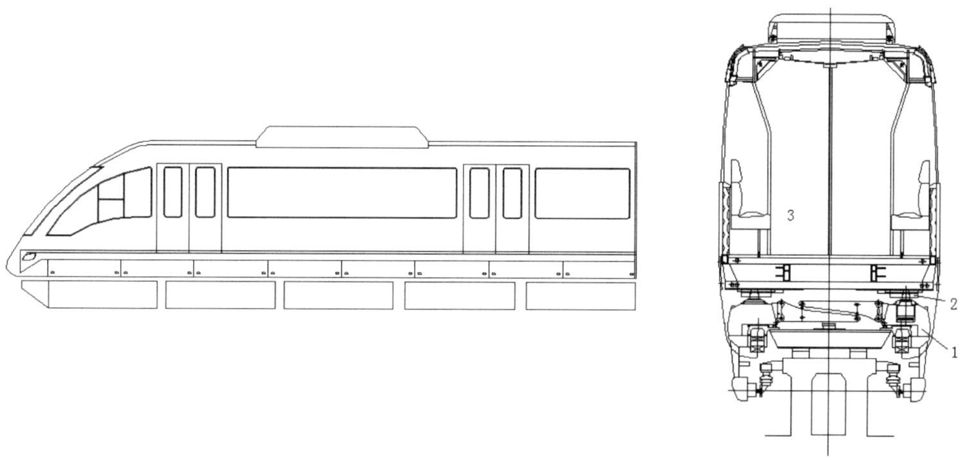

1. Maglev bogie, 2.Secondary Suspend system, 3.Car body.

Fig. 5. HSST-100

2.2 Characteristics of EMS maglev train structure

The structure of maglev trains has several extraordinary characteristics: 1) as light as possible; 2) enough degrees of freedom; 3) special mechanically-braking mode; 4) unique lateral load way 5) vehicles fall on rail to slide under emergency. The vehicles are composed of three parts as shown in Fig.4: car body at the top, secondary suspension at the middle and running gear at the bottom. The wheel rail vehicles have only two bogies through wheel pair contact with rail, but bogies of the maglev trains distributed along the entire length of vehicles, so they are strikingly different in structure.

The two wheel pair of wheel rail vehicles is installed on a rigid frame in the same plane. The four points in the frame of maglev bogies, the detection points of gap sensors, should move independently. The bogies have two typical structures: the bogie with torsion longeron is shown as Fig.6 (Maglev Technical Committee, 2007), Fig.8 (Z.S. ZHAO & L.M. YING, 2007). Two levitation frame units 8 are connected by torsion longeron 7 to form The maglev bogies. In vertical direction, the bogies realize the independent motion of four points by reversed longeron (the bogies hereinafter referred as T-type bogies); and the bogie is assembled by connection tow module 8 with anti-rolling beam 1, as shown in Fig.7.

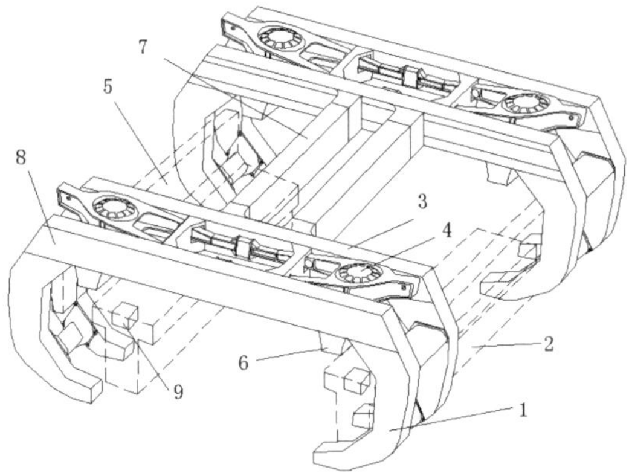

1. Support arm, 2.Levitation magnet, 3. Crossbeam, 4.Air spring & Pendulum arm, 5.Guidance magnet, 6.Support skid, 7.Torsional longeron, 8.Levitation frame unit, 9. Gap sensor

Fig. 6. A bogie of high-speed maglev vehicle

The bogies realize the independent motion of four points by relative torsion of two anti-rolling beams 1 (the bogies hereinafter referred as A-type bogies).

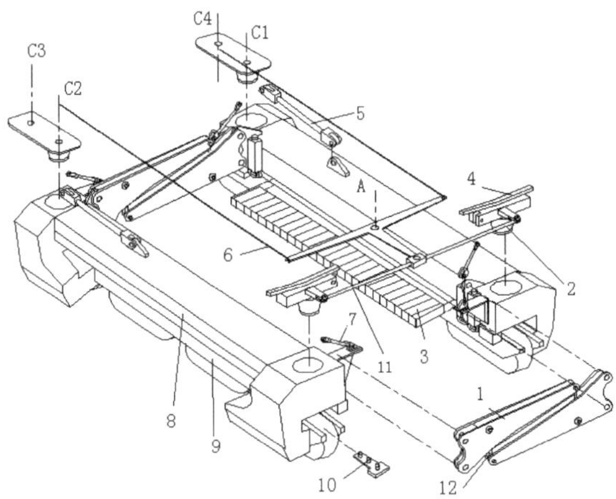

1. Anti-rolling beam, 2.Air spring, 3.Linear induction motor, 4.Linear rolling table, 5.Drive staff, 6.Forced steering mechanism, 7.transverse rod, 8.Module, 9.Lvitation magnet, 10.Gap sensor, 11.thrust rod, 12.Rocker.

Fig. 7. A bogie of middle-low speed maglev vehicle

Generally speaking, the running gear of maglev trains is composed of several bogies. The maglev trains and wheel rail trains also differ in the connection among bogies and between bogies and carriages. As shown in Fig.6, bogies are connected by overlap electromagnet 2 and spring hinges to form the maglev running gear (Fig.4), and joints with car body by the tilting suspension system 7. As shown in Fig.7, 9, 10, bogies are grouped in pairs by forced steering mechanisms 6 to make up the running gear, which is connected with the carriages by hinges A 、C1~C4 and rolling table 4. The linear rolling table is equipped at the end of bogie modules 8 which can rotate around the shaft C in a small angle. The forced steering mechanism 6 is composed of wire ropes and T-type rod. As it turns, the modules deflect to drive the air spring transverse rod 7, then the force is transmitted to thrust rod 8 whose motion drives the T-type rod to rotate, the rotation is passed to another T-type rod by linkage wire ropes, then thrust rod and transverse rod of next bogie drive its modules to deflect and so far the steering action is completed.

The secondary suspension system transmits three forces in different directions between car body and bogies and the transmission course is as follows: the vertical load transmits in maglev track←→ electromagnet ←→ modules ←→diaphragm air spring ←→ rolling table←→ car body.

The transverse load transmits in car body ←→ T-type rod ←→ wire rope, transverse link ←→ lower rolling table ←→ air spring tie rod ←→ modules ←→electromagnet ←→ track.

It can be seen that plenty of bogies distributed along the length of car body contribute to the relative complex joint of car body and bogies. If the tilting suspension system is adopted, the maglev bogie 4 has sixteen pendulum binding mechanisms; if the rolling table is adopted, there are ten point of junction for the steering mechanism.

3. Kinematic characteristics of EMS maglev trains

Although EMS maglev trains fly at a zero height, it still needs exercise along maglev guideway necessary. The position vectors can be divided along guideway (longitudinal), perpendicular guideway surface (vertical), perpendicular guideway side(lateral) three components.

The vertical motion is controlled by the system composed of gap sensor, levitation controller and levitation electromagnet with limitation. The transversal motion is restricted by transversal electromagnetic force and the longitudinal motion is related to the transversal motion and the constraint between all parts of vehicles. According to last paragraph, the vehicle is composed of running gear, secondary suspension system and car body and it's kinematic analysis includes the analysis on the spatial positions of all parts and the relative positions of all parts.

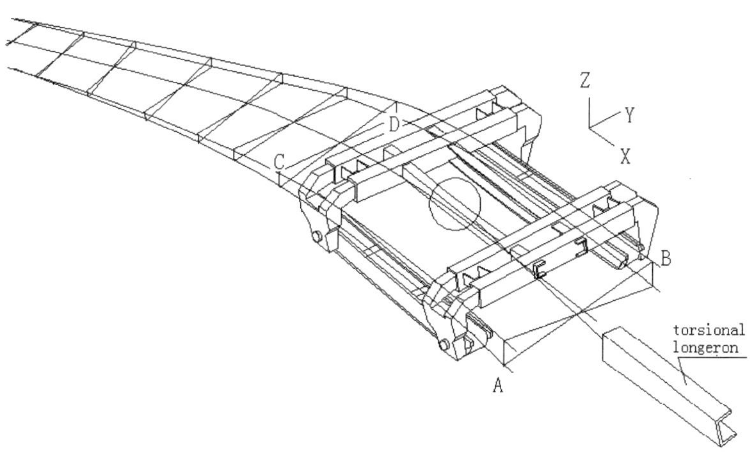

Fig. 8. Bogie Decoupling by torsion beam

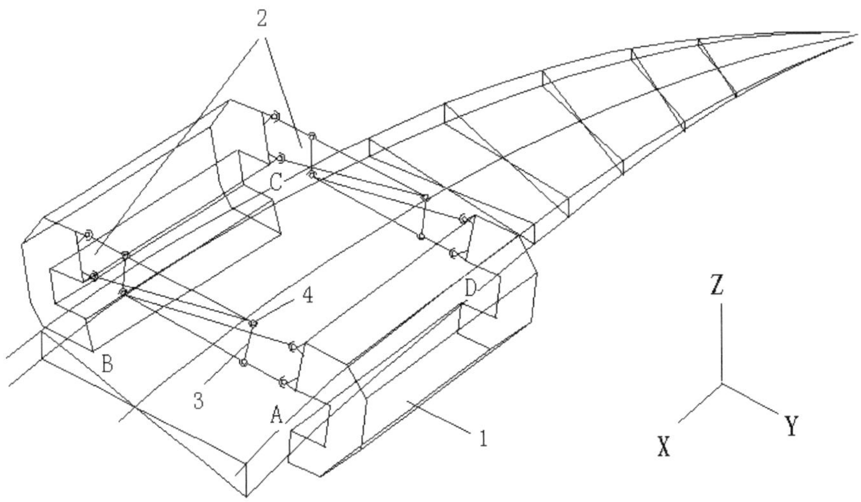

1. Module, 2.anti-rolling beam, 3.pedulum rod, 4. Sphere joint.

Fig. 9. Bogie decoupling by anti-rolling beam

The maglev bogies are the basic components of running gears and their displacements are crucial for the determination of vehicle motion. Their kinematic characteristic is that A, B, C and D points（Fig.8, 9）should move independently (uncoupling). Four straight lines can be drawn by the four points. When the maglev bogies are running along curved path, the four rectilinear motion space surfaces is the Coons surface. When the maglev bogies are passing the transition curve, the four points are not in the same plane. Both A-type bogies and T-type bogies can realize this motion. T-type bogies realize the motion by torsion beams and A-type bogies by the torsion of two anti-rolling beams. However, A-type bogies and T-type bogies have big differences in their transversal motions. The bogie as shown in Fig.6 can only make lateral movement as a whole. In addition, because its secondary suspension system is pendulous and there will produce a big transverse component of gravity force acting on the bogie by pendulum suspension system when the vehicle is passing the curve, the bogie has a bad ability to follow the guideway transversally and can not pass the curve with a small curvature radius and need an active guidance force provided by guidance electromagnet 5 as shown in Fig.6.

Two modules 8 of the bogies as shown in Fig.7 should move independently. Each module has three translational (X, Y, Z) and two rotational (Y, Z) degrees of freedom. It can pass the curves with a small curvature radius and there is hardly any limit in its lateral motion in a small range, so by adopting levitation electromagnet 9 as shown in Fig.6 it can provide a passive guidance force which is a component of levitation force and only exists when the electromagnet is deviating from the guideway and thus it is called as passive guidance force. By now it seems that the motion problem of vehicles has been solved. The track curve can determine the instantaneous position of the bogie, then the relative positions between bogies, bogie and second suspension system and car bodies by connection relationship and the absolute spatial positions of all parts, all of which only involve the deduction of geometric relationships.

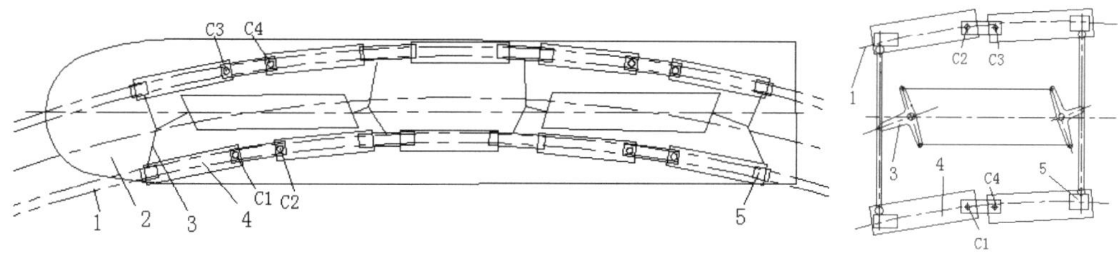

1. Railway, 2.Carbody, 3. Forced steering mechanism, 4.Module, 5.Linear bearing.

Fig. 10. Connection between Secondary suspend system, car body & running gear, and two bogies is connected by forced steering mechanism

But the problem is far from simple. For example, the relative position of car body and running gear of vehicle as shown in Fig.4 must be calculated based on the sixteen pendulum suspension mechanisms for the joint of car body and bogie. If the relative position of car body and bogie in the curve changes, the rocker deflects and the weight W of car body transmitted by the sixteen suspenders to the bogie is decomposed into two component forces W_i, W_j, so the transversal relative position of car body and bogie involves the balance of sixteen transversal forces W_i but not a simple calculation of geometric relationships.

For the vehicle as shown in Fig.5, the constraint of electromagnetic restoring force in the relative position of bogies and track is described in the preceding paragraph. It is easy to calculate the relative position of single bogie and guideway, namely the instantaneous position or locus, then the relative positions or topological relations among all components can be deduced by electromagnetic balance. However, owing to the complexity of connection relationships between several bogies and car body (Fig.10), this calculation method can not be extended to the vehicle. A typical case is when a bogie enters into the transition curve and the following bogie is still in the straight-line guideway, the front bogie rotates around the points C_1, C_2 and the following bogie is droved to rotate around the points C_3, C_4 owing to the effect of forced steering mechanism 3, so the following bogie doesn't move along a straight line. The reason lies in that there is a balance relationship of restraining force between the lateral electromagnetic restoring force and components and it should not considered simply that the bogies are pulled to the track by electromagnetic restoring forces. Therefore, different from the wheel rail vehicles, the passive guidance EMS maglev trains may not run in the track curve. The vehicle electromagnetic restoring force, constraint among all components and track geometry curve must be considered comprehensively to deduce the instantaneous position or trace of a bogie in absolute coordinate by the force balance relation and geometrical relation of vehicle in any position, then the relative positions between the rigid bodies or topological relations of all components are deduced by connection relations. However, it brings big difficulties in solving this problem.

4. Kinematic modeling and analysis of maglev trains (Z.S. Zhao and C. Ren, 2009)

The kinematic characteristics of EMS maglev trains illustrated in the preceding paragraph show that the motion of maglev trains can not be deduced simply by geometrical relations. Based on the passive EMS maglev trains, the following kinematic analysis includes kinematic modelling and analysis. The vertical position of vehicle is controlled by the levitation gap between electromagnet and guideway. Because the gap is constant, the vertical position of vehicle can be determined by the track curve and the determination of lateral instantaneous position is the key of kinematic research on vehicles.

The research on instantaneous position adopts two methods based on track fitting: strict fitting track (two endpoints of the bogie on the track) (ZHAO Z.S. & YING L.M., 2000; MEI Z. & LI J., 2007; JIANG H.B., et al., 2007) and balanced lateral electromagnetic restoring force of single bogie (ZENG Y.W. & WANG S.H. 2003; ZHANG K. & LI J., 2005; ZHAO C.F & ZAI W.M., 2005). The former is obviously an unproved hypothesis and the later doesn't consider the influence of constraint among all components in the motion.

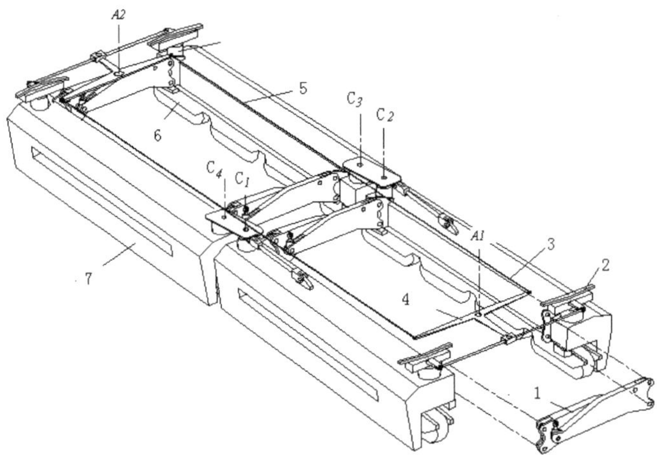

1.Anti-rolling beam, 2.Linear rolling table, 3.forced steering mechanism, 4.T type rod, 5.Wire rope, 6.Levitation magnet, 7.module.

Fig. 11. Running gear sketch of the passiveness guidance EMS maglev train

4.1 Kinematic modelling of EMS mid-low speed maglev trains

To simplify the problem without loss of generality, in this article derivations is made based on the following conditions: 1) because the Z-directional motion of vehicle has a little influence on its lateral motion, its mathematical deduction is based on X-Y plane; 2) the model is established for the vehicle with four bogies; 3) the axis C_1-C_4 are combined into two axis P_4, P_{11} (Fig.12); 4) the kinematic modelling is only based on the central line of track; 5) the carriages and bogies are rigid bodies with the lengths of L_C、L respectively.

4.1.1 Kinematic modeling of maglev trains based on geometrical relations

In the instantaneous position sketch of maglev trains as shown in Fig. 12, P_i (x_i, y_i) represent bogies' end point and intersection point of bogies and track curve Y(x). If P_i is definite, the instantaneous position or motion locus of vehicle and the relative positions (topological relations) among the components of vehicle and between vehicle and track may be determined. The section aims to establish the equations with the unknown quantities x_i, y_i

accordingly based on geometrical relations. P_i is in the straight line representing bogies respectively and should satisfy the following relations:

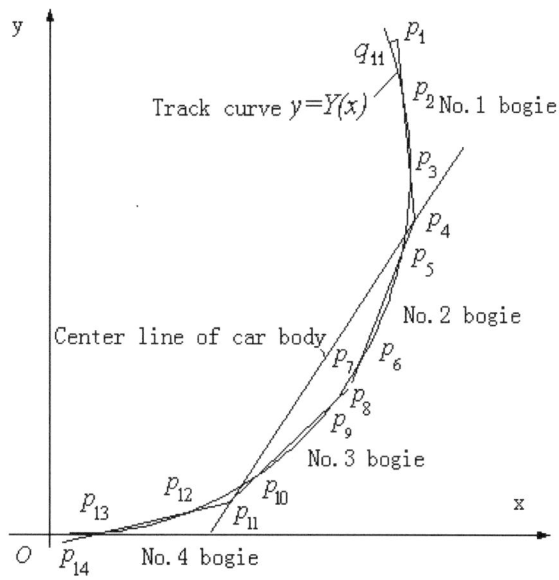

Fig. 12. Instantaneous Position of the maglev vehicle with four bogies

$$\left.\begin{aligned}&\frac{(y_i-y_{i+2})}{(x_i-x_{i+2})}-\frac{(y_{i+1}-y_{i+2})}{(x_{i+1}-x_{i+2})}=0\\&\frac{(y_{i+3}-y_{i+1})}{(x_{i+3}-x_{i+1})}-\frac{(y_{i+2}-y_{i+1})}{(x_{i+2}-x_{i+1})}=0\\&(x_i-x_{i+3})^2+(y_i-y_{i+3})^2=L^2\end{aligned}\right\} \tag{4-1}$$

There i=1, 4,8,11. The geometrical relation between carriage and bogie is:

$$\sqrt{(x_4-x_{11})^2+(y_4-y_{11})^2}=0.5L_c \tag{4-2}$$

The intersection relation of curve and straight line is:

$$\left.\begin{aligned}&\frac{y_i-Y(x_{i+2})}{x_i-x_{i+2}}-\frac{Y(x_{i+1})-Y(x_{i+2})}{x_{i+1}-x_{i+2}}=0\\&\frac{y_{i+3}-Y(x_{i+1})}{x_i-x_{i+1}}-\frac{Y(x_{i+2})-Y(x_{i+1})}{x_{i+2}-x_{i+1}}=0\end{aligned}\right\} \tag{4-3}$$

In the above equation, i=1, 4,8,11. The straight line representing the centre line of carriage is:

$$\left(\frac{y_4-y_{11}}{x_4-x_{11}}\right)x-y-\left(\frac{y_4-y_{11}}{x_4-x_{11}}\right)x_4+y_4=0$$

The offset distance from the bogie endpoints to the car body obtained by the distance from point to line: $\Delta_i = \dfrac{(y_4 - y_{11})(x_i - x_4) - (x_4 - x_{11})(y_i - y_4)}{0.5L_c}$, in which i=1,7,8,1. Follow equation (ZHAO Z.S. et al., 2000) is given by the structural symmetry and the constraint of forced steering mechanism,

$$\frac{\Delta_1}{\Delta_7} = \frac{\Delta_{14}}{\Delta_8} \tag{4-4}$$

there are twenty-two equations with twenty-eight unknown quantities Pi(xi、yi) in the above (4-1)-(4-4), it is obvious that the kinematic problem of vehicle can not be solved only by geometrical relations and other equations should be founded by the balance relations of lateral forces.

4.1.2 Kinematic modeling of maglev trains based on the constraint of lateral electromagnetic restoring force and mechanism constrain

The passive guidance EMS maglev train keeps a lateral position from electromagnetic restoring force. To seek balance of lateral force, it should be considered that the calculation of electromagnetic resilience generated by the linear bogie units fitting the curved track; the influence of constraint such as the binding force produced among the bogies owing to interconnection of the forced steering mechanisms and carriages an bogies. In this section, other equations shall be sought for by the balance of lateral forces, the calculation formula of lateral forces (Sinha P. K., 1987) is:

$$F_u = K_m L_w \tan^{-1}\left(\frac{\Delta u}{\delta}\right)$$

in which $K_m = \dfrac{\mu_0 N^2 I^2}{4\pi\delta}$, $L_w = \dfrac{A}{W_m}$, L_w is length of magnetic pole, μ0, N, A, I represents vacuum permeability and turns, effective area of magnetic pole, coil current respectively, other parameters can refer to Fig.13. Taking the first bogie for example, the electromagnetic restoring force of any infinitesimal curve unit d_s in the track is:

$$dF_{u1} = K_m \tan^{-1}\left(\frac{\Delta u_1}{\delta}\right) dL_w$$

Δu_1 represents the distance from a point q(x, y) in the curve to the line: $\Delta u_1 = \dfrac{(y_1 - y_4)x - (x_1 - x_4)Y(x) + x_1 y_4 - x_4 y_1}{L}$, seeing to (Fig. 13), $dL_w = \cos(\alpha_s - \alpha_1)ds$, $ds\cos\alpha_s = dx$, $k_s = \tan\alpha_s$, $k_1 = \tan\alpha_1 = \dfrac{y_1 - y_4}{x_1 - x_4}$, dF_{u1} can be written as：

$$dF_{u1} = \frac{1 + k_s k_1}{\sqrt{1 + k_1^2}} K_m \tan^{-1}\left(\frac{\Delta u_1}{\delta}\right) dx$$

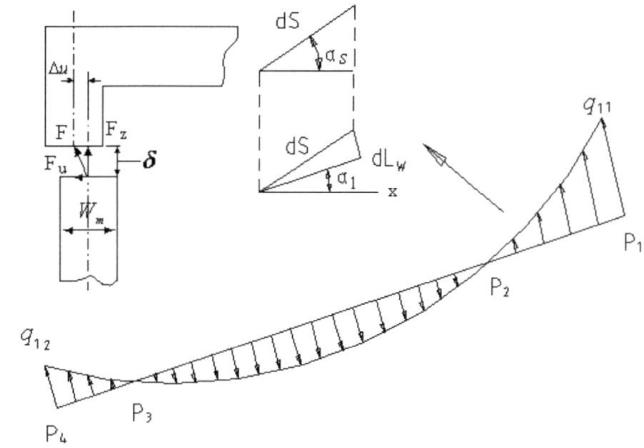

Fig. 13. Magnet lateral reversion force of the bogie module

In the same way, the calculation formula of the electro magnetic restoring force differential unit of other bogies can be deduced:

$$F_{ui} = \int_{x_{i1}}^{x_{i2}} K_m \frac{1 + k_s k_i}{\sqrt{1 + k_i^2}} \tan^{-1}\left(\frac{\Delta u_i}{\delta}\right) dx \quad (i=1,2,3,4) .$$

1. Balance equation of lateral restoring force and moment of the vehicle:

$$\left.\begin{aligned} \sum F_x &= \sum_i^4 F_{ui} \sin \alpha_i = 0 \\ \sum F_y &= \sum_i^4 F_{ui} \cos \alpha_i = 0 \end{aligned}\right\}$$

$$\left.\begin{aligned} &\sum_i^4 \int_{x_{i1}}^{x_{i2}} K_m \frac{(1 + k_s k_i) k_i}{1 + k_i^2} \tan^{-1}\left(\frac{\Delta u_i}{\delta}\right) dx = 0 \\ &\sum_i^4 \int_{x_{i1}}^{x_{i2}} K_m \frac{1 + k_s k_i}{1 + k_i^2} \tan^{-1}\left(\frac{\Delta u_i}{\delta}\right) dx = 0 \end{aligned}\right\} \tag{4-5}$$

Balance equation of moment of lateral restoring force (taking P_4 as the pivoting point)

$$\left.\begin{aligned} &\sum_{i \neq 3}^4 \int_{x_{i2}}^{x_{i1}} K_m \frac{1 + k_s k_i}{1 + k_i^2} \tan^{-1}\left(\frac{\Delta u_i}{\delta}\right) (-1)^i (x_4 - x) dx + \\ &\quad + \int_{x_{32}}^{x_{31}} K_m \frac{1 + k_s k_3}{1 + k_3^2} \tan^{-1}\left(\frac{\Delta u_i}{\delta}\right) (x - x_4) = 0 \\ &\sum_{i \neq 3}^4 \int_{x_{i2}}^{x_{i1}} K_m \frac{(1 + k_s k_i) k_i}{1 + k_i^2} \tan^{-1}\left(\frac{\Delta u_i}{\delta}\right) (-1)^i (y_4 - Y(x)) dx + \\ &\quad + \int_{x_{32}}^{x_{31}} K_m \frac{(1 + k_s k_i) k_3}{1 + k_3^2} \tan^{-1}\left(\frac{\Delta u_i}{\delta}\right) (Y(x) - y_4) dx = 0 \end{aligned}\right\} \tag{4-6}$$

2. Balance of constraint force of forced steering mechanism

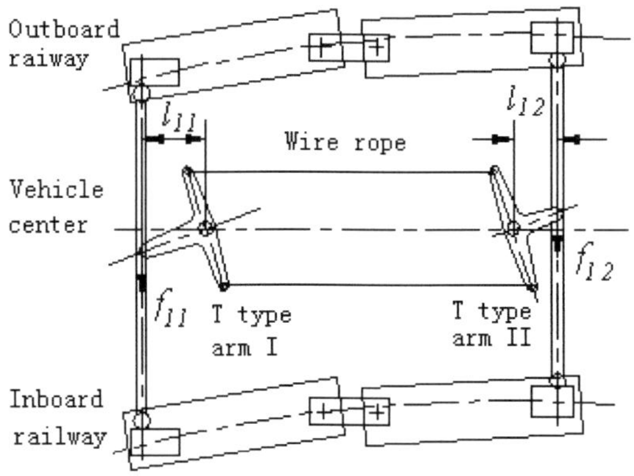

Fig. 14. Balance of two bogies linked by compelling guided mechanism

3. According to Fig.14, the balance equation of constraint force of forced steering mechanism linking the No.1,2 bogies is $f_{11}l_{11}=f_{12}l_{12}$, and the following equation can be obtained:

$$f_{12} = f_{11}\frac{l_{11}}{l_{12}} = \eta f_{11} \tag{4-7}$$

The moment arm length from the point P_4 to No.1 bogie's any point dFu1 on which electromagnetic resilience is exerted is $\frac{x-x_4}{\cos\alpha_1} = (x-x_4)\sqrt{1+k_1^2}$, the balance relation of No.1 bogie module's moment is $0.5 f_{11} L \cos\alpha_1 = \int_{x_{11}}^{x_{12}} K_m(1+k_s k_1)(x-x_4)\tan^{-1}\left(\frac{\Delta u_1}{\delta}\right)dx$, namely, $f_{11} = \frac{2}{L}\int_{x_{11}}^{x_{12}} K_m(1+k_s k_1)\sqrt{1+k_1^2}(x-x_4)\tan^{-1}\left(\frac{\Delta u_1}{\delta}\right)dx$, similarly f_{12} can be obtained, and substituted in (7) :

$$\sum_{i=1,2} \eta^{2-i} \int_{x_{i2}}^{x_{i1}} K_m(1+k_s k_i)\sqrt{1+k_i^2}(x_4-x)\tan^{-1}\left(\frac{\Delta u_i}{\delta}\right)dx = 0 \tag{4-8}$$

By the above method the constraint forces' balance equation of forced steering mechanisms connecting No.3, 4 bogies can be obtained:

$$\sum_{i=3,4} \eta^{i-3} \int_{x_{i2}}^{x_{i1}} K_m(1+k_s k_i)\sqrt{1+k_i^2}(x-x_{11})\tan^{-1}\left(\frac{\Delta u_i}{\delta}\right)dx = 0 \tag{4-9}$$

The upper or lower limit of integration are the coordinate x of the intersection points q_{i1} 、q_{i2} of two straight lines perpendicular to the endpoints P_i 、P_{i+3} of the No.1 bogie and the curve Y(x). Its expression is written as (corresponding to four bogies, j=1, 4, 8, and 11):

$$\left.\begin{aligned} k_i(x_{i1}-x_j)+(Y(x_{i1})-y_j)=0 \\ k_i(x_{i2}-x_{j+3})+(Y(x_{i2})-y_{j+3})=0 \end{aligned}\right\}$$

Substituting x_{11} 、 $y_{11}=Y(x_{11})$ in the above first equation, the following equation can be obtained:

$$k_1(x_{11}-x_1)+(Y(x_{11})-y_1)=0 \tag{4-10}$$

In the above six equations in (4-5) (4-6) (4-8) (4-9) obtained by the balance between lateral electromagnetic resilience and structural constraint force, the equations (8), (9) introduce a unknown quantity η. Hereby the equation (4-10) is introduced and a reference point q_{11} (x_{11} 、 y_{11}) to instantaneous position of vehicle is given. The equations (4-1)-(4-6) and (4-8)-(4-10) are the non-linear equation set with twenty-nine unknown quantities, namely twenty-nine unknown quantities P_i (x_i 、 y_i) and η can be resolved. This is the general formula for kinematic analysis on passive EMS maglev trains with four bogies which can be used to resolve the absolute position (motion trace) of any bogie and the relative position or topological relation of any component of the vehicle at any time. In the same way, kinematics equations of maglev train with other number bogies can be deduced.

4.1.3 General kinematic characters of passive guidance EMS maglev trains

Following kinematic characters of vehicles can be deduced by the above general kinematic formulas:

Character 1: kinematic the static determinacy or indeterminacy of vehicles is determined by the forced steering mechanism, namely the topological relation between a bogie and carriage (formula (4-4), (4-7)) must be given and if not, there will be multiple solutions of motion trace.

Character 2: n, namely the number of intersection points of the modules (straight line) and track (convex curve), $1 \le n \le 2$, two geometric equations will be reduced for each reduced crossing point and in the straight-line segment of track, the bogies are coincident with the track.

Character 3: the motion trace of vehicle is determined together by the topological relations between bogie and carriage and bogie and track but not only by the track.

Character 4: The steering characteristic and yawing characteristic of vehicle with transverse interference depend on the balance relation between the lateral electromagnetic restoring force and the constraint force of forced steering mechanism.

4.2 Solution and analysis of kinematic equations of EMS mid-low speed maglev trains

Given that N=320，W_m=28 mm，A=3360×28 mm2，L=3.4 m，L_c=14.5 m , circular curve radius R=100 m，superelevation is 60 mm，transition curve length l_0=12 m, the easement trace curve is the clothoid generally, the curvature of easement curve $k = s \div Rl_0$ and the high-order small quantities are ignored, the projection of x-y plane of trace curve is:

$$y=Y(x)=\begin{cases} 0 & x\le 0 \\ \dfrac{x^3}{6l_0R} & 0<x<x_e \\ y_c-\sqrt{R^2-(x-x_c)^2} & x_e\le x \end{cases}$$

in which q_0 (x_0, y_0) 、q_e (x_e, y_e) 、q_c (x_c, y_c) represent the bonding points of the straight segment and curved segment of trace curve Q(x) and transition curve and the center of bend circular curve respectively. $x_c = x_e - \dfrac{k_eR}{\sqrt{1+k_e^2}}$, $x_c = y_e + \dfrac{k_eR}{\sqrt{1+k_e^2}}$, $k_e = Y'(x)|_{x=x_e}$, q_e (11.9956, 0.2397) , q_c (11.6364, 100.06) . In the interval defined by track curve, a series of $q_{11}(x_{11}, y_{11})$ are valued according to the step length of 0.1m to resolve the kinematic equations obtained in the preceding paragraph, the motion trace of vehicle and the relative positions of all vehicle components can be derived. Some primary results obtained through numerical calculation by MATHEMATICA are given below.

1. Motion trace

To express the kinematic characters, the motion trace of vehicle is shown by the offset of bogies to the track but not the coordinate figure P_i. The fig.15 gives the fitting figure of computed results and the table 1 shows the computed results when q11 is valued as four typical coordinate points.

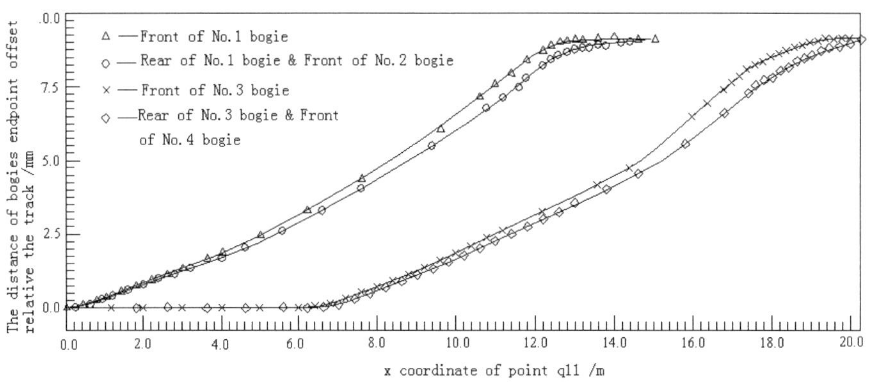

Fig. 15. Curves of bogies endpoint offset relative to the track

Bogie / q_{11}(x,y)	1		2		3		4	
	FrontΔu_{11}	RearΔu_{12}	FrontΔu_{21}	RearΔu_{22}	FrontΔu_{31}	RearΔu_{32}	FrontΔu_{41}	RearΔu_{42}
3.4, 0.0055	1.94	1.89	1.89	1.91	0	0	0	0
6.8, 0.0437	3.52	3.49	3.49	3.5	0	0	0	0
10.2, 0.1474	6.52	6.46	6.46	6.48	1.91	1.89	1.89	1.92
13.6, 0.3494	8.05	7.94	7.94	8.01	3.52	3.51	3.51	3.53

Table 1. Δu_{ij}, Amount of bogie endpoint offset relative to the track

2. Relative positions of all components of vehicle

The relationships of relative positions exist among the bogies and between bogies and carriages. For the sake of intuition, the computed results of relative positions among bogies are transformed into the included angles. The figure 16 gives the fitting figure of computed results of relative positions among bogies and the Tab. 2 gives the computed results of two position relations when q_{11} is valued as four typical coordinate points respectively

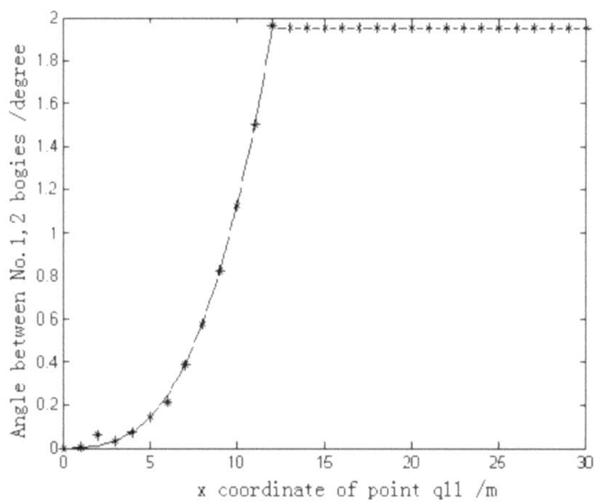

Fig. 16. Alteration curve of angle between No.1,2 bogies

Bogie P11(x,y) / q_{11}(x,y)	1	2	3	4	1-2	2-3	3-4
	FrontΔ_1	RearΔ_7	FrontΔ_8	RearΔ_{14}	θ_{12}	θ_{23}	θ_{34}
3.4, 0.0055	3.24	1.05	0.84	0.84	0.034°	0.02°	0°
6.8, 0.0437	13.78	4.52	1.02	1.02	0.424°	0.215°	0°
10.2, 0.1474	95.75	31.81	27.24	83.92	1.072°	0.556°	0.039°
13.6, 0.3494	157.8	52.54	48.05	146.03	1.771°	1.091°	0.411°

Table 2. Angle θ_{ij} between bogies & Bogie endpoint offset Δi relative to the car body

4.3 Conclusions

Based on the derivation and computational analysis of above kinematic kinematics mathematical formulas and test results, the following conclusions on relevant kinematic researches on passive guidance EMS maglev trains can be obtained.

1. The absolute position or trace of vehicle is not equal to track curve and their relation (offset Δu_{ij}) is also not constant. The change rule is: straight segment (zero) → easement curve segment (Monotone increasing) →bend segment (a maximum constant)
2. The kinematic static determinacy or indeterminacy of vehicles depends on the forced steering mechanism. If no forced steering mechanism, the kinematic relation of vehicles is indefinite.
3. The bogie and carriage, bogie between any two are restricted geometrically and the bogie and track is constrained by lateral electromagnetic restoring force, so the absolute

position Δu_{ij} of bogies should be determined by the electromagnetic balance of vehicle and geometrical constraint relations of all its components. Δu_{ij} of different bogies is different at the same time and position.

4. The parameter η=3 expressing the characteristic of relative position between bogies and carriages is applicable to all line segments.
5. The change law of relative position (θ_{ij}) among bogies is: straight segment (zero) → easement curve segment (Monotone increasing) →bend segment (a maximum constant).
6. The absolute position of vehicle and the relative position among all its components including the bonding points of all line segments change smoothly in the motion.
7. The results obtained by kinematics formulas are consistent with the past research results in circular curve and straight-line segment
8. The article gives that the mathematical models can be used in the kinematic analysis of vehicle by the transverse interference.

5. Kinematic analysis on the secondary suspension system of maglev trains

The composition and kinematic characteristics of secondary suspension system for the joint of car body and bogies has been illustrated in the paragraph 1, 2. The distinctive mid-structure of active and passive guidance maglev trains lie in pendular suspension mechanism and forced steering mechanism respectively. In this paragraph, their kinematic characteristics analysis and the calculation method is given and other kinematic analyses can see the references (ZHAO Z.S. et al., 2000).

5.1 Kinematic analysis on the forced steering mechanism of passive guidance ems maglev trains

The functions of forced steering mechanism are to connect two bogies to form the running gear (Fig.10, 11, 14), keep a proper geometric position between the running gear and car body (Fig.10) and transmit the transversal force between the running gear and car body. When realizing these functions, the uncoupling of bogies can not be affected by the mechanism. According to Fig.14, the transverse thrust rod of the forced steering mechanism may affect the uncoupling of bogies, which can be obtained by analyzing some motions of bogie as it goes through the curve.

The relative height and distance between the ends of two bogie modules in motion may change. The transverse thrust rod are equipped at the end of bogies, so it is possible to add spherical hinges at the end of links to adapt to the change of relative height between the ends of two modules and it is hard to change the length of rigid rods. Take the vehicle with five bogies Shown as Fig.17 for example. Setting：$\delta_1=\alpha_2-\alpha_1$，$\delta_2=\varphi_2-\varphi_1$, h, β , L represent width of track and angle of a、R2 , length of module respectively.

$$\phi_i = \sin^{-1}\left(\frac{L}{2R_i}\right), \quad \alpha_i = 5\phi_i ,$$

outside track radius is R_1 =R+h/2, inside track radius is R_2 =R-h/2, distance between two module endpoints of bogie in the curve： $a_i = \sqrt{R_1^2 + R_2^2 - 2R_1R_2\cos\delta_i}$, there i=1,2.

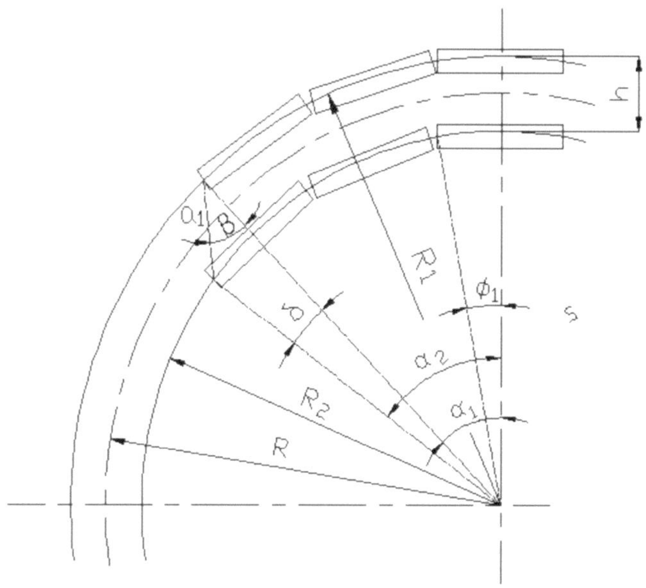

Fig. 17. Distance between two module endpoints of bogie in a curve

Obviously the distance between two module endpoints of bogie in a curve is enlarged relatively to in straight. Its size can be examined by an instance.

Given that R=50m，h=2.02m, L=2.24m

φ_1=1.25811
φ_2=1.31

The distance between two module endpoints of front bogie is:

$$a_1 = \sqrt{48.99^2 + 51.01^2 - 2 \times 48.99 \times 51.01 \times \cos(0.2594)} = 2.03264\text{m}$$

The distance between two endpoints of back bogie：a_2=2.0205m, a_1, a_2 represents the distance between endpoints of front and back bogies in a curve respectively, and the change of distance between two module endpoints of bogie is:

$D_{a1}=a_1-h$=12.64mm
$D_{a2}=a_2-h$=0.5mm

Therefore, the change of distance between two module endpoints of front bogie is bigger and transverse thrust rod must be able to extend 13mm at most when the vehicle is in motion. To solve this problem, the transverse thrust rod may be arranged in a V type (Fig.18) and the calculation of physical dimension of the forced steering mechanisms and their mathematical models based on kinematics principle is given below.

As shown in Fig.17, L、t、l、d、f represent length of module and transverse thrust rod and T-type arm, the horizontal distances from rotation center of module to rotation centre of T-type rod, the offset distance between the hinge point of transverse thrust rod and the center of air spring respectively. θ_1, θ_2 represent the oscillation angle of two modules respectively.

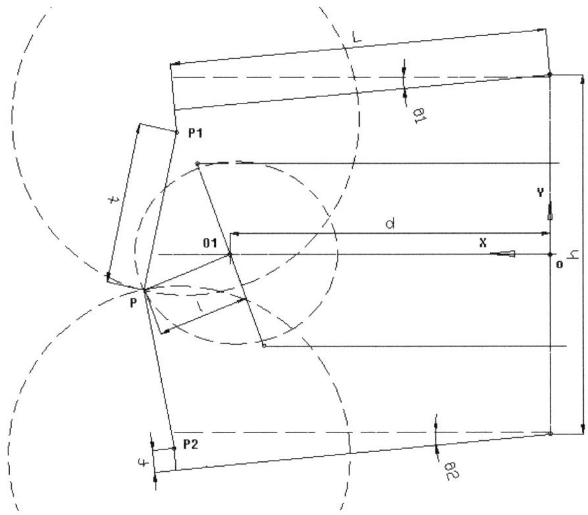

Fig. 18. Kinematic sketch of forced steering mechanism

When the running gear is in motion, two transverse thrust rod and T-type rod rotate around the point $P_1(x_1,y_1)$, $P_2(x_2,y_2)$ and O_1 respectively and the traces of their endpoints are three circles in the same plane with the radius of t, 1. Three circles intersect in the point P(x, y). Thereout the following equation set is given.

$$\left.\begin{array}{l}(x-x_1)^2+(y-y_1)^2=t^2\\(x-x_2)^2+(y-y_2)^2=t^2\\(x-d)^2+y^2=l^2\end{array}\right\} \tag{5-1}$$

among which,

$$\left.\begin{array}{l}x_i=L\cos 4\alpha_i+(-1)^i f\sin 4\alpha_i\approx L\sqrt{1-\dfrac{4L^2}{R_i^2}}\\y_i=(-1)^{i-1}(h-f\cos 4\alpha_i)-L\sin 4\alpha_i\approx(-1)^{i-1}(h-f)-\dfrac{2L^2}{R_i}\end{array}\right\} \tag{5-2}$$

In the above equation set, i=1, 2, and by the third equation in (1), $y=\sqrt{l^2-(x-d)^2}$ is obtained.

In consideration of $x_1\approx x_2$, by the first and second equations in (1), $y=-L^2(\frac{1}{R_1}+\frac{1}{R_2})$ is obtained and is substituted into second equations in (5-1) :

$x_2+\sqrt{t^2-[h-f+L^2(\frac{1}{R_2}-\frac{1}{R_1})]^2}=x$, thus the x, y is obtained and substituted respectively into third equation in (5-1):

$$(\sqrt{t^2-[h-f+L^2(\frac{1}{R_2}-\frac{1}{R_1})]^2}+L-d)^2+L^4(\frac{1}{R_1}+\frac{1}{R_2})^2=l^2 \quad (5\text{-}3)$$

Without considering of high-order small quantities, the equation (3) can be simplified as:

$$(\sqrt{t^2-(h-f)^2}+L-d)^2=l^2 \text{ or } t^2=(l+d-L)^2+(h-f)^2$$

There are two unknown quantities t and 1 in the above equation. One of them is set, the other can be obtained. A calculation sample is given below.

Given that L=2.24m, h=2m, R=50m, f=90mm, d=1.9m and l=550mm, t≈1.39m can be obtained.

5.2 Kinematic analysis on tilting suspension system of maglev train

5.2.1. Mathematical description of turing characteristic of tilting suspension system maglev train (Zhao Z.S., 2009)

The figure 18 show the motion state of high-speed maglev train with tilting suspension system goes around the curve, in which Δ_{ij}, θ_{ij} represent the lateral displacement and oscillation angle of rocker respectively and w、T_{ij}、f_{ij} represent the car body gravity, tension and lateral force of carriage acting on the rockers of tilting suspension system respectively. The train is composed of carriage, bogies, suspension system. four bogies and three overlapping modules are connected alternately to form the running gear (see the Fig. 18 left) and four set of pendular suspension systems are configured in the interval between four bogies and carriage respectively (see the Fig. 18 right). As the vehicle enters the curve, the bogies move along the track curve under the effect of active electromagnetic guiding force and produce relative displacement Δ_{ij} to the carriage which is driven by the oscillation of rockers of tilting suspension system. The sixteen pendular rod of tilting suspension system will produce the lateral force acting on the carriage and bogies. Δ_{ij} is determined by the balance of the forces f_{ij} acting on the carriage, the active electromagnetic guiding force can be obtained by the force f_{ij} acting on single bogie. Therefore the solution of the steering characteristic of maglev train with tilting suspension system lies in resolving the displacement of rockers and the force acting on them. From the viewpoint of design, it might as well make a hypothesis that the sixteen rockers receive the weight of carriage equally.

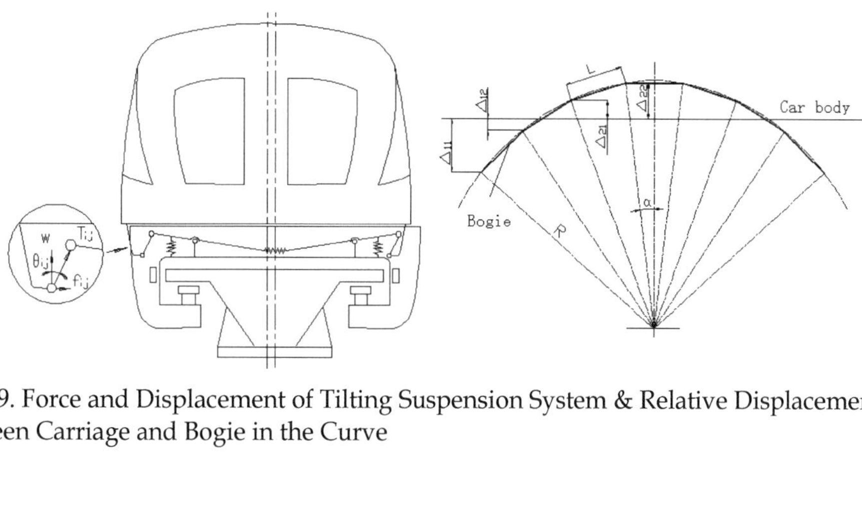

Fig. 19. Force and Displacement of Tilting Suspension System & Relative Displacement between Carriage and Bogie in the Curve

The lateral balance equation of carriage in the curve is:

$\sum_{i=1}^{4}\sum_{j=1}^{2}\vec{f}_{ij}=0$, because of the symmetry, $2\sum_{i=1}^{2}\sum_{j=1}^{2}\vec{f}_{ij}=0$,

in the above equations, the bilateral balances are considered similarly and i、j represent the number of bogie and its ends respectively. The above equation can be written as:

$$w(\tan\theta_{11}+\tan\theta_{12}-\tan\theta_{21}-\tan\theta_{22})=0 \tag{5-4}$$

namely:

$$\frac{\Delta_{11}}{\sqrt{l^2-\Delta_{11}^2}}+\frac{\Delta_{12}}{\sqrt{l^2-\Delta_{12}^2}}=\frac{\Delta_{21}}{\sqrt{l^2-\Delta_{21}^2}}+\frac{\Delta_{22}}{\sqrt{l^2-\Delta_{22}^2}} \tag{5-5}$$

among which l represent the length of rocker. The following geometrical relationships can be shown in Fig. 18:

$$\Delta_{11}=\Delta_{12}+L\sin 6\alpha$$

$$\Delta_{22}=\Delta_{21}+L\sin 2\alpha$$

Substituted into (2):

$$\frac{\Delta_{12}+L\sin 6\alpha}{\sqrt{l^2-(\Delta_{12}+L\sin 6\alpha)^2}}+\frac{\Delta_{12}}{\sqrt{l^2-\Delta_{12}^2}}=\frac{\Delta_{21}}{\sqrt{l^2-\Delta_{21}^2}}+\frac{\Delta_{21}+L\sin 2\alpha}{\sqrt{l^2-(\Delta_{21}+L\sin 2\alpha)^2}} \tag{5-6}$$

Likewise from the geometrical relationships, the following equation can be obtained:

$$\Delta_{12}+\Delta_{21}=L\sin 4\alpha \tag{5-7}$$

From trigonometric functional relations and in considering of R>>L：

$$\sin 2\alpha=2\sin\alpha\cos\alpha=\frac{L}{2R}\sqrt{1-\left(\frac{L}{2R}\right)^2}\approx\frac{L}{2R}$$

$$\sin 4\alpha=4\sin\alpha\cos^3\alpha=\frac{2L}{R}\left(1-\left(\frac{L}{2R}\right)^2\right)^{\frac{3}{2}}\approx\frac{2L}{R}$$

$$\sin 6\alpha=2(3\sin\alpha-4\sin^3\alpha)(4\cos^3\alpha-3\cos\alpha)=\frac{L}{R}\left(3-\left(\frac{L}{R}\right)^2\right)\left(1-\left(\frac{L}{R}\right)^2\right)\sqrt{1-\left(\frac{L}{2R}\right)^2}\approx\frac{3L}{R}$$

From the above three equations, given that m=L2/R and substituted in (3), (4) :

$$\left.\begin{aligned}&\frac{\Delta_{12}+3m}{\sqrt{l^2-(\Delta_{12}+3m)^2}}+\frac{\Delta_{12}}{\sqrt{l^2-\Delta_{12}^2}}=\frac{\Delta_{21}}{\sqrt{l^2-\Delta_{21}^2}}+\frac{\Delta_{21}+0.5m}{\sqrt{l^2-(\Delta_{21}+0.5m)^2}}\\&\Delta_{12}+\Delta_{21}=2m\end{aligned}\right\} \tag{5-8}$$

among which L represent the length of bogie. From the equations (5),

$$\frac{5m-\Delta_{21}}{\sqrt{l^2-(5m-\Delta_{21})^2}}+\frac{2m-\Delta_{21}}{\sqrt{l^2-(2m-\Delta_{21})^2}}=\frac{\Delta_{21}}{\sqrt{l^2-\Delta_{21}^2}}+\frac{\Delta_{21}+0.5m}{\sqrt{l^2-(\Delta_{21}+0.5m)^2}}$$

Given that $\Delta_{21}=\mu m$ and substituted in the above equations:

$$\frac{5-\mu}{\sqrt{n^2-(5-\mu)^2}}+\frac{2-\mu}{\sqrt{n^2-(2-\mu)^2}}=\frac{\mu}{\sqrt{n^2-\mu^2}}+\frac{\mu+0.5}{\sqrt{n^2-(\mu+0.5)^2}} \tag{5-9}$$

among which n=l/m 。

$$\left.\begin{aligned}\Delta_{11}&=(5-\mu)m\\\Delta_{12}&=(2-\mu)m\\\Delta_{21}&=\mu m\\\Delta_{22}&=(0.5+\mu)m\end{aligned}\right\} \tag{5-10}$$

The equations (5-4) (5-5) (5-9) (5-10) are the calculation formulas of steering characteristics of maglev train with tilting suspension system.

5.2.2 Calculation of steering characteristic parameters of maglev train with tilting suspension system

Structural parameters of vehicle is given, L=4.096m; l=0.24m; R=350m、400m; gauge is 2.2m; weight of carriage W=30T; w=W÷16=1.875T. Valuing the convergence accuracy as 0.005, μ can be obtained by solution of the equation (5-9) with numerical method, then the lateral force fij and lateral displacement Δij of rocker ends derived from equations (5-4) and (5-10), it is not hard to obtain the tension Tij of suspension rocket and the vertical displacement of its ends. The calculation results are as follows:

When the vehicle is passing the curve of 350m, R=350m ; R1=351.1m ; R2=348.9m.

Item / Position	Displacement of suspensor rod tip								Lateral force put on car body (KN)			
	Transverse (m)				Vertical (m)							
	Δ_{11}	Δ_{12}	Δ_{21}	Δ_{22}	Z_{11}	Z_{12}	Z_{21}	Z_{22}	f_{11}	f_{12}	f_{21}	f_{22}
Outside track	0.154	0.0106	0.0846	0.1085	0.056	0.0002	0.015	0.026	15.69	0.83	-7.09	-9.54
Inside track	0.156	0.0108	0.0857	0.1098	0.057	0.0002	0.0156	0.027	15.86	0.84	-7.14	-9.37

Table 3. Displacement of suspensor rod tip & Lateral force put on car body

From equation (5-9) it can obtain μ=1.777 and the above parameter table 3. The electromagnetic guiding forces acting on the bogie 1 and 2 are 33.22K and -33.15KN respectively, which is the reason why this kind of vehicles must adopt the active guidance structure.

6. Research on mechanisms and kinematics of maglev bogies

In this paragraph, the mechanism analysis and kinematic calculation methods of maglev bogies are introduced. As described in the paragraph 1 and 2, the bogies of EMS maglev trains have two structures. T-type bogies (Fig.6, 8) are decoupled by the torsion of longerons and A-type bogies (Fig.6, 8) are decoupled by anti-rolling beams. The vertical uncoupling of both kinds of bogies is based on the principle of relative torsion of modules. Their mechanism sketches are shown respectively in Fig.19 and Fig.20.

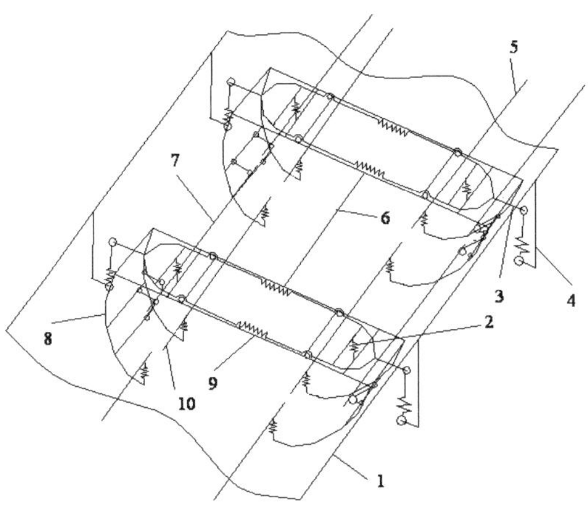

1.Car body, 2.Secongdary system spring, 3.Rocker arm, 4.Z support for car body, 5.Linkage levitation magnet, 6.Longeron, 7.Guidance magnet, 8.Suppot arm, 9.Levitation frame unit, 10.Levitation magnet.

Fig. 20. Mechanism sketch of T-type bogie

The Levitation frame unit of T type bogies is distributed both front and back and may be connected with a torsional elastic longeron, and the Levitation and guidance electromagnet is installed on the bracket arms of front and back modules. It is obvious that other relative motions of the front and back Levitation frame unit of T-type bogies are limited. The two modules of A-type bogie have three translational degrees of freedom and two rotational degrees of freedom. It can be seen from the sketch that the analysis on their X, Y-directional translational degrees of freedom and Z-directional rotational degree of freedom is much simple, and X-directional rotational degree of freedom is limited by the anti-rolling beams, so in this section, the analysis and calculation focus on Z-directional translation and Y-directional rotation of modules of A-type bogies.

Take the kinematic analysis on the right module in Fig. 19 for example. When the endpoint P of right electromagnet 1 elevates D_{11}, the corresponding points M, M′ to electromagnets in the same plane with anti-rolling beams 3, 4 elevate d_{11} 、d_{12}, the angle between the magnet 1 and the horizontal plane is α_1 and the module 2 rotates in the Y direction, namely twist relatively to the left module 9. As the motion of the module 2, the front and back anti-rolling beams 3, 4 move d_{11} 、 d_{12} upward in the Z direction. Owing to the immovability of the left module 9, as the motion of the module 2, the right and left pairs of anti-rolling beams 3-5,4-6 should stagger d_{11} 、 d_{12} in the Z direction, but the anti-rolling beams are connected by suspenders 1—2 which tend to stop this motion.

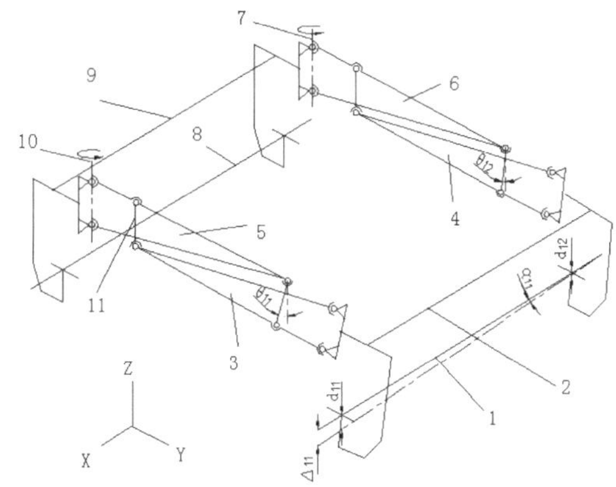

1. R. levitation magnet, 2.R. module, 3.R. Front Anti-rolling beam, 4.R.rear Anti-rolling beam, 5.L. Front Anti-rolling beam, 6.L.Rear Anti-rolling, 7.Rear axis of rotation, 8.L. levitation magnet, 9.L. module, 10.Front axis of rotation, 11. Pendular rod

Fig. 21. Module uncoupling movement of a type bogie mechanism

For the sake of further analysis, it may emphasize the analysis on the relative motion of front two anti-rolling beams 3-5. It is obvious that the anti-rolling beam 5 can not move in the Z direction but rotate around the shaft 10. When the module 2 is moving upward, the anti-rolling beam 3 exerts a press force on the Pendular rod 1-1′. Because there are the ball hinges at the ends of the rod, the bearing anti-rolling beam 3 is instability and will deflect to drive the anti-rolling 5 to move around the shaft 10. At this moment, the anti-rolling beam 3 can move upward, namely one end of the module 2 move upward and the other anti-rolling beam moves similarly. It can be seen that the analysis on the motion of modules focuses on the calculation of kinematic parameters of connecting two module ant-rolling beams. The relevant computational formulas are given below.

For the convenience of analysis, the mechanism sketch Fig.21 of ant-rolling beam is given separately. The sketch shows the position relations of motion of all points elevated by one end of the right module. Proposed that the length of ON is L1 、 the length of OP is t12 、 the length of OP′ is t11 ， the length of RM is H1 、 the length of RM′ is h11, lij represents the length of four rocker respectively and the first and second subscripts represent the number of modules and anti-rolling beams respectively.

$$d_{ij} = \frac{t_{ij}}{L_i}\Delta_{ij} \qquad i=1, j=1, 2 \tag{6-1}$$

$$\alpha_{ij} = Sin^{-1}\left(\frac{\Delta_{ij}}{L_i}\right)$$

$$\theta_{ij} = Cos^{-1}\left(1 - \frac{d_{ij}}{l_{ij}}\right) \tag{6-2}$$

$$S_{ij} = l_{ij}Sin\theta_{ij} = l_{ij}\sqrt{1 - Cos^2\theta_{ij}} \tag{6-3}$$

$$\beta_{ij} = Sin^{-1}(\frac{S_{ij}}{H_i})$$

$$s_{ij} = \frac{h_{ij}}{H_i}S_{ij} \tag{6-4}$$

$$\phi_{ij} = Sin^{-1}\left(\frac{s_{ij}}{l_{ij}}\right)$$

In above equation (6-3) (6-4), S_{ij} and s_{ij} represent transverse motion of end of pendular rod 1j and 2j respectively, from the above equations：

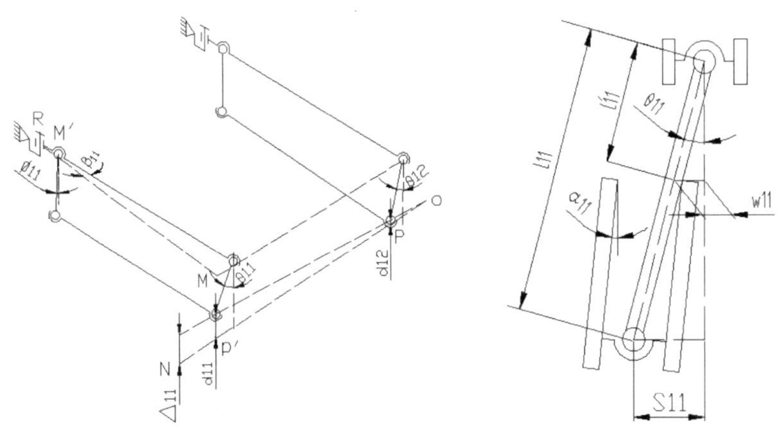

Fig. 22. Z-directional decoupling movement of anti-rolling beam mechanism & oscillation compensation of suspender

$$S_{ij} = l_{ij}\sqrt{1 - \left(1 - \frac{t_{ij}\Delta_{ij}}{L_i l_{ij}}\right)^2} = \frac{t_{ij}\Delta ij}{L_i}\sqrt{\frac{2L_i l_{ij}}{t_{ij}\Delta_{ij}} - 1}$$

$$s_{ij} = \frac{h_{ij}t_{ij}\Delta_{ij}}{H_i L_i}\sqrt{\frac{2l_{ij}L_i}{t_{ij}\Delta_{ij}} - 1}$$

$$\theta_{ij} = Cos^{-1}\left(1 - \frac{t_{ij}}{L_i l_{ij}}\Delta_{ij}\right)$$

$$\beta_{ij} = Sin^{-1}\left(\frac{t_{ij}\Delta ij}{L_i}\sqrt{\frac{2L_i l_{ij}}{t_{ij}\Delta_{ij}}} - 1\right)$$

$$\phi_{ij} = Sin^{-1}\left(\frac{h_{ij}t_{ij}\Delta_{ij}}{H_i L_i l_{ij}}\sqrt{\frac{2l_{ij}L_i}{t_{ij}\Delta_{ij}}} - 1\right)$$

The above equations are the computational formulas of relevant parameters to the Y-directional rotation and Z-directional translation of the right module. When the right module is translating in the Z-direction, $\Delta_{11}=\Delta_{12}$ and the calculation of connecting two pairs of anti-rolling beams is identical. The calculation of X, Y-directional translation and X-directional rotation is comparatively simple and the analysis and calculation of the left module are similar. About these it is unnecessary to go into details.

An example of calculation is given below. Given all relevant geometric dimensions：ON=L=2700，OP′=t_{11}=2320，
l_{ij}=200，OP=t_{12}=380，RM=H_1=1200，RM′=h_{ij}=26 and supposed that one end of module elevates Δ_{11}=8mm, calculation from the above formula, S_{11}=52, s_{11}=11.3, S_{12}=21.2, s_{12}=4.6, θ_{11}=15.1°, θ_{12}=6.1°, β_{11}=2.48°, β_{12}=1.01°, Φ_{11}=3.24°, Φ_{12}=1.32°.

If the anti-rolling beams and rocker are assembled as sandwich (Fig.7), the oscillation of pendular rod may be limited, so the width between two anti-rolling beams should be enough. Take the anti-rolling beam 11 for example (Fig.20 right) and it is not difficult to conclude that:

$w_{11} = \frac{l'_{11}}{l_{11}}S_{11}$ width between two anti-rolling beams：$W_{11} = \frac{2l'_{11}}{l_{11}}S_{11} + C_{11}$, among which C_{11} is the diameter of suspender. If $l'_{11} = 75$ mm, C_{11} =20mm，and others are same as the above instance，it is given that W_{11}=59mm.

It should be pointed out that when four rockers are oscillating, connection of four endpoints of the rocker l_{1j}、l_{2j} can form a pair of spatial quadrangles. It has the following two circumstances:

If the module translates in the Z direction, this pair of spatial quadrangles will be in two planes separately and they are parallelogram.

If one end of the module elevates or rotates in the Y direction, this pair of quadrangles will be spatial.

It can be seen that the motion of pendular rod is spatial and the above formulas based on simplified to the plane is approximate one in the circumstance 2. However the error is small and the results are conservative, so there is no problem to apply in the engineering design.

7. Prospects for structure and kinematic analysis on maglev trains

The research and application of maglev trains has gone for more than half century, the study of vehicle structures, focusing on the running gears and secondary suspension system, has undergone the replacement of many generations. Great strides have also been made in the kinematic analysis which is closely related to design. However, it is to be so regretted that contents of this section is involved in the core of structure and competitiveness and this kind of references are rare, so an brief introduction is given below according to the author's work.

7.1 Prospects for research on vehicle structures

The most feature parts of maglev vehicle structure are the bogies and secondary suspension system for the joint of bogies and car body on which the study touches upon the analysis methods of design and innovation of mechanisms.

1. The research on the mechanism of bogies focuses on the innovation of mechanism which requires providing at most five degrees of freedom for single levitation module. Now the mechanism and its developmental direction are focusing on the spatial linkages mechanism. The number of kinematic pairs and component and joints type are two mainstream research directions, for example, at the longeron's middle of T-type bogie two hinged rods are changed into one rod and more linkage rods are set at the junction part of two modules of A-type bogies. The number of kinematic pairs and component is closely related to degrees of freedom of bogie levitation unit (reduced to connecting rods), and T-type bogies are equipped with more elastic connecting pieces to add the degrees of freedom, which will produce some additional forces and affect their structural life and motion range of component. A-type bogies with plenty of kinematic pairs and component are much complicated in structure and the operation and maintenance work are also increased. Therefore it is an important direction of research on vehicle structure how to constitute the bogie mechanisms with minimum kinematic pairs and components to realize the maximum degrees of freedom now.
2. The innovation of mechanism is still the direction of research on secondary suspension system, but the mechanism of secondary suspension system is closely related to the bogies and is contrary to the bogies in the complexity. This is not hard to understand because the degrees of freedom of bogies are more and the matched secondary suspension system must satisfy its requirements but not limit its degrees of freedom. Therefore an important direction of the research on structure lies in the analysis and innovation of the whole mechanism formed by secondary suspension system and bogies. Of course, the difficulties are obvious.
3. As the advance in the research, the design analysis method is an important branch. It is a trend to apply the development achievements of mechanism in recent years into the structural design of maglev trains. In a nutshell, the topological structure of kinematic chain is represented by graph theory, namely the topological graph represented by points and edges is further represented by matrix. The formulation of experiences and imitation design methods may be very important to the synthesis of bogie mechanism and secondary suspension system. The optimization of mechanisms is another trend, including the objective functions such as scale of motion and degree of freedom and the parameters such as length of linkage rod and connection pair.

7.2 Progress in kinematic analysis on vehicles

The kinematic analysis on vehicles includes kinematic analysis methods, modelling and solutions of kinematics mathematical model, etc.

1. Progress in kinematic analysis methods

At present, the simulation method is widely applied and much mature. The analytic method is still developing and its main direction is to apply the mechanism kinematics theory into the kinematic analysis of maglev trains, for example, in multi-rigid-body kinematical analysis on robots, the traces and relative positions of all rigid bodies can be obtained successively by the determination of the motion trace of input end and D-H transformation, which is method of open chain analysis. However the problem is that the maglev trains have no trace of input end which is conveyed in the fourth section of this chapter, so it is Inappropriate to apply the above method into maglev trains. Another analytic method is to found an analytic equation set of the whole kinematic chain by combining geometrical analysis (traces, topological relations among rigid bodies) and equilibrium of internal with external forces, then the equation set is solved to derive traces (instantaneous positions) and topological relations of all bodies (relative positions including the relative positions with traces), which may be called as method of "closed-loop" analysis. That is to say, the traces of the whole kinematic chain and its any component are unknown and all unknown quantities are included in a non-linear equation set. This analytic method is proper to maglev trains and also universal. In this chapter, the analytical process on two kinds of EMS maglev trains introduced.

The further studies include that the dynamics vector equations of vehicles can be obtained by establishing the position vectors equations of spatial traces of all rigid bodies and derivation of the equations on time. In addition, considering the vehicle is composed of rigid-elastic bodies, its method of multi-body kinematic analysis is another important and difficult task.

2. Establishment of kinematics mathematical model

The analytic method is closely related to modelling, Transformation of areal model into space model becomes an important branch even though its sense may be restricted in theoretical category. If the kinematic analysis model of bogies stated in the sixth section is established based on the theory of spatial mechanism, the motion of binding mechanism of modules can be understood clearly and more accurate structural design may be guided if necessary. In addition, the model in the third section can establish the model with the width of vehicle and track by the method of offset curve.

More accurate models are also the pursuit of researchers, for example, considering the influence of change of the module Z-directional displacement caused by the adjustment of electromagnet and elastic elements which may change the kinematic models of maglev trains.

3. The solution should not differ greatly from that of mathematic and numerical solution without much further ado. For maglev trains, their unique features are the simplification of equations, setting of boundary conditions and precision of calculation.

8. References

Yoshio Hikasa, Yutaka Takeuchi.(1980). Detail and Experimental Results of Ferromagnetic Levitation System of Japan Air Lines HSST-01/=02 Vehicles[C]//IEEE. *IEEE Transactions on Vehicular Technology*, VOL. VT-29, No. 1, February, pp35-41.

J.L. He, D.M. Rote, and H.T. Coffey (1992). Survey of Foreign Maglev Systems[R]. Center for Transportation Research, Energy Systems Division, Argonne National Laboratory, 9700 South Cass Avenue, Argonne, Illinois 60439, July, pp13-14

Tejima Yuichi, et al., (2004). Aichi High-speed Traffic HSST-100 Type Vehicle[J]. *Vehicle Technology*, 227(3), pp86-97.

Seki, Tomohiro.(1995). The Development of HSST-100L[A]. In: *Proceedings MAGLEV'95 14th International Conference on Magnetically Levitated Systems* [C]. VDE-Verlag, ISBN-10 3800721554, Berlin, pp51-55.

Maglev Technical Committee.(2007). Vehicle Part I General Requirements, In: *Rapid Maglev System Design Principles* [R]. White paper, 12, pp18- 19.

Z.S.Zhao, L.M.Ying.(2007) One Running Gear of Maglev Vehicle: China, ZL03130750.7[p]. 24. 10.

Zhao Zhisu, Ren Chao.(2009). Modeling of Kinematics of EMS Maglev Vehicle[J]. *Journal of The China Railway Society.*,Vol.31, (4), pp32-37，ISSN 1001-8360.

Zhao Zhi-Su, Et al.,(2000) Motion Analysis and Design for Yawing Mechanism of Maglev Vehicle [J].*Electric Drive for Locomotive*, (6), pp11-13,30, ISSN 1000-128x.

Mei Zu, Li Jie.(2007). Dynamics Simulation for Yawing Mechanism of Maglev train Based on Virtual Prototype [J]. *Journal of System Simulation*, 19 (18), pp 4199-4203, ISSN 1004-731x

Jiang Haibo et al.(2007). A Study on Forced Steering Mechanism of Low-speed Maglev Train[J]. *Diesel Locomotive*, (4), pp15-18, ISSN 1003-1839.

Zeng You-Wen,Wang Shao-Hua.(2003). Research on geometri- cal curve nigotiating of three-truck maglev vehicle [J]. *Journal of Southwest Jiaotong University*, 38(3), pp282-285，ISSN 0258-2724.

Zhang Kun, LI Jie.(2005). CHANG Wensen. Structure de-coupling analysis of maglev train bogie[J]. *Electric Drive for Locomotive*, (1), pp 22-39, ISSN1000-128x.

Zhao Chun-Fa, Zai Wan-Ming.(2005). Guidance Mode and dynamic lateral characteristic of low-speed maglev vehicle[J]. *China Railway Science*, (1), pp28-32, ISSN 1001-4632.

Sinha P. K.(1987). *Electromagnetic Suspension Dynamics & Control* [M]. Perter Peregrinus Ltd., ISBN 10-0863410634, London.

Luo Kun, Yin Li-Ming, XIE Yun-de.(2004). Analysis on location parameters of line for mid-low speed maglev Train Calculation and analysis of gradient[J]. *Electric Drive for Locomotive*, (4), pp17-19, ISSN 1000-128x.

Zhao Zhisu.(2009). Researches On Turing Characteristic Of Tilting Suspension High-Speed Maglev Train[J]. *Electric Drive for Locomotive*, (1), pp43-45, ISSN 1000-128x.

6

Maglev

Hamid Yaghoubi[1,2], Nariman Barazi[2] and Mohammad Reza Aoliaei[3]
[1]Iran Maglev Technology (IMT), Tehran,
[2]Civil Engineering Division, Department of Engineering,
Payame Noor University (PNU), Tehran,
[3]Civil Engineering Division, Azad University, Ramsar,
Iran

1. Introduction

Magnetic levitation (maglev) is a highly advanced technology. It is used in the various cases, including clean energy (small and huge wind turbines: at home, office, industry, etc.), building facilities (fan), transportation systems (magnetically levitated train, Personal Rapid Transit (PRT), etc.), weapon (gun, rocketry), nuclear engineering (the centrifuge of nuclear reactor), civil engineering (elevator), advertising (levitating everything considered inside or above various frames can be selected), toys (train, levitating spacemen over the space ship, etc.), stationery (pen) and so on. The common point in all these applications is the lack of contact and thus no wear and friction. This increases efficiency, reduce maintenance costs and increase the useful life of the system. The magnetic levitation technology can be used as a highly advanced and efficient technology in the various industrial. There are already many countries that are attracted to maglev systems.

Among above-mentioned useful usages, the most important usage of magnetic levitation is in operation of magnetically levitated trains. Magnetically levitated trains are undoubtedly the most advanced vehicles currently available to railway industries. Maglev is the first fundamental innovation in the field of railroad technology since the invention of the railroad. Magnetically levitated train is a highly modern vehicle. Maglev vehicles use non-contact magnetic levitation, guidance and propulsion systems and have no wheels, axles and transmission. Contrary to traditional railroad vehicles, there is no direct physical contact between maglev vehicle and its guideway. These vehicles move along magnetic fields that are established between the vehicle and its guideway. Conditions of no mechanical contact and no friction provided by such technology makes it feasible to reach higher speeds of travel attributed to such trains. Manned maglev vehicles have recorded speed of travel equal to 581km/hr. The replacement of mechanical components by wear-free electronics overcomes the technical restrictions of wheel-on-rail technology. Application of magnetically levitated trains has attracted numerous transportation industries throughout the world. Magnetically levitated trains are the most recent advancement in railway engineering specifically in transportation industries. Maglev trains can be conveniently considered as a solution for transportation needs of the current time as well as future needs of the world. There is variety of designs for maglev systems and engineers keep revealing new ideas about such systems. Many systems have been proposed in different parts of the worlds, and a number of corridors have been selected and researched (Yaghoubi, 2008).

Rapid growth of populations and the never ending demand to increase the speed of travel has always been a dilemma for city planners. The future is already here. Rapid transit and high-speed trains have always been thought of and are already in use. This is the way further into the future. Trains with magnetic levitations are part of the game. Conventional railway systems have been modified to make them travel at much higher speeds. Also, variety of technologies including magnetic levitation systems and high-speed railway (HSR) systems has been introduced. Rapid development of transportation industries worldwide, including railroads and the never ending demand to shorten travel time during trade, leisure, etc. have caused planning and implementation of high-speed railroads in many countries. Variety of such systems including maglev has been introduced to the industry. Maglev trains are a necessity for modern time transportation needs and vital for the future needs of railways, worldwide. This has resulted in the development of a variety of maglev systems that are manufactured by different countries. Maglev systems currently in use have comparable differences. The current models are also changing and improving.

Industries have to grow in order to facilitate many aspects of modern day life. This comes with a price to pay for by all members of socities. Industrial developments and widespread use of machineries have also increased risks of finanicial damages and loss of lives. Safety and needs to physically protect people against machineries may have not been a priority in the past but they are neccessities of modern times. Experts of industries have the task of solving safety and protection issues before implementing machineris. This is a step with high priority for all industrial assignments. While being fast, relaible and comfortable, maglev systems have found special places in minds of people. Running at such high speeds, maglev sytems have to be safe and need to be renown for safety. This puts much heavier loads on the shoulders of the corresponding experts and managers, compared to some other means of transportation. Safety is knowingly acting with proper functions to provide comfort and reduce dangers, as much as possdible. Risk management techniques have a vital role in organizing and implementing proper acts during incidents, accidents or mishaps in maglev systems operations. Effective management has a specific place in such processes. Obviously, such plannings put considereable finanicial load on the system. Implementation of internationally accepted standards is a fundamental step toward uplifting track safety. It will also serve to improve route quality, increase passenger loads and increase speed of travel. Maglev vehicle is one of the important transportation equipment of the urban track traffic system toward the future (Wang et al., 2007).

The overall plan for research and development and application of maglev technology should be made at the national level. This plan shall include the development plans as to research and development of key maglev technology, project implementing technology research and development of maglev project, plans of building maglev passage based on traffic demands, investment and financing system for the construction and operation of maglev system, research on implementing plans of high-density operational organization and maintenance of maglev route and so on.

It is very important to be vigilant about economical aspects of any major project during its planning and construction phases. Optimal use of local resources must be all accounted for. Technical and economical evaluation of the projects is a necessity to their success. It is necessary to have prior knowledge for investing into a project and then implementing its goals. Good planning makes it feasible to run the projects with reduced risks and increased return for the investment.

2. Vehicle

Maglev suspension systems are divided into two groups of ElectroMagnetic Suspension (EMS) and ElectroDynamic Suspension (EDS). There are varieties of vehicles that are manufactured based on these two types of systems. Vehicle path in EMS and EDS systems are called guideway and track, respectively. Basically, there are two main elements in a maglev system including its vehicle and the guideway. The three primary functions in maglev technology are levitation, propulsion, and guidance. Magnetic forces perform all of these. Magnets are used to generate such magnetic forces. For EMS systems, these magnets are located within the vehicle while for EDS systems magnets are located in the track. Performance of EMS system is based on attractive magnetic forces, while EDS system works with repulsive magnetic forces. In EDS system, the vehicle is levitated about 1 to 10 cm above the track using repulsive forces as presented in Fig. 1. In EMS system, the vehicle is levitated about 1 to 2 cm above the guideway using attractive forces as presented in Fig. 2. In EMS system, the electromagnets on the vehicle interact with and are attracted to levitation rails on the guideway. Electromagnets attached to the vehicle are directed up toward the guideway, which levitates the vehicle above the guideway and keeps the vehicle levitated. Control of allowed air gaps between the guideway and vehicle is achieved by using highly advanced control systems. Figs. 1, 2 show the components of the guideway and track including levitation and guidance systems in aforementioned maglev systems.

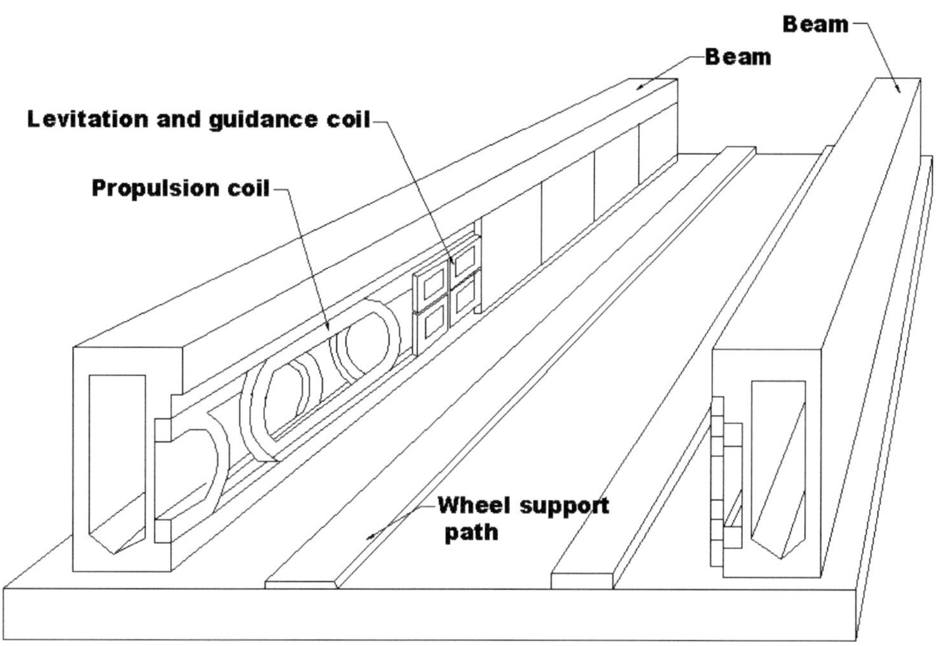

Fig. 1. Schematic diagram of EDS maglev system

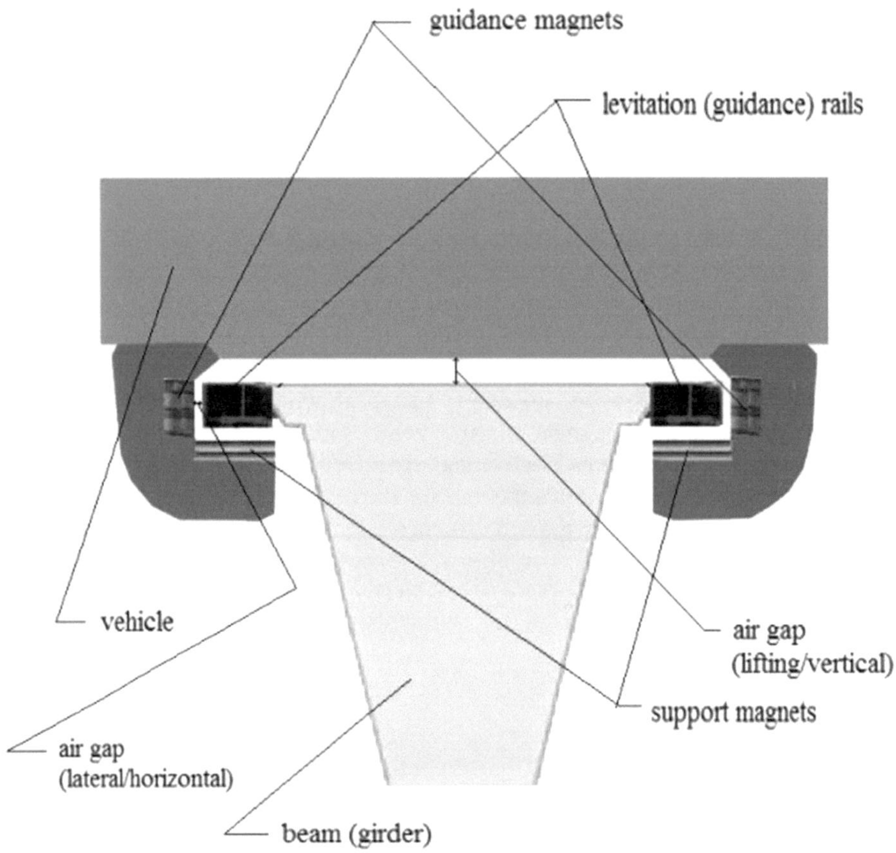

Fig. 2. Schematic diagram of EMS maglev system

Germany and Japan are clearly the front runners of the maglev technology. German's Transrapid International (TRI), a joint venture by Siemens AG and ThyssenKrupp, with EMS system has presented ninth generation of its maglev vehicles namely TR01 to TR09. TRI has been investigating electromagnetic levitation since 1969 and commissioned TR02 in 1971. The eighth generation vehicle, TR08 operates on 31.5 km of the guideway at Emsland test track in northwest Germany. The contract for implementing the world's first Transrapid commercial line was signed in Shanghai in January 2001. Construction work of the Shanghai Transrapid line began in March 2001. After only 22 months of construction time, the world's first commercially operated Transrapid train made its successful maiden trip on December, 31 2002. On December, 2003, the world's first commercial Transrapid line with a five section train started scheduled operation in Shanghai. TR08 and TR09 vehcles are used for the Shanghai Maglev Train (SMT) and TR09 Munich project, respectively. TR08 consists of 2 to 10 car bodies. SMT consists of 5 car bodies and travels on a 30km double-track elevated guideway, connecting the LongYang Road station (LYR), served by Metro Line 2 situated in the Pudong trade centre in Shanghai, to the Pudong International Airport (PIA). High-speed signifies operation of at least at $250_{km/hr}$. SMT has reached to the record speed of $501_{km/hr}$, a average speed (peak operating speed) of $431_{km/hr}$ and average speed of $268_{km/hr}$.

In 2005, China built its own maglev train. This train reached to the test speed of $150_{km/hr}$ over a track length of 204m. In February 2006, Chinese government announced that they decided to extend Shanghai maglev to Hangzhou city the capital of Zhejiang province. It would create the world's first intercity maglev line. The project will be managed by a German consortium leaded by Siemens Company. This route is of 170 to 175 km in length. The Ministry of Railways chief planner said in March 2010 that China had agreed to build a maglev line between Shanghai and Hangzhou. The line will start construction this year, Xinhua news agency reported. The new link will be 199.5 kilometers, about 24 kilometers longer than that included in the 2006 plan. The top speed of the maglev will be 450 kilometers per hour. It will take about half an hour to travel from Shanghai to Hangzhou, a trip which usually takes one and an half hours on the current service. The new line will also contain a downtown section of about 34 kilometers which is expected to connect the city's two international airports, Pudong and Hongqiao.

Maglev transport system features its potential development in a region with fast growing demand of intercity travel, such as the Shanghai maglev transport system (Yau, 2009). Growth of maglev technologies originated from human's pursuit of travel speed. Since the past 80 years, a number of scientists have made several researches on the feasibility of applying this transport technology. Eventually, they have realized commercial operation in Shanghai, China. Since China has a large population, the demand of applying this technology not only comes into being in the intercity long-distance transport but also in the city traffic field, which is mainly materialized in the low-speed technology and light vehicles (Siu, 2007). The Shanghai maglev line solved many important problems concerning the practical use of maglev transportation system. It has proved that the maglev technology is mature and can be put into practical application with good safety and reliability (Luguang, 2005). The construction data and operational experience of Shanghai maglev route create quite advantaged conditions for the application of maglev technology in China. It is also a blessed advantaged condition for research and development of maglev technology of China. Therefore, to share and make full use of the experiences and technical data of this operational route at national level may promote the research and development progress of maglev technology in China (Baohua et al., 2008).

In field of low-speed maglev systems, the National Defense University and the South South-West Jiaotong University worked for a long time for the development of the system similar to Japanese HSST. The Beijing Enterprises Holdings Maglev Technology Development Co. together with the National Defense University built a CMS-03 test vehicle and a 204m long test line with minimum radius of 100m and maximum climbing of 4% in 2001 in Changsha. Up to now, the vehicle traveled over 7000 km with over 20,000-test run and 40,000 times start and stop operations, its safety and reliability are proved. Recently, based on the test results a new engineering prototype vehicle has been constructed. It is planned to build a 2 km test and operation line in Kunming, after all necessary testing is finished. The whole system can be accepted for real urban application in 3-5 years (Luguang, 2005).

Technical specifications of high-speed and low-speed maglev trains are presented in Table 1 and 2, respectively (Yaghoubi & Sadat Hoseini, 2010).

<table>
<tr><th>Country</th><th>System</th><th>Suspension</th><th>Performance</th><th>Levitation</th><th colspan="2">Vehicle</th><th>Car-body</th><th>Speed</th><th>Year</th></tr>
<tr><td rowspan="5">German</td><td rowspan="5">TRI</td><td rowspan="5">EMS</td><td rowspan="5">Attractive force</td><td rowspan="5">At low speed and even at standstill</td><td colspan="2">TR06</td><td>2</td><td>392</td><td>1987</td></tr>
<tr><td colspan="2">TR07</td><td>2</td><td>450</td><td>1993</td></tr>
<tr><td rowspan="2">TR08</td><td>TR08</td><td>3</td><td>500</td><td>1999 (German)</td></tr>
<tr><td>SMT</td><td>5</td><td>501</td><td>2003 (Shanghai)</td></tr>
<tr><td colspan="2">TR09</td><td>3</td><td>350</td><td>2008</td></tr>
<tr><td rowspan="7">Japan</td><td rowspan="7">Railway Technical Research Institute (RTRI) and JR (Japan Railways) Central</td><td rowspan="7">EDS</td><td rowspan="7">Repulsive force</td><td rowspan="7">At speeds higher than 100km/hr</td><td colspan="2">ML-500R</td><td>1</td><td>517</td><td>1979</td></tr>
<tr><td colspan="2">MLU001</td><td>2</td><td>405</td><td>1980-1982</td></tr>
<tr><td colspan="2">MLU001</td><td>3</td><td>352</td><td>1980-1982</td></tr>
<tr><td colspan="2">MLU002</td><td>1</td><td>394</td><td>1987</td></tr>
<tr><td colspan="2">MLX01</td><td>3</td><td>550</td><td>1997</td></tr>
<tr><td colspan="2">MLX01</td><td>5</td><td>552</td><td>1999</td></tr>
<tr><td colspan="2">MLX01</td><td>3</td><td>581</td><td>2003</td></tr>
</table>

Table 1. Characteristics of high-speed maglev trains

Country	U.S	U.S	U.S	U.S	U.S	Korea	Indonesia	Japan
System/ Project	Magne Motion	GA (General Atomics)	CDOT[a]	AMT[b] / ODU[c]	M2000	MOCIE[d]	Jakarta	HSST[e]
Vehicle	M3	-	HSST-200 Colorado 200	-	-	-	-	HSST-100L
Suspension	EMS	EDS	EMS	EMS	EDS	EMS	EMS	EMS
Max. Operation Speed (km/hr)	160	80	200	64	-	110	110	100
Max. Initial Acceleration (m/s^2)	2	1.6	1.6	-	2	-	1	1.1
Capacity (pphpd[f])	12000	12000	6000	-	-	-	-	-
Passenger Capacity (One Car)	-	100	Seated: 103	Standin: 100	Seated: 50-100	100	Seated: 33 Standing: 67 Total:100	100
Air gap (mm)	20	25	-	10	100	-	10	-
Service Brake Max. Deceleration (m/s^2)	1.6	1.6 (standing) 2.5 (seated)	1.25	-	2	-	1	1.1
Emergency Brake Max. Deceleration (m/s^2)	-	3.6	3.1	-	4	-	1.25	1.1
Car-body	-	1	2	1	-	2	-	2
Number of Bogies in each Car body	-	2	5	2	-	-	-	5
Number of Magnets in each Bogie	-	-	4	6	-	-	-	-
length of each car body (m)	-	13	24.3	13.5	20- 30	13.5	13.5	15
Car width (m)	-	2.6	3.2	-	3.3	28.5	28.5	-
Car height (m)	-	3	3.65	-	3	3.5	3.53	-
Vehicle weight (ton)	-	Empty: 12 75% Loaded: 17.6	44	Empty: 11.5	Empty: 19.5-27 75% Loaded: 23.5- 36	28.5	Empty: 21 75% Loaded: 27.5	-

(a) Colorado Department of Transportation
(b) American Maglev Technology
(c) Old Dominion University
(d) Ministry of Commerce, Industry and Energy
(e) High Speed Surface Transport
(f) pphpd: passengers/hr/direction

Table 2. Characteristics of low-speed maglev trains

3. Guideway

The guideway is the structure that maglev vehicles move over it and are supported and guided by it. Its main roles are: to direct the movement of the vehicle, to support the vehicle load, and to transfer the load to the ground. It is the function of the guideway structure to endure applied loads from the vehicle and transfer them to the foundations. It is the main element in maglev system and holds big share of costs for the system. It is vital for maglev trains. The cost of the guideway structure is expected to be 60-80 percent of the overall initial capital investment cost (Zicha, 1986; Uher, 1989; Cai et al., 1994; FTA, 2004; Ren et al., 2009). Maglev train levitates over single or double track guideway. Guideway can be mounted either at-grade or elevated on columns and consists of individual steel or concrete beams. Elevated guideways occupy the least amount of land on the ground. Moreover, with such systems there is guarantee of meeting no obstacle while along the route. To guarantee safety for maglev trains necessitates guarantee that there will be no intersection between guideway and other forms of traffic routes. To serve the purpose, general proposition is to have elevated guideways.

Guideway provides guidance for the movement of the vehicle, to support the vehicle load, and to transfer the load to the ground. In maglev guideways contrary to traditional railroad tracks, there is no need to ballast, sleeper, rail pad and rail fastenings to stabilize the rail gauge. A guideway consists of a beam (girder) and two levitation (guidance) rails. Guideways can be constructed at grade (ground-level) or elevated including columns with concrete, steel or hybrid beams. Maglev elevated guideways minimize land occupation and prevent collision with other forms of traffic at-grade intersections. Guideways are designed and constructed as single or double tracks. Guideways can be U-shaped, I-shaped, T-shaped, Box, Truss and etc. Majority of cross-sections of guideway girders are also U-shaped. The rail gauges (track gauges) and spans are mostly 2.8 m and 24.8 m (Type I), respectively.

During the past three decades, different guideways have been developed, constructed and tested. Technical specifications of guideways for Federal Transit Administration (FTA) in U. S. Department of Transportation and TRI in Germany are presented in Table 3 (FTA, 2004, 2005a) and Table 4 (Schwindt, 2006), respectively. The guideway for the Transrapid in the Shanghai project was realized as a double-track guideway in 2001 and 2002. This Hybrid guideway is generation H2, type I as single-span (24.8 m) and two-span (2 x 24.8 m) girders. The Shanghai guideway I-shaped hybrid girder is 24.8m long, 2.8 wide, 2.2m high with a reinforced concrete girder (Schwindt, 2006; Dai, 2005).

Guideway consists of superstructures and substructures. Fig. 3 shows components of guideway's superstructures including beam and levitation (guidance) rails in an EMS maglev system where L is span length in meters and H is girder height in meters.

Depending on height of the guideway, it is separated in:

- At-grade guideway: $1.45 \leq h \leq 3.50$
- Elevated guideway: $h > 3.50$

Where h is guideway gradient height in meters.

The standard guideways are (Figs. 4, 5):

- Type I: L= 24.768 and $H \leq 2.50$
- Type II: L= 12.384 and $H \leq 1.60$
- Type III: L= 6.192 and plate construction, construction height ≤ 0.40

System	Guideway	Girder	Column	Span	Length of Span (m)	Cross-section	Width of Girder (m)	Height of Girder (m)
Magne Motion	Elevated	Concrete	Concrete	Two-span	36	Box	1	1.6
ODU [(a)]	Elevated	Concrete-Steel	Concrete	Single-span	25-27.5	Inverted-T	-	-
Colorado	Elevated	Concrete	Concrete	Single-span	25	U-shaped	-	-
Colorado	Elevated	Concrete-Steel	Concrete	Single/Two span	20 to 30	Box	2.972	3.66 (at mid-span) 5.49 (at the supports)
Colorado	Elevated	Steel	Concrete	Single-span	30	Truss	-	-
GA [(b)]	Elevated/ at-grade	Hybrid	Concrete	Single/Two span	36	Box	1.7	1.98, 1.22

(a) Old Dominion University
(b) General Atomics

Table 3. Technical specifications of guideways for FTA

No.	Guideway	Girder	Column (elevated)/ Support (at-grade)	Generation	Type	Span	Length of Span (m)	Year of Installation
1	Elevated	Concrete	Concrete	C 1	I	Single-span	24.8	1981-83
2	Elevated	Steel	Concrete	S 1	I	Single-span	24.8	1981-83
3	Elevated	Concrete	Concrete	C 2	I	Single-span	24.8	1984-86
4	Elevated	Steel	Concrete	S 2	I	Single-span	24.8	1984-86
5	Elevated	Steel	Concrete	S 4	I	Two-span	24.8	1995
6	Elevated	Concrete	Concrete	C 4	I	Single-span	24.8	1995
7	Elevated	Steel	Concrete	S 4	II	Two-span	12.4	1997
8	Ground-level (at-grade)	Steel	Concrete	S 4	II	Two-span	12.4	1997
9	Ground-level (at-grade)	Steel	Concrete	S 4	III	Two-span	6.2	1997
10	Elevated	Concrete	Concrete	-	I	Single-span	24.8	-
11	Ground-level (at-grade)	Concrete	Concrete	-	-	Two-span	-	-
12	Ground-level (at-grade)	Concrete	Concrete	C 4	III	Two-span	6.2	1998
13	Elevated	Hybrid	Concrete	H 1	I	Two-span	31	1999
14	Elevated	Hybrid	Concrete	H 2	I	Two-span	24.8	2001
15	Elevated	Hybrid	Concrete	H 2	I	Single-span	24.8	2002
16	Ground-level (at-grade)	Concrete	Concrete	C 5	II	Single-span	9.3	2005
17	Ground-level (at-grade)	Concrete	Concrete	C 8	III	Two-span	6.1	2006-2007
18	Ground-level (at-grade)	Hybrid	Concrete	H 3	II	Single-span	12.4	2006

Table 4. Technical specifications of guideways for TRI

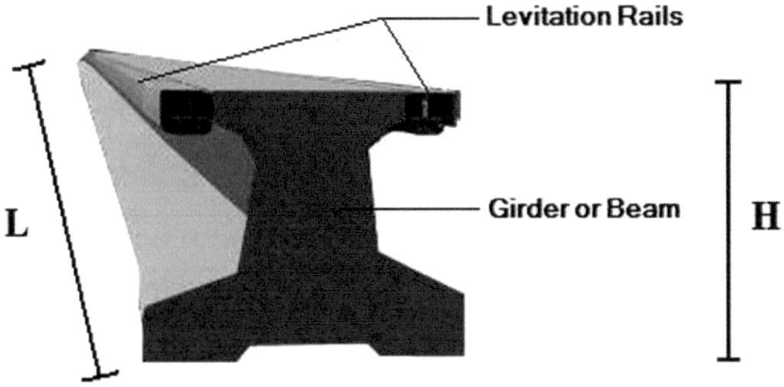

Fig. 3. Components of guideway in an EMS maglev system

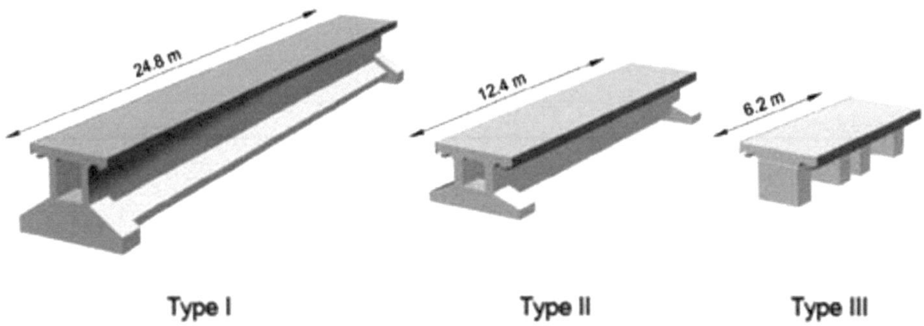

Fig. 4. Standard guideway types

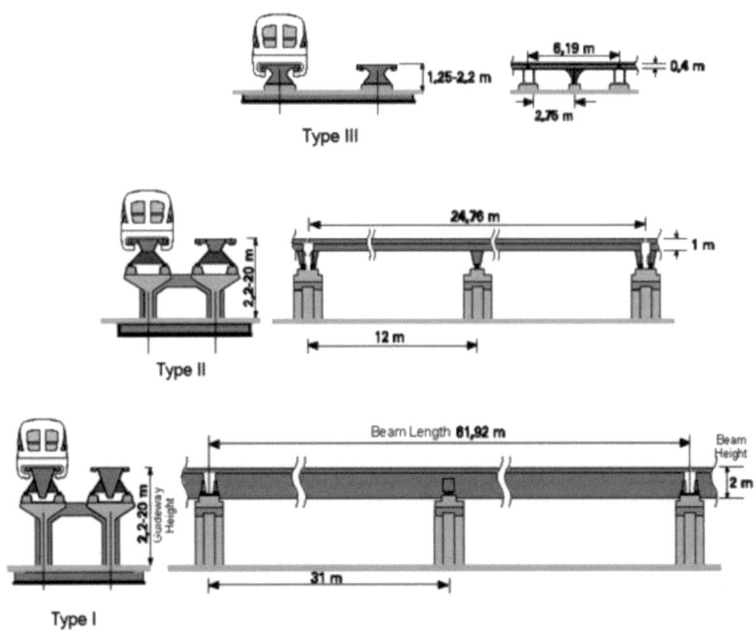

Fig. 5. Standard guideway types

The guideway height varies smoothly between 1.45 m and about 20 m. For greater guideway heights or span lengths larger than 40 m primary structures are needed in the form of conventional bridges. For substructures such as columns or foundations, reinforced concrete is proposed. The substructures for the guideway girders consist of several components. These are, depending on guideway type and gradient height, the column heads with bearing supports, the columns, tie beams and intermediate beams and the foundation slabs. They are built onto the natural soil, soil with soil improvement and/or on piles. The dimensions of the reinforced concrete substructures result from the high demands on the permissible deformations of the substructures (Grossert, 2006).

Different types of existing maglev magnetic suspension systems and technical specifications of existing guideways are presented in Table 5. As seen in this table, the majority of the maglev suspension systems are of electromagnetic suspension type. This table shows the most commonly used guideway structures and suspension systems. As indicated in the table, majority of guideway are elevated, double-track and U-shaped. The track gauges and spans are also mostly 2.8 m and 24.8 m, respectively.

Maglev systems	Shanghai China	Transrapid Germany	HSST Japan	JR Japan	U.S	Korea
Suspension	EMS	EMS	EMS	EDS	Different types (mostly EMS)	EMS
Section	I-shaped	Different Types (mostly U-shaped)	U-shaped	U-shaped Inverted T-shaped	U-shaped Box Truss	U-shaped
Track (rail) gauge	2.8 m	2.8 m	1.7 m	2.8 m	1-2.972 m	2.8 m
Guideway	Elevated	Elevated Ground-level (at grade)	Elevated	Elevated Ground-level (at grade)	Elevated	Elevated
Span length (elevated)	24.8 m	mostly 24.8 m	30 m	-	mostly 25 m	25-30 m
Maximum number of tracks in a route (guideway structure)	Double track	Double track	Double track	Double track	Double track	Double track
Maximum percent of tunnel in a route	0	22%	15%	87%	0	0

Table 5. Guideway structures and suspension systems

In recent years, different designs of guideways have been developed, constructed and tested among which U-girder guideways happens to be the most popular. In Korea, since 1980, prestressed concrete U-girder guideway for straight route applications has been proposed (Jin et al., 2007). In Germany, during 1981-1983, a U-girder guideway was built at the Transrapid Test Facility (TVE) by TRI and was used in the TR07 project (Lever, 1998). This guideway was further optimized and improved during 1984-1986 (Schwindt et al., 2007). An elevated concrete U-girder guideway was installed in TVE in 1995 (Schwindt, 2006). One of the three kinds of girders considered for Colorado Maglev Project (CMP) in the Colorado Department of Transportation (CDOT) is a concrete U-girder (FTA, 2004).

Frequent studies on the technical characteristics of beams shows that the following are among the reasons for frequent use of U-shaped cross-sections in majority of projects (Yaghoubi & Ziari, 2010):

- In general, in beams with closed cross-sections like box or U-shaped with continuous deck, the torsion is reduced. Also, structural continuity in cross-section reduces deflections due to vertical loads and possibly allow for higher speeds.
- The U-shaped beams have less deflection compared to other types of cross-sections.
- Cost of design, construction, installation and operation of U-shaped concrete beams compared to welded steel box girders and tubular steel space truss guideways is lower. Welding in steel box girders will cause a cost hike. Truss guideways also require many full penetration welds to insure the truss integrity under loadings, and this in turn would cause another cost hike.
- The continuity of the girder and the deck and the lack of need for installation of horizontal shear connectors between the concrete girder and the deck, in contrast to the railroad bridges.
- The U-shaped cross-sections are ideal as far as structural and strength (ultimate strength) are concerned.
- More centroidal moment of inertia and the section modulus of U-shaped cross-sections among cross-sections of equal sectional-area, including the I-shaped.
- Lower torsion of U-shaped cross-sections (as a closed cross-section) relative to other cross-sections including the I-shaped (as an open cross-section).
- Lower weight and volume (concrete used and lower dead load) of U-shaped cross-sections relative to other sections including the I-shaped.
- And generally, the U-shaped cross-sections are technically, operationally, economically, more satisfying.

It is geometrically simpler to design railway tracks without horizontal or vertical curves and without longitudinal or lateral inclinations. Practically, this does not happen very often. When tracks have to be laid in mountain ranges, engineers have to design horizontal and vertical curves and axial and lateral slops. While passing through horizontal curves centrifugal forces are added to other forces already present. Centrifugal forces are generated due to curves and tend to push the vehicle further away from the centre of the curve. The centrifugal forces also transmit efforts to the track pillars. In fact these forces are generated in the track and the main components to resist them are the track pillars. If the track or part of it is located in a horizontal curve, the effect of centrifugal forces needs to be included for the calculation proposes. These forces act horizontally and in a direction perpendicular to the tangent to horizontal axis of the track. Normally, track superelevation is added to the guideway to compensate these centrifugal forces. Regarding the structure, presence of centrifugal forces on the horizontal curves disturbs the balance of magnetic forces acting on the guideway. Therefore, it is necessary to make allowance for such effects when analyzing and designing for guideways on the curves. An important effect of introducing centrifugal forces on the horizontal curves is the unsymmetrical distribution of vertical loads on the guideway. This causes different calculation procedures for guideways on the curves compared to the straight routes.

The maglev can easily handle tight curves and steep grades of up to 10 %, resulting in fewer tunnels and other encroachment on the terrain (Siemens AG, 2006). The main task of route alignment is to stipulate the geometry of the guideway's function planes in such a way that the passenger enjoys maximum travel comfort when a vehicle travels on the guideway. Apart from acceleration, however, a consideration of changes in acceleration is also an important aspect of comfort. An exception to this is the track changing equipment, where, on the basis of the beam theory, the transition curve of the turnout position is also in the form of a clothoid in the horizontal plane. When route alignment, including determination of the spatial curve, is carried out, these or other aspects are taken into account, as well as the system's characteristics (Schwindt, 2006).

It is well known that torsion has particular significance on the curved bridges. A box section has a special advantage for a curved guideway because of its high torsion rigidity. A curved steel box girder guideway can provide longer curved spans with fewer supports than would be required for I girders, thereby creating greater cost savings in the substructure. For a given design speed and superelevation, the minimum radius of the circular portion of the horizontal curve may be determined based on either the passenger comfort criteria (lateral acceleration) or the vehicle stability criteria, depending on which criterion results on the smallest curve radius.

In the Colorado system, the passenger comfort criterion is based on the American Society of Civil Engineers (ASCE) People Mover standards. These standards provide a maximum recommended lateral acceleration on the passenger. The lateral acceleration is a function of the velocity and the radius of curvature. In addition to passenger comfort, the stability of the vehicle itself needs to be considered in the relationship between allowable curve radius and superelevation. Fabrication of the curved guideway sections is not widely discussed in maglev system literature. However, it is a central element of the guideway construction technique and can become a major consideration in the guideway cost (FTA, 2004).

When the vehicle travels on a straight piece of route, gravity is the most influential load acting on it. If the vehicle has geometrical symmetry and is loaded symmetrically, the gravity load passes through guideway axis of symmetry. When the vehicle travels on the curves, the centrifugal force will also be added to this effect.

Ideally, if there is adjustment for superelevation on the curve, loads exerted from vehicle to the guideway are symmetric and loading pattern will not be different from that of a straight route. However, on the curves with insufficient superelevation loading pattern will be different. While the principals of calculating the guideway loading on the curves and the straight routes are basically the same the main differences arise due to insufficient superelevation on the curves. This results in different amount of loads being applied to the guideway for both cases. Eccentricity caused by such effects, makes load on internal and external levitation rails different. It is clear that as a result of eccentricity due to insufficient superelevation the portion of load transmitted from vehicle to each one of the internal and external levitation rails will not be equal. It is vital for the guideway loading calculations to depict a proper pattern for its loading.

3.1 Loading

The most important part in the analysis and design of guideway is structural loading. The loading of the maglev vehicle is an important parameter in the practical application. It is related to the magnetic forces (He et al., 2009). The guideway must carry a dead load due to its own weight, and live loads including the vehicle loads. To incorporate the dynamic interaction between the guideway and the vehicle, the live load is multiplied by a dynamic amplification factor. Lateral and longitudinal loads including wind and earthquake loads may also need to be considered. The guideway loadings are modeled as dynamic and uniformly distributed magnetic forces to account for the dynamic coupling between the vehicle and the guideway. As maglev vehicle speeds increase to 300-500 km/h, the dynamic interactions between vehicle and guideway become an important problem and will play a dominant role in establishing vehicle suspension requirements. Magnetic forces are generated by the maglev vehicle and cause structural loading that transmits to the guideway. This can happen whilst such a vehicle is stationary or in motion. In order to prevent contact between the vehicle and the guideway and maintain the required gap between them, the system is continuously under Operation Control System (OCS) command.

Some decisive factors for the design of maglev guideways are listed as being constructible, durable, adaptable, reliable, readily maintained, being slim in accordance with urban environment and being light to be constructed more efficiently (Jin et al., 2007; Sandberg et al., 1996a, b). In this regard, one of the main challenges to guideway designers is to produce a structure that will be easily maintainable to the narrow tolerances and precise alignment required for practical high-speed maglev operation, to achieve a structure which is economically and financially justifiable and attractive (Plotkin at al., 1996a ,b). Besides satisfying the above conditions, important parameters in the design of guideway include vertical live loads and its pursuant dynamic amplification factors (DAF), plus deflection due to this load. These parameters constantly govern the design process, and they play a major determining role in the structural optimization of the guideway girder systems.

Vehicle/guideway interaction of the maglev system is an important and complicated problem. It is influenced by the levitation system, guideway structure, vehicle mechanical structure, running speed, etc. So the investigation of it should be launched out in many aspects (Wang et al., 2007). Among the various parameters which affect on design of maglev guideway, dead and live loads, dynamic amplification factor and deflection have major importance. Assessment of deflections due to the vertical loads for guideway beam during operation of maglev vehicle is very important. It is the most influential parameter in design of guideway (Lee et al., 2009).

While there are routine processes for the calculation of the guideway dead loading, there is a need for special treatment in the calculation of its live loads. Live load intensity and its distribution patterns are highly dependent on the structural behavior. According to AREMA (American Railway Engineering and Maintenance-of-Way Association) and UIC (International Union of Railways) regulations live load models for conventional railway track are based on a combination of concentrated and distributed loads. This is compatible with the use of wheels and the behavior of locomotives in conventional trains. In the case of trains with magnetic levitation with no wheels and added complexity of lifting magnetic forces due to support magnets, the analysis is much more complicated.

The forces are of attractive magnetic forces and can be categorized as lifting magnetic forces and lateral magnetic forces. The lateral magnetic forces include the restoring magnetic forces, the impact forces, etc. on the straight route and the restoring magnetic forces, the wind forces, the earthquake forces, the centrifugal forces, the impact forces, etc. on the curved route. Lateral magnetic forces due to interaction of the guideway and the guidance magnets ensure the lateral stability of the vehicle. Lateral guidance is provided by the configuration of the vehicle-guideway interface and by the levitation electromagnets. The "horseshoe" configurations of the electromagnet and levitation rails provide strong lateral restoring forces when perturbed from equilibrium (FTA, 2004). Guidance magnets are located on both sides along the entire length of the vehicle to keep the vehicle laterally stable during travel on the guideway. Electronic control systems assure the preset clearance.

The mechanical load at a specific point of the structure depends on its location within the car body, but not on the overall length of the vehicles or the position of the car body within the vehicle set. The interaction forces are the magnetic forces that can be derived from magnetic suspension models. Fig. 6 schematically presents locations for interactions between the vehicle and guideway. In this figure, (a) presents location for interaction between support magnets of the vehicle and the guideway levitation rail; (b) presents location for interaction between guidance magnets of the vehicle and guideway levitation rail.

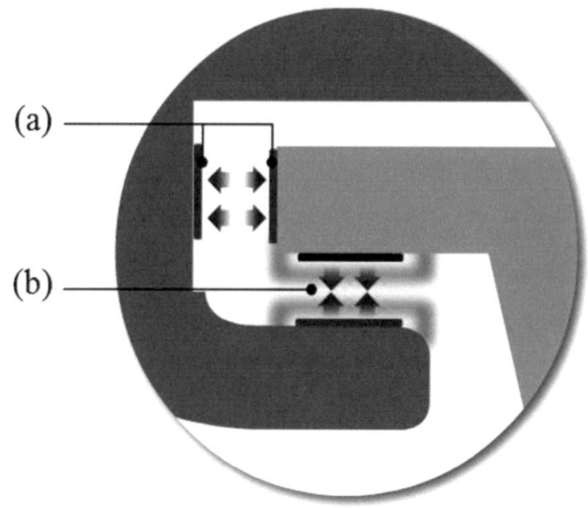

Fig. 6. Schematic of Interacting Maglev Vehicle and Guideway

Static (dead) load on the guideway generally consists of structural weight. It is important to remember that same type of guideway beams are selected for the straight and the curved routes. Even though, they may be selected with different types and shapes for the two types of routes (Jin et al., 2007).

3.1.1 Lifting magnetic forces (vertical loads)

Vertical loads imposed on maglev guideway can be categorized as dead loads due to the weight of the guideway divided by the length of the span, and live loads due to the interaction between guideway and the vehicle. Table 6 presents dead loads on guideways for some different maglev systems. Also, presented in this table, is the calculated dead loads on a railroad bridge.

Items	Span Length (m)	Girder Height (m)	Dead load (ton/m)
Urban Maglev Program, Korea (Jin et al., 2007)	25	1.51	2.948
	25	1.99	3.460
	25	1.402	2.68
	30	1.625	3.10
	25	1.515	2.95
	30	1.837	3.28
	25	1.794	3.16
	30	2.183	3.63
	25	1.991	3.46
	30	2.320	4.05
Linimo, Japan (Jin et al., 2007)	30	2.5	7.733
AGT, Korea (Jin et al., 2007)	30	1.92	5.531
Transrapid, Germany (Jin et al., 2007)	25	2.20	5.732
KIMM, Korea (Jin et al., 2007)	25	2.06	3.544
Expo Park Korea, Korea (Jin et al., 2007)	25	1.90	4.256
(Dai, 2005)	30	-	1.5[(a)], 3.5[(b)]
(Cai, et al., 1996)	-	-	1.82
Railroad Bridge [(c)]	18	1.45	14.8

(a) single-car maglev
(b) three-car maglev
(c) Consisting of four precast prestressed concrete girders of type 3 AASHTO.

Table 6. Dead loads on some typical maglev guideways

The magnetic force between the guideway and the supporting magnets causes the vehicle to levitate. The maglev guideway live loads consist of both static and dynamic loads. The static live load is the load due to the weight of the vehicle. In this case, the vehicle rests directly on the guideway. When the vehicle rises, an air gap appears between the vehicle and the guideway. The interaction force between the guideway and the i-th bogie in each car body is equal to $1/n_b$ of the total vehicle weight, including the car body (wagon), bogies, magnets and passengers. n_b is the number of bogies in each car body.

The movement of the vehicle over the guideway amplifies the static loads. Dynamic amplification factor (DAF) is a non-dimensional ratio of dynamic magnetic force to the static magnetic force. Incorporating a dynamic amplification factor, the dynamic lifting magnetic force between the guideway and the i-th bogie in each car body. DAF is the most influential parameter in design of guideway. The DAF of the guideway girder caused by the maglev vehicle is generally not severe compared with that caused by a traditional railway load, and is not significantly affected by vehicle speed. The effects of the deflection ratio and span continuity on the DAF of the guideway are negligible (Lee et al., 2009). DAF for variety of maglev systems is presented in Table 7. The DAF defined as the ratio of the maximum dynamic to the maximum static response of the guideway under the same load plus one, are used to evaluate the dynamic response of the guideway due to the moving vehicular loads.

In general, DAF is not a deterministic value and must be estimated through probabilistic methods. The amount of DAF depends on several parameters including the geometry of the guideway such as length of span, type of span (single-span or multi-span), etc.

Items	Span Length, L (m)	Maximum Speed (Km/h)	Maximum DAF	
Bechtel (Lever, 1998)	24.82	500	1.4 (a)	
TR07, Transrapid, Germany (Lever, 1998)	24.82	500	1.56	
Maglev Transit (Lever, 1998)	-	-	1.4	
Grumman (Lever, 1998)	27	500	1.2	
New (corrected) Grumman (Lever, 1998)	27	500	1.36(b)	
(Dai, 2005)	30	500	1.37	
UTM01, Korea (Yeo et al., 2008; Lee et al., 2009)	-	100	1+15/(40+L)	
Urban Maglev, Korea (Yeo et al., 2008)	25	110	Steel girder: 1.15 Concrete girder: 1.1	
Linimo, Japan (Yeo et al., 2008)	-	100	Steel girder: 1.15 Concrete girder: 1.1	
AASHTO LRFD Bridge Design Specifications (Dai, 2005)	-	-	1.33(c)	
Conventional Railroad Bridge (d)	5	-	δ_1=1.35	δ_2=1.53
	10	-	δ_1=1.2	δ_2=1.31
	15	-	δ_1=1.14	δ_2=1.21
	20	-	δ_1=1.10	δ_2=1.16
	25	-	δ_1=1.08	δ_2=1.125
	30	-	δ_1=1.06	δ_2=1.09
General Atomics (FTA, 2005 a, b)	36	200	1.5	

(a) The Bechtel report indicates that this is a conservative value is used to design the girder (Lever, 1998).
(b) Calculated using diagrams of static vehicular loading and dynamic vehicular passage over guideway (Lever, 1998).
(c) In AASHTO LRFD Bridge Deign Specifications, the dynamic allowance (IM) for highway bridge design is 0.33 (the corresponding dynamic amplification factor is 1.33) (Dai, 2005).
(d) With track maintenance to accurate standards and criteria. δ_1: For Shear Force and δ_2: For Bending Moment.

Table 7. Dynamic Amplification Factor (DAF) for some typical maglev guideways

The interaction force (dynamic lifting magnetic force) between the i-th bogie in each car body and the guideway is transferred to two levitation rails. Due to the uniform distribution of the load on the levitation rails, the loading pattern on the guideway spans can be considered as a uniform distributed load.

The amount of live load of some different maglev systems are presented in Table 8.

Items	Span Length (m)	Live Load (ton/m)	Deflection Regulation (m)
Urban Maglev Program, Korea (Jin et al., 2007)	25	2.3	L/2000
	25	2.3	L/4000
Urban Maglev Program, Korea (Yeo et al., 2008)	25	2.6	L/2000
Linimo, Japan (Jin et al., 2007)	30	1.78	L/1500
Linimo, Japan (Yeo et al., 2008)	-	2.3	$20<L \leq 25m$: L/1500 $25<L$: $(L/25)^{1/2} \times L/1500$
UTM01, Korea (Yeo et al., 2008)	-	2.2	L/4000
AGT, Korea (Jin et al., 2007)	30	-	L/1000
Transrapid, Germany (Jin et al., 2007)	25	2.4	L/4000
KIMM, Korea (Jin et al., 2007)	25	1.86	L/4000
Expo Park, Korea (Jin et al., 2007)	25	2.5	L/3000
CHSST, Japan (FTA, 2004)	30	2.3	-
Colorado, U.S. (FTA, 2004)	30	2.3	-
TR08, Transrapid, Germany (Schach et al., 2007)	-	2.2	-

Table 8. Live loads on some typical maglev guideways

In the static position or while maglev vehicle is resting on its guideway, the thickness of the air gap between vehicle and guideway is nil. Therefore, the total load of the vehicle weight will be transmitted to the guideway. As a result, the interaction force (total static lifting magnetic force of each car body) between each car body and the guideway is the static weight of the vehicle. Each car body is equipped with n_b bogies. Thus, the total interaction force between each car body and the guideway is the summation of interaction forces (static lifting magnetic forces) between the i-th bogie in each car body and the guideway. Maglev trains achieve a weight reduction in reaching the design speed (FTA, 2004).

As presented in Fig. 3, each guideway includes one beam (girder) and two levitation rails. Therefore, the total interaction force between each car body and each of the levitation rails, is equal to one-half of the total interaction force between each car body and the guideway. In other words, the interaction force between the i-th bogie in each car body and each of the levitation rails, is equal to one-half of the interaction force between the i-th bogie in each car body and the guideway. Dynamic magnetic lifting forces are the forces generated while the vehicle moves. The interaction force (total dynamic lifting magnetic force of each car body) between each car body and the guideway, is the sum of the interaction forces (dynamic lifting magnetic forces) between the i-th bogie in each car body and the guideway. Also, considering the fact that each guideway consists of two levitation rails, the interaction force between the i-th bogie in each car body and each of the levitation rails, is equal to one-half of the interaction force between the i-th bogie in each car body and the guideway. Interaction force between each bogie in each car body and the guideway is a uniformly distributed live load. Load uniformity comes from the absence of the wheels and presence of lifting magnetic forces with uniform intensity that is generated by support magnets. Interaction force between each bogie in each car body and each of the levitation rails is also uniformly distributed. If bogie lengths in each car body are the same, as normally is the case, then the total interaction force intensity between each car body and the guideway is equal to the interaction force intensity between the i-th bogie in each car body and the guideway over the length of bogie. Also, in such case, the total interaction force intensity between each car body and each of the levitation rails over the length of each car body is equal to the interaction force intensity between the i-th bogie in each car body and each of the levitation rails over the length of each bogie. Each maglev vehicle involves some (one to ten) car bodies with different lengths. Hence, maximum interaction force intensity between car bodies and the guideway can be considered as maglev live load. In general, maglev live loading is evenly and uniformly distributed. The amount of maglev live load is generally less than the dead load of its guideway. Also, the uniformly distributed live load of maglev applied to each levitation rail over the length of live loading.

3.1.2 Lateral magnetic forces (lateral loads)

In the static case, lateral (guidance) magnetic forces do not exist. However, during vehicle movements and while it moves to the sides, interaction of guidance magnets and levitation rails brings the vehicle back to its central stable position. This causes lateral magnetic forces. These lateral forces act in lateral and normal directions to the levitation rails and transmit to the guideway. When the vehicle deviates to the right, guidance magnets on the right side of horseshoe shaped section of the vehicle and levitation rail on the right side of guideway attract each other while guidance magnets on the left side of horseshoe shaped section of the vehicle and levitation rail on the left side of guideway repulse each other. This brings the vehicle back to its stable position. At the location of interaction between guidance magnets and levitation rails, forces in the left and right zones are of the same size and act on the same direction to the guideway.

One of the main advantages of the elevated transportation systems such as maglev is the high resistance of their tracks in dealing with the earthquake forces. Earthquake forces are included in the guideway design for Shanghai in China (Dai, 2005) and in Japan. There is no report of major earthquake in central Europe. Therefore, German Transrapid TR07 has ignored such effects, all together (Lever, 1998). Earthquake lateral forces imposed on maglev guideway are less than that of the railroad bridges.

Irregular earth movements generate such forces that can be capable of damaging the man made structures. The size of such forces depends on the nature of the earthquake, the natural period for the bridge structural vibrations and the natural period for vibrations of the soil under the foundation. For the design of the exceptional bridges with very large spans or for the bridges that are near the earth's fault lines, calculations for the earthquake forces depend on some detailed studies. One may use the static analysis for the design of small to medium size bridges. Dynamic bridge analysis however, needs huge number of calculations that are economically formidable and sometimes turn to be impossible. On the other hand, the quasi static approach uses a load (or an impact) factor that converts the dynamic loads into the static loads. Therefore, such method assumes static equilibrium when determining the structural behavior. The load (or the impact factor) comes from the experiences, engineering judgment and from mathematical models.

Guideways must endure the earthquake lateral forces in two perpendicular directions. They need to also transfer the lateral forces to the guideway foundations in both directions. These two directions normally include the guideway longitudinal axis and the direction perpendicular to it. The guideway columns must endure the earthquake forces caused by the guideway weight in addition to enduring the earthquake forces that are related to the columns weight. The later force comes from multiplying the earthquake factor by the weight of the columns. The earthquake factor is the same factor that is also used for the calculation of the earthquake force. For the calculation of the earthquake lateral force, if the size of the live load is less than half of the size of the deal load, the live load will be ignored. Otherwise, two third of the summation of the dead and live loads on the guideway needs to be accounted for. While calculating the earthquake lateral force for urban maglev guideways, at least half of the live load must be included.

Generally, the wind effect depends on the geographical position of the district, its altitude from the sea level, the local topography and to some geometrical characteristics. For the guideway static calculations, regardless from the number of the tracks the wind force affects only one maglev vehicle.

The interaction force (dynamic lateral magnetic force) between the i-th bogie in each car body and each of the levitation rails is defined by the summation of the interaction forces (dynamic lifting magnetic force) between the i-th bogie in each car body and each of the levitation rails and the wind or the earthquake lateral force, whichever that turns to be bigger. The earthquake lateral force also includes a DAF.

Lateral forces on the maglev guideway can be caused by the vehicle sliding, particularly on curves. Lateral guidance is provided by guidance magnets. The dynamic lateral magnetic force imposed on the guideway can be considered as a uniformly distributed load. Centrifugal forces, in equal speed and curve radius, are less in maglev due to lower weight of the vehicle than in rail tracks.

3.1.3 Longitudinal loads

In recent years, with increasing traveling speed of the rail systems, aerodynamic load problems became very important. From the system point of view, aerodynamical topics which affect and define the interface between rolling stock, infrastructure and operation are of paramount importance and the corresponding loads increase with the vehicle speed. If maglev vehicles pass in close proximity to each other or move close to fixed objects such as barriers or buildings, the aerodynamic interactions can produce significant loads on the vehicle or the fixed object. The magnitude and duration of the load depends on the velocity and geometry of the vehicles and also on the ambient wind speed and direction. For high-speed railroads several studies have examined the loads produced by passing trains and their potential for causing an accident. The results of these studies show an important pressure load acting on the object which can have serious consequences. The experiments were carried out on conventional railroad vehicles but from the system point of view, in principle, the aerodynamics of a maglev and a high-speed railroad system do not differ. Although the safety aspect does not concern the maglev vehicle as strongly as it concerns conventional railroads, because maglev is guided by magnets on both sides and cannot derail, many aspects are similar. In both cases, the interaction of vehicles and infrastructure implies aerodynamic system issues, e.g. that of train induced aerodynamic loads leading to structural vibrations and a decrease of ride comfort. The pressure load caused by passing maglev vehicles has an important aerodynamic effect on the sidewall motion and therefore on the ride comfort (Tielkes, 2006). While two vehicles are passing each other at high relative speed, the quasi-static pressure distribution along each vehicle presents a dynamic load on the other vehicle. The dynamic pressure load strongly depends on the velocity of the oncoming vehicle, the geometry of the bow-part of the oncoming vehicle and the distance between the two tracks. The time behavior is given by the relative velocity between the two vehicles. The mechanical load on the car body depends mainly on:

i. the amplitude of the pressure wave, given by
   - the velocity of the oncoming vehicle
   - the bow-shape of the oncoming vehicle
   - the distance between the two tracks
ii. the relation between the propagation speed of the structural Eigenmode with the corresponding wavelength and the relative velocity between the two vehicles
iii. the load at a specific point of the structure depends on its location within the car body, but not on the overall length of the vehicles or the position of the carriage body within the vehicle set.

In general, the aerodynamic forces play an important role in affecting the interaction response of maglev-vehicle/guideway system due to their velocity-dependent characteristics, especially for the higher speeds over $600_{km/h}$ (Yau, 2009). Further development of the ground transport calls for solution variety of problems among which aerodynamic problems are very important. The up-date state of high-speed ground transport problem shows that the use of aerodynamic effects will make it possible to optimize the technical and economic performances of vehicles.

Longitudinal force can be applied to the guideway through braking and acceleration of the vehicle, vehicle weight when the guideway has a longitudinal slope, and air pressure (aerodynamics). Since maglev vehicles have no wheels, axles and transmission, they weigh less then a conventional railroad train. The lack of wheels also means that there is no friction between the vehicle and the guideway. These factors result in a reduction in energy consumption. Therefore, the vehicle requires a lesser force for braking and stopping it. For example, the attractive force due to braking in the Colorado maglev vehicle equals to 4.2-4.5 ton, which amounts to about 10% of its loaded vehicle mass of 44 ton (FTA, 2004). In conventional rail tracks, brake force is usually equal to 1/7 of the weight of the part of the train which is located on the bridge.

3.2 Analysis

During the past four decades, many maglev models have been proposed. In 1974, Katz proposed two simplified one dimensional maglev vehicle and suspension models. A simple two degree-of-freedom (DOF) vehicle system with one car body was used in his study (Katz et al., 1974). In 1993, Cai studied a multi-car vehicle model traversing on a guideway. Concentrated loads and distributed loads were compared. The coupled effects of vehicle and guideway interactions over a wide range of vehicle speeds with various vehicle and guideway parameters were investigated. Only vertical vehicle motion is considered in their study. A beam model with a uniform-cross-section was used. They found that a distributed load vehicle model was better than a concentrated load vehicle model which might result in vehicle accelerations in simulations. They concluded that multi-car vehicles had less car body acceleration than a single-car vehicle, because of the inter-car vertical constraints. However, a magnetic suspension model is not included in their study. The interface between the vehicle and the beam was modeled with an elastic spring and dashpot, which is not the case in a real maglev system (Cai et al., 1996). In 1995, Nagurka and Wang developed a dynamic maglev model which includes a five DOF vehicle model. The effects of the vehicle speed on the system performance were studied (Nagurka et al., 1997). In 2005, Huiguang Dai influenced by German TR08 maglev, defined a vehicle model, a magnetic suspension model and a beam roughness model. He studied dynamics of a single-car vehicle model with 4 bogies and a three-car vehicle model with 12 bogies. He used an elevated guideway with multiple concentrated moving loads. A total number of 500 simulations were performed to study the dynamic behavior of maglev vehicle and guideway beam (Dai, 2005).

Although extensive simulations and analyses have been performed, the development of design criteria for maglev guideways will require additional studies. Aerodynamic forces must be considered. Effects of horizontal curves should be considered. Maglev trains may be extended to 4 or more cars (Dai, 2005). Maglev vehicle/guideway interaction problem bothers the investigators and engineers for years. No well-accepted interpretation has been reported, yet. Vehicle/guideway interaction of the maglev system is an important and complicated problem. It is influenced by the levitation system, guideway structure, vehicle mechanical structure, running speed, etc. So the investigation of it should be launched out in many aspects (Wang et al., 2007).

During the past four decades, research and development have been performed in the areas of magnetic levitation, interaction of vehicle with guideway, and optimization of vehicle suspensions. The results of these efforts are useful in providing appropriate criteria for the design of maglev systems. The dynamic response of magnetically levitated vehicles is important because of safety, ride quality and system cost. As maglev vehicle speeds increase to 300-500$_{km/h}$ the dynamic interactions between vehicle and guideway become an important problem and will play a dominant role in establishing vehicle suspension requirements. Different dynamic responses of coupled vehicle/guideway systems may be observed, including periodic oscillation, random vibration, dynamic instability, chaotic motion, parametric resonance, combination resonance, and transient response. To design a proper vehicle model that provides acceptable ride quality, the dynamic interaction of vehicles and guideways must be understood. The coupled vehicle/guideway dynamics are the link between the guideway and the other maglev components. Thus, reliable analytical and simulation techniques are needed in the design of vehicle/guideway systems. Furthermore, a coupled vehicle/guideway dynamic model with multiple cars must be developed to meet system design requirements.

For a dynamic analysis of vehicle/guideway interactions, an understanding of the effects of distributed loads is essential. The maglev vehicle is the source of magnetic forces and loading starts from this vehicle. These forces transfer to the guideway while the vehicle is stopped or when it moves. Each car body model can be considered as a uniform rigid mass. It is supported by two to eight springs and two to eight dashpots that form the secondary suspension for maglev vehicle. The primary suspension consists of two to eight magnetically supported bogies. Maglev vehicle can be single-car or multiple-car.

Magnetic levitation is caused by magnetic forces that transmit to guideway by maglev vehicle. In fact, these forces are the consequence of interactions between vehicle and guideway caused by magnets. For EMS systems, these magnets are installed within the vehicle. The forces are of attractive magnetic forces. Lifting magnetic forces due to interaction of guideway and support magnets cause the levitation of the vehicle. Support magnets are located on both sides along the entire length of the vehicle. The attractive force produces inherently unstable vehicle support because the attractive force increases as the vehicle/guideway gap decreases.

The interaction forces are the magnetic forces that can be derived from magnetic suspension models. Static load on guideway generally consists of the vehcile weight. In either case, the dead load is uniformly distributed along the full length of the beam. Calculations of live load need more attention. Dynamic lifting forces are derived from static lifting forces. Therefore, accuracy of these models is vital to the accuracy of live load models. Combination of these models plus the live load models leads to the analysis and design of guideway. In static position or while maglev vehicle is resting on its guideway, thickness of the air gap between vehicle and guideway is nil. Therefore, total load of vehicle weight will be transmitted to the guideway. As a result, the interaction force (total static lifting magnetic force of each car body) between each car body and the guideway is the static weight of the vehicle. Each car body is equipped with n_b bogies. Thus, the total interaction force is summation of interaction forces (static lifting magnetic forces) between the i-th bogie in each car body and the guideway. Interaction force (dynamic lifting magnetic force) between each bogie in each car body and the guideway is a uniformly distributed live load. Load uniformity comes from absence of the wheels and presence of lifting magnetic forces with uniform intensity that is generated by support magnets. Maglev live loading is evenly and uniformly distributed. Amount of maglev live load is generally less than dead load of its guideway.

These forces transfer to the guideway while the vehicle is stopped or when it moves. Each car body model can be considered as a uniform rigid mass. It is supported by two to eight springs and two to eight dashpots that form the secondary suspension for maglev vehicle. The primary suspension consists of two to eight magnetically supported bogies. The electromagnets are mounted on the rigid bogies and generate attractive magnetic forces while interacting with ferromagnetic stator packs. Connections between magnets and bogies are assumed to be rigid. Two dimensional motions of the vehicle include heave motion and rotational motion. Maglev vehicle can be single-car or multiple-car. For example, a single-car vehicle with four bogies has 6 DOF including one translational and one rotational displacement at the center of mass of the car body, and one translational displacement for each of the four bogies. By the same token, a three-car vehicle with four bogies in each car body has 18 DOF. A dynamic simulation for maglev vehicle/guideway interaction is essential to optimize the vehicle design.

A variety of these parameters are presented in Table 9 for different types of maglev systems.

Maglev System/ Model	General Atomics (GA), U.S. (FTA, 2005a, b)	Old Dominion Uni. (ODU), U.S. (FTA, 2005)	(Wang et al., 1997)	(Cai et al., 1996)	TR08, German (Schach et al., 2007)	Shanghai, China (Guangwei et al., 2007)	(Dai, 2005)	HSST-100L, Japan (FTA, 2005b)	CHSST, Japan (FTA, 2002, 2004)	Colorado, U.S. (FTA, 2002, 2004)
Number of car body in the vehicle	1	1	1	3	5	5	3	2	2	2
Length of each car body, in meters	13	13.7	18	25	24.8	24.8	24	15	24.38	24.38
Number of bogies in each car body	2	2	2	8	4	4	4	5	5	5

Table 9. Parameters of some maglev vehicles

3.3 Design

Guideways are designed and constructed with concrete or steel girders. Concrete guideway girders can be as reinforced or prestressed. Guideway girder is evaluated for different load cases. As example, the Shanghai guideway girder was evaluated with respect to as many as 14,000 load cases by consideration of the deflection, dynamic strength and thermal expansion. The guideway girder for Urban Maglev Program in Korea was also evaluated for five load cases that are combinations of the dead load, live load and the prestressing forces of the tendon (Jin et al., 2007).

Guideways are usually made as single-span or two-span elevated or at-grade. But for larger spans the use of continuous two span supports is recommended. This can reduce deflection and the effect of temperature variations (FTA, 2004). Guideways are modeled as a single or multi span beam with uniformly distributed dead and live loads. Analyses are aimed at obtaining maximum stresses and deflections in guideway spans. The design criteria have deflection regulation on live load and concrete strength condition on top and bottom ends of the girder. The design criteria of the maglev guideway can be summarized as the deflection regulations due to live load in the sense of serviceability and the stresses limits of the girder due to the combination of the dead load and live load. Any classical beam analysis or finite element methods can be adapted in order to obtain maximum stresses and deflections of the beams. Design methods of guideway beam should satisfy the design criteria regarding loading conditions (live load and dead load) and deflection conditions due to live load. The stresses are controlled according to regulations such as AREMA or AASHTO specifications. As example (Jin et al., 2007), the allowable stresses for prestressed concrete compressive strength f_{ck} are described in Table 10.

Parameter	Description	Unit
Compressive strength at initial prestressing	$0.8f_{ck}$	MPa
Compressive strength just after prestressing	$0.55f_{ci}'$	MPa
Tensile strength just after prestressing	$0.75\sqrt{f_{ci}'}$	MPa
Compressive strength under design load	$0.4f_{ci}'$	MPa
Tensile strength under design load	$1.50\sqrt{f_{ci}'}$	MPa

Table 10. Allowable stresses of concrete guideway girder

Till now, variety of design methods has already been used. The Allowable Stress Method was used for design of GA maglev system foundations in U.S (FTA, 2005b) and the Urban Maglev Program in Korea (Jin et al., 2007). The AASHTO Standard Specifications for Highway Bridges were used for design of the Colorado maglev system in U.S (FTA, 2004), the maglev system of GA in U.S (FTA, 2005b) and Transrapid TR08 maglev system in German (Dai, 2005). The AREMA Standard Specifications were used for design of the tensile stress in prestressed concrete for the maglev systems of Colorado in U.S and CHSST in Japan (FTA, 2004). The Service Load Design Method was used for preliminary design of the Colorado special guideway to obtain a reasonable proportioning of members and for estimating material quantities (FTA, 2004).

The maximum allowable total deformation of the guideway can come from the settlements caused by consolidation or creep, by dead load, by cyclic loads from the vehicles or by dynamic loads during operation. Due to the importance of the geometry deviation in the serviceability and safety of the maglev guideway, tighter control over the deflection due to live load is required. In other words, there should be very strict limit adopted for the deflection in order to provide the required serviceability and safe parathion. Main contributors to guideway beam deflection is its' live load. Deflection due to dead load of guideway beam is usually very small and time-dependant. The maglev systems of CHSST and Colorado are no exceptions (FTA, 2002, 2004).

Lower deflection in guideway brings the possibility of reaching at higher speeds of travel. Structural continuity, reduction in span length, reductions in live load and DAF, load combinations, much concrete characteristics compressive strength, use of prestressed concrete, section modulus and etc. are among effective factors in the reduction of the deflection due to the live load. In general, utilization of prestressed concrete, increase of required concrete compressive strength and modules of elasticity reduces guideway beam deflection.

Structural continuity reduces live load and dead load deflections and possibly allow for higher speeds. Deformations due to creep and joint bearing costs also reduce with the use of structural continuity. Over time, precast girders get considerable variation in cambers and early creeps, but very little time deflection after continuity and composite behavior is achieved. The design with AREMA specifications results in the relatively high live load to total load ratio combined in comparison to other load combinations (because of the high effect of this type of load). It should be noted that this loading combination in comparison to other regulations consist of the highest load factors (LF), and at the same time the current regulation scheme with the use of the allowed tensile stresses applies a more accurate control over the deflection due to live load (FTA, 2004).

Up till now, different proposed regulations for deflection ratios due to live load such as L/500, L/1000, L/1500, L/1750, L/2000, L/2500, L/3000, and L/4000 have been proposed. In the Transrapid maglev systems generally beams are designed for the deflection ratios due to live load of L/4000 which is the optimum in design terms and in terms of economic efficiency (Jin et al., 2007; Lever, 1998; Schwindt, 2006). The allowable deflection ratios due to live load of some different maglev systems, a high-speed railway and conventional railroad bridges are presented in Table 11 where L is the span length.

Items	Span Length (m)	Dynamic Deflection Regulation (m)
Korea(a) (Jin et al., 2007)	25	L/1500
	30	L/1500
	25	L/2000
	30	L/2000
	25	L/3000
	30	L/3000
	25	L/4000
	30	L/4000
Proposed Girders(b) (Jin et al., 2007)	25	L/2000
	25	L/4000
A Proposed Girder(c) (Yeo et al., 2008)	25	L/2000
Linimo, Japan (Jin et al., 2007)	30	L/1500
Linimo, Japan (Yeo et al., 2008)	-	$20<L \leq 25m$: L/1500 $25<L$: $(L/25)^{1/2} \times L/1500$
AGT, Korea (Jin et al., 2007)	30	L/1000
Transrapid, Germany (Jin et al., 2007)	25	L/4000
KIMM, Korea (Jin et al., 2007)	25	L/4000
Expo Park, Korea (Jin et al., 2007)	25	L/3000
UTM01, Korea (Yeo et al., 2008)	-	L/4000
TGV(d) -Atlantique (Lever, 1998)	-	L/4000
TR07 (Lever, 1998)	25	L/4000
Bechtel (Lever, 1998)	25	L/2500
Foster-Miller (Lever, 1998)	27	L/2300
Grumman (Lever, 1998)	27	L/2500
Magneplane (Lever, 1998)	9.1	L/2000
Colorado, U.S. (FTA, 2004)	30	L/1750
CHSST, Japan (FTA, 2004)	30	L/1750
(Dai, 2005)	30	L/4000
Railroad Bridge	-	L/800

(a) Structural optimization results of Korea guideway girder systems
(b) Two types of proposed U-type girder systems for Urban Maglev Program in Korea (Jin et al., 2007)
(c) A proposed U-type girder system for Urban Maglev Program in Korea (Yeo et al., 2008).
(d) The French Train a Grand Vitesse (a high-speed railway train)

Table 11. Allowable deflection due to live load for some typical maglev systems

4. Station

Stations have emerged as a new central place in metropolitan cities and have become hub of networks due to their high accessibility by different modes of transport in high scale level. Furthermore, they produce movements which offer sufficient opportunity for the development of commercial land use. Railway stations entered a new age again in the late 20th century after the introduction of high-speed trains. Stations play a very important and influential role in the maglev transport system. The efficacy of the maglev system over the national and regional development depends on the stations. The development hub of maglev system mainly formed around stations.

Transportation facilities are both collectors and distributors. The overall goal of these transit stations is to collect and distribute as many passengers as possible with a minimum amount of confusion and inconvenience. Stations should have the capacity to accommodate large concentrations of passengers at various times throughout the day. The stations activities consist of everything from passenger service to the maintenance of the building. It is important to provide the traveler with a pleasant experience and atmosphere that will hopefully lead to repeat business in the future. The station should be able to provide for all of the modern conveniences to better serve the employees as well as the weary travelers. The important idea is to be able to get the people to their next destination as quickly as possible, and if a wait happens to occur then the station should be equipped to accommodate the passengers' needs (Stone, 1994).

Maglev stations are key regional transportation facilities designed to provide access for high volumes of passengers. The Maglev stations will provide regional and local intermodal connections, as well as national and international connections to passenger facilities. The aesthetic features of the stations are intended to reflect the intrinsic values of the Maglev system: advanced technology, movement, and speed. The conceptual design calls for open-air stations with natural light and ventilation.

Fundamentally, a maglev station is equivalent in planning, design, and operation to an inter-city or commuter railroad station. There is only one technical aspect of maglev that constrains station design: unlike railroad tracks, the maglev guideway cannot be crossed by passengers and vehicles at grade. As a result, maglev station designs must provide grade-separated passenger access to the station platforms. This form of access requires "vertical circulation" (stairs, elevators, escalators) to connect the platforms with tunnels under or bridges over the tracks. Stations should provide the proper functions of typical transit stations, including platforms, Shelter, Vertical and Horizontal circulation, Amenities and Services, Climate controlled waiting room, Public restrooms, Snack service, Public telephones, Changeable message display, Safety. All the station designs are planned to be consistent with the character of the buildings in the area of operation or predicated on the community standards of the local area where each station is located. The station must support the safe movement of passengers at specified flow rates and must also support particular levels of vehicle traffic. Based on the patron markets the following elements, features, and design standards should be common to all maglev system stations, regardless of location or patronage volumes. The expression of these standards will vary and additional features may be added, depending on station location.

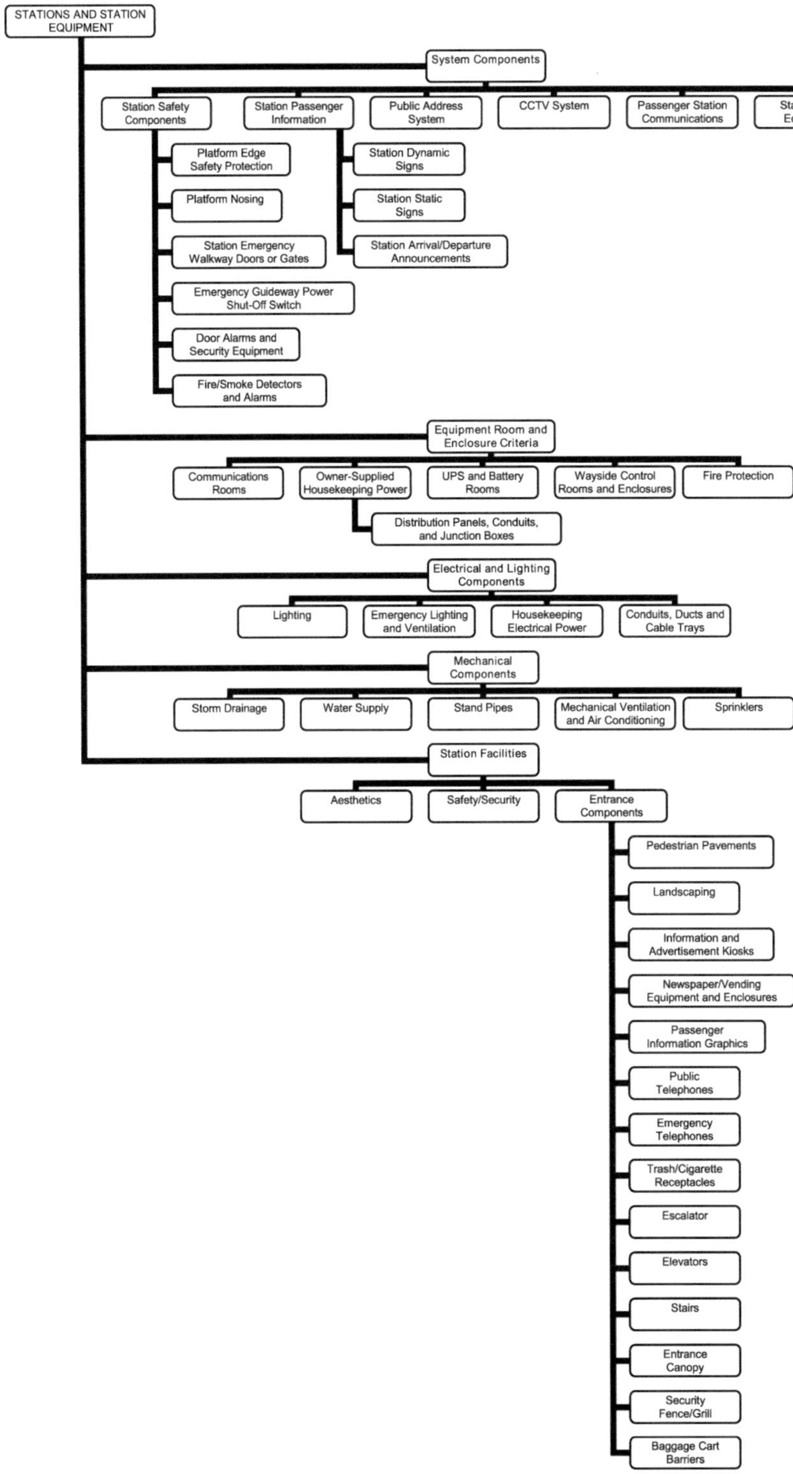

Fig. 7. Maglev station equipment

Platforms will be elevated, allowing direct access through train doors without steps or ramps. No free passenger access to the guideway will be permitted, for safety reasons. This is mandatory due to the speed and low noise profile of maglev systems. The use of docks and in-station transfer switches means that passing trains, while not necessarily in close proximity to platforms, could injure anyone who strayed into the active main guideway. For vertical circulation all maglev system stations will provide escalators and elevators as the primary elements and stairs as the backup.

The station services, including public rest rooms, snack service, newsstand, staffed ticketing and information center, and public telephones should be provided. Stations should also provide facilities (shops, changing rooms, luggage storage, etc.), access to traveler services such as station cars, and advertising displays. All stations would feature public art appropriate to their locations. Public art is an excellent adjunct to station design and a popular feature. Train and station operations require the station personnel and security. Station managers and ticket agents control the station activities, providing passenger assistance and information as well as inspecting train sets when in station. Armed security personnel are provided at every station. Large stations have multiple security personnel, and parking garages are policed (FTA, 2004).

5. Operation

5.1 Performance

The most important task or essential aim when designing the alignment is to specify the geometry of the guideway's functional planes so that the passenger traveling in the vehicle on the guideway experiences optimum comfort during the journey. The geometry defines the limit values for accelerations in the three spatial directions (X, Y, and Z direction). However, apart from the acceleration, the consideration of the change in acceleration (jerk) is also an important aspect for comfort. Therefore, various mathematical formulae were discussed for the transition curves and lengths, with the result,

- the horizontal transition curves are designed as sinusoidal curves and
- the vertical transition curves are designed as clothoids.

An exception are the track switching devices which, based on beam theory, are also designed using clothoids for the horizontal transition curves in the turn-out position. The alignment is designed and the space curve established taking into consideration the aspects given above as well as the system characteristics, e.g.

- climbing capability up to 10% and
- cant (superelevation) in curves up to 12%.

The space curve data are used in the next design phase as the design criteria for

- specifying the substructures,
- height of the columns,
- geometry of each individual beam,
- location of the track switching devices and for the
- precise location of the functional components on the beam (Schwindt et al., 2004).

The reductions in speed in the track course result from slopes, where the residual acceleration abilities do not maintain a high speed. Based on faster acceleration, the operation speed of the maglev can be smaller than that of the ICE3 in order to achieve the same running time. The primary energy demand that is relevant for the comparison between different means of transport averages under the examination of the current power mix as 2.5 times the secondary needs (Witt et al., 2004).

5.2 Propulsion system

Electronic control systems control the clearance (nominally 10 mm). The levitation system uses on-board batteries that are independent of the propulsion system. The vehicle is capable of hovering up to one hour without external energy. While traveling, the on-board batteries are recharged by linear generators integrated into the support magnets.

A synchronous, long stator linear motor is used in the Transrapid maglev system both for propulsion and braking. It functions like a rotating electric motor whose stator is cut open and stretched along under the guideway. Inside the motor windings, alternating current is generating a magnetic traveling field that moves the vehicle without contact. The support magnets in the vehicle function as the excitation portion (rotor). The speed can be continuously regulated by varying the frequency of the alternating current. If the direction of the traveling field is reversed, the motor becomes a generator which brakes the vehicle without any contact.

In accordance with Lenz's Law, the interaction of the levitation field with the current in the slots of the rail results in propulsion or braking force. During the motion of the magnet along the rail, the linear generator winding of the main pole is coupled with a non-constant flux, which induces a voltage and reloads the on-board batteries. The generation process begins in the range of 15 km/h and equals the losses of the magnetic suspension systems at 90 km/h. The whole energy losses of the vehicle are compensated at a velocity of 110 km/h and the batteries are reloaded. Thus the levitation magnet integrates three tasks: levitation, propulsion and transfer of energy to the vehicle (Dai, 2005).

The superhigh-speed Transrapid magnetic levitation system is powered by a synchronous, ironcored long stator linear motor which - in contrast to the classic railroad - is not installed on board the vehicle but in the guideway along the route. The special features of the long stator linear propulsion system enable its dimensions to be individually adapted to the running requirements of the route as well as to specific operating concepts.

The structure of the propulsion system developed for revenue service comprises a number of components, which are located along the guideway. These drive components are temporarily switched together to form the propulsion units necessary to permit maglev operation over the guideway. A propulsion unit remains in the switched configuration for as long as a vehicle is operating within the corresponding control range (drive control zone). It is capable of driving, accelerating and retarding one maglev train. A prolusion unit comprises the line section itself and, depending on the type of power supply selected, one or two propulsion blocks. The propulsion blocks are housed in substations, the latter being situated beside the guideway and spaced at a maximum distance of 50 kilometers. A substation for a single guideway contains one or two propulsion blocks, the necessary power supply and the decentralized operations control equipment (Fig. 8). A substation for a dual guideway simply is composed of two single-guideway substations.

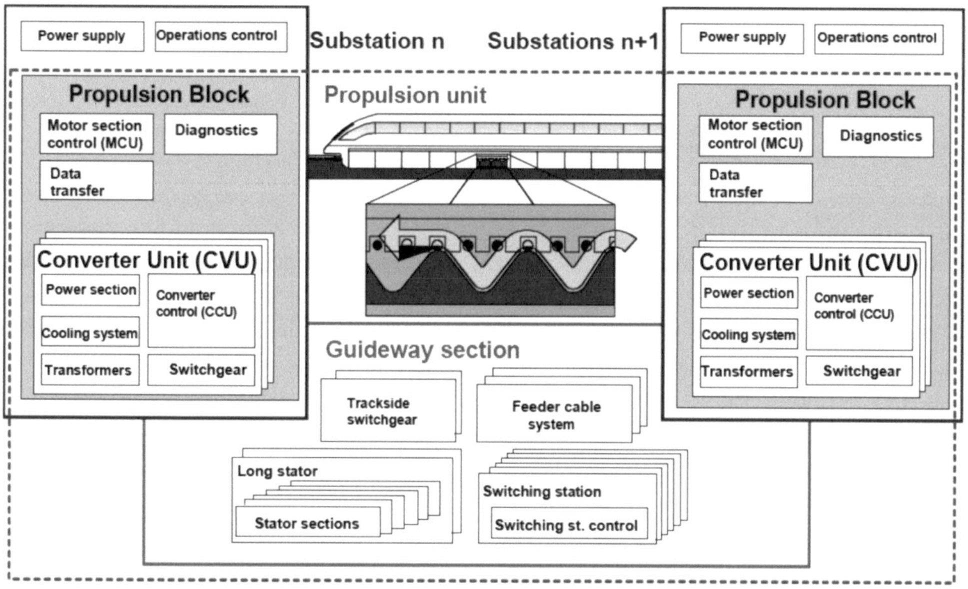

Fig. 8. Structure of the propulsion system

A guideway section consists of two long stators, each comprising the necessary stator sections and switching stations to cover a distance of up to 50 km, the feeder cable systems (two or three according to the selected mode of stator section switching) and the trackside switchgear. A propulsion block is made up of the converter units as well as the motor section control, diagnostics, and components of the data transfer system. In turn, one converter unit comprises a converter power section, rectifier- and output transformers, a closed-loop/open-loop converter control system, a converter cooling system and converter switchgear.

In a double-end feeding configuration, power is supplied to both ends of a guideway section from the two propulsion blocks of adjacent substations. If each substation has only one propulsion block per guideway, there must be at least one clear drive control zone between two maglev vehicles running in the same direction. However, if each substation has two propulsion blocks per guideway, the second maglev vehicle may enter a zone just cleared by the first. Data exchange between the components of a drive control zone as well as between adjacent control zones and external subsystems is made possible by a powerful data transfer system (Henning et al., 2004).

Propulsion in the maglev system is achieved by a linear synchronous motor (LSM). The linear synchronous motor comprises three-phase stator windings mounted on the underside of the guideway and producing a traveling magnetic field BS (its velocity being proportional to the frequency of the input signal) along the guideway. The second component of the LSM is the onboard excitation system. The excitation system made of the levitation electromagnets produces an excitation magnetic field BR. Propulsion is achieved when the excitation magnetic field BR synchronizes and locks to the travelling magnetic field BS. As a consequence, the speed of the vehicle is proportional to the input frequency of the three-phase stator windings.

The task of the power supply system is to supply all components of the Transrapid system with the demanded power. The main consumer naturally is the propulsion system; others are the power rail supply for the on-board supply of the vehicle, the auxiliary power supply for the propulsion control system as well as the operation control system, the guideway switches and the reactive power compensation.

The components of the Power supply system (PS) are installed in substations - where the main components of the propulsion system and the decentralized operation control system are installed, too - and in transformer stations, which are both located along the guideway. The distance between the substations and transformer stations mainly depends on the characteristic data of the operating program and system layout, such as speedtime-diagramm, minimum interval between maglev vehicles, stations, and auxiliary stopping areas. Furthermore, the availability of the power supply system is a very important stipulation for the power supply system layout.

The propulsion system and power supply for Shanghai Maglev Transrapid Project is based on the structures described above and is designed according to the requirements of the transportation system. The main requirements for a transportation system are:

- track length
- passenger capacity
- travel time to destination
- maximum waiting time for passengers at the stations

Therefore the main design parameters for a transportation system are:

- alignment
- comfort criteria
- speed profiles
- operation concept including headway times
- availability, reliability

In relation to these parameters and the necessary input data

- number of vehicles and number of sections per vehicle
- vehicle data, e.g. aerodynamic resistance
- comfort criterias, e.g. max acceleration, max. jerk
- restrictions regarding size or location of power supply or propulsion equipment the propulsion system and power supply was designed (Hellinger et al., 2002).

5.3 Optimal design speed

The optimal design speed of transportation project relates not only to the national integrated transportation system structure, but also to the energy consumption structure of national economy and the traveling quality of passengers. Starting from the analysis to technical and economical characteristics of the maglev system, this paper tries to find the optimal design speed of high-speed maglev transportation system in different aspects such as the speed structure of integrated transportation system and the project benefit. As a result, it gives reference to the planning of high-speed maglev transportation project.

The determination of the design speed is a strategic decision-making for a transportation model. It relates to the compatibility with social economic development. The design speed of a transportation model has remarkable influence on its construction and operation cost, the ability of its competition in transportation system, then its survivability further. The design speed of high-speed transportation system is a basic precondition for its line-planning, developing and manufacturing of vehicles and other equipments, forecast of the market demand, the assessment of economical and social benefit. It is the most important parameter to develop a high-speed transportation system.

The maximum operating speed the train may be raised step by step along with the market demand and the technical development. Therefore, the optimal design speed of mobile equipment of a project should be considered according to the conditions in the near and far future.

The commercial service speed refers to actual operating speed of the train under synthetic consideration of the market demand and economic benefit of the project. It can be determined according to many factors such as the function of a project in the whole transportation system, the competitive ability, the operating cost, the ticket price, the paying ability and the payment wish of passenger and so on. To adapt the market demand and obtain the best economic benefit, including national economic benefit and social benefit, is the principle to determinate the commercial service speed. Along with the development of economy and society, the best commercial service speed will therefore change. Therefore, in different period there is different optimal commercial service speed.

The optimal design speed of infrastructure will be affected by the natural conditions; The optimal commercial service speed will be affected by the social and economic environment; The optimal design speed of mobile equipments will be affected by the related industry technical level.

The following points should be taken into account:

1. The technologically suitable speed range of high-speed maglev system;
2. The best speed to improve the speed structure of integrated transportation system;
3. The requirement on travel speed of passenger;
4. The influence to the optimal design speed for a certain project situation, such as the line length or travel distance, the local economic development level and the natural conditions (Wanming et al., 2006).

5.4 Transrapid propulsion system

The propulsion system structure meets all the requirements for commercial operation in Shanghai, such as a modular design, high reliability, high availability as well as low maintenance expenditure. The outstanding advantage of the modular structure introduced is that individual components can be replaced in accordance with project requirements without affecting the rest of the system. For example, three different converter power sections are being used in the Shanghai project in order to adapt the converter output to the route's particular requirements regarding acceleration and speed. The double-track route is 30 km long. Consequently, at a maximum operating speed of 430 km/h, the travel time is only 7.5 minutes. Three 5-section maglev vehicles operate in round-trip mode at intervals of

10 minutes. The propulsion and power supply system has been specially configured for this service frequency. Although SMT is only 30km long, the test results show it has excellent characteristics of power-energy consumption/speed and it is the best tool for long distance transportation.

6. Safety and risk assesment

6.1 Safety concept

Despite high speeds, passengers are safer in maglev vehicles than in other transportation systems. The electromagnetically suspended vehicle is wrapped around the guideway and therefore virtually impossible to derail. Elevated guideways ensure that no obstacles can be in the way (Dai, 2005). Maglev systems are required by law to guarantee construction and operation of a system that meets proper safety standards. The responsibility of maglev systems are schematically shown in Fig. 9.

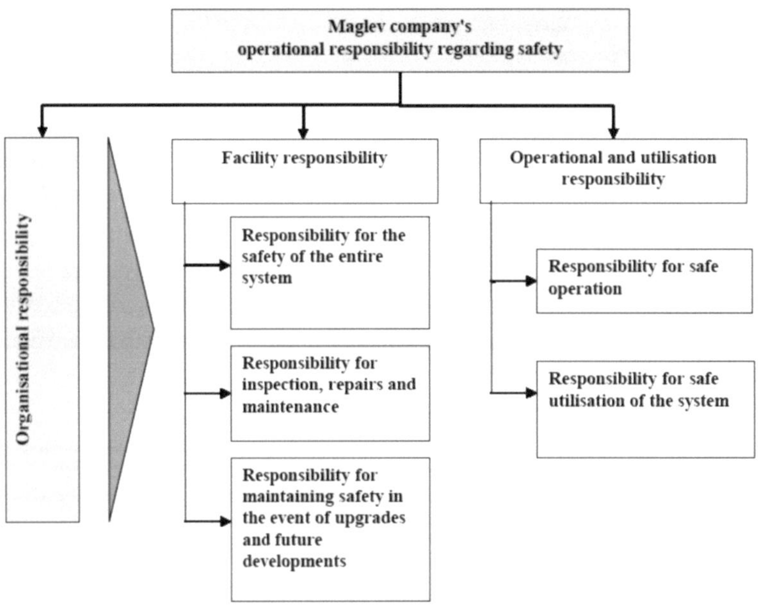

Fig. 9. Maglev system's responsibility

The requirements resulting from safety concepts have effects on all system components and on the whole planning and approval process. Certain internal and external dangers, which affect a maglev system installation, may only be limited regarding their risk potential through constructional measures.

6.2 Rescue concept

An essential component of the safety concept is the rescue concept. The maglev vehicle operator has to explain in this concept with which measures self and external rescue shall be guaranteed. Depending on conception self and/or external rescue measures require different sizes of escape routes, places for emergency stops and accessibilities. Therefore, the rescue concept influences the extent of the required properties so that the effects on the planning approval procedure are given immediately. The examples of protection against going off and rescue concept clearly show how safety concept and planning approval are connected with each other. This means that the development of a safety concept must be at the beginning of the planning process of a maglev system. However, changes of the route course may occur because of others than for safety reasons, so that corresponding customizations of the safety concept can become necessary at a later date.

The factors affecting transportation safety and security are various, among which, the physical structure and guideway security patrols play significant roles. Elevated guideways can be operated safety and efficiently (Liu & Deng, 2003). A means will be required to transfer passengers from the emergency walkway to the ground unless rescue vehicles are used to remove passengers from the walkway. The proposed method of egress from the emergency walkway is a pair of hinged stairways located within one guideway span where the walkway beam would be discontinuous. The stairways would be hinged at the end of the walkway beam and would be attached to dampers that would control the lowering of the stair. The passengers would need to activate a manual release mechanism and then the stair would lower by gravity, slowed by the dampers. The stairways would need to be located at intervals that are a reasonable walking distance. An interval of 0.40 kilometers has been assumed for cost estimating purposes. Signs would be mounted on the emergency walkway that direct passengers to the stairways and indicate the distance from their present location. Figs. 10 to 13 and Fig. 14 show required facilities while emergency situations for Colorado maglev project in the Colorado Department of Transportation (CDOT), U.S. and MOCIE maglev project in the Ministry of Commerce, Industry & Energy of Korean Government (MOCIE), respectively (FTA, 2004, 2005a).

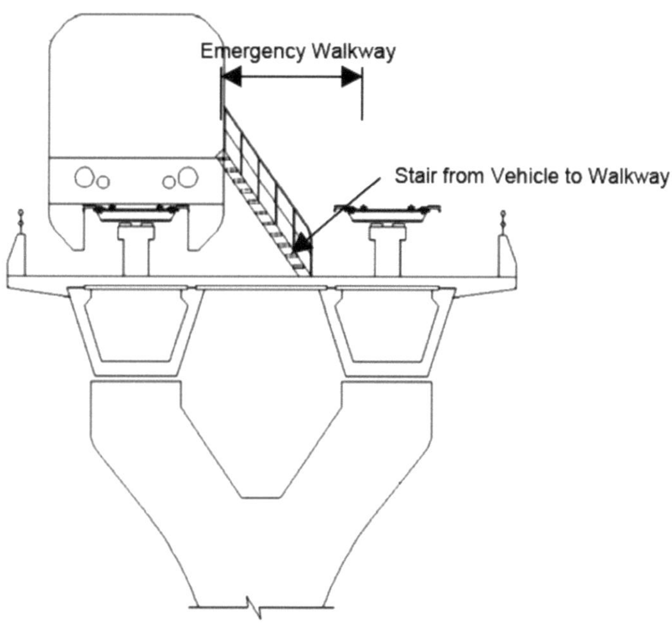

Fig. 10. Double-track guideway

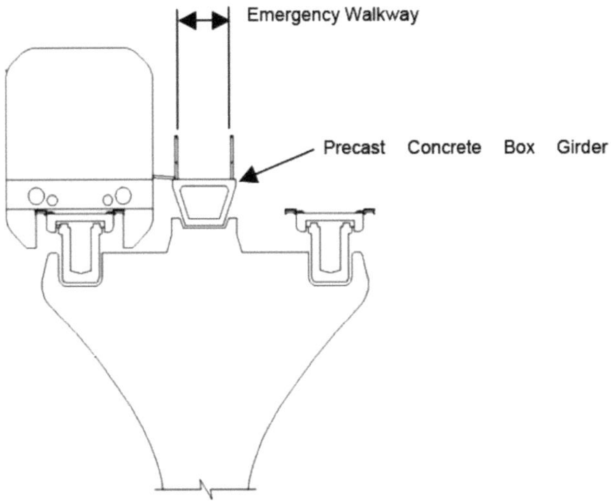

Fig. 11. Separated walkway beam

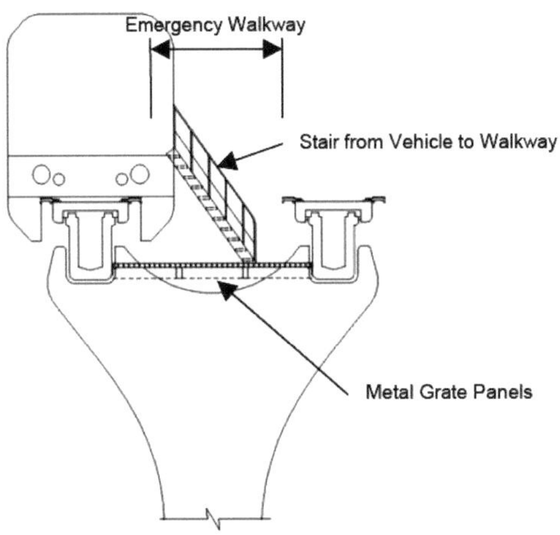

Fig. 12. Metal grate panels

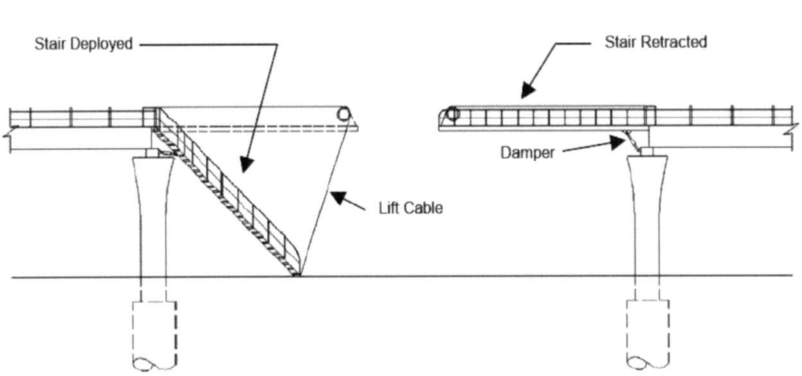

Fig. 13. Stairs from emergency walkway to ground

Fig. 14. Emergency door and ladder

6.3 Operation control system (OCS)

The OCS comprises all technical facilities for planning, monitoring and safeguarding of vehicle operation which means a combination of automatic vehicle operation (ATO) and automatic vehicle protection (ATP) functions like e.g. providing a safe vehicle travel path in order to avoid collisions and the monitoring of vehicle travel speed range in order to assure stopping only at predefined stopping points. The OCS consists of central, wayside and mobile components with interactions to other sub-systems respectively operational and maintenance staff (Fig. 15).

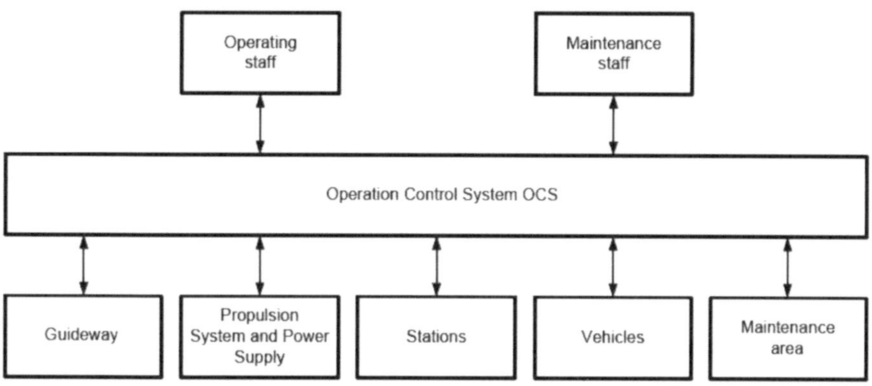

Fig. 15. Structure of the OCS

The system is involved in the assessment of the sub-systems Operation Control System, maglev vehicle and guideway switches are responsible for the overall system safety assessment for the safety concept, rules and regulations for operation and maintenance during commissioning and commercial operation and effectiveness of staff training. The Guideline Mü8004 (the traditional German signaling guide-lines for main lines) distinguishes safety relevant (vital) and not safety relevant (non-vital) requirements (Sawilla & Otto, 2006).

Operation control system (OCS) is the part of an overall maglev system that integrates all subsystems like operation control center, guideway elements, stations, maintenance areas, propulsion and power supply, and vehicles. An OCS contains all components and functions to control and monitor the safe maglev operation. OCS allows control of the vehicle movements and guideway elements both manually and automatically. On the base level, OCS provides all the safety functions generally known in railway signaling, e.g. vehicle locating, guideway switch control, route protection (interlocking), and automatic vehicle control including speed profile monitoring. There are some crucial differences between OCS and most existing railway signaling systems. All vehicle control and vehicle detection (vehicle locating) functions are purely communication-based, using a highly available radio system. Only the safe vehicle brake is used by OCS for emergency braking if the service brake is failed. Emergency systems are mechanical. They act simultaneously if there is an emergency. Each system is controlled by separate component of on-board computer. Emergency systems are independent on each bogie. Each component in the system checks the others. Each component controls at least one of the braking systems. The interior has been designed to concentrate upon the urban commuters' convenience and safety. Whole interior fittings such as panel, floor and seats are made of non-combustible material comply with international fire and safety standards, (Fig. 16) (FTA, 2005a). There are also some innovative safety functions like minimum speed profile monitoring which guarantees the availability of designated stopping points in the event of power shut-offs, transmission failures or hardware faults.

Fig. 16. Emergency landing and guidance wheel

The OCS functions comprise of:

- - Ensure safe movement of vehicles
- - Ensure safe route
- - Ensure safe separation of vehicles
- - Ensure safe speed
- - Authorize vehicle movement
- - Ensure detection and management of emergency situations
- - Handle emergency situation

(Kron Hans, 2006a).

The operation control system (OCS) monitors and controls the various subsystems, integrating them to form a safe, automated overall system. (Kron Hans, 2006b).

6.4 Safety life-cycle

The risk analysis pertaining to the safety concept for maglev vehicles, which is a key document, is an important criterion in the implementation process for the entire project in accordance with the DIN EN 50 126 life-cycle model, (Fig. 17). The European railway life-cycle standard DIN EN 50126 defines a process, based on the system life-cycle including RAMS management. It is applicable to modifications of existing systems in operation prior to the creation of the standard, although it is not generally applicable to other aspects of the existing system (Steiner & Steinert, 2006).

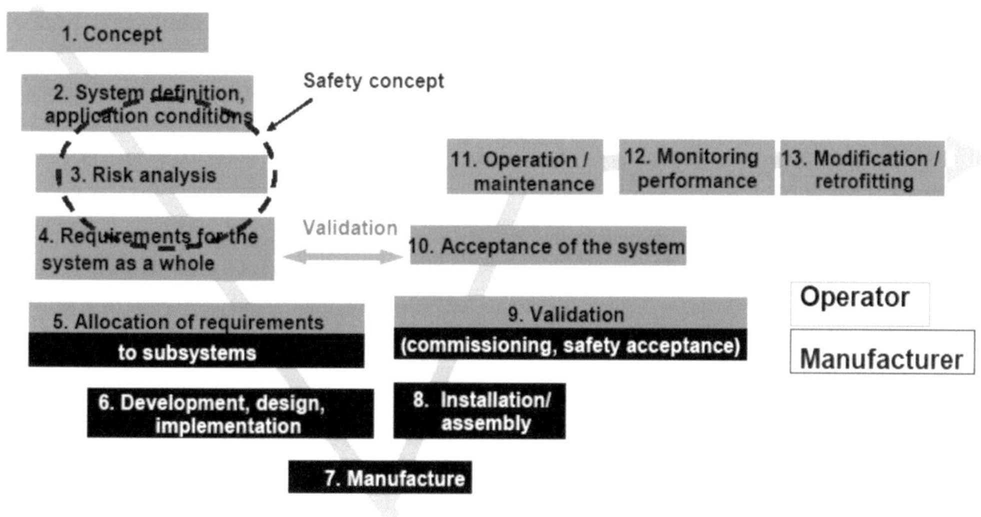

Fig. 17. Life-cycle model

An OCS may only be approved by the safety authority and accepted by the maglev transport authority if both the generic subsystem and the corresponding application data have successfully passed the safety life-cycle, including (Kron Hans, 2006b):

1. Verification - to determine by analysis and test that the output of each life-cycle phase meets the requirements of the previous phase
2. Validation - to demonstrate by analysis and test that the system meets in all respects its specified requirements
3. Safety assessment

7. Technical comparison of maglev and HSR

The need for rapid transit systems has become vital in both urban and intercity travels. There are two technologies for these systems, high-speed rail (HSR) and magnetic levitation (maglev). They are dramatically different in lots of terms. This section focuses only on the technical comparison of these technologies. For a comprehensive comparison, many criterions are included. In fact, this part surveys technical advantages of the maglev systems over the HSR systems.

Mobility and transportation infrastructure is a primary need for the population. They guarantee a high grade of freedom and quality for the citizens, for their work and leisure time. Infrastructure is an important location factor in the regional and global sense. It strongly influences the development of the society and the growth of the national economies. The mobility of individuals is impossible without an equivalent volume of traffic and transportation infrastructure. Against the background of increasing energy requirement, limited fossil resources and ever-growing CO2-loads, the road traffic may not be the adequate answer for the challenges of the future developments. It is necessary to establish integrated and sustainable traffic systems for the effectively and environmentally acceptable handling of traffic (Naumann et al., 2006). Cities' developments lead to a considerable increase of the road, a capacity overloads of road traffic network, and an increase of stresses for people and environment. The transport policy must be faced up to this challenge and take appropriate measures in time. A major vision is the development and implementation of rapid transit systems, which can relocate certain parts of road and air traffic to these systems and to enhance growth of congested urban areas and coalescence of the area (Schach & Naumann, 2007).

The congestion in transportation modes associated with increased travel has caused many problems. These problems include the public concern, among which are prolonging travel time, growing accident rates, worsening environmental pollution, and accelerating energy consumption. On the contrary, high-speed ground transportation, characterized by high speed, operating reliability, passenger ride comfort, and excellent safety record, is considered one of the most promising solutions to alleviate the congestion. There are two distinguished technologies, HSR and maglev. Both provide higher operating speed. However, they have dramatically different technical specifications. Various organizations in the world are facing difficult decisions, when choosing or settling on a specific technology, in a particular corridor. Due to the complexities of HSR and maglev technology, it is not an easy task to select the most efficient technology in any given corridor.

A new rapid transit system influences the society, the industry and the ecology in various manners. A HSR or maglev system must prove its advantages. Therefore, extensive and detailed studies must be carried out. It must be examined in an intense planning process, with feasibility studies. The criterions for the decision must be evaluated in a multi-criteria procedure. This process delivers a master plan for new construction of the transportation network. The plan for the research and development of a rapid transit technology should be made at the national level. The study focuses only on the technical comparison of these technologies. For a comprehensive comparison, a lot of criterions are included. It leads to a wider consideration and the development of the technical comparison. It comprehensively compares the characteristics of HSR and maglev in detail in different aspects. These aspects include geometrical requirements, speed, acceleration, RAMS, environmental impacts, energy consumption, noise emission, vibration level, land use, loading, etc. The obtained results clearly indicate that the maglev generally possesses better technical advantages over HSR.

Rapid transit system is a definition that covers both HSR and maglev. It is defined as an intercity passenger transit system that is time-competitive with air and/or auto on a door-to-door basis. This is a market-based, not a speed-based, definition: it recognizes that the opportunities and requirements for high-speed transportation differ markedly among different pairs of cities (Liu & Deng, 2004). The fundamental reason for considering the implementation of rapid transit systems is higher speed, which can easily equate to shorter travel time. Therefore, there is a need to look at the technical specifications of each technology. This examines the potential improvement of each technology in terms of speed, travel time and other advantages.

HSR trains represent wheel-on-rail passenger systems. These trains currently operate at maximum speeds of about 350 km/h in China, and have been tested at 574 km/h in France. Examples of HSR trains include the French Train à Grand Vitesse (TGV), the Japanese Shinkansen, the German Intercity Express (ICE), the Spanish AVE, etc. Maglev is an innovative transportation technology. It is the first fundamental innovation in the field of railway transportation technology.

HSR and maglev systems are each developed for specific purposes. Selection of the appropriate technology will depend primarily on acceptable funding levels, transportation objectives, and implementation schedule (Najafi & Nassar, 1996). Rapid transit systems must fulfill the major elements of the transport politics. The main aims consist in the increase of speed in the transportation corridors, flexibility, environmental acceptance, ride comfort, stresses (noise, pollutions, and vibrancies), etc. The two existing rapid transit systems must be evaluated and compared against the background of these requirements and the traffic demands.

HSR and maglev are guided ground transportation modes with very large capacity, and both use electric power from the utility grid for propulsion. They also exhibit some fundamental differences that distinguish them as very separable transportation modes. Maglev systems offer the unique combination of technical attributes. These include light weight vehicles, centralized and fully automated control of propulsion systems, non-reliance on adhesion for vehicle acceleration and braking forces, and the ability to operate with consists of as little as single cars. These cars carry fifty to one hundred passengers without the need for highly-skilled operators. The ability to use single or double-car allows even relatively small markets to be given high frequency, reliable service. This together with frequent, highly reliable service, are required to attract new ridership and divert passengers away from their cars. The maglev technology attracts a significantly greater ridership and provides more benefits than HSR systems.

Fig. 18 shows a classification to compare the different parameters for the rapid transit systems in this research. The paper focuses only on the technical comparison of the maglev and HSR systems. For a comprehensive comparison, a lot of criterions are included. It leads to a wider consideration and the development of the technical comparison. The purpose of this research is not to recommend one technology over the other. Actually, both technologies are highly advanced and have some advantages. However, this research surveys technical advantages of the high-speed maglev systems over the HSR systems (Yaghoubi, 2011; Yaghoubi et al., 2011).

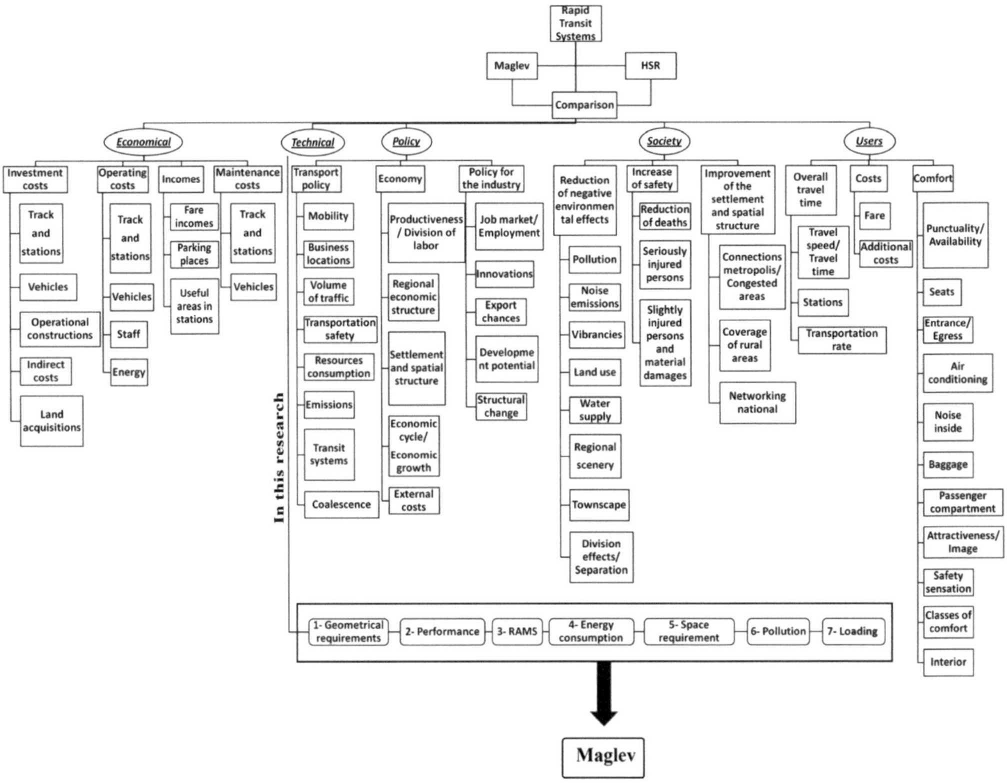

Fig. 18. Classification to compare different parameters

In general, there are many good reasons to turn to magnetically levitated trains. By lower levels of consuming energy, pollution, less noise emission and vibration level, maglev vehicles cause fewer disturbances to the nature and have increased compatibility with environmental issues. Possibility of traveling on elevated guideways means less land occupation. In addition, maglev guideway has lower dead loading. These vehicles can travel at steeper gradients and are capable of traveling at higher speeds with increased accelerations and higher braking, more effective use of regenerative as opposed to dynamic electrical braking, and lower staff and maintenance costs. Maglev vehicles have lower static and dynamic loading, higher passenger capacity and increased passenger comfort and convenience. Such vehicles can travel along routes with lower curve radiuses. They are reliable, reasonably safe, and convenient. Other benefits of maglev systems include travel time, health, flexibility, frequency, operational and schedule reliability (weather and equipment delays), accessibility, safety and security, system availability (origin and destination). Amongst the most important aspects of using maglev vehicles is the possibility of traveling at 10% grades while for high-speed trains such as German ICE this grade angle reduces to 4%. This important aspect considerably reduces the total length of the routes for maglev vehicles. As a further bonus, the cost of constructing and establishing maglev routes at grades and hilly areas considerably reduces. Maglev is obviously the most attractive and powerful transportation system. On the other hand, it is particularly suitable for long-distance transportation of passengers. Maglev is very competitive with air transportation at long distances and against passenger cars at distances starting of 100 kilometers. In contrast to maglev, HSR is only conditionally able to compete with passenger road and air traffic at shorter distances between approx. 150 and 350 kilometers (Naumann et al., 2006).

7.1 Geometrical requirements

Although the guideway has the different procedure with the manufacturing and examination, its geometrical requirements and criteria can be compared with railway tracks. The engineering rules of guideway geometry specification define the requests at the function planes of the guideway and their permissible deviations from the nominal values. These tolerances are valid for a guideway girder, finished equipping and under load of dead weight of the girder. The geometrical examination occurs to the outfit of the girders with the functional components in the manufacturing plant. Based on the defined space curve geometry, the deviations to that can be represented graphically. A comparable criterion of the wheel-on-rail system is the internal, shortwave geometry. This is with 2 mm related to 5 m length indicated in each case for layout (y-direction) and height (z-direction). Standardized onto a consideration length of 1 meter the comparative value turns out 1.5 mm/m at the maglev and 0.4 mm/m at the wheel-on-rail-system. It results from that this tolerance request is significantly higher at the wheel-on-rail system. The tolerance requests at the geometry are approximately identical with both systems. The comparison of the geometrical requests between the maglev and wheel-on-rail shows that similar tolerance requests are made. During the change of the inclination at the wheel-on-rail, track system is approximately 4-times higher as the maglev guideway (Suding & Jeschull, 2006).

7.2 Performance

Based on little wear and tear, the maintenance of the maglev system is less than that of the HSR systems. Due to high operating speed and acceleration, abilities and the low maintenance expenses' maglev can reach very high operation performances (Köncke, 2002). Maglev generally has an advantage over HSR in terms of travel speed. The operating speed of maglev is about 45% higher than that of the HSR trains (Liu & Deng, 2004). The limited speed of HSR is always the main concern of railway professionals. Resistance increases as the speed increases, which limits the increase of speed of HSR. On the contrary, high-speed potential is an inherent characteristic of the maglev technology.

If the speed of each mode plays a key role in the travel time comparison, acceleration and deceleration rate is an even more important factor in terms of safety spacing and average travel speed over certain distances. The maglev vehicle accelerates quickly to higher speeds. Acceleration and braking capabilities of the maglev system result in minimal loss of time for station stops. The vehicles reach high operating speeds in a quarter of the time and less than one quarter of the distance of HSR systems (AMG, 2002).

A maglev vehicle with acceleration/deceleration rate of 1 m/s^2 can obtain the maximum speed in much less time and space than HSR trains. For example, the distance required for the maglev vehicle to accelerate to 300 km/h from a standing start is just about 4-5 kilometers, while HSR trains require about 20-23 kilometers and over twice the time to reach the same speed. Therefore, this advantage of the maglev system results in much less loss of the time for the station stops. The German TR08 maglev vehicle takes 265 s and 19.3 km for the acceleration to achieve the speed of 500 km/h, which are less and shorter than the corresponding values 370 s and 20.9 km for ICE03 train to achieve 300 km/h. The deceleration time and distance via maglev are both shorter so it can maintain ideal speed much longer. The eventual travel time via HSR doubles that of maglev even though the analysis only presented about 50% difference (Liu & Deng, 2004; Witt & Herzberg, 2004; Baohua et al., 2008).

The maglev vehicles can easily overcome uphill gradients and slopes with inclinations up to 10 % comparing to a maximum 3.5 % - 4 % for the HSR trains. In general, the maglev vehicle can climb grades from 2.5 to 8 times steeper than HSR trains with no loss of speed. Embankments and incisions are necessary for the compensation of the small ability of climbing and the constructive design of the guideway. This can lead to a considerable land use. The maglev vehicles can negotiate 50-percent tighter curves (horizontal and vertical) at the same speeds as HSR trains. They can travel through a curve of the same radius at much higher speeds than HSR trains. For example, the maglev vehicle can cant up to 16°. The minimum curve radius of the maglev guideway under the speed of 300 km/h is also 1590–2360 m, which is smaller than 3350 m of HSR tracks (AMG, 2002; Liu & Deng, 2004; Dai, 2005; Jehle et al., 2006; Stephan & Fritz, 2006; Baohua et al., 2008).

Resulting from the greater propulsion performance, the maglev systems offer not only a higher travel speed but also a higher acceleration and deceleration level. The maglev accelerates very well and almost constantly with 0.9 m/s^2. Its maximum speed of 450 km/h is reached within 3 min. The ICE train requires nearly 5 min until it reaches its maximum speed of 300 km/h. Moreover, the maglev vehicle may run approaches to the stop stations in urban surrounding with a speed of 250 km/h due to its low noise emissions and vibrations. The pure running time difference of both systems regarding a line length of approximately 300 km from Berlin to Prague amounts of 29 minutes (50 % more) (Stephan & Fritz, 2006).

Table 12 shows the results of comparison between a maglev train and a HSR train from operational viewpoint (Schach & Naumann, 2007; Liu & Deng, 2004; Witt & Herzberg, 2004; Köncke, 2002; Baohua et al. 2008).

7.3 Reliability, availability, maintainability and safety

An important issue in the proper operation of rapid transit systems is the reliability, availability, maintainability and safety (RAMS). RAMS is the item that needs to be considered in any new rapid transit system establishment. This item is the factor that affects the passenger's mode choice decisions and is important for project evaluation. Safety is amongst most important factors for ensuring operational of integrity high-speed trains. Maglev is one of the safest means of transportation in the world. The concept of maglev has essentially eliminated the safety risks associated with the operation of HSR systems. The use of a dedicated and separated guideway without intersections with other transportation modes such as roads and highways ensures no safety conflicts and allows uninterrupted maglev operations. The maglev technology has essentially eliminated the safety risks associated with the operation of rapid transit systems. Compared to the operating experiences of HSR, the maglev technology has a scarce record. On the other hand, the German Transrapid Test Track in Elmsland has been operating for more than 20 years and

Parameter	Unit	InterCityExpress (ICE) 3 ICE-03 the type series 403		Transrapid Shanghai Maglev TR-08 SMT the type series TR 08				
Operational maximum speed	km/h	until 300		until 450				
Sections per vehicle		8		5 (from 2 to 10 possible)				
Seats (on average)		415		446				
Length (total)		200		128.3				
Capacity		8: 850		10: 1192				
Maximum engine performance	kW	8.000		approx. 25.000				
Power Requirement at Constant Speed of	MW			Train Sections				
				2	6	10		
200	km/h			0.9	2.2	3.6		
300	km/h			2.2	5.0	7.9		
400	km/h	-		4.4	10.3	16.1		
500	km/h			8.2	18.7	-		
Net weight vehicle	ton	409		247				
Weight / Seat	kg	Approx. 930		Approx. 550				
Maximum longitudinal gradient	%	3.5		10				
Acceleration	m/s^2	maximum 1,0		constant 1,5				
Acceleration	m/s^2	Distance (m)	Time (s)	Distance (m)	Time (s)			
0- 100	km/h			424	31			
0- 200	km/h	4400	140	1700	61			
0- 300	km/h	20900	370	4200	97			
0- 400	km/h			9100	148			
0- 500	km/h			22700	256			
Train Configuration		Driving Trailer/ End Car	Trailer Car	End Section	Middle Section			
Train Size		2	6	2	0-8			
Section Length	m	25.68	24.78	26.99	24.77			
Section Width	m	2.95	2.95	3.70	3.70			
Section Height	m	3.84	3.84	4.16	4.16			
Payload / Section	ton	-		10.3	13.9			
Seats / Section		-		62-92	84-126			
Floor Space / Section	m	-		70	77			
Weight / Seat	kg	Approx. 920 to 1000		500 – 700	400 – 600			
Number of Sections		8		2	4	6	8	10
Seats (high density)		408 to 418		184	436	688	940	1192
Seats (low density)		-		124	292	460	628	796
Passengers	ton	-		20.6	48.4	76.2	104	131.8
Curve Radii	m							
Minimum	km/h	300		350				
200	km/h	1400		705				
250	km/h	2250		1100				
300	km/h	3200		1590				
350	km/h	-		2160				
400	km/h	-		2825				
450	km/h	-		3580				
500	km/h	-		4415				

Table 12. Comparison between two German trains of ICE-03 HSR and TR-08 maglev

close to a million passengers have ridden around the 40-kilometer closed loop. The maglev vehicle wraps around the guideway beam so it is virtually impossible to derail. Redundancies achieved through the duplication of components as well as the automated radio-controlled system ensure that operational safety will not be jeopardized. The principle of synchronized propulsion on the guideway makes collisions between vehicles virtually impossible. In general, no other obstacles can be in the way. If two or more vehicles were ever placed simultaneously in the same guideway segment, they would be forced by the motor in the guideway to travel at the same speed in the same direction. The vehicles are also designed to withstand collisions with small objects on the guideway. Energizing only the section of the guideway on which the train is traveling enhances operational safety and efficiency. The maglev vehicle is absolutely weatherproof and masters wind and adverse weather easily. Regarding the aspect of fire protection the maglev vehicle meets the highest requirements of the relevant standards. No fuels or combustible materials are on board. All used materials within the vehicles are PVC-free, highly inflammable, poor conductors of heat, burn-through-proof and heat-proof. The fire proof doors can be optionally used in order to separate the vehicle sections. The system is controlled in all the directions of the movement to ensure ride comfort throughout all the phases of the operation. The seat belts are not required, and passengers are free to move about the cabin at all speeds (AMG, 2002; Köncke, 2002; Liu & Deng, 2004; Dai, 2005).

7.4 Energy consumption and space requirement

With non-contact technology, there is no energy loss due to the wheel-guideway friction. The vehicle weight is lower due to the absence of wheels, axles and engine (low mass of approx. 0.5 t per seat). In terms of energy consumption, the maglev vehicles are better than HSR trains. The maglev consumes less energy per seat-mile than HSR trains due to the utilization of lightweight materials and improvement in the advanced technology. The energy consumption of the maglev system with its non-contact levitation and propulsion technology, highly efficient linear motor and low aerodynamic resistance is very economical when compared to other transportation modes. The high-speed maglev system consumes 20 to 30 percent less energy per passenger than the very modest railroad. With the same energy input, the performance of the maglev system is substantially higher than HSR systems (Liu & Deng, 2004; Köncke, 2002).

As consumers of energy, the transportation sectors are vulnerable to environmental and global warming concerns and the increasing volatile oil market. Reducing dependency on foreign oil is also an important criterion. The system of the external power supply over the contact rail causes higher investment and operational costs. The energy costs of the maglev vehicle despite higher design speed, is lower than that of ICE3 train (Witt & Herzberg, 2004). The maglev vehicles running at 400 km/h has lower environmental impact indicators, such as system energy consumption, waste gas discharges, site area and the like, then the ICE trains running at 300 km/h (Baohua et al., 2008). They also have low running resistance of approx. 0.2 kN per seat at 400 km/h (Köncke, 2002).

Maglev is one of the first transportation systems to be specially developed to protect the environment. The system can be co-located with existing transportation corridors and needs a minimum amount of land for the support of the guideway beams. Use of the elevated guideway minimizes the disturbance to the existing land, water and wildlife, while flexible alignment parameters allow the guideway to adapt to the landscape. Compared to roads or railway tracks, especially the elevated guideway does not affect wildlife movement. Even the ground-level guideway allows small animals to pass underneath due to the clearance planned under the guideway. Compared to all other land-bound transport systems, the maglev requires the least amount of the space and the land. The land area required for a ground-level double-track by either maglev or HSR systems is about similar so it is 14 m^2/m and 12 m^2/m, respectively. But for an elevated double-track guideway, approx. 2 square meter of land is needed for each meter of guideway (Schwindt, 2006). Considering the densely populated and limited land resources, an elevated structure is a preferred choice. The traffic effects on the land-use have been always considered by urban planner and transportation engineers. In the center of metropolitan areas with large economic activities, such as Mashhad, the increase of traffic volume has indirectly cost. It includes wasting time and damages such as environmental pollution.

7.5 Pollution

As maglev is electrically powered, there is no direct air pollution as with airplanes and automobiles. The maglev causes lower CO2 emissions. It is also easier and more effective to control emissions at the source of electric power generation rather than at many points of consumption. Maglev is the quietest high-speed ground transportation system available today. Due to its non-contact technology, there is neither rolling nor gearing or engine noise. The frictionless operation of the maglev vehicle reduces vibration and maintenance resulting from wear. Comparing the noise levels at different speeds, the maglev vehicle is much quieter than the HSR trains. For example, The German TR07 maglev vehicle can travel about 25 percent faster than existing HSR trains before reaching the peak noise restrictions of 80 to 90 dBa. Such an advantage in speed will yield reduced the trip times along the noise-limited routes, which is most urban areas. At the speeds up to 200 km/h, the noise level compared to other noises from the surroundings can hardly be heard. At 250 km/h, the pass-by noise level is 71 dB(A), and from 250 km/h upwards, the aerodynamic noises begin to dominate the noise level. The result is that, at the speed of 300 km/h, the system is no louder than a light rail vehicle, and at 400 km/h, the noise level can be compared to a conventional train traveling at around 300 km/h. Even when at respective high speeds, data also indicates that maglev vehicle is 5 to 7 dBa quieter than the HSR train (Liu & Deng, 2004; Dai, 2005; Schwindt, 2006). The American JetTrain HSR train is almost twice as noisy as the maglev vehicle at the similar operational speeds (AMG, 2002). The results of the noise measurements of the TR08 Maglev System may be compared with similar data, documented by the Federal Railroad Administration (FRA, 1998), for other high-speed ground transportation systems (FRA, 2002a). The noise analysis associated with the Shanghai maglev train shows that the system is quieter than high-speed railway trains for comparable distances from the track (Chen et al., 2007).

A field experiment was conducted, to investigate the possible differences in perceived annoyance of noise caused by high-speed trains, both HSR and maglev. These results were evaluated for the TGV train at speeds of 140 km/h & 300 km/h and for the maglev vehicle at speeds of 200 km/h, 300 km/h and 400 km/h. The LAeq-annoyance relationships determined for the HSR and for the maglev train did not differ significantly. This study has shown that the noise annoyance caused by different types of trains at the same average outdoor façade exposure level is not significantly different. In particular, the magnetic levitation systems are not more annoying than the HSR trains, which is in agreement with earlier research (Coensel et al., 2007).

Whatever the kind of transport system, a passing maglev vehicle always creates ground vibrations due to dynamic loading of the track. Depending on the speed, load transfer, load dispersion and the nature of the ground, these vibrations are transmitted through the ground to different degrees and may thus be felt as shocks in neighboring buildings. For especially sensitive areas, technical solutions are currently being investigated, which minimize the dynamic loads that are transferred from the vehicle to the guideway and then to the bearings in the supports and foundations (Schwindt, 2006).

TR08 vibration levels for both the concrete elevated and concrete at-grade (AG) guideways are compared with those of the TGV, the Italian Pendolino, the Swedish X2000, and the Acela at 240 km/h. The vibration levels for the TR08 traversing the at-grade guideway structures are comparable to those from HSR trains measured in Italy (Pendolino) and France (TGV), whereas the levels for the elevated structure are considerably lower for the distances measured. Vibration levels measured at 15 m for the TR08 traversing the at-grade guideway at 400 km/h are less than those previously measured at 15 m for the Acela traveling 240 km/h. These comparisons, however, are representative of data collected at various sites and are generally typical of local geological conditions. In general, ground-borne vibration levels from trains on elevated structures tend to be lower than those from at-grade operations (FRA, 2002b). The curves for European HSR trains are taken from the FRA high-speed ground transportation guidance manual (FRA, 1998), and for the Acela from measurements conducted by HMMH (FRA, 2000).

7.6 Loading

In this part of research, maglev guideways and road and railroad bridges are compared from loading and design aspects. The optimal design of all bridges, including road, railroads and maglev elevated guideways is really vital. Majority of the existing maglev guideways are elevated and completely built on the bridge. In fact, a maglev elevated guideway is one kind of bridges. Therefore, it can be compared with any bridge, including railroad and road.

According to the AREMA regulations and the UIC leaflets, the live loading models for the rail tracks, is a combination of the concentrated and distributed loads. However, the live loading models for the maglev trains, in the absence of wheels and pursuant to uniformity in the intensity of magnetic forces due to the magnets, are uniformly distributed on the guideways. The lateral magnetic force in maglev is less than the lateral force in the rail tracks. The low level of this force in maglev is due to the absence of the rails and wheels, lower weight of the vehicle and the presence of lateral restoring and equilibration magnetic force.

In general, vertical loadings (dead and live) in the spans of maglev guideways are much lower than those of the railroad bridges. The intensity of the uniform distributed load in live loading of the railroad bridges is almost four times that in maglev. One reason for this difference is the lower weight of the maglev vehicle due to the absence of wheels, axles and transmission parts plus the overall short length of the vehicle. The amount of the earthquake lateral force on the maglev guideway is less than one third of its value for the road bridge.

The loading of the guideway is almost equal to the loading of each one of the four girders of the railroad bridge. In other words, taking into account the fact that the bridge consists of four girders, comparison of the results indicates that the load on the railroad bridge deck is four times greater than the load on the maglev guideway. This means that the guideway by itself can play the role of each one of the girders of the railroad bridge (Yaghoubi & Ziari, 2011; Yaghoubi & Rezvani, 2011).

8. Conclusion

Rapid increase in traffic volume in transport systems plus the need for improving passenger comfort have highlighted the subject of developing new transport systems. The recent required increases in the traffic volume in transport systems, as well as a need for the improvement of passengers' comfort, and required reductions in track life cycle costs, have caused the subject of the development of a new transportation system. One of the important systems which have attracted industries is maglev transport system. In this regard, maglev transport system turns out to be a proper choice for transportation industries around the world. Maglev systems have been recently developed in response to the need for rapid transit systems. The maglev system comes off clearly better and surpasses the HSR systems in almost most fields. These include the pollution, noise emission, vibration level, environmental issues, land occupations, loading, speed, acceleration and deceleration, braking, maintenance costs, passenger comfort, safety, travel time, etc. With the maglev guideway it is also possible to reach to the minimal radiuses for the horizontal and vertical curves. A maglev vehicle can as well travel at the steeper gradients compared with the HSR systems. This considerably reduces the total length of track for the maglev routes compared to the HSR systems. The possibility of traveling with the higher grade angles also reduces the number of tunnels that are required to travel through the mountainous areas. This can also shorten the total length for the maglev route. Therefore, construction of the maglev routes in the hilly areas, in addition to many other advantageous of these systems, can be considered as an attractive choice for the transportation industries. The lower energy consumption of the maglev vehicles in comparison with the HSR systems is also among major characteristics of the magnetically levitated trains. This can be easily associated with the absence of the wheels and the resulting situation of no physical contact between the maglev vehicle and its guideway. Therefore, the energy loss due to the unwanted friction is out of the equations. Furthermore, the vehicle weight is lower due to the absence of wheels, axles and engine. On the other hand, reduction in the travel time considerably reduces the energy consumption. The limited energy resources that are currently available to the nation have highlighted the fact that every individual has to be the energy conscious. The government had to take steps, and it started by setting the preventative rules and the tightening access to the cheap energy resources. Clearly, the widespread application of the magnetically levitated trains for the public transport, in short and long distances, can provide the nation with huge saving in the energy consumption. This is not a fact that can be easily ignored nor can it be bypassed.

Effective parameters in the design of guideways including dead and live loads, dynamic amplification factor and deflection, and structural analysis and design criteria were investigated. According to AREMA regulations and UIC leaflets, live loading models for loading of rails, is a combination of concentrated and distributed loads. However, live loading models for maglev trains, in the absence of wheels and as a result of uniformity in the intensity of lifting magnetic forces due to support magnets, are uniformly distributed on the guideways. The guideway loading is modeled as dynamic and uniformly distributed magnetic forces to account for the dynamic coupling between the vehicle and the guideway. In general, vertical loadings (dead and live) in the spans of maglev are much lower than those of the railroad bridges. The railroad bridge dead load is four times larger than the maglev guideway dead load, and the intensity of the uniformly distributed live load on the railroad bridge is almost four times that of the maglev guideway. Moreover, loading of guideway is four times that in the railroad bridge. One reason for this difference is the lower weight of the maglev vehicle due to the absence of wheels, axles and transmission parts plus the overall short length of the vehicle. The lateral force on the maglev guideway is also much lower than that on the railroad bridge. Also, it is predicted that on the straight routes as a result of negligible lateral magnetic force, there is no considerable amount of torsion created in the cross-section. Therefore, if the beam cross-section and the vertical loading are symmetrical, special design of guideway cross-sections to overcome torsion, is not necessary. Moreover, there usually is no need for the design of deck-shaped cross-sections to care for tension lateral magnetic forces and for the moments due to vertical magnetic forces. Compared to the road and railway bridges, the amount of lateral earthquake force on maglev guideway is lower. Maglev guideways have high resistance against earthquake forces. Maglev vehicles are lighter compared to conventional railway vehicles. These lighter vehicles cause less centrifugal force. The absence of wheels and wheel/rail contact, lighter vehicles and presence of compensating magnetic forces opposing any lateral deviation are the main reasons behind the lower centrifugal forces. A distributed-load vehicle model is better than a concentrated-load model. Multicar vehicles have less car-body acceleration than does a single-car vehicle, because of intercar constraints. This indicates that the multicar vehicle would provide better ride comfort. Weight of required longitudinal bars of guideway is also one-fourth that in the railroad bridge. Deflections due to the vertical loads (dead and live) are also lower in guideways than in rail tracks. Torsion reduction, deflection reduction due to vertical loads, reduction in the costs of construction and operation, increase in resistance and technical justification, possibility of motion in higher speeds are among main reasons to utilize beams with a U-shaped cross-sections and structural continuity in the guideways. Therefore, as noticeable improvements and developments are made in structural optimization of these cross-sections, they could be considered as a good choice among other sections and could be used with a relatively high safety factor. Also, it is shown that the lower loads on the maglev guideway lead to lower bending moments and sheer forces in comparison with the railroad bridge. This indicates that the maglev support structure requires less mechanical strength than the railroad bridge support structure for the same loading pattern. A dynamic simulation for maglev vehicle/guideway interaction is essential to optimize the vehicle design and reduce the cost.

9. Acknowledgment

This chapter of book was performed and financially supported completely by Iran Maglev Technology (IMT).

10. References

American Magline Group (AMG). (2002). Technology comparison: high speed ground transportation, Transrapid superspeed maglev and bombardier JetTrain.

Baohua, M., Rong, H. & Shunping, J. (2008). Potential applications of maglev railway technology in China. *Journal of Transpn Sys Eng & IT*, Vol. 8, No. 1, pp. 29–39.

Behbahani, H. & Yaghoubi, H. (2010). Procedures for safety and risk assessment of maglev systems: a case-study for long-distance and high-speed maglev project in Mashhad-Tehran route. *The 1st International Conference on Railway Engineering, High-speed Railway, Heavy Haul Railway and Urban Rail Transit*, Beijing Jiaotong University, Beijing, China, China Railway Publishing House, pp. 73-83, ISBN 978-7-113-11751-1.

Cai Y., Chen, S. S., Rote, D. M., & Coffey, H. T., (1996). Vehicle/guideway dynamic interaction in maglev systems. *ASME, Journal of Dynamic Systems, Measurement, and Control*, Vol. 118, pp. 526-530.

Coensel, B. D., Botteldooren, D., Berglund, B., Nilsson, M. E., Muer, T. D. & Lercher, P. (2007). Experimental investigation of noise annoyance caused by high-speed trains. *Acta Acustica united with Acustica*, Vol. 93, No. 4, pp. 589-601.

Dai, H. (2005). Dynamic behavior of maglev vehicle/guideway system with control. Ph.D. Thesis, *Department of Civil Engineering, Case Western Reserve Uni*.

FRA (Federal Railroad Administration). (1998). High speed ground transportation noise and vibration impact assessment. Report No. 293630-1, *U.S. Department of Transportation*.

FRA. (2000). Acela trainset noise and vibration measurements on the northeast corridor. Report No. 295450-3, *U.S. Department of Transportation*.

FRA. (2002a). Noise characteristics of the Transrapid TR08 maglev system. Report No. DOT-VNTSC-FRA-02-13, *U.S. Department of Transportation*.

FRA. (2002b). Vibration characteristics of the Transrapid TR08 maglev system. Report No. DOT-VNTSC-FRA-02-06, *U.S. Department of Transportation*.

Federal Transit Administration (FTA), Office of Research, Demonstration, and Innovation, (2002). Assessment of CHSST maglev for U.S. urban transpn., *U. S. Department of Transpn.*, Final Report.

FTA, Office of Research, Demonstration, and Innovation, (2004). Urban maglev technology development program Colorado maglev project, *U. S. Department of Transportation*, Final Report.

FTA, Office of Mobility Innovation, (2005a). Proceedings of the Federal Transit Administration's urban maglev workshop, *U. S. Department of Transpn.*, Washington, DC.

FTA, Office of Research, Demonstration, and Innovation, (2005b). General Atomics low speed maglev technology development program (supplemental #3), *U.S. Department of Transpn.*, Final Report.

Grossert, E. (2006). Actual development in guideway constructions at the example of the Transrapid Munich project, *The 19th Int. Conf. on Magnetically Levitated Sys. and Linear Drives*, Dresden, Germany.

Guangwei, S., Meisinger, R., & Gang, S. (2007). Modeling and simulation of Shanghai maglev train Transrapid with random track irregularities. *Sonderdruck Schriftenreihe der Georg-Simon-Ohm-Fachhochschule Nürnberg*, Nr. 39.

He, Q., Wang, J., Wang, S., Wang, J., Dong, H., Wang, Y. & Shao, S. (2009). Levitation force relaxation under reloading in a HTS Maglev sys. *J. Physica C*, Vol. 469, pp. 91–94.

Hellinger, R., Engel, M. & Nothhaft, J. (2002). Propulsion system and power supply for Transrapid commercial lines, *The 17th International Conference on Magnetically Levitated Systems and Linear Drives*, Lausanne, Switzerland.

Henning, U., Hoke, D. & Nothhaft, J. (2004). Development and operation results of Transrapid propulsion system, *The 18th International Conference on Magnetically Levitated Systems and Linear Drives*, Shanghai, China.

Jehle, P., Schach, R. & Naumann, R. (2006). Comparison of at-grade or elevated guideway construction and railroad tracks. *The 19th International Conference on Magnetically Levitated Systems and Linear Drives*, Dresden, Germany.

Jin, B. M., Kim, I. G., Kim, Y. J., Yeo, I. H., Chung, W. S., & Moon, J. S. (2007). Proposal of maglev guideway girder by structural optimization, *Proceeding of International Conference on Electrical Machines and Systems*, Seoul, Korea.

Köncke, K. (2002). Technical and economical aspects of the Transrapid compared to traditional HSR systems. *The 17th International Conference on Magnetically Levitated Systems and Linear Drives*, Lausanne, Switzerland.

Kron Hans, H. (2006a). Commissioning operation control systems, *The 19th Int. Conf. on Magnetically Levitated Sys. and Linear Drives*, Dresden, Germany.

Kron Hans, H. (2006b). Application design assessment engines, *The 19th Int. Conf. on Magnetically Levitated Sys. and Linear Drives*, Dresden, Germany.

Lee, J. S., Kwon, S. D., Kim, M. Y. & Yeo, I. H. (2009). A parametric study on the dynamics of urban transit maglev vehicle running on flexible guideway bridges. *J. Sound and Vibration*, Vol. 328, pp. 301–317.

Lever, J. H. (1998). Technical Assessment of maglev system concepts, Final Report, The Government Maglev System Assessment Team, U.S. Army Corps of Engineers (USACE) Cold Regions Research and Engineering Laboratory Hanover NH, Report No. A392853.

Liu, R. & Deng, Y. (2004). Engineering comparison of high-speed rail and maglev systems: a case study of Beijing-Shanghai corridor. *The Transportation Research Board*.

Luguang, Y. (2005). Progress of the maglev transportation in China. *MT'19 Conf.* Genoa, Italy.

Najafi, F. T. & Nassar, F. E. (1996). Comparison of high-speed rail and maglev systems. *ASCE, Journal of Transportation Engineering*, Vol. 122, No. 4, pp. 276-281.

Naumann, R., Schach, R. & Jehle, P. (2006). An entire comparison of maglev and high-speed railway systems. *The 19th International Conference on Magnetically Levitated Systems and Linear Drives*, Dresden, Germany.

Sawilla, A. & Otto, W. (2006). Safety assessment for the maglev operation control and overall sys. - Experience Gained and Lessons Learned, *The 19th Int. Conf. on Magnetically Levitated Sys. and Linear Drives*, Dresden, Germany.

Schach, R. & Naumann, R. (2007). Comparison of high-speed transportation systems in special consideration of investment. *Journal of Transport*, Vol. 12, No. 3, pp. 139–147.

Schwindt, G. (2006). The guideway. The 19th International Conference on Magnetically Levitated Systems and Linear Drives, Dresden, Germany.

Schwindt, G., Hauke, U. & Fried, A. (2004). Interaction vehicle/guideway, guideway design aspects for the Munich airport link, *The 18th International Conference on Magnetically Levitated Systems and Linear Drives,* Shanghai, China.

Siemens AG (2006). Transrapid, be forward-looking, experience the future today, transportation systems, a joint project of *Siemens, ThyssenKrupp and Transrapid International,* Germany.

Steiner, F. & Steinert, W. (2006). Safety assessment for the maglev vehicle TR09 - an approach based on CENELEC railway standards, *The 19th Int. Conf. on Magnetically Levitated Sys. and Linear Drives,* Dresden, Germany.

Stephan, A. & Fritz, E. (2006). Operating concept and system design of a Transrapid maglev line and a high-speed railway in the Pan-European corridor IV. *The 19th International Conference on Magnetically Levitated Systems and Linear Drives,* Dresden, Germany.

Stone, A. L. (1994). High speed maglev station - Los Angeles, Thesis in Architecture, Bachelor of Architecture in Design, *The Architecture Faculty of the College of Architecture of Texas Tech University.*

Suding, A. & Jeschull, B. (2006). Comparison of the geometrical requirements for guideways of Transrapid and wheel-on-rail. *The 19th International Conference on Magnetically Levitated Systems and Linear Drives,* Dresden, Germany.

Tielkes, T. (2006). Aerodynamic aspects of maglev systems, Proceedings of the 19th International Conference on Magnetically Levitated Systems and Linear Drives, Dresden, Germany.

Wang, S. K., & Nagurka, M. L. (1997). A superconducting maglev vehicle/guideway system with preview control: part II—controller design and system behavior. *ASME, Journal of Dynamic Systems, Measurement, and Control,* Vol. 119, pp. 644-649.

Wanming, L., Jinbin, Y. & Baofeng, Z. (2006). Study of optimal design speed of high-speed maglev project, *The 19th International Conference on Magnetically Levitated Systems and Linear Drives,* Dresden, Germany.

Witt, M. & Herzberg, S. (2004). Technical-economical system comparison of high speed railway systems. *The 18th International Conference on Magnetically Levitated Systems and Linear Drives,* Shanghai, China.

Yaghoubi, H. (2008). *Magnetically Levitated Trains, Maglev.* Vol. 1, Pooyan Farnegar Publisher, ISBN 978-600-5085-05-1, Tehran, Iran.

Yaghoubi, H. (2011). The most important advantages of magnetically levitated trains, Towards Sustainable Transportation Systems Proceedings of the 11th International Conference of Chinese Transportation Professionals (ICCTP2011), Nanjing, China, American Society of Civil Engineers (ASCE) Publisher, 3038 pp., Stock No. 41186 / ISBN 978-0-7844-1186-5.

Yaghoubi, H., Barazi, N., Kahkeshan, K., Zare, A. & Ghazanfari, H. (2011). Technical comparison of maglev and rail rapid transit systems, *The 21st International Conference on Magnetically Levitated Systems and Linear Drives (MAGLEV 2011),* Daejeon Convention Center, Daejoen, Korea.

Yaghoubi, H. & Sadat Hoseini, M. (2010). Mechanical assessment of maglev vehicle - a proposal for implementing maglev trains in Iran. *The ASME 10th Biennial Conference on Engineering Systems Design and Analysis (ESDA),* Yeditepe University, Istanbul, Turkey, Vol. 2, pp. 299-306, ISBN: 978-0-7918-4916-3.

Yaghoubi, H. & Ziari, H. (2010). Assessment of structural analysis and design principles for maglev guideway: a case-study for implementing low-speed maglev systems in Iran. *The 1st International Conference on Railway Engineering, High-speed Railway, Heavy Haul Railway and Urban Rail Transit*, Beijing Jiaotong University, Beijing, China, China Railway Publishing House, pp. 15-23, ISBN 978-7-113-11751-1.

Yaghoubi, H. & Ziari, H. (2011). Development of a maglev vehicle/guideway system interaction model and comparison of the guideway structural analysis with railway bridge structures. *ASCE, Journal of Transportation Engineering*, Vol 137, No. 2, pp. 140-154.

Yaghoubi, H. & Rezvani, M. A. (2011). Development of maglev guideway loading model. *ASCE, Journal of Transportation Engineering*, Vol. 137, No. 3, pp. 201-213.

Yau, J.D. (2009). Aerodynamic response of an EMS-type maglev vehicle running on flexible guideways, *10th International Conference on Flud Control, Measurements, and Visualization*, Moscow, Russia.

Part 2

Modelling for Railway Infrastructure Design and Characterization

7

Power System Modelling for Urban Massive Transportation Systems

Mario A. Ríos and Gustavo Ramos
Universidad de los Andes, Bogotá, D.C., Colombia

1. Introduction

Urban Massive Transportation Systems (UMTS), like metro, tramway, light train; requires the supply of electric power with high standards of reliability. So, an important step in the development of these transportation systems is the electric power supply system planning and design.

Normally, the trains of a UMTS requires a DC power supply by means of rectifier AC/DC substations, know as traction substations (TS); that are connected to the electric HV/MV distribution system of a city. The DC system feeds catenaries of tramways or the third rail of metros, for example. The DC voltage is selected according to the system taking into account power demand and length of the railway's lines. Typically, a 600 Vdc - 750 Vdc is used in tramways; while 1500 Vdc is used in a metro system. Some interurban-urban systems use a 3000 Vdc supply to the trains.

Fig. 1 presents an electric scheme of a typical traction substation (TS) with its main components: AC breakers at MV, MV/LV transformers, AC/DC rectifiers, DC breakers, traction DC breakers. As, it is shown, a redundant supply system is placed at each traction substation in order to improve reliability. In addition, some electric schemes allow the power supply of the catenaries connected to a specific traction substation (A) since the neighbour traction substation (B) by closing the traction sectioning between A and B and opening the traction DC breakers. In this way, the reliability supply is improved and allows flexibility for maintenance of TS.

So, an important aspect for the planning and design of this electric power supply is a good estimation of power demand required by the traction system that will determine the required number, size and capacity of AC/DC rectifier substations. On the other hand, the design of the system requires studying impacts of the traction system on the performance of the distribution system and vice versa. Power quality disturbances are present in the operation of these systems that could affect the performance of the traction system.

This chapter presents useful tools for modelling, analysis and system design of Electric Massive Railway Transportation Systems (EMRTS) and power supply from Distribution Companies (DisCo) or Electric Power Utilities. Firstly, a section depicting the modelling and simulation of the power demand is developed. Then, a section about the computation of the placement and sizing of TS for urban railway systems is presented where the modelling is based on the power demand model of the previous section. After that, two sections about the power quality (PQ) impact of EMRTS on distribution systems and grounding design are presented. Both subjects make use of the load demand model presented previously.

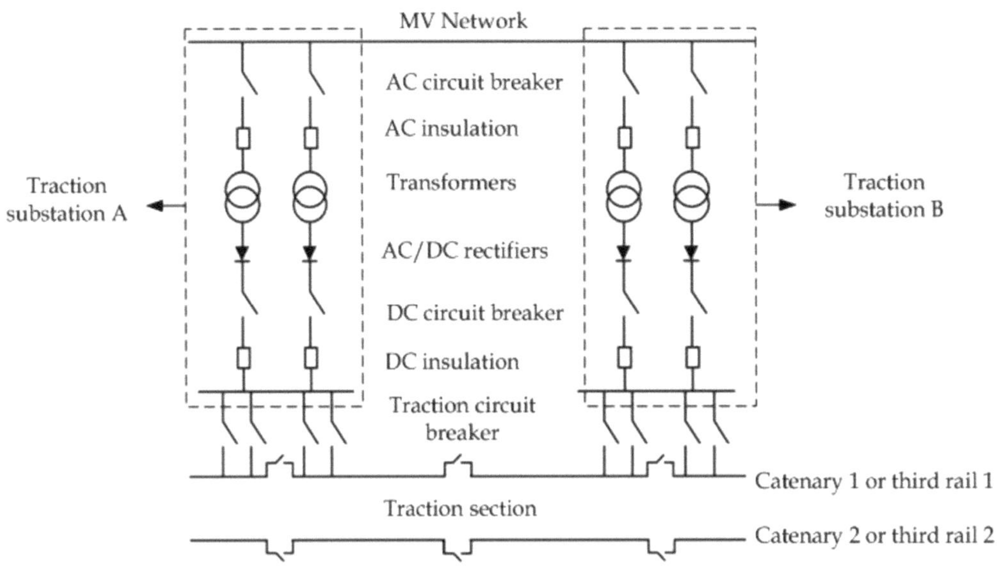

Fig. 1. A Typical Traction Substation (TS)

2. Power demand computation of electric transportation systems

This section presents a mathematical model useful to simulate urban railway systems and to compute the instantaneous power of the Electric Massive Railway Transportation Systems (EMRTS) such as a metro, light train or tramway, by means of computing models that take into account parameters such as the grid size, acceleration, velocity variation, EMRTS braking, number of wagons, number of passengers per wagon, number of rectifier substations, and passenger stations, among other factors, which permit to simulate the physical and electric characteristics of these systems in a more accurate way of a real system.

This model connects the physical and dynamic variables of the traction behaviour with electrical characteristics to determine the power consumption. The parametric construction of the traction and braking effort curves is based on the traction theory already implemented in locomotives and urban rails. Generally, there are three factors that limit the traction effort versus velocity: the maximum traction effort (F_{max}) conditioned by the number of passengers that are in the wagons, the maximum velocity of the train (or rail), and the maximum power consumption. Based on these factors, a simulation model is formulated for computing the acceleration, speed and placement of each train in the railway line for each time step (1 second, for example). So, the power consumption or re-generation is computed also for each time step and knowing the placement of each train in the line, the power demand for each electric TS is calculated.

2.1 Power consumption model of an urban train

The power consumed by one railway vehicle depends on the velocity and acceleration that it has at each instant of time. Its computation is based on the traction effort characteristic (supplied by the manufacturer of the motors), the number of passengers and the distances between the passengers' stations (Vukan, 2007), (Chen et al., 1999), (Perrin & Vernard, 1991). The duty cycle of an urban train between two passengers' stations is composed by four operation states: acceleration, balancing speed, constant speed and deceleration. Fig. 2 shows the behavior of the speed, traction effort and power consumption of a traction vehicle during each operation state elapsed either time or space (Hsiang & Chen, 2001).

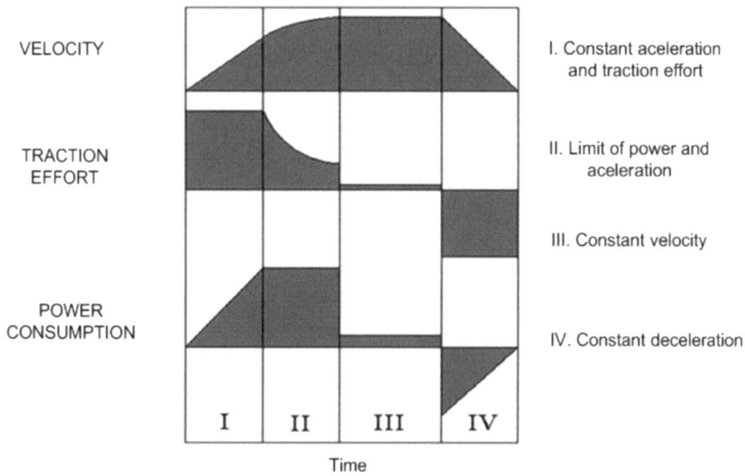

Fig. 2. Velocity, Traction Effort, and Power Consumption of an Urban Train Travel between adjacent Passenger Stations (Hsiang & Chen, 2001)

During the first state (I), the vehicle moves with constant positive acceleration, so the speed increases. When the vehicle reaches a determined speed lower than the constant speed, the second operation state starts. In this state, the acceleration decreases, but the speed keeps increasing. In the third state (III), the cruise speed is reached and the acceleration is zero. In the fourth state (IV), the braking operation starts with negative acceleration until the moment it decelerates with a constant rate and finally it stops at the destination station (Vukan, 2007), (Chen et al., 1999), (Perrin & Vernard, 1991), (Hsiang & Chen, 2001).

2.1.1 Net force of a traction vehicle

The parametric construction of the traction and braking effort curves is based on the traction theory already implemented in locomotives and high speed rails. Three factors limit the traction effort versus velocity: the maximum traction effort F_{max} conditioned by the number of passengers that are in the wagon, the maximum velocity of the vehicle, and the maximum power consumption. The maximum traction effort used by the acceleration, and then transferred to the rail, is limited by the total weight of the axles given by:

$$m_m = TM - \left(n_{axis} - n\right) \times w_{axle} \tag{1}$$

where TM is the total vehicle mass, n is the number of motor drives, n_{axis} the number of axles in the vehicle, and w_{axle} the weight per axle (Buhrkall, 2006). The total vehicle mass is:

$$TM = w_v + \left(n_p \times w_{pas}\right) \times M_{DYN} \tag{2}$$

where w_v corresponds to the weight per wagon without passengers, n_p the number of passengers per wagon, w_{pas} the average weight per passenger, and M_{DYN} the dynamic mass of the railway, which represents the stored energy in the spinning parts of the vehicle, typically of 5-10% (Buhrkall, 2006). Then, the maximum traction effort is calculated as:

$$F_{\max} = \mu \times m_m \times g \tag{3}$$

Where μ corresponds to the friction coefficient between the wheels and the rail, which is about 15% for the ERMTS, and gravity g equals to 9.8 m/s^2 (Buhrkall, 2006). The force needed to move a traction vehicle (TM times the acceleration (a)) is:

$$F = TM \times a = TM\frac{dv}{dt} = TE(v) - MR(v) - B_e(v) \tag{4}$$

Where $TE(v)$ is the traction effort in an EMRTS that provides the necessary propulsion to exceed inertia and accelerate the vehicle, $MR(v)$ is the movement resistance as an opposite force to the vehicle movement, $B_e(v)$ is the braking effort used to decelerate the vehicle and stop it permanently (Vukan, 2007).

The traction and braking effort act directly in the vehicle wheels edges. The movement resistance is given by:

$$MR(v) = 10^{-3} \times \left(2.5 + 10^{-3} \times k\left(v + \Delta v\right)^2\right) \times TM \times g \tag{5}$$

Where $k \approx 0.33$ for passengers' vehicles, $\Delta v \approx 15$km/h is the wind velocity variation, TM is the total mass of the vehicle, and g the gravity. Table 1 presents the action forces in an EMRTS that makes its path between two passengers' stations. As a result, there are four regimens of operation: stopping, acceleration, constant velocity, and deceleration. This is how the difference between the traction effort, the movement resistance, and the braking effort, which are not velocity variants, represent the net force of the vehicle (Jong & Chang, 2005b).

Operative Regimen	Net Force	Velocity
Stopping	$TE(v) - MR(v) - B_e(v) = 0$	$v=0$
Acceleration	$TE(v) - MR(v) - B_e(v) > 0$	$0 < v < v_{max}$
Constant Velocity	$TE(v) - MR(v) - B_e(v) = 0$	$v > 0$
Deceleration	$TE(v) - MR(v) - B_e(v) < 0$	$0 < v < v_{max}$

Table 1. Net Force and Velocity as function of the Operative Regimen (Jong & Chang, 2005b)

2.1.2 Computation of dynamic variables

The incremental acceleration (a_i) is obtained from the net force and the total mass of the vehicle (Jong & Chang, 2005b) computed for each instant t, as:

$$a_i(t) = \frac{F(t)}{TM(t)} \tag{6}$$

The velocity is assumed an independent variable, which determines the path time of the traction vehicle, with steps fixed by velocity and acceleration (Jong & Chang, 2005b). So, the time steps and the incremental travelled distance are given by:

$$t_{i+1} = t_i + \frac{v_{i+1} - v_i}{a_i} \tag{7}$$

$$s_{i+1} = s_i + v_i \left(t_{i+1} - t_i\right) \tag{8}$$

2.1.3 Power consumption computation

The motor torque and the velocity for an EMRTS are linear functions of the acceleration and the angular velocity. So, the instantaneous power consumption by the EMRTS, for the first three operative states (Chen et al., 1999), (Perrin & Vernard, 1991), (Hsiang & Chen, 2001), is:

$$P(t) = \left(TM(t) \times a_i(t) + MR(v)\right) \times v \tag{9}$$

For the last operative state where the braking acts, the consumption is given by:

$$P(t) = B_e(v) \times v \times \eta_B \tag{10}$$

which describes the braking effort multiplied by the velocity in the range of $0 \leq v \leq v_{max}$ and a multiplicative factor η_B which describes the efficiency of the regenerative braking which it is considered of 30% for this type of systems (Perrin & Vernard, 1991), (Jong & Chang, 2005a), (Hill, 2006).

2.2 Simulation model

The model presented at section 2.1 allows the computation of the power consumption and travel time characteristics (t, x) for each train i in the railway line. Naturally, a railway line simulation must include a number n of passengers' stations and k trains travel in the line (go and return).

The integration of these characteristics requires modelling the mobility of passengers associated at each train. It can be simulated in a probabilistic way, computing the number of passengers coming up and leaving the train (i) in each passenger's station (j) and the stopping time of the train in each station. This first part, stated here as Module 1, uses the following parameters: the passengers' up (r_{up}) and down (r_{down}) rates, and up (t_{up}) and down (t_{up}) times per passenger.

The number of passengers in the first station and the number of passengers waiting in each station (pax_{wait}) are modelled as random variables of uniform distribution. As, the railway line simulation includes a number n of passengers' stations; Module 1 computes for each train i the number of passengers that the train transport between station j and $j+1$ as:

$$pax(i,j) = (1 - r_{down}) \times pax(i,j-1) + pax_{wait}(j) \times r_{up} \quad (11)$$

The number of passengers is constrained to be less or equal than the maximum capacity of passengers at the train. In addition, this module gives the stopping time for each train at each passenger's station ($t_{stop}(i,j)$) based on passengers up and down times, as:

$$t_{stop}(i,j) = t_{down} \times (1 - r_{down}) \times pax(i,j-1) + t_{up} \times pax_{wait}(j) \times r_{up} \quad (12)$$

The second part of the model, called Module 2, simulates the overall travel of train *i*. This means, the simulation gives the power consumption of train *i* for each instant of time *t* for a complete travel (go and return). At the same time, the placement ($x(t)$) of the train is get for each *t*. If the line railway has a length *L*, then the total travel of one train is *2L*, and *x* will be between *0* and *L* in one sense and between *L* and *0* in the another sense.

So, Module 2 computes the train's time of travel between passengers' stations and the instantaneous power demand for one train based on equations (1) to (10) and the number of passengers and stopping time obtained from (11) and (12), respectively; as Fig. 3 shows. As, it is shown, the simulation considers the initial dispatch time and computes the initial value of passengers using the second term of equation (11).

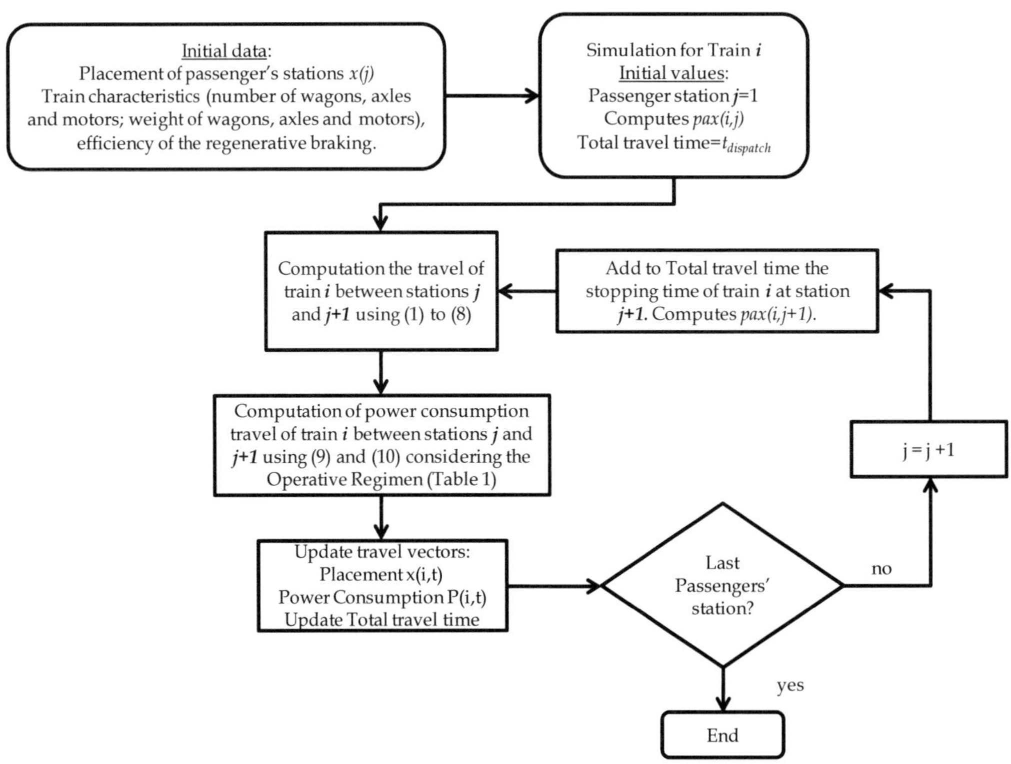

Fig. 3. Simulation of Train *i* Travel - Module 2

On the other hand, Module 2 considers the maximum velocity, the braking and traction effort curves as input variables. These curves are parameterized by means of (1), (2), and (3) and are given by manufacturers of traction equipment. Each curve is used to establish the net force at each operative regime, I to IV in Fig. 2. Fig. 4. shows an example of the simulation of placement and power consumption for a train in a metro line using a power demand simulator reported at (Garcia et al., 2009).

Finally, the simulation of Module 2 is run for the total number of k vehicles in the railway line, taking into account the dispatch time of each one. Then, the power consumption at each TS is computed as Fig. 5. shows. Each TS supplies the power to trains (going or returning) placed for its specific portion of the railway line (the DC section connected to the TS).

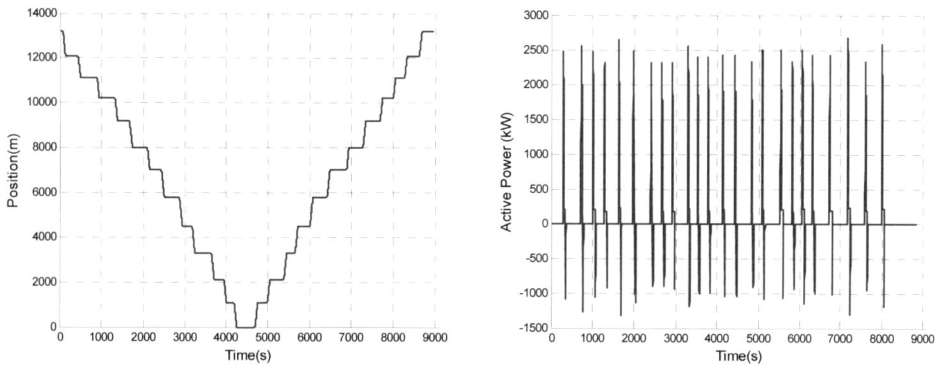

a) Graphical Interpretation of x(i,t) b) Graphical Interpretation of P(i,t)

Fig. 4. Example of Simulation of Train *i* Travel - Module 2

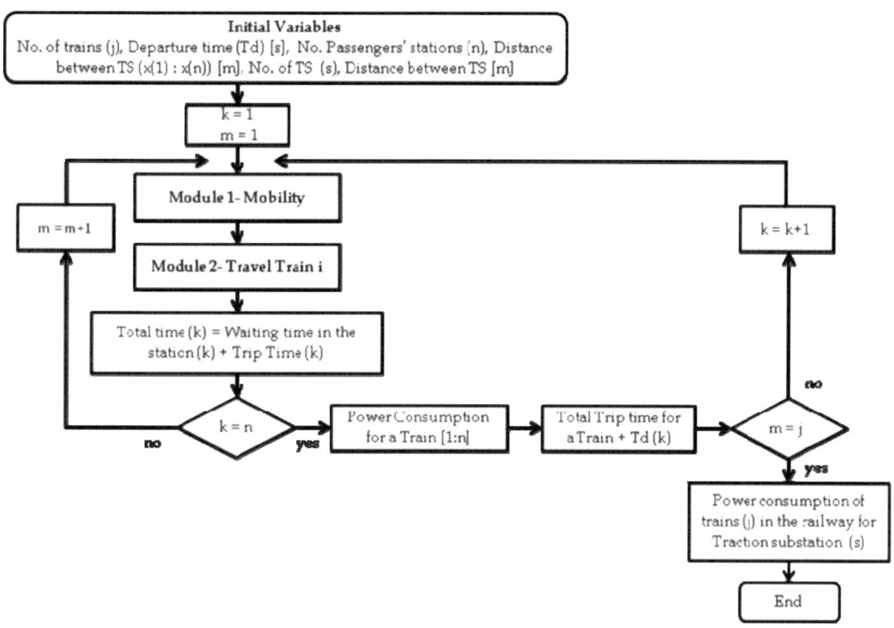

Fig. 5. Simulation of Power Consumption of a Railway Line

2.3 Simulation example

This section illustrates the application of the power consumption mathematical and simulation model in a possible metro line for the city of Bogota of 13.2 km and 13 passenger stations. Fig. 6 shows one section of the possible line 1 to be developed in Bogotá. Fig. 7. presents the results of a simulation of the Metro Line of Fig. 6 using the previous algorithms.

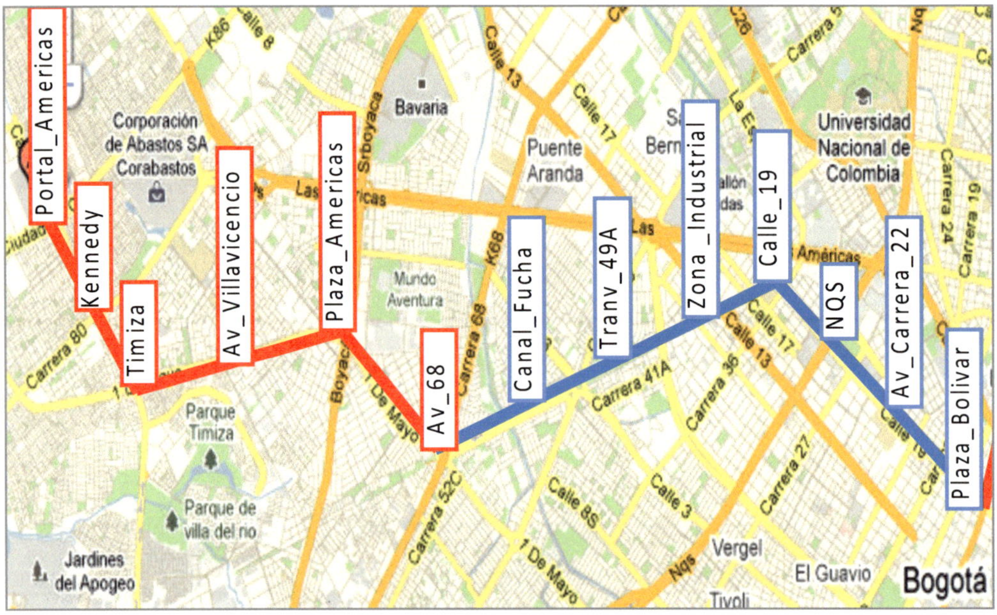

Fig. 6. Example Case of Power Consumption for a Metro Line

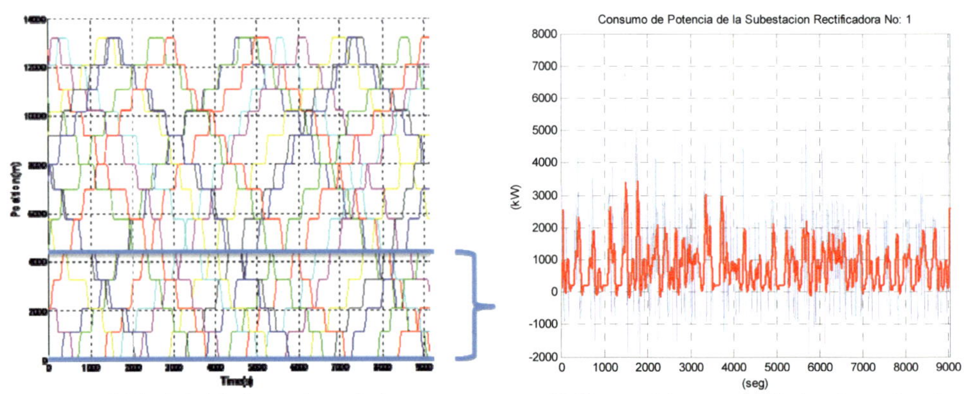

a) Train's Trajectories x(i,t) b) Power Demand P(i,t) substation 1

Fig. 7. Power Demand of a Traction Substation - Estimation by Simulation

The simulation establishes the trajectories of 17 trains-vehicles at the Metro Line (Fig. 7.a). The power is supplied by 3 TS. Fig. 7.b shows the power demand at the first traction substation that supplies all trains placed between position 0<x(t)<4400 m.

3. Placement and sizing traction (rectifier) substations in urban railway systems

In this section, a methodology of placement and sizing of traction substations under an electric connection scheme, in which high reliable levels are guaranteed, is presented. In this scheme, each traction substation (TS) is able to support the load of each adjacent substation. That means that in the case when a fault occurs in one TS, there is a support system based on automatic switches normally opened that close and allow the two neighbour substations to supply the power to the associated load with the faulted substation (each one would feed half of the load of the faulted one). The input data to calculate the sizing of substation is obtained from the power demand computation, explained in the previous section.

On the other hand, the placement of each TS is obtained by a heuristic optimization problem. This problem minimizes the total cost of a given configuration, that is composed of investment costs (rectifiers, transformers, and protection and control cells), the cost of energy losses composed by AC losses (associated with the transformer) and DC losses (associated to rectifiers) and the failure cost, that represents the cost of the annual expected energy not supplied (EENS).

3.1 Traction substation (TS) configurations

A scheme of supply of an urban railway system must satisfy electric conditions, such as: operating limits, voltage drops through the catenaries or third rail (called here, in general, DC section), and maximum capacity of transformers. These conditions must be satisfied for supplying the power demand independently of the operating state of the system, i.e., normal state or a post-contingency state after a fault of a HV/MV substation, or TS, or one DC section. So, the TS location and configuration's selection are strongly linked problems. Fig. 8 shows three possible schemes of connection of the MV network to a set of TS. Each TS is designed to supply (in normal operation state) a DC sector of length L.

The way of behave in a fault condition determines the following three possible configurations:

1. **One transformer-rectifier unit with possibility of power supply from the adjacent TS**. Each TS acts as a support of its adjacent TS. This implies that the substations must be able to supply at least 1.5 times the length of the normal DC section length (3L/2).
2. **Two transformer-rectifier units in each traction substation**. This configuration means the redundancy in the main equipment of the TS. In case of a fault in one transformer and/or rectifier, the parallel unit must supply the total power demand of the TS. This scheme assumes that there is not possibility of support of adjacent substations. The wide dotted line if Fig. 8 remarks the parallel unit of transformer-rectifier unit.
3. **Two transformer-rectifier units in each TS and support of adjacent DC section**. This is the combination of configurations 1 and 2. This means that there is redundancy in each traction substation and there is also possibility of support of adjacent DC section feeder.

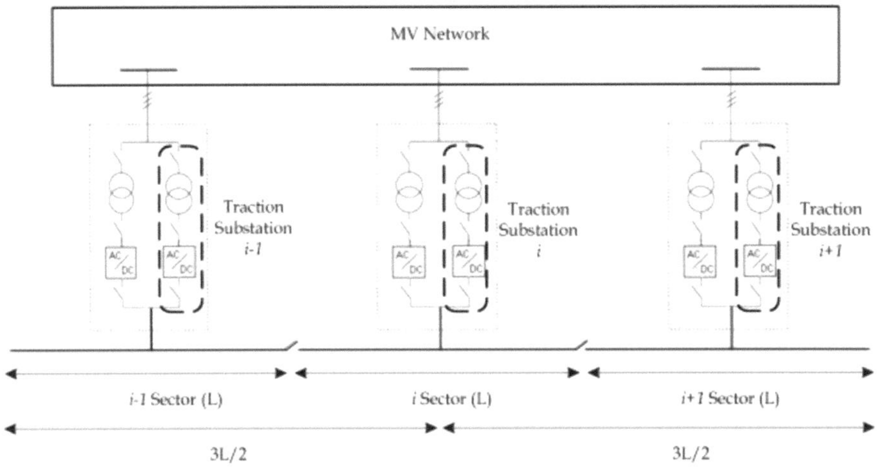

Fig. 8. Configurations of Traction Substations' Connection

3.2 Optimization problem

A minimization of the total project cost is solved for determining the quantity of traction substations, their connection configurations, and their locations. The optimization is a constrained problem that guarantees the electrical requirements, like voltage levels and high reliability requirements. The distance between TS is assumed to be equal, and each TS is located at the middle point of the DC section that it supplies, as Fig. 8 shows.

The cost function (*TC*) includes investment costs, operation costs C_{op} (associated with losses) and reliability cost or cost of energy not supplied (C_{ENS}). So, the total cost for each configuration is given by:

$$TC = C_{inv} + \sum_{j=1}^{T} \left(\frac{C_{op}(j)}{(1+r)^j} + \frac{C_{ENS}(j)}{(1+r)^j} \right) \tag{13}$$

Where T is the number of years of the project, and r is the discount rate of the project. The investment cost depends on the length of the DC network, and is given by:

$$C_{inv} = (L_{cat} \times C_{cat}) + \sum_{i=1}^{m} \left(N_{Mod}(i) \times C_{Mod}(i) + C_{Place}(i) \right) \tag{14}$$

Where, L_{cat} and C_{cat} are the total length and the unitary cost of the catenaries or rail (DC section, in general). N_{Mod} is the number of modules in one traction substation (1 or 2). C_{Mod} is the cost associated to one module; C_{place} is the cost of the terrain where the substation is built; m is the number of substations.

The annual operation cost ($C_{op}(j)$) is computed as the sum of annual AC and DC losses in the year j multiplied by the energy cost. Transformer losses are defined as the sum of the instantaneous iron losses ($AC_{iron\text{-}loss}$) and copper losses ($AC_{copper\text{-}loss}$) during the year. Then, the total losses cost for m substations is:

$$C_{op}(j) = Cost_{Energy}(j) \times \sum_{i=1}^{m} \left(AC_{loss}(i,j) + DC_{loss}(i,j) \right) \tag{15}$$

$$AC_{loss}(i,j) = \frac{1}{T_o} \times \int_0^{To} \left(AC_{iron-loss}(i,j) + AC_{copper-loss}(i,j,t) \right) dt \tag{16}$$

T_o is the operation time of the urban rail during the year. The "iron losses" in the transformer are constant (Institution of Electrical and Electronic Engineers [IEEE], 2007) and computed as it is established in (IEEE, 1992). The "copper losses" are directly proportional to the square of the utilization factor (*UF*), and the constant of proportionality is the nominal copper losses of the transformer (P_{nom_Cu}) (IEEE, 2007). The *UF* is defined as the ratio of instantaneous demanded power ($P_{dem}(i,j,t)$) and the transformer rating (P_{nom}).

On the other hand, the DC losses are power losses in the rectifiers (AC/DC converters):

$$AC_{cooper-loss}(i,j,t) = P_{nom\_Cu}(i) \times (UF(j,t))^2 = P_{nom\_Cu}(i) \times \left(\frac{P_{dem}(i,j,t)}{P_{nom}(i)} \right)^2 \tag{17}$$

$$DC_{loss}(i,j) = \frac{1}{T_o} \times \int_0^{To} (1 - eff(i,j,t)) \times P_{dem}(i,j,t))dt \tag{18}$$

The $eff(i,j,t)$ is the DC efficiency of the rectifier of the TS *i* and depends on the instantaneous power demanded at year *j* hour *t*. It varies from 95.4% to 95.7% (Hill, 2006).

The third term of the objective function of the minimization problem is the cost of energy not supplied (C_{ENS}), computed as function of the time of no-supply in hours/year (T_{NS}), the unitary cost of fault in USD$/kWh ($C_{fault}$) and the average power not supplied by TS (P_{av}):

$$C_{ENS} = T_{NS} \times P_{av} \times C_{fault} \tag{19}$$

3.3 Technical constraints

The voltage drop between a supply point and a utilization point must not be more than 15% in normal operation and as maximum 30% in special cases (Arriagada & Rudnick, 1994). These specials cases may be the outage of a substation or the last DC section in the route. Table 2 presents the voltage margins according to the different used DC system voltages.

DC system voltage (V)	600	750	1500	3000
Lowest voltage. Undefined duration (V)	400	500	1000	2000
Nominal design system voltage (V)	600	750	1500	3000
Highest voltage. Undefined duration (V)	720	900	1800	3600
Not-permanent highest voltage. Duration of 5 minutes (V)	770*	950**	1950	3900

* In the case of regenerative braking, 800 V is admissible.
** In the case of regenerative braking, 800 V is admissible.

Table 2. Voltages in DC Traction Systems (White, 2009).

A voltage drop of 30% between the TS and the last vehicle can be tolerated in a suburban system, where the vehicles are constantly accelerating, but a voltage drop over the principal line of a *metro* during any time interval might exceed all the established limits. Therefore, the maximum voltage drop allowed is limited to 15% on nominal voltages under normal conditions. A voltage drop in the farthest point of a section supplied by TS is defined as:

$$V_T = V_S - L_2 ZnI - Z_x(n+n')I - \frac{1}{4}(Zn \times I \times LCat(i) - Z \times L_2) \tag{20}$$

$$Z_x = (R_u + R_T)\cos\phi + (X_u + X_T)\sin\phi \quad and \quad Z = R\cos\phi + X\sin\phi \tag{21}$$

Where V_s is the DC voltage at the TS (p.u), V_T is the minimum DC voltage in the DC section for correct vehicle operation (p.u.); I is the current demanded by a vehicle (p.u.); R_u and X_u are the equivalent resistance and reactance, respectively (p.u.); R_T and X_T are the transformer resistance and reactance, respectively (p.u.); ϕ is the angle of power factor (zero for DC systems); R and X are the DC section resistance and reactance, including the return way, in p.u./mi; L_2 is the distance between the TS and the nearest vehicle at the right; n and n' are the number of vehicles at the right and the left, respectively, of the TS.

Voltage drop in the farthest point is determined by the maximum length of the sector supplied. In normal conditions, this value is the length L (see Fig. 8). However, when a contingency occurs, the sector length must be modified to almost twice the original length. Then, for normal conditions, the voltage must satisfy:

$$V_T \le V_S(i) - Z(i, L/2) \times I_{TS}(i) \tag{22}$$

Where I_{TS} is the current delivered by the TS depending on the number of vehicles in sector i supplied in a determined time by the substation. Under a contingency of the TS, the voltage must satisfy the constraint for the sector *i-1* and sector *i+1* (adjacent sectors):

$$V_T(i \pm 1) \le V_S(i \pm 1) - Z(i \pm 1, 3L/2) \times I_{TS}(i \pm 1) \tag{23}$$

The minimum capacity of transformers and rectifiers is calculated from the maximum demanded current in each TS. The transformers and rectifiers size must be chosen as the nearest superior value to the demanded power, depending on the commercial capacities. As previously, normal conditions and post-contingency operation must be considered. In normal operation with 2 transformers, the power capacity of transformers must satisfy:

$$CapT(MW)(i) \ge I_{TS}(i, L) \times V_{DC} + P_{Loss}(L) \tag{24}$$

Meanwhile, when the traction substation i is unavailable, the capacity of active power of the 2 transformers in the *i-1* and *i+1* sector must satisfy:

$$CapT(MW)(i \pm 1) \ge I_{TS}(i \pm 1, 3L/2) \times V_{DC} + P_{Loss}(3L/2) \tag{25}$$

The capacity in MVA of the transformer is computed dividing the capacity in MW by the power factor (p.f.). As shown in (24) and (25), the power loss (P_{loss}) in the DC section feeder for the maximum demand must be determined for each section. The total power loss in DC section associated to the TS for a round trip is:

$$P_{LOSS} = \sum_{t=0}^{To} P_{loss}(t) = \sum_{t=0}^{To}\left[\sum_{j=1}^{n'} I_j(t)^2 \times \rho \times L_j(t) + \sum_{j=1}^{n} I_j(t)^2 \times \rho \times L_j(t)\right] \quad (26)$$

I_j is the current in each DC section that is defined as the catenaries/rail between two vehicles or between a vehicle and the TS (in the case of the nearest vehicle to the feeding point of the TS). The total losses at the DC section takes into account all vehicles placed at left and right of the TS. ρ is the resistivity of the DC section [Ω/km or Ω/mi]; T_o is the total annual operation time.

3.4 Application to the study case

The analysis was developed for the metro line showed in Fig. 6 corresponding to the study case of section 2.3. The study was developed as function of the number of substations and the three possible configurations explained in section 3.1.

The unitary cost of fault was assumed 1074 US$/kWh, from reliability analysis. Simulations were done for three levels of load: high (the maximum number of vehicles in service), medium (half of the total vehicles in service), and low (with no vehicles in service). The simulator allows the calculation of power losses in N-0 state, and the demand of each substation for N-0 and N-1 contingencies state.

Simple contingencies (N-1) at the maximum load were made in order to sizing the TS when configurations 1 and 3 are used, to give support of adjacent TS. While, normal state operation was used for sizing TS in configuration 2.

Table 3 presents the total cost computed as function of the number of TS and configuration of connection. Additionally, the investment cost (C_inv) and the net present value of the operation cost (NPV_Oper) is shown. The fault cost was of 155.000 USD$/year.

The investment cost (without the cost of catenaries/rail that is common for all alternatives), noted C_inv, includes the switchgear in SF6, rectifiers, transformers, having into account the number of each equipment depending on the configuration (see Fig. 8). The NPV_Oper includes the operation cost for a useful life of the project of 20 years.

In the second column, in brackets, the rating commercial capacity of each substation is shown, based on the results of simulations and the algorithm for finding catenaries/rail losses. The capacities of each substation for configurations 1 and 3 are the same, due to the high electrical similitude between both schemes.

#TS's	Configuration (rating/TS)	Maximum length of catenary/rail (km)	Millions of dollars		
			C_inv	NPV_Oper	Total Cost
3	1 (5MW)	8.8	6.11	0.81	7.95
	2 (4 MW)	4.4	9.75	1.63	12.4
	3 (5MW)	8.8	11.9	1.62	14.5
4	1 (5 MW)	6.6	6.73	0.90	8.65
	2 (3.75MW)	3.3	12.3	1.79	15.1
	3 (5 MW)	6.6	13.0	1.79	15.8
5	1 (3.75MW)	5.28	7.97	0.98	9.97
	2 (3.75MW)	2.64	14.3	1.97	17.3
	3 (3.75MW)	5.28	15.4	1.96	18.3

Table 3. Cost Comparison of Several Configuration of TS's Connections - Study Case

The third column shows the maximum length of the DC section that each TS can supply. TS in configurations 1 and 3 must have a capacity to supply even twice the total length of the line divided by the number of considered substations. Instead, TS in configuration 2 supply the maximum length of catenaries/rail, just the normal operation length because this configuration is not able of supporting of adjacent substation in case of fault.

The lowest total cost at Table 3 is presented in the case of three 5 MW TS because, in the study case of section 2.3, the investment cost weights more in the final cost than the operation cost. That is, looking just the configuration 1, it is evident that even though the operation costs do not grow up linearly as more TS are considered, the difference between investment costs is higher than operation costs, so the optimal solution is the location of 3 TS of 5 MW, under the configuration 1.

4. Power quality impact of urban railway systems on distribution systems

Power quality phenomena originated in power distribution systems impacts on the electrical power supply system of UMTS and, at the same time, power electronics used in the traction system impacts on the power quality (PQ) service of the distribution system.

In addition, the power demand of UMTS presents high and fast variations as consequence of the operation cycles of each train-vehicle and the non-coincidence of operational cycles among several vehicles. So, PQ phenomena are time variable (Singh et al., 2006).

4.1 PQ Phenomena and railways' electrical system components

Fig. 9 shows the existing relationships between the different PQ phenomena and the railways' electrical system components. As it is shown, the main electrical components in the railway system are: the train-vehicle as an electric load that involves a great use of power electronics, rectifier substations, the electric HV/MV substation, and the distribution network system (White, 2008).

On the other hand, the main PQ phenomena involved in the interaction between the railways' electrical systems and the power distribution system are: electromagnetic interference (EMI/RFI) at high frequency (HF); harmonics, flicker, and voltage regulation at low frequency (LF) (Sutherland et al., 2006). Also, PQ phenomena include sags at instantaneous regime, unbalance of the three-phase power system, and transients' phenomena (Lamedica et al., 2004).

Fig. 9 (Garcia & Rios, 2010) presents also where the cause of the phenomena is, what are the affected or perturbed systems, and where a solution of the problem can be implemented. For example, the electromagnetic transients occur in microseconds and they are caused by capacitor switching or lightning. Hence, they can be generated in the distribution network, MV side of the rectifier substation or in the train (represented by X in Fig. 9). The main problems are related to the rectifier substation or the train (represented by circle in Fig. 10) where the electronic sensitive equipment are susceptible to misuse or damage due to the transient overvoltage. An effective overvoltage transient protection could be located at the rectifier substation and, finally, at the train (represented by triangle Fig. 9).

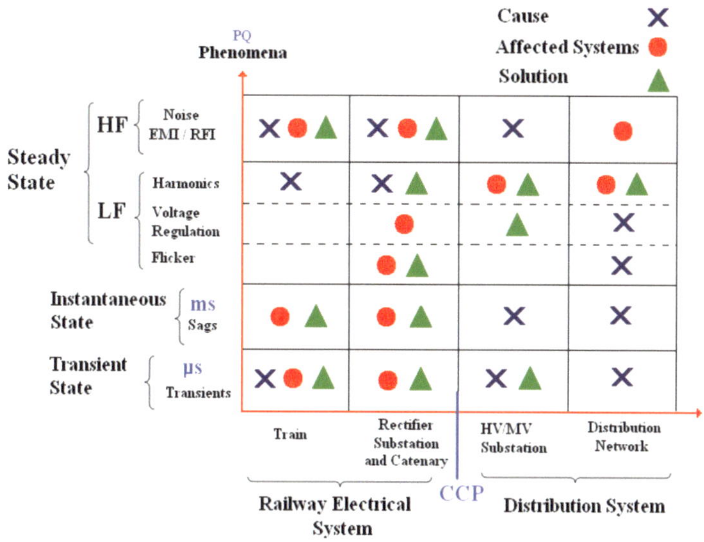

Fig. 9. PQ Phenomena and Railways' Electrical System Components Relationships

4.2 Harmonic distortion analysis

The identification of PQ problems in power systems represents an important issue to the distribution utilities. The harmonic distortion is one of the main PQ phenomena in the electrical system feeding an EMRTS because the injection of harmonics by its nonlinear loads flows through the network and affects other consumers connected to the distribution system. According to the conceptual diagram of Fig. 9, the production of harmonics in the EMRTS is a PQ phenomenon at steady state caused by the rectifier substations, normally, a controlled rectifier of 6 or 12 pluses.

In addition, the computation of the total harmonic distortion (THD) in the AC side of the rectifier substation at the railway system must take into account the time load variability at each TS. So, the instantaneous power load must be computed as function of time and distance as it was explained at section 2. Once the current consumption in each TS is obtained, it is possible to identify the variation of the THD during the time.

4.2.1 Probabilistic model

Generally, deterministic models have been adopted for network harmonic analysis; however, these models can fail for modelling the load variation in systems such as the railways' electrical system (Chang et al., 2009). So, a probabilistic analysis to characterize the harmonic current loads properly must be used in order to obtain an accurate model.

An EMRTS is characterized by fluctuating loads due to the different operation states of the trains in the traction system (See Fig. 7 b). Thus, the harmonics injection from the rectifier substations to the MV network causes that the current harmonic spectrum at the distribution system's connection point (PCC) varies over time. So, each traction substation can be represented as a harmonic current source that provides a probabilistic spectral content at the PCC (Rios et al., 2009).

Then, it is necessary to perform the vector sum of several harmonic sources (i.e. traction substations) at the distribution system's connection point to determine the total harmonic distortion. There are two methods to evaluate the effect of different non-linear loads: the analytical method and Monte Carlo simulation method. The complex implementation of analytical methods for large power systems studies involves little practical application in real systems. By contrast, Monte Carlo simulation has proved to be a practical technique (Casteren & Groeman, 2009) based on the low correlation between different harmonic loads (independence of the sources). Fig. 10 presents the methodology useful for probabilistic harmonic distortion analysis of railways' electrical systems with different harmonic sources.

The methodology for probabilistic analysis of harmonic starts from values obtained from deterministic simulations. Once the different conditions of loads are defined in the behaviour of the traction system, it is possible to use probability distribution plots to evaluate the harmonic level in the system during the travel time. So, the next step is to determine the probability density function to fit the harmonic components of each harmonic source and its phase angle.

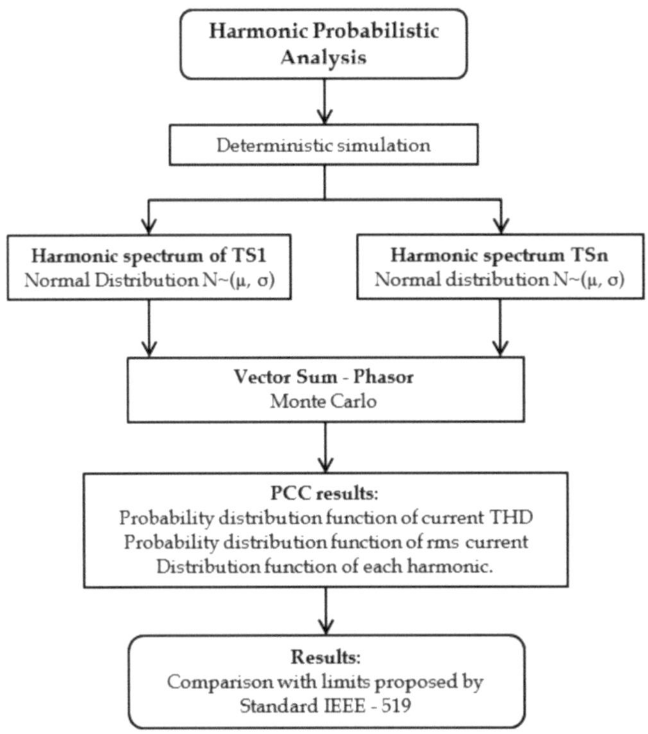

Fig. 10. Methodology for Harmonic Distortion Probabilistic Analysis of EMRTS

Many studies agree that the normal function is suitable as probability density function to use in the case of a random behaviour (Wang et al., 1994). In addition, according to the Std. IEEE - 519 (IEEE, 1993) the recommended window time to evaluate the harmonic distortion is 15 or 30 minutes. Therefore, it is recommended the selection of random time intervals of 15 minutes to make a probabilistic characterization of the THD distortion.

Then, a process called "Vector Sum - Phasor" is run through Monte Carlo simulations. Finally, the probabilistic characterization is obtained; where the probability distribution function of current THD, the probability distribution function of rms current and the probability distribution of each harmonic component are obtained.

Table 10.3 of Std. IEEE - 519 (IEEE, 1993) contains the current distortion limits in the voltage range of 120 V to 69 kV, which applies for typical railways' electrical systems connected to distribution systems at MV. So, based on this standard, a comparison between the current distortion levels at 95% and 50% of probability and the given limits must be realized to assess if the current distortion must be reduced or not. If a current THD distortion must be reduced, it could be used several filters methods. The next section presents the application of active power filtering to reduce THD distortion.

4.2.2 Active power filter allocation methodology

The harmonic distortion produced by railways' systems at the distribution system's connection point can be reduced using passive or active power filters (APF). However, due to the random and time variability of the harmonic distortion in traction systems, it is required an active power compensation with the ability of adaptation to different load conditions. Passive filters are designed with fixed parameters and for specific harmonics, so this type of filter does not have the required ability. By contrast, APFs based on the p-q theory became an effective solution in traction systems; normally, they are used for dynamic harmonic suppression (Xu & Chen, 2009). This type of compensation presents the advantage of eliminating a wide range of harmonics simultaneously.

On the other hand, the traction system has several rectifier substations and from the economic point of view it is difficult to install an APF in each TS due to its high cost. Then, it is necessary to allocate APFs in the most sensitive positions in the own power system of the EMRTS using the least number of filters and minimizing their size. An important factor to be considered in the decision of harmonic compensation in traction system is the sudden fluctuation of traction load because this dynamic behavior is also observed in the harmonic distortion, as it has been explained in the previous section.

The allocation methodology of APFs in distribution systems supplying a traction load is based on probabilistic data of harmonic distortion presented in all traction substations. According to the Std. IEEE - 519 (IEEE, 1993), using a 15 minutes time interval it is enough to understand the dynamic behavior of the traction load because in this interval there are 900 different data of the load behaviour in each TS. Fig. 11 shows the proposed methodology of allocation of APF in urban railways systems.

As, it was shown in section 3.4, for the study case of this Chapter, the metro line can be supplied by three TS at MV. The total harmonic distortion in the distribution system is analyzed with and without active compensation. The APF is allocated in the low voltage side of the transformer in the TS. As, the railway line has three TS, there are seven possible allocations of APF, as Table 4 shows at the first column.

Table 4 shows the THD distortion at levels of 50% and 95% of probability when the system is without active power compensation and when APFs compensation is used according to the seven different configurations. This table shows the effectiveness of the APFs to reduce the current THD distortion. Although the reduction is achieved with one filter, the amount of reduction is low because the two rectifier substations without active power filter present high variability and distortion. It is also observed that when an additional filter is used the amount of reduction in the THD is higher. Obviously, if three APF are used (one at each TS), the higher THD reduction is obtained.

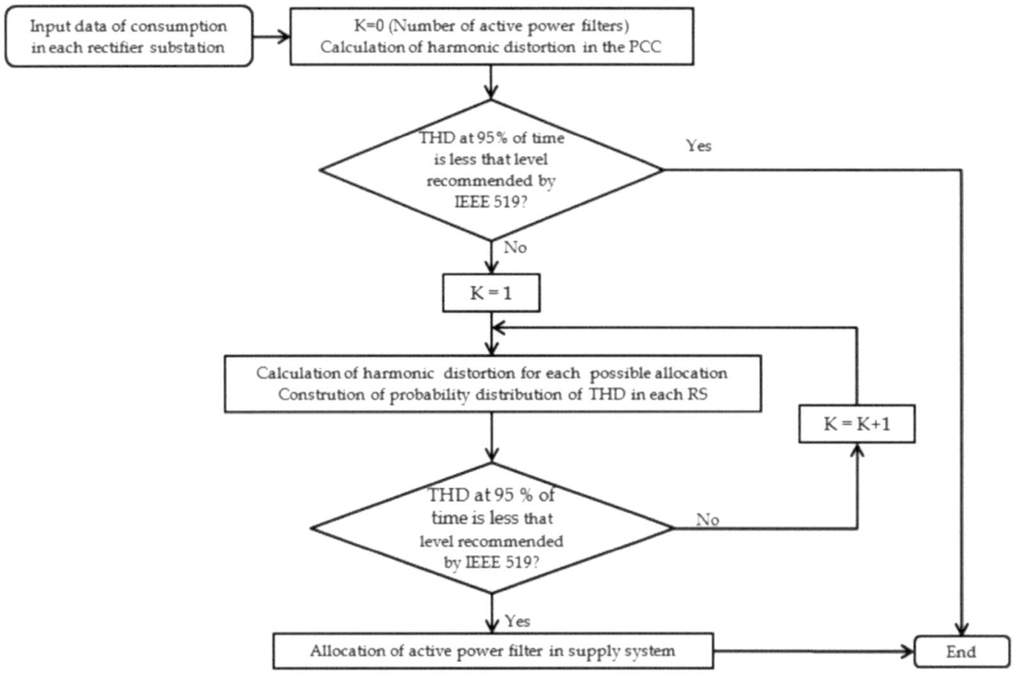

Fig. 11. Methodology for Allocation of Active Power Filters in Urban Railway Systems

The final decision about what configuration selects depends on the short circuit level of the system; for example, if the short circuit level is lower than 50 MVA a placement of one APF at each TS is required to satisfy Std. IEEE-519. By contrast, if the short circuit level is between 50 and 100 MVA, the best option is to place APF at TS1 and TS2.

Case	MEAN THD (%)		THD of 95% of time (%)	
	SUPPLY 1	SUPPLY 2	SUPPLY 1	SUPPLY 2
Without filter	22.96	22.98	23.66	23.66
APF in TS1	14.59	14.84	19.31	19.25
APF in TS2	13.81	13.82	18.94	18.94
APF in TS3	13.75	13.50	17.80	17.77
APF in TS1 and TS2	3.64	3.69	5.44	5.56
APF in TS1 and TS3	3.81	3.82	5.21	5.17
APF in TS2 and TS3	3.68	3.64	5.18	5.06
APF in all TS	3.01	3.00	3.60	3.59

Table 4. Total Current Harmonic Distortion – Active Power Filter Allocation

5. Grounding in DC urban railway systems

A primary requirement to ensure the appropriate operation of any electrical system is to guarantee personnel and system safety, either under normal and fault conditions. So, grounding is the most important component to control electrical system failures.

Grounding in electric traction systems requires a different treatment than in typical AC electrical systems, because of the existence of traction substations AC/DC of high capacity, the high variable load characteristic in time and distance, the direct contact of the rails with the earth, the current flow through the ground during normal operating conditions that can cause corrosion of underground metallic elements, the appearance of step and touch voltage that can jeopardize the integrity of persons.

The grounding system is composed by two subsystems. The first one (subsystem 1) assures the personnel safety and the protective device operation; while, the second one (subsystem 2) is used to ground the negative pole in the DC side of the railway's traction substation.

The grounding subsystem 1 is used to ground all metallic structures: boxes, protective panels, pipeline, bridges, passenger platforms, etc. There are two ways to connect this subsystem:

- High Resistance Grounding Method (HRGM): A constant voltage of 25 Vdc is applied between the TS's housing and the ground, in order to energize a relay to send the opening order to the protection equipment. When the voltage level decreases, other relay is set to send the opening order to the protection if a big current flows through the module. This path is supplied with a resistance of 500 Ω.
- Low Resistance Grounding Method (LRGM): A constant voltage of 1 Vdc is applied between the TS's housing and the ground. In this case no resistance is used, but a direct connection is made to the ground system. In addition, when the relays and protections detect the voltage's absence, they will send the opening order to the protection system.

So, Table 5 presents a comparison of the performance of these two methods.

Technical Features Description	HRGM	LRGM
Monitor constant voltage	25 Vdc	1 Vdc
Relay circuit resistance	High (200-700 Ω)	Low (< 1 Ω)
Current fault-ground structure	Low (1-2 A)	High (70-1500 A)

Table 5. Comparison of HRMG and LRMG performance.

The second subsystem is used to ground the negative conductor of the TS (Paul, 2002) (Lee & Wang, 2001) which corresponds physically to running rails. In DC traction systems, the rails are used as return conductor current, which could cause corrosion problems in underground metallic structures. There are three options to connect this subsystem:

- Solid-grounded system: This system keeps under control touch voltage but it permits the corrosion of the elements grounded to the earth.
- Ungrounded system (Floating): This system keeps under control stray currents but it permits high touch and step voltages.

- Diode-grounded system: Its purpose is to maintain the system without grounding while operating conditions are normal. But in the case of a failure, it quickly makes a change that provides a physical connection between the negative pole and grounding. When it returns to normal conditions, connection with grounding is suppressed. The diode is able to perform this function, as it is complemented by a security relay. The disadvantage is that under normal operating conditions small voltage differences may occur between the negative pole and grounding, forcing the diode to enter in function mode which increases stray currents and their associated effects.

Table 6 compares the main characteristics of the three options of railway's grounding system.

Grounding method	Riel to ground voltage (Vehicle touch voltage)	Stray current level
Solid-grounded system	Low	High
Ungrounded system (Floating)	High	Low
Diode-grounded system	Middle-Low	Middle-High

Table 6. Comparison of the Railway's Grounding Systems

5.1 Generalized grounding model in DC to EMRTS

Fig. 12 illustrates a grounded scheme for a railway line, in which for general purposes there are k trains, m substations and a total rail length *l*.

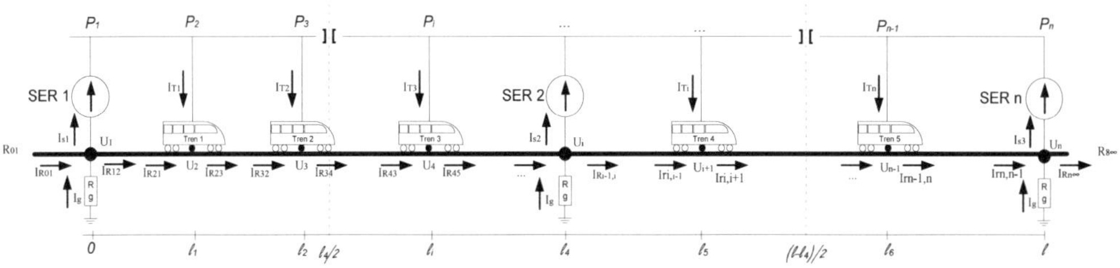

Fig. 12. Grounded System - General Scheme

The behavior of the current and voltage on the rail for each section between points P_i and P_{i+1} for i = 1, 2, 3..., n, is modeled by:

$$I_{Ri,i+1}(x) = c_{(2xi-1)}e^{\gamma x} + c_{(2xi)}e^{-\gamma x} \tag{27}$$

$$U_i(x) = -R_0(c_{(2xi-1)}e^{\gamma x} - c_{(2xi)}e^{-\gamma x}) \tag{28}$$

For $0 \le x \le l$, where *n* is equal to the number of trains running (*k*) plus the number of TS that are in operation (m), n = k + m. $U_i(x)$ is the rail to ground potential [V], and $I_{Ri,i+1}(x)$ is the stray current in the rail conductor [A].

The constant values $c_{(2xi-1)}$ and $c_{(2xi)}$ can be determined from the solution of a linear system of 2x(n-1) equations with 2x(n-1) unknowns obtained from the boundary conditions of each point P_i applying Kirchhoff's laws and assuming that the magnitudes of the currents delivered or absorbed by the trains and the substations are known as well as the location of each of the trains at the moment that these currents are delivered. Different scenarios can arise during the operation, which can be described as: railway starting point (P_1); railway ending point (P_n); point where a train is passing (P_2, P_3, P_4, ... , P_{n-1}); and point where a traction substation is located, for example (P_i).

5.2 DC grounding algorithm model in time

The model uses information on the train location and current consumption or delivered by the traction substations, for all time t. The power demand simulator (section 2) gives the power consumed and delivered by each train and TS, as well as the location of each train along the rail for each time instant. Fig. 13 shows the flowchart of the algorithm.

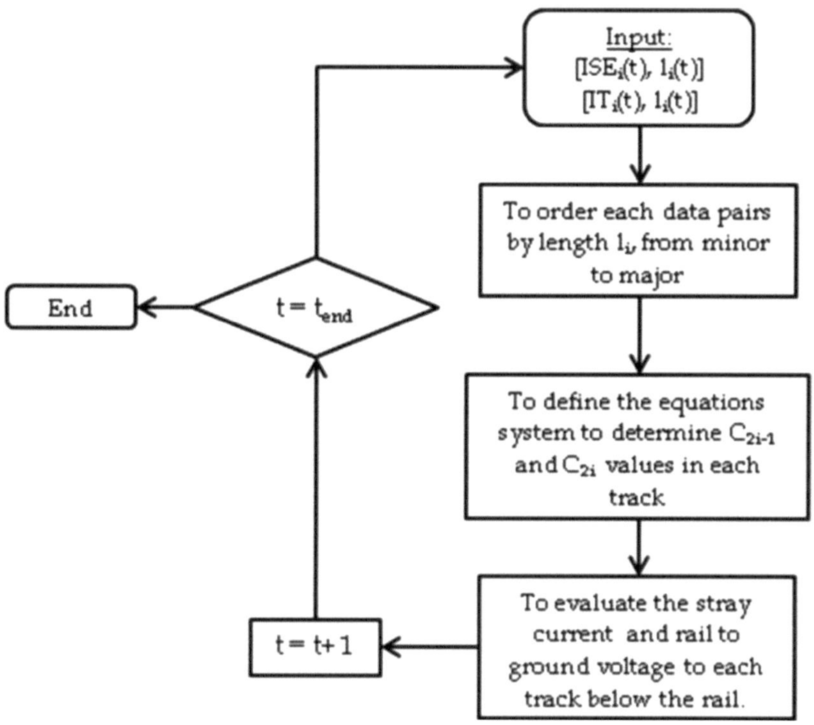

Fig. 13. DC Grounding Algorithm Model in Time

This algorithm has the following characteristics:

- The input data consist of arrays of pairs with the current supplied or absorbed by each TS or train and the respective train locations. This information is supplied by the model presented in section 2.
- As the trains are in constant motion the input for each instant of time is ordered from minor to major, in accordance to their location to the starting point of the track, in order to determine the track to be evaluated.
- After defining the tracks and points (P_i) on the total rail length, values are determined for each constant $c_{(2xi-1)}$ and $c_{(2xi)}$ respectively.

- Finally, using the constants obtained in the previous step for the instant of time the stray current and the voltage rail to ground are evaluated and the information obtained is saved. This process is repeated from the second step until all points in time.

5.3 Example

Let us consider a simplified study case similar to the system of section 2.3 (Fig. 6) with three TS located at 0, 2000 and 4000 meters and four trains moving along the 4 kilometers of rail. Constant system parameters are: R=0.04Ω/km, G=0.1S/km, R_g=0.01Ω/km and $R_{01}=R_{8\infty}=R_0$.

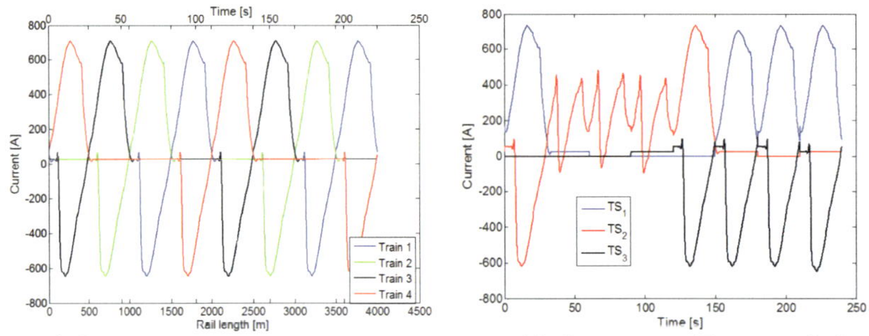

a) Currents at Trains b) Currents at Traction Substation

Fig. 14. Example of Simulation of Grounding

Fig. 14a shows the current magnitude of each train and its location on the rail for each moment. Likewise, Fig. 14b shows the current magnitude in each traction substations for each time instant.

Fig. 15 shows the voltage profile along of the rail length at different points in time obtained from the simulation for the case of diode-grounded system. With this system, it is possible to reduce the voltage difference presented in the ungrounded system as the solid-grounded system behaviour. The simulations results show that the diode-grounded system guarantees greater security because it control the step and touch voltage and reduces the stray currents that cause the deterioration of the physical installation.

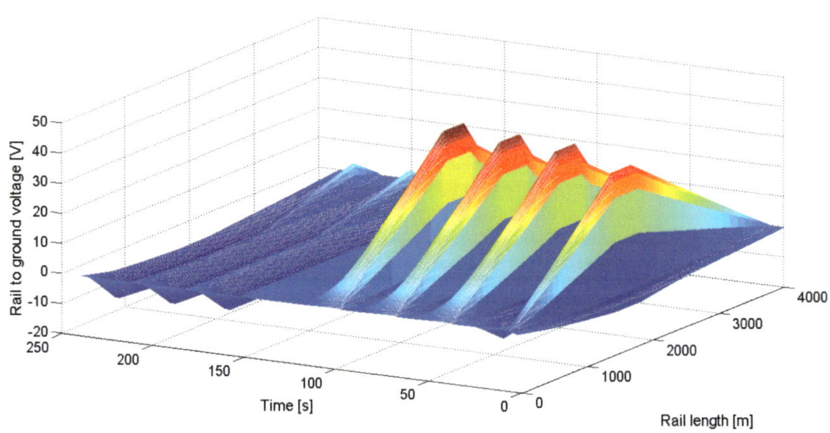

Fig. 15. Rail to Ground Time Voltage Profile

6. Conclusion

This chapter has presented useful tools for power systems modelling, analysis and system design of Electric Massive Railway Transportation Systems (EMRTS) and power supply from Distribution Companies (DisCo) or Electric Power Utilities. Firstly, a section depicted to present the modelling and simulation of the power demand was developed. Then, a section about the computation of the placement and sizing of traction substations for urban railway systems was presented where the modelling is based on the power demand model of the previously mentioned.

After that, two sections about the power quality impact of EMRTS on distribution systems and grounding design are presented. Both subjects make use of the load demand model presented at section 2.

These tools allow the optimization of the design scheme of railway electrification for UMTS, taking into account an adequate sizing and number of traction substations, and the number and location of harmonic filters to improve the power quality of the system.

7. Acknowledgment

The authors want to thanks to Ana María Ospina, Camilo Andrés Ordoñez, and Elkín Cantor for the support given in the preparation of the material for this Chapter.

8. References

Arriagada, A.; & Rudnick, H. (1994). *Reliability Evaluation in Electric Distribution Systems* (in spanish). Escuela de Ingeniería, Pontificia Universidad Católica de Chile.

Buhrkall, L. (2006). Traction System Case Study, In: *Electric Traction Systems*, IET (Ed.), pp. 53-71, ISBN 978-0-86341-9485, London, UK.

Casteren, J.V.; Groeman, F. (2006). Harmonic analysis of rail transportation systems with probabilistic techniques, *9th International Conference on Probabilistic Methods Applied to Power Systems*, ISBN 978-91-7178-585-5, Stockholm, Sweden, June 11-15, 2006.

Chang, G.W.; Hung-Lu, W.; Gen-Sheng, C.; Shou-Yung, C. (2009). Passive harmonic filter planning in a power system with considering probabilistic constraints, *IEEE Transactions on Power Delivery*, Vol. 24, No.1, (Jan. 2009), pp. 208-218, ISSN 0885-8977.

Chen, C.S.; Chuang, H.J.; & Chen, J.L. (1999). Analysis of dynamic load behavior for electrified mass rapid transit systems, *34th IEEE Industry Applications Conference*, Vol. 2, pp. 992-998, ISBN: 0-7803-5589-X, Phoenix, Arizona, USA, October 3-7, 1999.

Garcia, J.G.; Ríos, M.A.; Ramos, G. (2009). A power demand simulator of electric transportation systems for distribution utilities, *44th International Universities Power Engineering Conference UPEC 2009*, ISBN 978-1-4244-6823-2, Glasgow, Scotland, September 1-4, 2009.

Garcia, J.G.; Rios, M.A. (2010). PQ analysis in tramway systems, *2010 IEEE ANDESCON*, ISBN 978-1-4244-6740-2, Bogotá, Colombia, 15-17 Sept., 2010.

Hill, R.J. (2006). DC and AC Traction Motors, In: *Electric Traction Systems*, IET (Ed.), 33-52, ISBN 978-0-86341-9485, London, UK.

Hsiang, P.; & Chen, S. (2001). Electric Load Estimation Techniques for High-Speed Railway (HSR) Traction Power Systems, *IEEE Transactions on Vehicular Technology*, Vol. 50, No. 5, (September 2001), pp. 1260-1266, ISSN 0018-9545.

IEEE (1992). IEEE Loss Evaluation Guide for Power Transformers and Reactors, IEEE (ed.), ISBN 1-55937-245-1, New York, USA.
IEEE (1993). IEEE recommended practices and requirements for harmonic control in electrical power systems Standard 519 - 1992, IEEE (ed.), ISBN 1-55937-239-7, New York, USA.
IEEE (2007). IEEE standard general requirements for liquid-immersed distribution, power, and regulating transformers C57.12.00-2006, IEEE(ed.), ISBN 0-7381-5251-X, New York, USA.
Jong, JC..; & Chang, S. (2005a). Models for Estimating Energy Consumption of Electric Trains, *Journal of the Eastern Asia Society for Transportation Studies*, Vol. 6, (2005), pp. 278 - 291, ISSN 1881-1124.
Jong, J.C.; & Chang, S. (2005b). Algorithms for Generating Train Speed Profiles, *Journal of the Eastern Asia Society for Transportation Studies*, Vol. 6, (2005), pp. 356 - 371, ISSN 1881-1124.
Lamedica, R.; Maranzano, G.; Marzinotto, M.; & Prudenzi, A. (2004). Power quality disturbance in power supply system of the subway of Rome, *IEEE Power Engineering Society General Meeting*, pp. 924 - 929, ISBN 0-7803-8465-2, Denver, USA, 6-10 June, 2004.
Lee, C.H.; Wang, H.M. (2001). Effects of Grounding Schemes on Rail Potential and Stray Currents in Taipei Rail Transit Systems, *IEE Proceedings on Electric Power Applications*, Vol. 148, No. 2, (Mar. 2001), pp. 148–154, ISSN 1350-2352.
Paul, D. (2002). DC traction power system grounding, *IEEE Transactions on Industry Applications*, Vol. 38, No.3, (May/Jun 2002), pp. 818 - 824, ISSN 0093-9994.
Perrin, J.P.; & Vernard, C. (1991). *Urban Electric Transportations* (in French), D5-554 Techniques de l'Ingénieur, France.
Rios, M.A.; Ramos, G.; Moreno, R. (2009). Evaluación de Calidad de la Potencia en la Interacción del Sistema de Distribución y los Sistemas Eléctricos Ferroviarios Urbanos, *8th Latin-American Congress on Electricity Generation and Transmission CLAGTEE*, ISBN 978-85-61065-01-0, Ubatuba, Brazil, October 18-22, 2009.
Singh, B.; Bhuvaneswari, G.; & Garg, V. (2006). Improved power quality AC-DC converter for electric multiple units in electric traction, *2006 IEEE Power India Conference*, ISBN 0-7803-9525-5, New Delhi, India, 2006.
Sutherland, P.; Waclawiak, M.; & McGranaghan, M. (2006). Harmonic Impacts Evaluation for Single-Phase Traction Load, *International Journal on Energy Technology and Policy*, Vol. 4, No. 1, (2006), pp. 37-59, ISSN 1472-8923.
Vukan, R. (2007). *Urban Transit Systems and Technology*, John Wiley & Sons, Inc., ISBN 978-0471758235, NJ, USA.
Xu, X.; Chen, B. (2009). Research on Power Quality Control for Railway Traction Power Supply System, *Pacific-Asia Conference on Circuits, Communications and Systems*, ISBN 978-0-7695-3614-9, Chengdu, China, May 16-17, 2009.
Wang, Y.J.; Pierrat, L.; Wang, L. (1994), Summation of harmonic currents produced by AC/DC static power converters with randomly fluctuations loads, *IEEE Transactions on Power Delivery*, Vol. 9, No. 2, (April, 1994), pp. 1129 - 1135, ISSN 0885-8977.
White, R.D. (2008). AC/DC railway electrification and protection, In: *2008 IET professional development course on Electric Traction Systems*, IET (ed.), pp. 258 - 305, ISBN 978-0-86341-948-5, UK.
White, R.D. (2009). DC electrification supply system design, In: *4th IET professional development course on Railway Electrification Infrastructure and Systems*, IET (ed.), 44-69, ISBN 978-1-84919-133-3, UK.

8

Optimized Model Updating of a Railway Bridge for Increased Accuracy in Moving Load Simulations

Johan Wiberg, Raid Karoumi and Costin Pacoste
KTH Royal Institute of Technology
Sweden

1. Introduction

The moving load problem has been studied intensively since the first research by Willis in 1849 (Willis, 1850). Today's railway bridges are analyzed in more detail for moving loads due to increased speeds, axle loads and more slender bridge designs. Such analyzes are very time consuming as it involves many simulations using different train configurations passing at different speeds. Thus, simplified bridge and train models are chosen for time efficient simulations. However, these FE models are often called into question when they are in conflict with in-situ bridge measurements. Model updating has therefore been a rapidly developing technology and has gained a lot of interest in recent years. It is the popular name for using measured structural data to correct the errors in FE models. Clearly, the approach of numerical predictions to the behavior of a physical system is limited by the assumptions used in the development of the mathematical model (Friswell & Mottershead, 1995). Model updating, at its most ambitious, is about correcting invalid assumptions by processing test results.

Mottershead & Friswell (1993) provided a state of the art and addressed the problem of updating a numerical model by using data acquired from a physical vibration test (Friswell & Mottershead, 1995). Optimization has been used by many others since then, improving FE model predictions based on real measurements. This chapter highlights the importance and the potential of such optimization procedures for increased accuracy in moving load simulations. A large-scale simplified railway bridge FE model is used and the updating process is based on previously identified updating parameters in Wiberg et al. (2009). Natural frequency, static strain, static deflection and acceleration residuals are used, separately and combined, to optimize the values of modulus of elasticity, mass density and modal damping ratio. The updated FE model is finally used to identify and analyze the most critical moving load configuration in CEN (2002) concerning bending moment, vertical bridge deck deflection and acceleration.

The optimization algorithm was easily implemented for FE model updating and was shown to operate efficiently in a benchmark test and for the specific bridge. The optimization algorithm converges against reasonable values of the updating parameters. A previously questioned high-valued equivalent modulus of elasticity, found for a manually tuned FE model in Wiberg (2009), was proven to be reliable. Further, the difference in load effect between an initial manually tuned FE model and the optimized FE model is found most significant for vertical deflection. However, more measured dynamic characteristics (natural frequencies, mode shapes and modal damping ratios), together with complementing updating parameters and a more detailed FE model are considered necessary for dynamic load effect predictions with highest accuracy.

Finally, it should be given attention that the adopted methodology can not only be used for model updating based on measurements, but also introduced in the early design phase. The reasonable range of a typical modeling factor or parameter is then based on the drilled engineer's qualified guess and the risk of for example a resonance problem can be investigated by, e.g. letting the maximum allowed code limit for vertical bridge deck acceleration be "measured" response. Performing the optimization will then result in a model configuration, needed to fulfill the requirements in the code.

2. FE model optimization

2.1 General

The objective of FE model updating is to improve an FE model in order to reproduce the measured response of a structure. Model updating brings together the skills of the numerical analyst and the load test engineer, and requires the application of modern estimation techniques to produce the desired improvement (Friswell & Mottershead, 1995). Basically, an understanding of the updated model is necessary. The updated model may only reproduce physical test data but could lack physical meaning. It is therefore required to accurately know the application area of the updated model. Typically, the physical meaning of the model must be improved if the updated model is to assess the effect of changes in construction.

Optimization techniques are used to find a set of design parameters, $\mathbf{p} = \{p_1, p_2, \ldots, p_n\}$, that can somehow be defined as optimal. FE modeling procedures involve an optimization with respect to an objective function, i.e. finding an optimal model that behaves similarly to the real structure and represents the physical characteristics of it (Zárate & Caicedo, 2008). Thus, residuals of the response, as a nonlinear function of the input parameters, are established and accounted for in the objective function. Different types of objective functions are found in the literature and by their minimization an FE model may be optimally updated.

The optimization process is rather straightforward. More complex is the choice of updating parameters, i.e. those exerting an influence on the bridge model in question. It is reasonable to believe that an accurate representation of a structure depends on the type of FE model used to represent the structural members and the properties assigned to these elements. Therefore, relatively large differences can exist between the behavior of a FE model before updating and the real structure.

Considering the minimization problem as unconstrained nonlinear, i.e. finding a vector $\mathbf{p}$ that is local minimum to a scalar function $\Pi(\mathbf{p})$:

$$\min_{\mathbf{p}} \Pi(\mathbf{p}) \tag{1}$$

with no restriction placed on the range of $\mathbf{p}$, the Nelder-Mead simplex algorithm as described in Lagarias et al. (1998) can be used for optimization. The algorithm is capable of escaping local minima in some cases and can even handle discontinuities (Coleman & Zhang, 2009). Unlike gradient based optimization routines, facing ill-conditioning for the Jacobian and

Hessian matrices, the Nelder-Mead simplex algorithm is less prone to numerical difficulties at iteration steps. Also for noisy measurements the Nelder-Mead simplex algorithm has been proven effective, see e.g. the updating results of a simple beam in Jonsson & Johnson (2007) or the more extensive study of Schlune et al. (2009) to improve the FE model of the new Svinesund Bridge between Sweden and Norway. Further, the optimization algorithm is general, problem independent and can be implemented easily for FE model updating.

2.2 The objective function

The objective function is the crucial heart of FE model updating. It represents the magnitude of the error of the response vector, $\mathbf{z}$, defined as the difference between the observed responses and the expected responses $E\{\mathbf{z}\}$ (Friswell & Mottershead, 1995):

$$\Pi = E\left\{(\mathbf{z} - E\{\mathbf{z}\})^T (\mathbf{z} - E\{\mathbf{z}\})\right\} \tag{2}$$

Typically, the response residual vector is weighted to reflect the confidence in different measurements:

$$^{\mathbf{z}}\Pi = \left(\mathbf{z}_m - \mathbf{z}_j\right)^T \mathbf{W}_\mathbf{z} \left(\mathbf{z}_m - \mathbf{z}_j\right) \tag{3}$$

where $\mathbf{z}_m$ is the measured response vector, $\mathbf{z}_j$ is the FE response vector at iteration j and the response weighting matrix, $\mathbf{W}_\mathbf{z}$, is a diagonal matrix with corresponding reciprocals as diagonal elements depending on the type of objective function. Notations are also to be found in Appendix 5.1.

The selection of the objective function has a profound impact on the problem (Jaishi & Ren, 2005). A classical least squares approach fails to acknowledge that the observations are not recorded with equal confidence (Friswell & Mottershead, 1995). In reality, different error sources will also reduce the ability of the FE model to reproduce the experimental measurements. This can be systematic errors, experimental noise and modeling limitations. In a weighted least squares approach each squared measurement residual is therefore multiplied by a weight, w_i, and the sum of weighted squares of the residuals is calculated. When the weights are given by the inverse observation variances,

$$\mathbf{W} = \mathrm{diag}\left(\frac{1}{\sigma_1{}^2}, \frac{1}{\sigma_2{}^2}, \ldots, \frac{1}{\sigma_i{}^2}, \ldots, \frac{1}{\sigma_{N_z}{}^2}\right) \tag{4}$$

the minimization problem, $\min_{\mathbf{p}} {}_\sigma\Pi$, has the objective function:

$$^{\mathbf{z}}_{\sigma}\Pi = \sum_{i=1}^{N_z} \sqrt{\frac{(z_{mi} - z_i)^2}{\sigma_{z_i}{}^2}} = \sum_{i=1}^{N_z} \frac{|z_{mi} - z_i|}{\sigma_{z_i}} \tag{5}$$

This is the objective function used by Jonsson & Johnson (2007) and Schlune et al. (2009). To keep the least squares form of the objective function, the square root is omitted and the objective function reformulated as:

$$^{\mathbf{z}}_{\sigma}\Pi = \sum_{i=1}^{N_z} \frac{(z_{mi} - z_i)^2}{\sigma_{z_i}{}^2} = \left(\mathbf{z}_m - \mathbf{z}\left(\mathbf{p}_j\right)\right)^T \mathbf{W}_\mathbf{z} \left(\mathbf{z}_m - \mathbf{z}\left(\mathbf{p}_j\right)\right) \tag{6}$$

which corresponds to Eq. 3 with possibility to take the significance of different measurements into account and with dimensionless terms as a result. The normalized updating parameter vector is defined as

$$\mathbf{p}_j = \left(\frac{p_{1,j}}{p_{1,0}}, \frac{p_{2,j}}{p_{2,0}}, \ldots, \frac{p_{n,j}}{p_{n,0}} \right) \tag{7}$$

2.3 Optimization procedure

FE model updating becomes very efficient, neat and easy to implement by coupling the FE analysis software in question to a mathematical analysis software such as the Matlab® package. This also most conveniently facilitate the use of the Matlab® incorporated optimization toolbox and the updating process is therefore fully controlled from within Matlab®. In order to use a typical optimization solver, a function handle of the objective function together with an initial normalized updating parameter vector are sent to the optimization subroutine. The updated FE model code is then automatically generated by the optimization algorithm as it iterates. The Matlab® syntax for a general and problem independent optimization procedure is then:

```
% Response function as function handle @:
z=@(p)FEA(p);

% Objective function as function handle @:
obj=@(p)obj_func(zm,z,sigma);

% Initial normalized updating parameter vector:
p0=[1 1];

% Nelder-Mead simplex optimization based on functions Z and OBJ:
[p,objval,exitflag,output]=fminsearch(obj,p0,options);
```

In this case `FEA(p)` includes the appropriate code for initiation of the finite element analysis and calculation of the response vector as a function of the updating parameter vector `p` from the optimization solver `fminsearch` in the optimization toolbox of Matlab®. In this case the Solvia® FE system software was adopted.

The optimization algorithm starts at the point `p0` and attempts to find a local minimum `p` of the function described in `obj`, with measured response `zm`, standard deviations in responses `sigma` and optimization options specified in `options`. The algorithm returns in `objval` the value of the objective function `obj` at the solution `p`, in `exitflag` the exit condition of `fminsearch` and in `output` the user specified information about the optimization are found.

3. Benchmark test

A benchmark test was performed to verify the updating procedure implemented in Matlab®. The physical problem consisted of a 2D dynamic analysis of a moving vehicle across a ballasted railway bridge with vehicle-bridge interaction due to contact definitions, see Fig. 1. The I-beam steel bridge had two spans, assumed to be linearly elastic, and the vehicle speed was 30 m/s. The bridge surface and the neighboring rigid surface portions are assumed to initially form a horizontal straight line. Each span was modeled to consist of 20 beam elements and a mass-spring-damper system was used to model the vehicle. The mass density

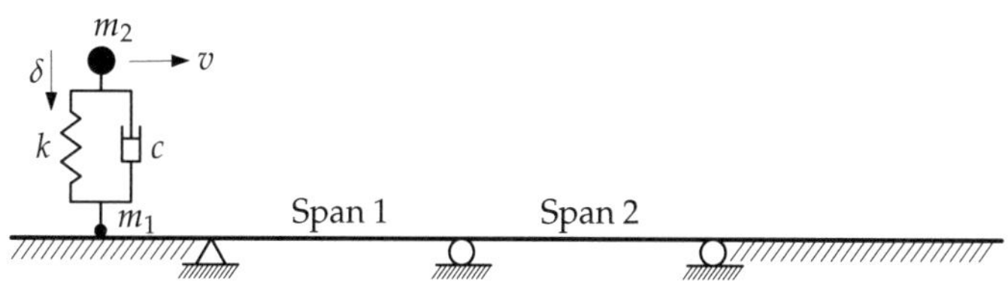

Fig. 1. The physical benchmark problem.

of the beam was increased to include the mass of the ballast. Direct time integration with the Hilber method was adopted. All necessary input data to the Solvia® FE system is given in Appendix 7.

To symbolize "measured" result, material properties of 210 GPa and 16000 kg/m^3 for modulus of elasticity and mass density, respectively, were used and the corresponding "measured" maximum deflection in each span was calculated to 2.152 mm and 2.165 mm. In addition to these "measured" responses, the predicted finite element analysis responses, as a function of the iteratively updated material properties, constituted the objective function expression. For simplicity, the Euclidean norm of the normalized response residual was chosen as objective function:

$$ {}^{\mathbf{z}}\Pi = \sqrt{\sum_{i=1}^{N_z} \frac{(z_{mi} - z_i)^2}{z_{mi}^{\;2}}} = \left\| \frac{\mathbf{z}_m - \mathbf{z}(\mathbf{p}_j)}{\mathbf{z}_m} \right\| \tag{8} $$

Using the modulus of elasticity and mass density as updating parameters, with initial values of 175 GPa and 20000 kg/m^3, i.e. corresponding to initial deflections of 2.658 mm and 2.607 mm, the `fminsearch` solver in Matlab® converged towards 210 GPa and 16000 kg/m^3 at the deflections 2.152 mm and 2.165 mm in 30 iterations and 62 objective function counts. Fig. 2 illustrates the iteration sequence, starting at the normalized input parameter coordinates $(\frac{175}{175}, \frac{20000}{20000})$ and ending at $(\frac{210}{175}, \frac{16000}{20000})$. Interestingly, the algorithm first seemed to localize a local minimum but proceeded to the global minimum.

4. Case study

The New Årsta Bridge in Stockholm, Sweden, was adopted for FE model updating (see Fig. 3). Previous research pointed out some of the difficulties in studying bridge dynamics resulting from moving traffic (Wiberg et al., 2009). Not only does the dynamic amplification depends on the considered load effect, but different modeling parameters, individually or jointly, influence the dynamic load effect or dynamic property in question. The use of statistically identified updating parameters as a step in more effective model optimization is highlighted in previous study by the author and typical results from a statistical parameter study on this specific bridge are exemplified in Fig. 4 and found in Wiberg et al. (2009) where the factorial experimentation technique was used. The type of information encountered in Fig. 4 is considered extremely important and valuable. Thus, the statistical method of factorial experimentation, in contrast to ordinary parameter sensitivity analyzes where parameters are varied one at a time, captures the synergy effects. Consequently, a modeling parameter can be significant even though it individually is found insignificant and an optimal amount of updating parameters to include in the optimization can therefore be identified. This leads to shorter solution times as the optimization algorithm itself is iterative and becomes

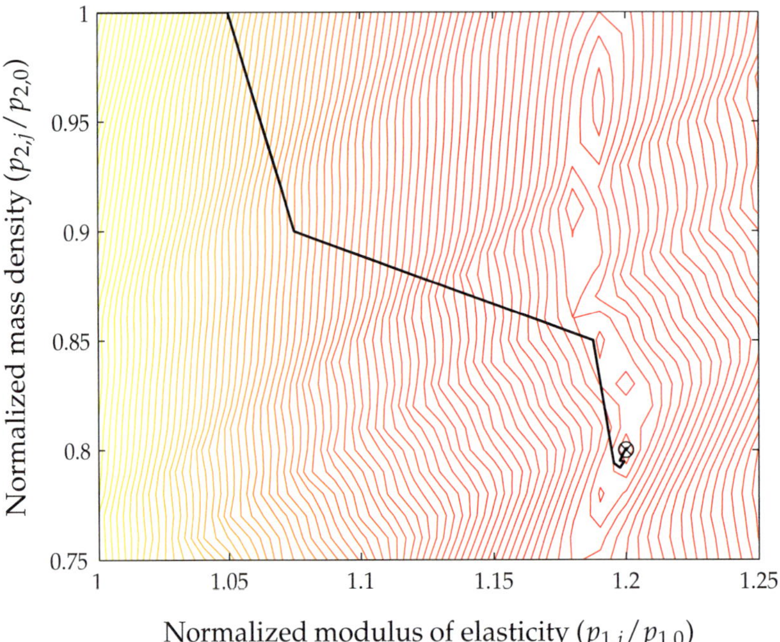

Fig. 2. Sequence of updating parameter points in the normalized updating parameter space. The contours represent the magnitude of the response objective function.

very time-consuming for large dynamic simulations with inappropriately many updating parameters causing unnecessarily many iterations.

A large-scale simplified bridge FE model in the Solvia® FE system was verified as reliable for global analysis and manually tuned concerning an equivalent modulus of elasticity and mass density by using operational modal analysis and static load tests (Wiberg, 2006; 2007; 2009; Wiberg & Karoumi, 2009). This 3D modified Bernoulli-Euler beam model was therefore used as a basis for the present study.

4.1 The bridge

The eleven span New Årsta Bridge of approximately 815 m has main spans of 78 m. Elevation and plan view with the monitoring sections is presented in Fig. 5. The cross section of the bridge is complex with a parabolic height variation. To make the slender design possible, the sections were extensively reinforced and prestressed. To use a simplified inclusion of tendons in the model, they were concentrated to the center of gravity along the bridge and not distributed within the cross section. Further, the UIC 60 rails of the double track bridge were modeled with rectangular beam elements, giving cross sectional properties corresponding to the actual rail cross section. The element length was at most 0.5 m (both for bridge and rail elements) and each rail node was connected to the corresponding bridge node with a rigid link. The FE model of the bridge consisted of linear, elastic and isotropic materials. Support conditions were assumed according to bridge design documents, but also verified as reliable in previous work (Wiberg, 2009). Fig. 6 represents the boundary conditions, where the legend *F* indicates that the bridge deck and pier are fixed in translation movement. The main girder

Fig. 3. The spectacular New Årsta Bridge in Stockholm.

was released for longitudinal movements at other supports. The totally 24 Swiss mageba pot bearings had the function of hinges for free rotation about the transverse bridge deck axis. Torsional rotation of the bridge over piers was constrained to follow the bending of the oval piers in their stiff direction. Bridge deck rotation about the vertical axis over each pier was prevented due to the support consisting of two bearings in the transverse direction. The lower basis of all piers were assumed to be clamped but in reality the foundation blocks of P8 and P9 rested on concrete filled steel piles. A thorough description of the bridge with all sensor locations is found in Wiberg (2006).

4.2 The loadings

4.2.1 General

In this study the FE model was updated using tests with Swedish Rc6 locomotives. The updated model was then used to study the effect of passing high speed trains (HSLM) as specified in design codes.

4.2.2 The Rc6 locomotive

The updating process considered a field test with two Swedish Rc6 locomotives positioned at different locations in a static load test according to Wiberg (2009) and, in a dynamic load test, one locomotive crossing the bridge at different speeds. The locomotive is visualized in Fig. 7(a). Each of the four axles was represented as a point load of 19.5 tons and a

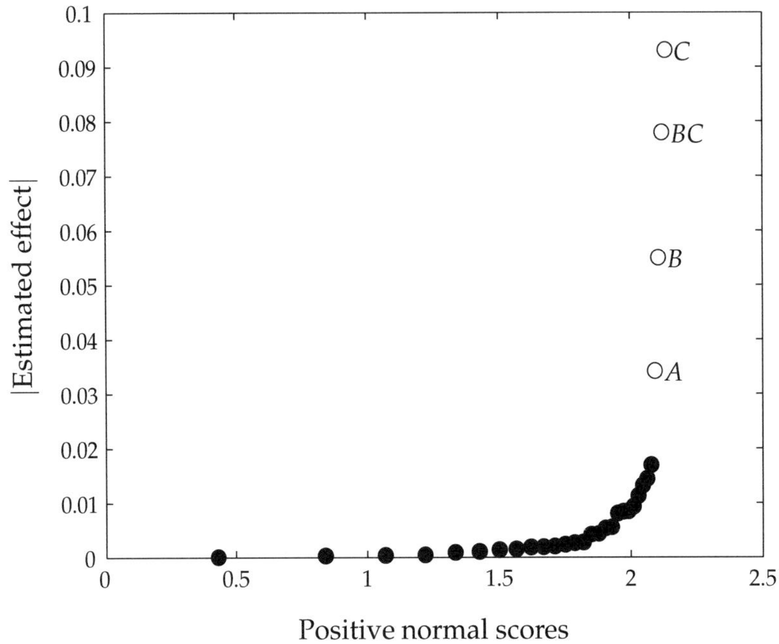

Fig. 4. Half normal plot with absolute values of estimated FE modeling parameter effect on vertical bridge deck acceleration. Factor definition: (*A*) damping ratio, (*B*) tendons and (C) vehicle speed.

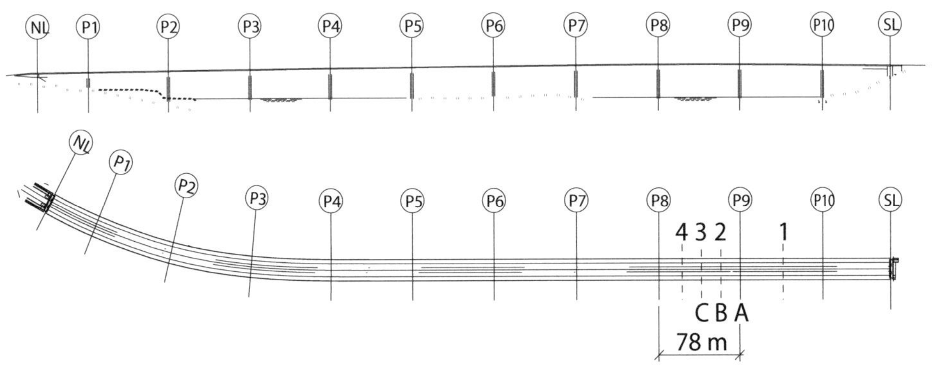

Fig. 5. Elevation and plan view of the New Årsta Bridge. Between the northern and southern abutment, NL and SL, respectively, the 10 piers are designated P1 to P10. Strain and acceleration sensor sections are marked A, B and C. Section 1, 2, 3 and 4 were used for vertical deflection measurements.

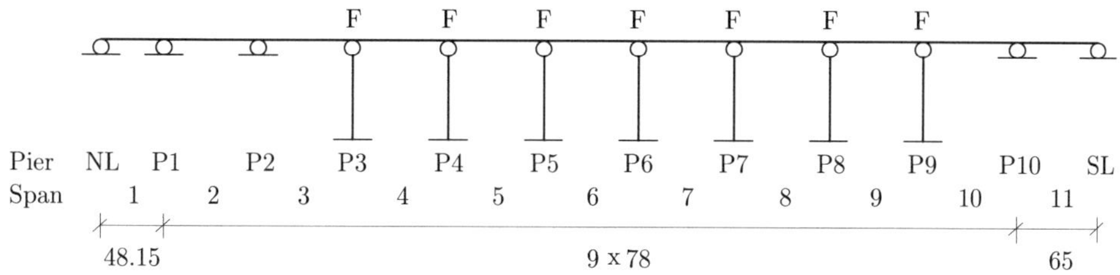

Fig. 6. Boundary conditions assigned to the bridge model.

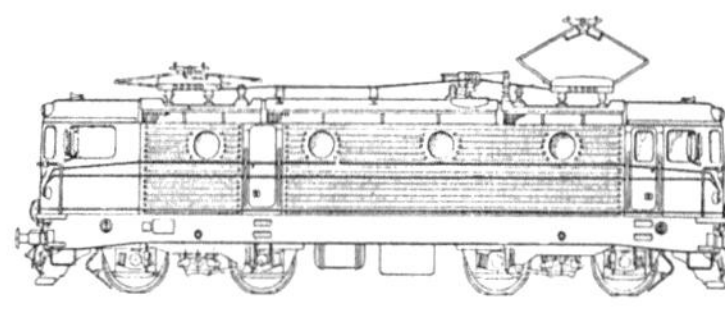

(a) Rc6 locomotive.

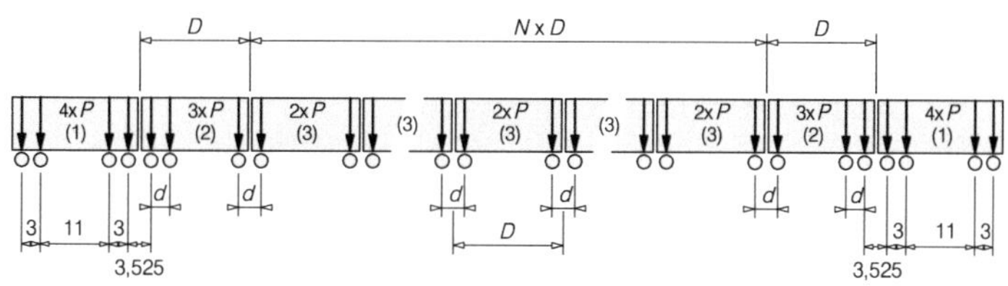

(b) HSLM-A configurations.

Fig. 7. Representation of vehicle loads.

representative distribution in the moving load case using amplitude functions. The internal distance between the axles in a bogie was 2.7 m and the bogie center to center distance was 7.7 m.

4.2.3 The HSLM-A configurations

The high speed load models, intended for railway bridge simulations above 200 km/h, were adopted here to subject the optimized FE model for more extreme dynamics than the current maximum speed limit of 140 km/h across the bridge. Fig. 7(b) is used in Eurocode to represent the HSLM-A configurations (CEN, 2002). Appendix 5.1 specifies the varying number of intermediate coaches, coach lengths, bogie axle spacings and point forces between the 10 different HSLM-A configurations.

4.3 Model optimization

In dynamic modeling, the dynamic characteristics of the bridge are of main concern, i.e. natural frequencies, mode shapes and damping ratios, why those should be focused on in detail. For that purpose, the objective function of Zárate & Caicedo (2008) would be optimal:

$$\Pi = \sum_{i=1}^{n} \left[\left[1 - \mathrm{MAC}\left(\phi_{mi}, \phi_i \left(\mathbf{p}_j \right) \right) \right] + \left| \frac{f_{mi} - f_i \left(\mathbf{p}_j \right)}{f_{mi}} \right| \right] \tag{9}$$

However, as the modal assurance criterion (MAC) values were unavailable the objective function in Eq. 6 was considered instead. No focuses was placed in evaluating different objective functions and the influence of variations in standard deviations (weight).

All 6 modeling parameters in Wiberg et al. (2009), i.e. damping ratios, modulus of elasticity, rails, tendons, vehicle speed and mass density, were significantly influencing typical dynamic load effects and the dynamic properties of the bridge. Therefore, they were all included in the optimization process. Damping and vehicle speed were obviously only considered in the dynamic analyzes. The importance of rails and tendons was analyzed based on their inclusion or exclusion in the FE model. The material properties of the rails and tendons were assumed as known. The prestress effect was included in a geometrically nonlinear large displacement analysis preceding each linear FE model restart execution for static and dynamic load effects. Modal damping was used in mode superposition of the moving load simulations. Thus, in the calculation of the mode shapes and frequencies, the effects of the axial compressive load on the modes and frequencies are included since the numerical calculation is based on the configuration at the start of the restart analysis. The linear mode superposition analysis that followed were then based on these mode shapes and frequencies, resulting in a dynamic response relative the prestressed bridge configuration.

In Table 1 the frequency columns from left to right are results from: an initial and manually tuned FE model in Wiberg (2006) but without rails and tendons, fast Fourier transforms of acceleration signals in Wiberg (2006), enhanced frequency domain decompositions from operational modal analysis in Wiberg & Karoumi (2009) and stochastic subspace identifications from operational modal analysis in Wiberg (2007). A dash only (see EFDD in Table 1) means undetected, while the dashes with parentheses (see SSI-PC in Table 1) stands for detected but unstable in the stabilization diagram as a result of operational modal analysis in Wiberg (2007). As can be seen from Table 1, already a simple manual updating resulted in a correct estimation of natural frequencies. However, the obtained high equivalent modulus of elasticity was questioned and therefore object of optimized updating. In addition, the initial manually tuned FE model used boundary conditions proven to be somewhat inaccurate according to Wiberg (2009). Henceforth, the notations differ between initial, $FE_{initial}$, initial manually tuned, FE_{tuned_1}, final manually tuned, FE_{tuned_2}, and optimized FE model, $FE_{optimized}$.

The optimization process was performed in the following two steps, based on the final manually tuned FE model:

1. Identification of updated material parameters (modulus of elasticity and mass density) from static load tests for strain and deflection residuals, together with frequency residuals.
2. Identification of modal damping ratio from dynamic load tests with maximum and root mean square (rms) acceleration residuals.

The frequency residuals were based on FE solutions using the subspace iteration method carried out for the structure linearized at the start of the restart analysis after prestress. The frequencies f_1 and f_5 at 1.3 Hz and 3.55 Hz were used, see Table 1. Strain residuals were based on axial beam stresses. The FE code can be modified to include the constrained warping effect on stresses. This was not considered here as it is based on torsional curvatures, manually given from separate analyzes. To include them as updating parameters was tested but made

No.	Type	Predicted		Measured		
		$FE_{initial}$	$FE^{a}_{tuned_1}$	FFT[a]	EFDD[b]	SSI-PC[c]
1	bending	1.03	1.30	1.30	1.29	(-)
2	coupled	1.13	1.44	1.45	1.45	(-)
3	coupled	1.91	2.45	2.43	2.43	2.43
4	coupled	2.53	3.26	3.26	-	3.24
5	coupled	2.78	3.55	3.55	3.50	3.55

[a] Wiberg (2006)
[b] Wiberg & Karoumi (2009)
[c] Wiberg (2007)

Table 1. Natural frequencies (Hz) from measured acceleration signals and for an initial and manually tuned FE model.

the optimizing algorithm unstable. However, the effect of unconstrained warping is relatively small in this case, see Wiberg (2009).

Frequency residuals only, strain residuals only, deflection residuals only, acceleration residuals only and their combinations were studied independently according to the principle:

$$\left\{ \begin{array}{c} \mathbf{p}_1 = \left(\frac{E_j}{E_0}, \frac{\rho_j}{\rho_0} \right) \\ \mathbf{z}_1 = (f_1, f_5, \varepsilon_1, \ldots, \varepsilon_{51}, v_1, \ldots, v_{20}) \end{array} \right\} \tag{10}$$

$$\left\{ \begin{array}{c} \mathbf{p}_2 = \left(\frac{\zeta_j}{\zeta_0} \right) \\ \mathbf{z}_2 = (a_1, \ldots, a_3) \end{array} \right\} \tag{11}$$

with the indexes 1 and 2 on **p** and **z** corresponding to the two different optimization steps. Thus, Eq. 10 was used with 2 frequencies, 51 strains and 20 deflections all together included, but also for frequencies, strains and deflections separately. Unfortunately, none of the installed axial strain transducers was active during the dynamic load testing. Therefore, according to Eq. 11, modal damping was tuned against acceleration residuals only, based on the three monitoring sections of Fig. 5. The location of all sensors within the monitoring sections were considered redundant information here but is to be found in for example Wiberg (2006).

An educated guess of the initial vector of updating parameters was necessary. Based on the results of Wiberg (2006) and Wiberg (2009), manually tuned start values of $E_0 = 55\,\text{GPa}$ and $\rho_0 = 2500\,\text{kg/m}^3$ were used for modulus of elasticity and mass density, respectively, and a constant damping ratio of $\zeta_0 = 0.01$ was assigned to all modes and used in the mode superposition procedure. This corresponds to the modal damping ratio found for prestressed bridges in design codes (see CEN (2002)).

All load testing used Rc6 locomotives and is described in detail in Wiberg (2009). Mode superposition was used to calculate the responses of the simulated Rc6 locomotive crossings with a time step of $\Delta t = 5\,\text{ms}$. The initial implicit time integration in the geometrically nonlinear axial load case operated on the basic equation of motion using the BFGS matrix update method algorithm (SOLVIA® Finite Element System, 2007).

Due to the restrictions of the beam FE model, i.e. using a beam element node to compare accelerations at the locations of accelerometers in the cross section, these signals were not comparable in the first place. Measured and modeled acceleration signals from the crossing Rc6 locomotive were therefore first low-pass filtered with a Butterworth filter at 5 Hz and then smoothed, using Savitzky-Golay filtering. Generally, a FE model is not optimal in reproducing high frequency content, especially not in representing a complex structure with a simple beam as is the case here. The low-pass filter at 5 Hz for reasonable acceleration comparison was therefore motivated. A Savitzky-Golay smoothing filter was chosen as they typically are used for a noisy signal whose frequency span (without noise) is large and they are considered optimal in the sense that they minimize the least-squares error in fitting a polynomial to frames of noisy data (The MathWorks, Inc., 2009).

To remove the rotational accelerations due to torsion, the measured signals from two accelerometers, 1 and 2, at the same distance from the center of gravity but on opposite sides were combined to compute the vertical translation acceleration only according to:

$$a_b = \frac{a_1 + a_2}{2} \tag{12}$$

In this way, assuming an infinitely stiff cross section, predicted vertical node accelerations were directly comparable with the measured bending acceleration a_b in monitoring section C, see Fig. 5. However, with only one accelerometer in monitoring section B, the predicted total vertical acceleration a_{tot} for comparison with measurements was calculated from beam node accelerations as:

$$a_{tot} = a_b + L \cdot a_r \tag{13}$$

with a_b the bending acceleration at center of gravity, a_r the rotational acceleration around the axial beam axis through center of gravity and L the distance perpendicular to the vertical axis, from center of gravity to the measuring position.

4.4 Relevant moving load simulations

After optimization, resulting in updated modulus of elasticity, mass density and modal damping ration, the FE model was finally subjected to all ten HSLM-A configurations for more reliable moving load simulations. These load configurations crossed the bridge as point loads with corresponding amplitude functions on the outermost track solely, moving from NL to SL, at speeds between 100 and 250 km/h. Typical results of interest were bridge deck deflection, acceleration and bending moment. These were all estimated and evaluated in more detail for the most critical HSLM configurations.

5. Results and discussion

5.1 Model optimization

The optimization algorithm operated efficiently but it was found unattainable to include all measurements in the response vector simultaneously. This was basically since the large-scale simplified model is incapable of predicting results based on all monitoring sections in Fig. 5

Parameter	FE_{init}	FE_{tuned_1}	FE_{tuned_2}	$FE_{optimized}$
Modulus of elasticity (GPa)	36	55	55	60
Mass density (kg/m^3)	2500	2400	2500	2700
Modal damping ratio (%)	-	-	1.0	0.92;2.10
Rail	excluded	excluded	included	included
Tendons	excluded	excluded	included	included
Boundary condition	state 1	state 1	state 2	state 2

Table 2. Differences between updating parameters of initial, manually tuned and optimized FE model. Modal damping ratios for the optimized FE model corresponds to rms and maximum acceleration, respectively.

with its restrictions as a beam model and the relatively few number of updating parameters chosen for the FE model. In addition, some sensors and deflection measurements resulted in result distortion, probably due to a difference in assumed sensor position or other sources of errors. Frequencies and load effects are also non stationary due to time dependent effects, not considered in the FE model and therefore influencing the optimization accuracy since the measurements took place at different occasions.

The mean result of adding the updating parameter vector from the frequency optimization procedure separately, the deflection optimization procedure separately and the strain optimization procedure separately, constituted the updating parameter values of modulus of elasticity and mass density in the optimized FE model. Consequently, these three different objective function contributors, separately gave different optimized updating parameter values of modulus of elasticity and mass density. Notice therefore, if the intention for example is superior dynamic characteristics, it would have been better to concentrate on the frequency residuals solely, complemented with mode shape information. However, the intention here was again to implement the algorithm and investigate the possibilities with a simplified FE model.

The results of the optimized updating parameters are summarized in Table 2 as parameter value or structural condition before and after optimization. Obviously, the optimized values of modulus of elasticity and mass density had a negligible influence concerning the frequencies. This was reasonable as Table 1 already indicated good agreement in frequencies between measurements and manually tuned FE model. Therefore, the bending stiffness to mass ratio for the final manually tuned FE model at iteration start (55/2500) was similar to the ratio of the converged optimized FE model at (60/2700) in the typical iteration sequence of Fig. 8. Still, frequencies were included in the objective function to account for the change in structural system concerning the inclusion of rails and tendons.

The increased values in modulus of elasticity and mass density were believed to have a larger effect in the optimization based on static strains and deflections. Table 3 summarizes results for static strains and deflections as initially predicted, predicted with the optimized FE model and measured. Observe that strain results are exemplified with the values of one single strain transducer and its position in that monitoring section (A, B or C) according to Fig. 5 for one of the six different static load test configurations in Wiberg (2009). Deflections were presented

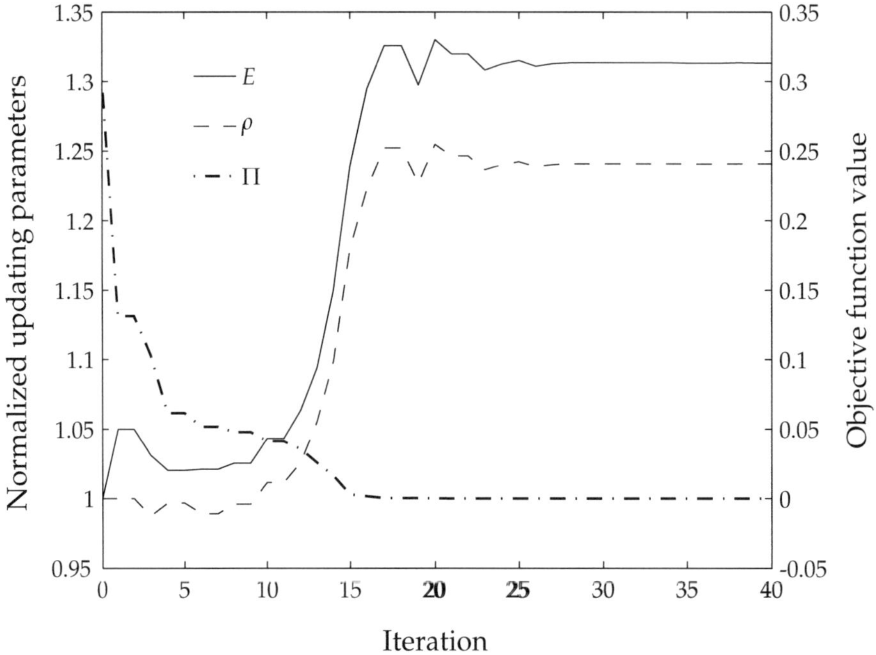

Fig. 8. Typical iteration sequence for optimization according to frequency residuals separately.

for the largest residual between initial FE model prediction and measured mean deflection in the point of interest (1, 2, 3 or 4) in Fig. 5. Notice that the initial FE model gave results on the safe side in all cases, while the updated FE model tended to be too optimistic, i.e. gave smaller load effect values than the measured ones.

Results for accelerations, based on the optimized values of modal damping ratios in Table 2, are presented in Table 4 for both measured, maximum and rms predicted acceleration. The corresponding acceleration signals are shown in Fig. 9. These signals represent a complete Rc6 locomotive crossing from SL to NL, traveling in outer curve at a speed of 120 km/h. Observe that the measured acceleration in the top of Fig. 9 is based on the signal manipulation according to Eq. 12. However, the way the measured signal looks may indicate that it still has some torsional acceleration included, i.e. that the beam element assumption of a rigid cross section with negligible in-plane stresses may not be completely satisfactory for the studied section. To be correct, a volume or shell element model of part of the bridge is necessary to include typical local flange modes, probably influencing the edge beam but are completely missed with the stiff cross section of the beam element. Consequently, it seems reasonable to base an optimized damping ratio on maximum acceleration for comparison with design codes, as those specify requirements on the maximum acceleration. However, in this case the predicted maximum acceleration may be too low due to the discretization and solution errors.

Even if the optimized model did not reproduce measured responses with highest accuracy in all cases it was considered reliable for the type of dynamic analysis assigned in design codes. At the same time, this is likely to be as far as one can get with a simplified FE model. It was not the intention with the simplified model in the first place to most accurately

Static load effect	$FEA_{initial}$	$FEA_{optimized}$	Measured
Strain A (10^{-6})	37.9	22.7	23.2
Strain B (10^{-6})	2.8	1.7	1.8
Strain C (10^{-6})	16.9	10.2	11.1
Deflection 1 (mm)	4.3	2.6	3.0
Deflection 2 (mm)	5.8	3.5	4.0
Deflection 3 (mm)	9.5	5.7	6.5
Deflection 4 (mm)	5.8	3.5	3.5

Table 3. Modeled and measured static strains and deflections.

$FEA_{initial}$ max	$FEA_{initial}$ rms	$FEA_{optimized}$ max	$FEA_{optimized}$ rms	Measured max	Measured rms
0.01630	0.00387	0.01195	0.00303	0.01195	0.00303

Table 4. Modeled and measured vertical accelerations (m/s^2) in the center of gravity of monitoring section C, filtered with a low pass filter at 5 Hz.

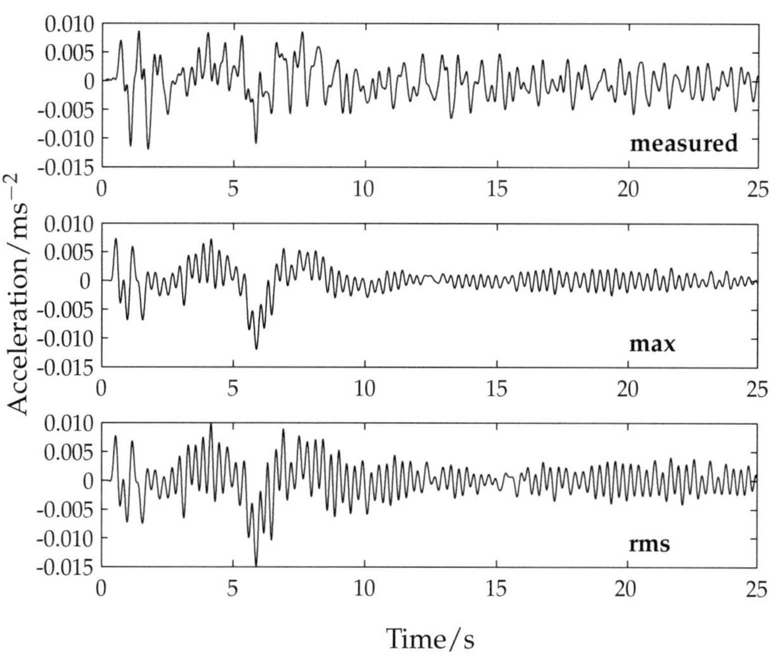

Fig. 9. Measured acceleration signal together with maximum and rms acceleration of the optimized FE model. All signals are filtered with a low pass filter at 5 Hz.

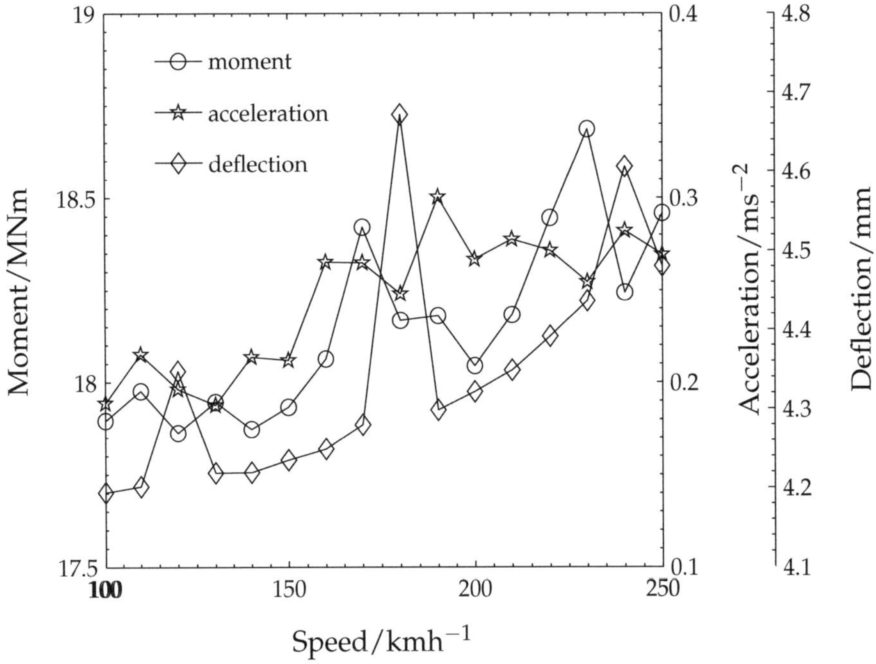

Fig. 10. Envelopes for bridge deck bending moment, vertical acceleration and deflection among all ten HSLM-A configurations with a speed increment of 10 km/h.

predict measurements. The main objective was to implement the optimization methodology and procedure in combination with statistical techniques to identify individually and jointly influencing FE modeling parameters to perform time efficient and relevant moving load simulations. In addition, the number and choice of updating parameters and the content of the objective function influence the possibilities of the adopted model.

Based on the optimized FE model, moving load simulations were performed with the ten HSLM-A train loads for increased accuracy in dynamic load effect predictions. Fig. 10 illustrates the envelope results for bridge deck bending moment, vertical acceleration and vertical deflection with a speed increment of 10 km/h between 100 km/h and 250 km/h. The most critical configurations were identified in Fig. 11 and found to be HSLM-A2, HSLM-A7 and HSLM-A10, respectively. For those three train loads, new results were presented in Fig. 12 for a speed increment of 5 km/h. Finally, the identified critical speeds in Fig. 12 were used to predict the load effects, at the critical location found in Fig. 13, from complete train crossings in time domain. The critical speeds corresponded to 165 km/h, 240 km/h and 180 km/h for HSLM-A2, HSLM-A7 and HSLM-A10, respectievely.

Moments and accelerations showed relatively small differences compared to results using the initial FE model. However, deflections were considerably smaller compared to the results of the initial FE model according to Fig. 14. The optimized FE model corresponded to a dynamic amplification factor of 1.15 in deflection, compared to 1.09 for the initial FE model. Observe that these results are given for the node with maximum vertical acceleration. Obviously, the dynamic amplification can be larger elsewhere.

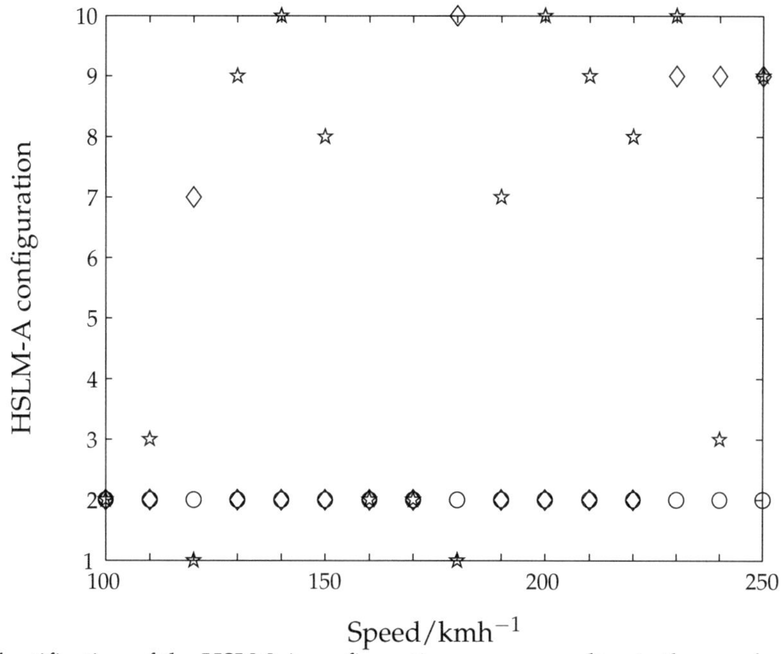

Fig. 11. Identification of the HSLM-A configurations corresponding to the envelope results. Bending moment is symbolized with (○), vertical acceleration with (⋆) and vertical deflection with (◇).

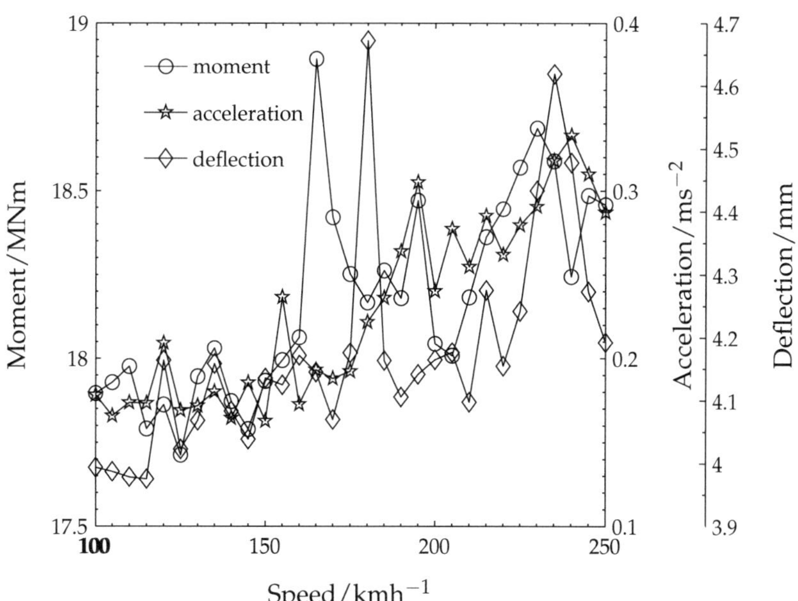

Fig. 12. The three most critical HSLM-A configurations, corresponding to HSLM-A2, HSLM-A7 and HSLM-A10, respectively, for bridge deck bending moment, vertical acceleration and deflection. A speed increment of 5 km/h was used.

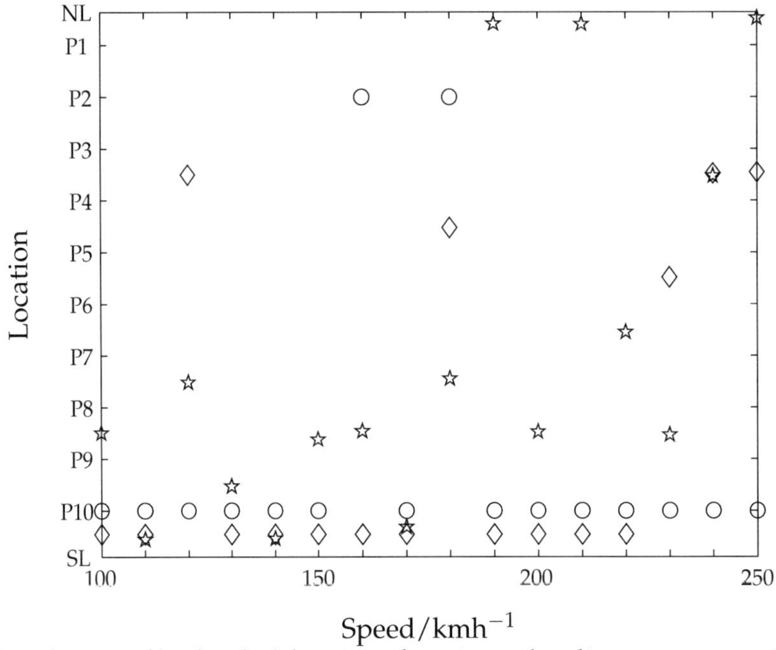

Fig. 13. Identification of bridge deck location of maximum bending moment, vertical acceleration and deflection of HSLM-A2, HSLM-A7 and HSLM-A10. Bending moment is symbolized with (○), vertical acceleration with (⋆) and vertical deflection with (◇).

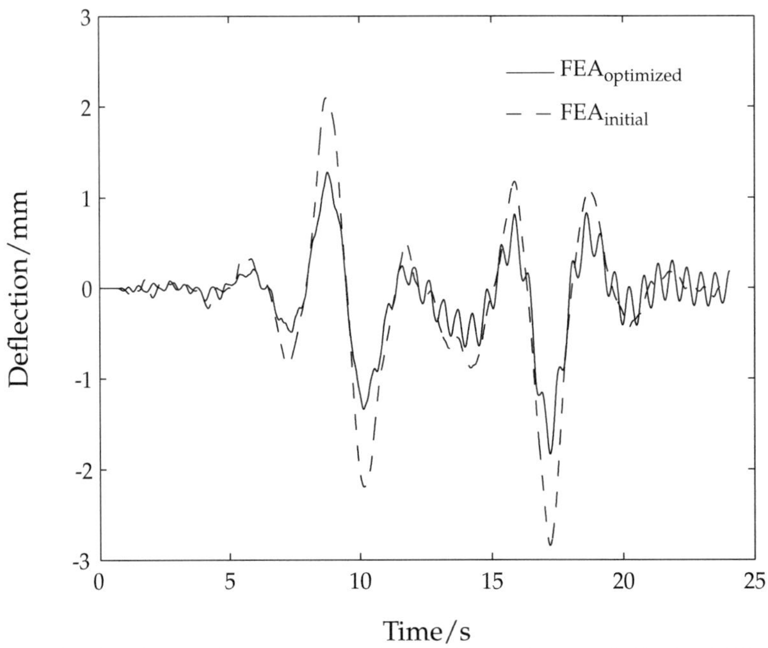

Fig. 14. Vertical deflection at the most critical node, i.e. midspan P4-P5 , for HSLM-A10 at 180 km/h.

6. Conclusions

Railway bridge design codes require detailed analyzes of passing trains at high speeds. Such analyzes are very time consuming as it involves many simulations using different train configurations passing at different speeds. Thus, simplified bridge and train models are chosen for time efficient simulations.

In this chapter a large-scale simplified railway bridge FE model for time efficient moving load simulations was optimized. The optimization uses natural frequencies from operational modal analysis and load effects from load testing, based on previously identified updating parameters. The optimization algorithm was easily implemented for FE model updating and was shown to operate efficiently in a benchmark test and for the specific bridge. The importance and the potential of optimization procedures in FE modeling for increased accuracy in moving load simulations is highlighted. Further, it was generally concluded that:

- Based on individually and jointly influencing factors, the optimized FE model predicted static and dynamic load effects more accurately.
- Even though the updated FE model predicted too optimistic load effects, i.e. not being on the safe side, the updated model resulted in larger dynamic amplification factor in comparison with the initial model.
- The optimized FE model predicted static load effects most accurately. The natural frequencies were already accurately calculated for the manually tuned FE model.
- The previously questioned high-valued equivalent modulus of elasticity was proven to be reasonable.
- Reliable modal damping ratios of 0.92% and 2.10%, for rms and maximum bridge deck acceleration, respectively, were predicted.
- Even if the simplified FE model in some sense is insufficient in load effect predictions, it stands for better predictions when optimization is used.
- More measured dynamic characteristics (natural frequencies, mode shapes and modal damping ratios), complementing updating parameters and a more detailed FE model are necessary for dynamic load effect predictions with highest accuracy.

To conclude, the author strongly recommend the working procedure of (1) manual FE model tuning, (2) updating parameter identification, and (3) final optimization focused on the result of interest and therefore based on a suitable objective function with that intention.

Finally, this chapter highlights the potential of the adopted optimization procedure. This methodology can not only be used for model updating based on measurements, but also be introduced and customized to work already in the bridge design phase for better (based on design code requirements) and more cost-effective designs.

7. Appendix

A. Benchmark input

The following input are used in the Benchmark test:

Masses:	$m_1 = 1000\,\text{kg}$
	$m_2 = 20000\,\text{kg}$
Initial velocity:	$v = 30\,\text{m/s}$
Initial displacement of m_2:	$\delta = 0.05\,\text{m}$
Stiffness:	$k = 4 \cdot 10^6\,\text{N/m}$
Damping:	$c = 1 \cdot 10^5\,\text{Ns/m}$
Time step:	$\Delta t = 0.5\,\text{ms}$
Beam section properties:	$A = 0.06128\,\text{m}^2$
	$I = 0.01573\,\text{m}^4$
Material properties:	$E = 210\,\text{GPa}$
	$\nu = 0.3$
	$\rho = 16000\,\text{kg/m}^3$

B. HSLM-A

Universal Train	Number of intermediate coaches N	Coach length D (m)	Bogie axle spacing d (m)	Point force P (kN)
A1	18	18	2.0	170
A2	17	19	3.5	200
A3	16	20	2.0	180
A4	15	21	3.0	190
A5	14	22	2.0	170
A6	13	23	2.0	180
A7	13	24	2.0	190
A8	12	25	2.5	190
A9	11	26	2.0	210
A10	11	27	2.0	210

C. Notation

The following symbols are used in this chapter

a = acceleration (m/s^2);
A = cross section area (m^2);
E = modulus of elasticity (Pa);
f = natural frequency (Hz);
n = mode number;
N = response vector length;
p = elements of $\mathbf{p}$;
$\mathbf{p}$ = updating parameter vector;
v = vertical deflection (m);
w = elements of $\mathbf{W}$;
$\mathbf{W}$ = weighting matrix;
$\mathbf{z}$ = response vector;
ε = axial strain;
ζ = equivalent modal critical damping ratio;
Π = scalar objective function value;
ρ = mass density (kg/m^3);
σ = standard deviation;
ϕ = natural vibration mode;
ω = circular frequency (rad/s);

8. References

CEN (2002). *Eurocode 1: Actions on structures – Part 2: Traffic loads on bridges (prEN 1991-2).*
Coleman, T. F. & Zhang, Y. (2009). *Optimization Toolbox™ 4 User's Guide*, The MathWorks, Inc.
Friswell, M. I. & Mottershead, J. E. (1995). *Finite Element Model Updating in Structural Dynamics*, Springer.
Jaishi, B. & Ren, W. X. (2005). Structural finite element model updating using ambient vibration test results, *J. Struct. Engrg.* 131(4): 617–628.
Jonsson, F. & Johnson, D. (2007). *Finite Element Model Updating of the New Svinesund Bridge. Manual Model Refinement with Non-Linear Optimization*, Msc thesis, Chalmers University of Technology, Sweden.
Lagarias, J. C., Reeds, J. A., Wright, M. H. & Wright, P. E. (1998). Convergence properties of the nelder-mead simplex method in low dimensions, *SIAM Journal of Optimization* 9(1): 112–147.
Mottershead, J. E. & Friswell, M. I. (1993). Model updating in structural dynamics: A survey, *Journal of Sound and Vibration* 167(2): 347–375.

Schlune, H., Plos, M. & Gylltoft, K. (2009). Improved bridge evaluation through finite element model updating using static and dynamic measurements, *Engineering Structures* 31(7): 1477–1485.

SOLVIA® Finite Element System (2007). Users Manual Version 03.

The MathWorks, Inc. (2009). *Signal Processing Toolbox*™ *6 User's Guide*.

Wiberg, J. (2006). *Bridge Monitoring to Allow for Reliable Dynamic FE Modelling. A Case Study of the New Årsta Railway Bridge*, Lic thesis, Royal Institute of Technology, Sweden.

Wiberg, J. (2007). Railway bridge dynamic characteristics from output only signal analysis, *Proc., Int. Conf. on Experimental Vibration Analysis for Civil Engineering Structures (EVACES'07)*, Porto, Portugal.

Wiberg, J. (2009). An equivalent modulus of elasticity approach for simplified modelling and analysis of a complex prestressed railway bridge, *Advances in Structural Engineering* . Submitted.

Wiberg, J. & Karoumi, R. (2009). Monitoring dynamic behaviour of a long-span railway bridge, *Structure and Infrastructure Engineering* 5(5): 419–433.

Wiberg, J., Karoumi, R. & Pacoste, C. (2009). Statistical screening of individual and joint effect of several modeling factors on the dynamic fe response of a railway bridge, *Structural Control and Health Monitoring* . Submitted.

Willis (1850). Deflexion of railway bridges under the passage of heavy bodies, *Journal of the Franklin Institute* 49(1): 7–8.

Zárate, B. A. & Caicedo, J. M. (2008). Finite element model updating: Multiple alternatives, *Engineering Structures* 30(12): 3724–3730.

9

Controlling and Simulation of Stray Currents in DC Railway by Considering the Effects of Collection Mats

Mohammad Ali Sandidzadeh and Amin Shafipour
School of Railway Engineering, Iran University of Science and Technology, Tehran, Iran

1. Introduction

Urban rail transit systems are mostly electrically DC type. Usually in these systems for reducing the costs, running rails are used as the return current paths. Because of the electrical resistance of rails against the flow of traction currents and also rail to ground conductivity (despite rail-to-ground insulations), parts of the return current that flow from trains to traction substations leak to the ground. These leaking currents are called stray currents (as shown in Fig. 1). Stray currents can enter their neighboring metallic infrastructures and, as a result of anodic interactions, cause electrochemical corrosions in their leakage path.

The amount of electrochemical mass reduction as a result of anodic interaction due to flow of current *i(t)* from a metal to an electrolytic environment can be gained as below

$$M = C\int_{t1}^{t2} i(t)d(t) \quad (1)$$

in which C is the electrochemical coefficient and is based on the type of the metal, electrolyte, and chemical calculations. For example, C is 9.11 $KgA^{-1}Year^{-1}$ for iron. It means that a current of 1 ampere can oxidize 9.11 kilograms of iron per year. This can reduce the safety and the life time of structures and infrastructures in tunnels. The corrosions caused by stray currents yield a total loss of $500 million each year to the American railway system [1]. When a train is running, especially during its accelerating time, the traction supply current can sometimes reach 6000 amperes. Since the resistance of rails is between 15 to 20 $m\Omega Km^{-1}$, the return current can face a voltage drop of up to 120 VKm^{-1}. This voltage, due to inadequate insulations of rails and their underlying structures, allows the current to leak to the ground. Stray current control is usually done via improvements performed to transportation systems or the neighboring ground structures. Increasing the resistance between rail and ground is a very effective method in reducing stray currents. The increase in resistance reduces the tendency of return current to flow in any path other than the rails. Other methods for avoiding corrosions include cathodic protection, rail insulation, traction voltage increase, employment of proper rails that have very low electrical resistance and usage of proper grounding systems in traction substations.

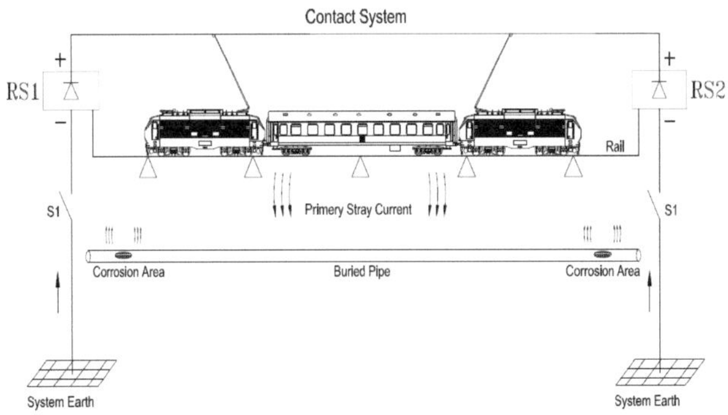

Fig. 1. Exposure of Stray Current in DC Railway Systems.

Despite application of the mentioned methods for controlling stray currents, some portion of the traction current would still flow through the ground instead of the rails to reach the negative terminal of the substation. In this paper, by discussing the ways for using stray current collection mat and grounding system via simulations, the effects of collector cables and stray current collection mats below the rails are described.

Many papers, published between 1995 and 2005, discussed and examined various grounding systems in railway transportation. In addition, general techniques for reducing voltage and stray current levels were discussed in these papers. "Paul" specifically examined grounding in power grid systems of subways [2]. "Goodman", for the first time, calculated the rail voltage and stray current profiles. His calculations were not computer aided and instead based on simple hypothesis [3]. Later, "Case" used Π model to investigate a diode grounding system and compared it with EN50122-2 standard [4]. "Lee" used the floating models instead of "Case"'s proposed Π model [5, 6]. Although Lee's model benefited from high accuracy, it showed no significant different results. In the mentioned papers, the effects of grounding systems on voltage and stray current profiles were investigated. Despite mentioning corrosion in the previous papers, it was "Cotton" who, for the first time, discussed the influence of soil structures and stray current collection mats on corrosion performance of metallic infrastructures. The outcome of his survey was software that analyzed and studied influences of soil and current collection mats on corrosion of metallic infrastructures [7].

"Cotton" discussed the importance of stray current reduction in DC rail transit systems and proposed usage of stray current collection mats. Also, he talked about the influences of his proposed system's performance and resistivity of soil on stray currents and the resulting corrosions [8].

"Lee" used the floating model to simulate and investigate stray currents and their effects on underground structures. He used a coefficient of 0.1V voltage increase of these structures as a high stray current leakage in these systems. In the end, he proposed some methods for reducing stray currents by improving railway systems and the neighboring structures [9].

In this research, after a brief introduction of Tehran Metro line 4, the present methods used for simulation and analysis of stray currents are investigated and later stray current and touch potentials are studied by simulating the current path between two substations of line

4. The simulations are performed for various grounding scenarios of traction substations and the influence of stray current collection mats, under various scenarios, and effective parameters on performance improvement of the collection system, in the simulated line, are discussed. In the end the stray current and touch potential of rails under presence and absence of the stray current collection mats, in the worst scenarios, and movement of four trains from three stations, with the middle station not having a traction substation, are investigated.

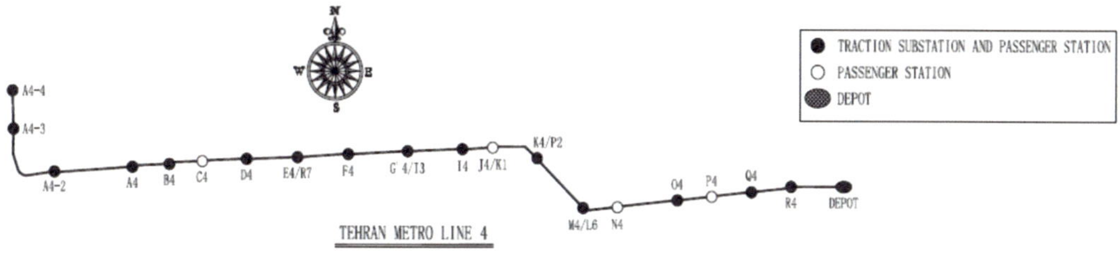

Fig. 2. Tehran Metro line 4 plan

2. An introduvtion of Tehran metro line 4

Tehran metro line 4 is the forth subway line in Tehran with a length of 25 Km and comprises 22 stations. It is stretched from the west to the east part of the city and is equipped with 18 traction substations. The details of the substations are shown in tables 1 and 2. The total trip time in line 4 is 42 minutes, which includes 25 seconds of dwell time in each station. At the most crowded hours, the train headways reach 2 minute cycles. Fig. 2 shows the line 4 plan.

Each traction substation is supplied with two rectifier transformers that have nominal powers of 2.5 or 3.3 MW. The required power is supplied directly by 63 KV power grid of Tehran Regional Electric Co. in B4, H4 and R4 substations, and is then stepped down to 20 kV via 3×2 (63 to 20kV) transformers (Fig. 3). After that, it is distributed in LPSs (Lighting and Power Substations) and RSs (Rectifier Substations) by means of separate 20 kV rings.

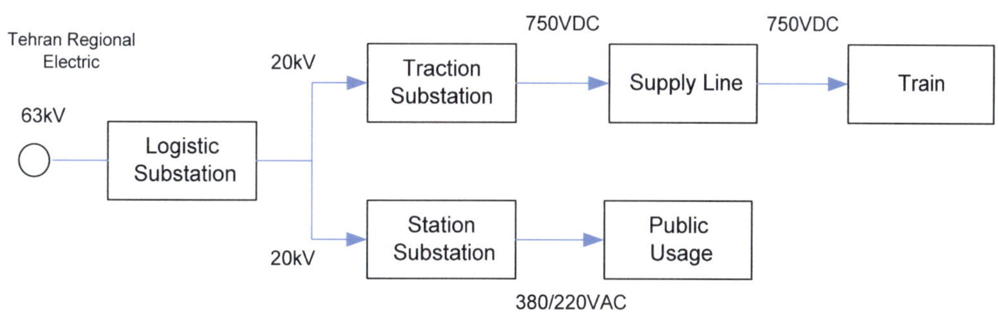

Fig. 3. General diagram of Tehran Metro Line 4 power system.

For a 750V power supply, under full load and setting of 6%, the substation output voltage has a voltage drop of 45V. Since the train requires high power amounts during its acceleration, the RS should be able to supply the extra required load. Based on class VI in compliance with IEC146 standard, the RS should be capable of a constant load supply of 100%, a nominal load supply of 150% for a period of 2 hours and a nominal load supply of 300% for a period of 1 minute. Based on its internal resistance, at 300% overload times, the voltage drop in the RS would be around 135V.

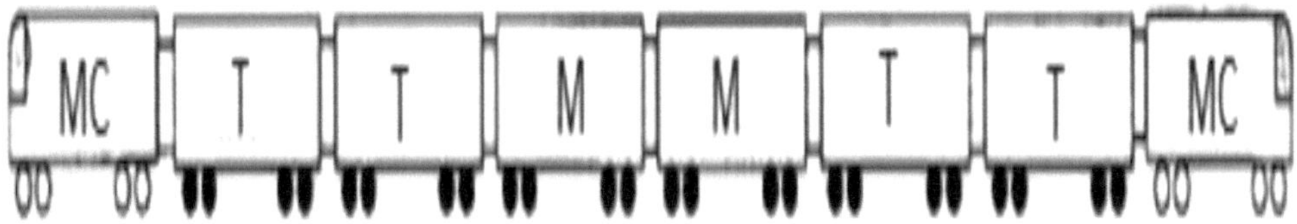

Fig. 4. The order of rolling stock cars

The trains comprise 8 cars and weigh around 274 tons without passengers and 379 tons in normal transit conditions (Fig. 4). The traction power is supplied from 750V voltage source and from the third rail. Each train includes 16×200kW traction motors. The maximum train speed and acceleration is 80 kmh^{-1} and 0.2 ms^{-2}, respectively. The rolling resistance equation (i.e., the Davis formula) is $1.57+0.00106V^2$ for the traction mode and $1.97+0.0025V + 0.00106V^2$ for the breaking mode. Also, the internal auxiliary power consumption of each train is 125 kW.

Item	Station Name	Station Characteristics		
		Begin	Center	End
1	A4-6			
2	A4-5	-3+645	-3+566	-3+487
3	A4-4	-1+611	-1+532	-1+453
4	A4-3	-0+15	0+063	0+142
5	A4-2	1+661	1+740	1+819
6	A4	2+811	2+890	2+969
7	B4	3+971	4+050	4+129
8	C4	4+926	5+005	5+084
9	D4-H2	5+645	5+724	5+803
10	E4	7+064	7+143	7+222
11	F4	8+153	8+232	8+311
12	I3G'4	9+427	9+506	9+585
13	I4	10+716	10+795	10+874
14	K1J4	11+310	11+389	11+468
15	P2K4	12+461	12+540	12+619
16	M4	13+773	13+852	13+931
17	N4	14+537	14+616	14+695
18	O4	15+593	15+671	15+750
19	P4	16+489	16+568	16+647
20	Q4	17+443	17+522	17+601
21	R4	18+793	18+872	18+951
22	S4	20+661	20+740	20+819

Table 1. Characteristics of line 4 stations

Item	Station Name	Type of Substation	Capacity
1	A4-6	RS	
2	A4-5	RS	2×3300
3	A4-4	RS	2×2500
4	A4-3	RS	2×3300
5	A4-2	RS	2×2500
6	A4	RS	2×2500
7	B4	RS	2×2500
8	C4	RIC	-
9	D4-H2	RS	2×2500
10	E4	RS	2×2500
11	F4	RS	2×2500
12	I3G'4	RS	2×2500
13	I4	RS	2×3300
14	K1J4	RIC	-
15	P2K4	RS	2×3300
16	M4	RS	2×3300
17	N4	RIC	-
18	O4	RS	2×3300
19	P4	RIC	-
20	Q4	RS	2×3300
21	R4	RS	2×2500
22	S4	RS	2×3300

Table 2. Characteristics of line 4 substations and their capacities

3. Stray current modeling

For simulation of a DC traction network and analysis of stray currents, two different methods can be used. The first method is based on the floating model and distributed elements [5, 6]. In this method, distributed elements like rail conductance and resistance per unit length are used for the calculations, and by utilizing the floating equations, the equations for stray current and touch voltages of rails are determined. Equations (2) and (3) show the current and voltage in various parts of the rail, respectively as below: [6]

$$i(x) = c_1 \exp(\gamma x) + c_2 \exp(-\gamma x) \tag{2}$$

$$v(x) = -R_0(c_1 \exp(\gamma x) + c_2 \exp(-\gamma x)) \tag{3}$$

in which *i(x)* is the rail current, *v(x)* is the rail voltage, γ is the propagation constant ($=\sqrt{RG}$), R_0 is the characteristic resistance of the rail ($=\sqrt{R/G}$), c_1 and c_2 are the equitation constants and determined based on the boundary values, *R* is the resistance per unit length of the rail and *G* is the leakage conductance between the rail and the ground. Although this method has the ability of determining quantities in each part with high accuracy, it has a low flexibility and therefore for various structures, many of its equations should be altered.

The second method is based on using concentrated elements or Π line model in the DC mode. In this method, the line is divided into sections with equal lengths and each section is modeled with a rail to ground resistance and conductance (Fig. 5). In most applications, selecting the length of each section as 100 meter creates a good accuracy for stray current determination. Each substation is modeled via its Thevenin equivalent (i.e., a voltage source and an equivalent resistance) or Norton equivalent circuit. The train is also modeled via a time-varying current source. Finally, based on proposed model and node analysis rules, rail surface potential and the current amount profiles are resolved. The linear equation for node analysis is as below: [4]

$$\overline{Y}_n \cdot \overline{e} = \overline{i}_s \qquad (4)$$

in which $\overline{Y}_n$ is the node admittance matrix, $\overline{e}$ is the node voltage matrix and $\overline{i}_s$ is the node current sources matrix. In the node admittance matrix, y_{ii} entry is the sum of all conductances tied to the ith node and y_{ij} is the negative of the sum of all conductances between the ith and the jth nodes. Also, i_{si} is the sum of all currents entering the ith node. The currents entering the positive and exiting the negative nodes are also considered here. The node voltage matrix is the unknown matrix that is determined by solving the above equation. This matrix actually represents the amount of rail surface voltage.

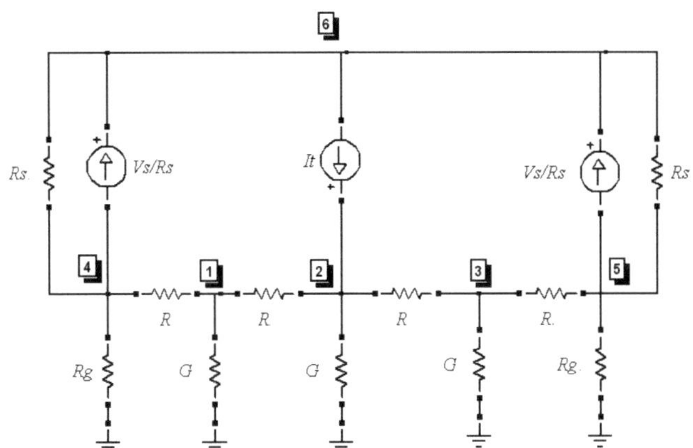

Fig. 5. Simplest form of discrete Norton line model for two stations and rail

Because of the existence of resistance network, the admittance matrix is symmetrical and has a positive determinant. The voltage of the station rectifier can be determined from the DC load distribution of the traction network. The current amount in the I_T current model, which is the train model, can be determined from the train's current - velocity equations.

Fig. 6 shows the calculation steps based on this model in a DC railway network. According to the presented flowchart, at first, based on the train time schedule, the locations of various trains are determined and then each substation voltage is gained by distributing the load through the system. Also, the traction current of each train is determined according to the velocity of that specific train. Finally, by solving the node equations, the touch voltage and stray current for each node are calculated. [10]

Although this method is not as accurate as the first one, selection of 100-meter sections creates the acceptable preciseness required for analysis of stray currents. Besides, this method allows simulating and modeling different structures with minimal changes in them. Since all the equations in this method are linear, the simulations have an acceptable performance.

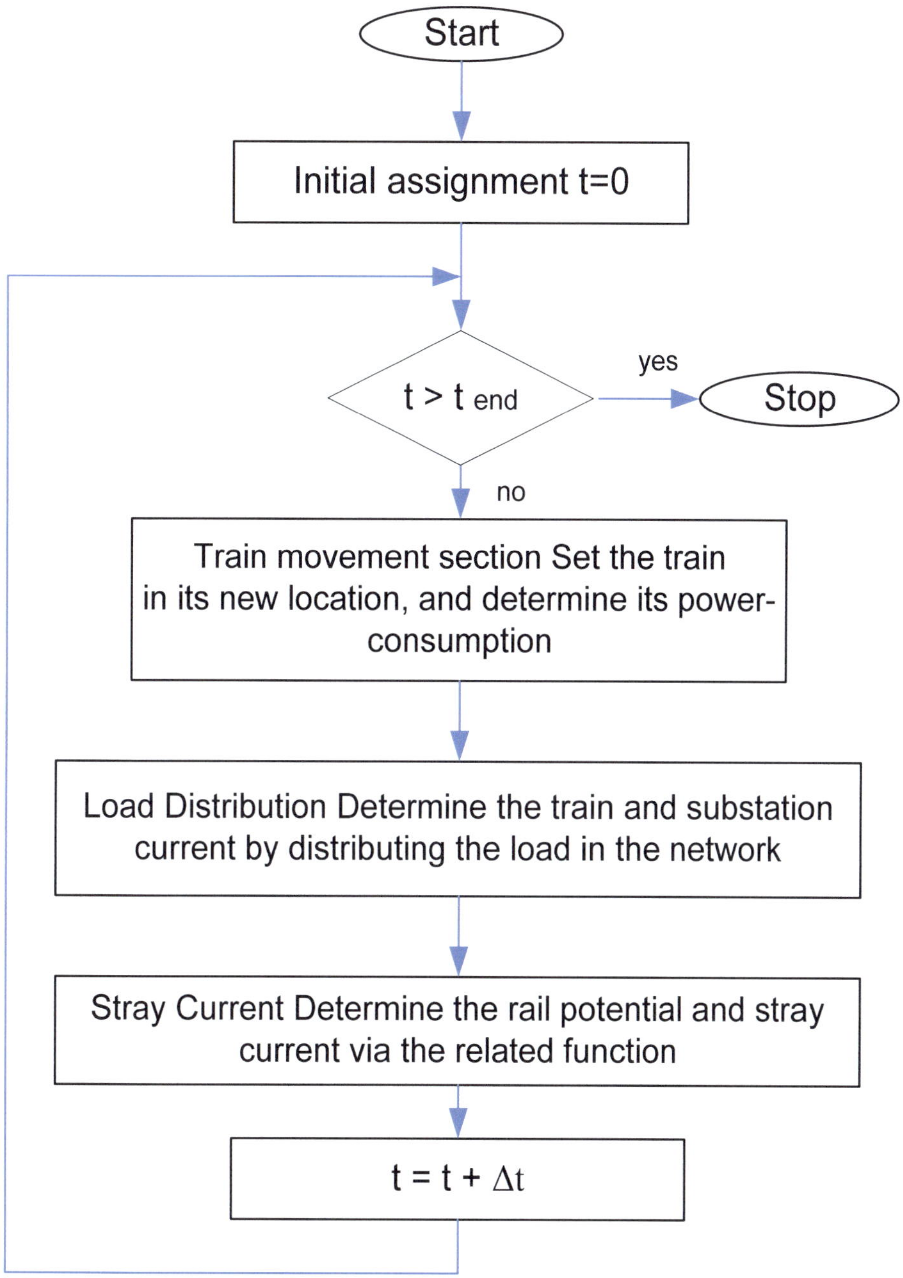

Fig. 6. Flowchart for determining the touch potential and stray current in the discrete model

4. Modeling the line and determining the parameters

For line modeling, Fig. 7 schematic is used, in which V_{dc} is voltage of the DC busbar, I_t is the train model, R_{rr} is the resistance per unit length of the rail, R_{rm} is the rail to mat resistance, R_{mm} is the resistance per unit length of the mat and R_{mg} is the mat to ground resistance. Also, R_g is the resistance of the substation ground. The type of grounding can be defined based on the way the negative substation busbar is connected to the ground; if, as shown in Fig. 7, S_1 key is closed, the system is grounded and if this key is open, the system is floating. If a diode is used instead of S_1 key, then the system is diode grounded.

The amount of I_t current varies based on the train velocity and is calculated from the power consumption of traction motors and the Davis formula. Fig. 8 shows the relation between the current and speed of the traction motors.

The parameters used in the simulation are determined from the available line 4 characteristic data, the available standards, references [8], [11] and division of the line into equal 100 meter cells. According to the manufacturers' data, R_{rr} is 1650 μΩ (i.e., with 5% corrosion, the resistance per unit length of the rail is 16.5 $\mu\Omega m^{-1}$.) The rail to soil conductivity, based on the type of foundation, is 10 Ωkm^{-1} and therefore R_{rm} is assumed to be 100Ω. Assuming the special resistance of steel as 15.9 $\mu\Omega cm^{-1}$, a corrosion coefficient of 5% and a cross sectional area of 1800mm^2 for the underground stray current collection mat, the amount of R_{mm} is calculated as 12mΩ. The stray current collector cable with a cross sectional area of 185mm^2 is assumed to have a resistance per unit length of 180 $\mu\Omega m^{-1}$. The R_{mg} is assumed to be 1Ω.

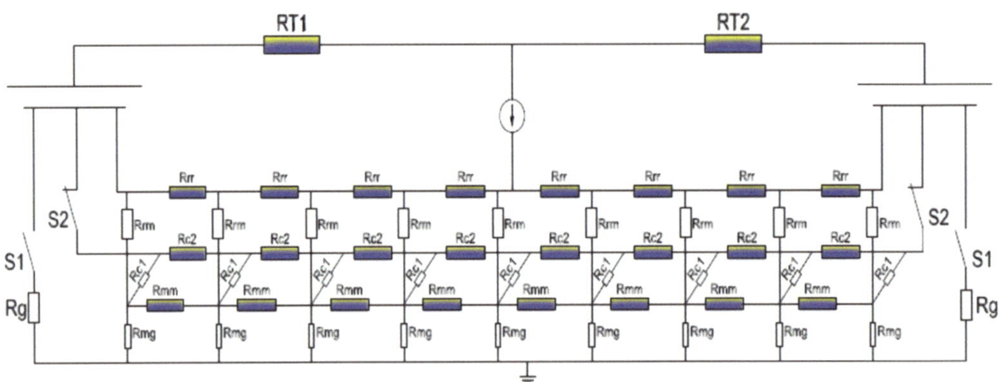

Fig. 7. The schematic used for line modeling

Since the grounds in I3G4 and F4 stations have the lowest resistance values, the line between the two stations, which is 1274 meters long, is selected for the simulation. Fig. 9 shows the time-speed relation for this line.

As shown here, the train starts its movement with acceleration of 0.75 ms^{-2} and after 330 meters reaches a speed of 80kmh^{-1}. Then the supply from the power source is cut and the train runs the rest of the line as the result of its inertia, and its speed starts to decrease. After 1150 meters the dynamic brakes are engaged and the train begins to stop.

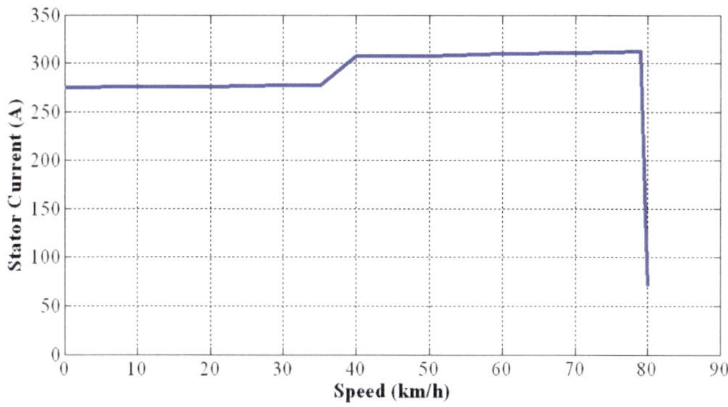

Fig. 8. The current-speed relation of train traction motors in the power consumption and return modes

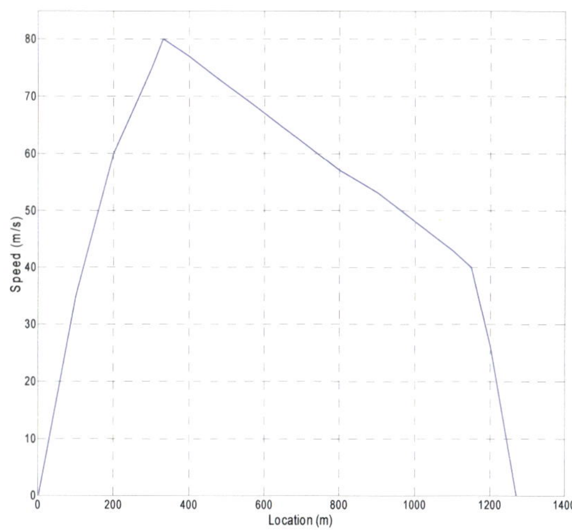

Fig. 9. The train speed-location curve

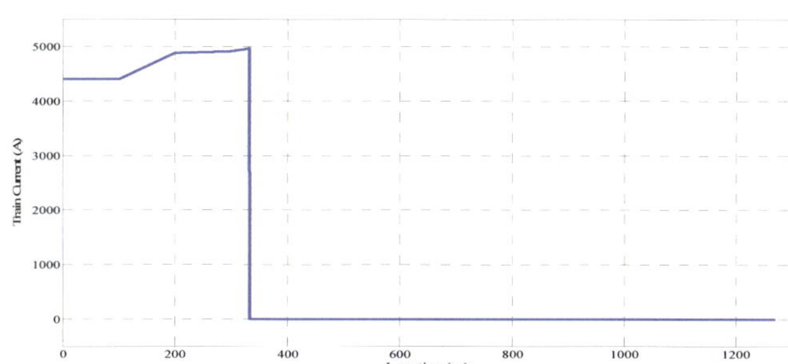

Fig. 10. Current - location curve of traction motor

Fig. 10 shows the current-speed curve in the line. It shows the total train consumed current for all the 16 traction motors. Since stray current effect on metallic structure in long term, in this analysis transition currents and the current related to train auxiliary power have been ignored. Also, the train regenerative current is lossed on the resistors located on the train.

5. Simulation results and analysis

For analyzing level of the stray current for Tehran Metro line 4, the simulations are performed under various scenarios. In the first scenario, while no collection system is utilized, effect of grounding systems on current leakage from rail is studied; then in the second scenario while the reinforcement grid exit in the second stage of concrete under rail, effect of electrical continuity of this grid on controlling and collecting stray current is investigated. In the next scenario, stray current collector cable is added to collection mat and in the last part, previous scenarios are considered on the passenger stations (without traction substations).

5.1. Effect of grounding systems

This part is when no stray current collection system is utilized and train is assumed to be running and 2 below cases are studied:

At first, while the train is in a specific location, the stray current and rail potential is observed along the whole line (case I) later, the stray current is studied for a specific location, while the train moves along the whole line (case 2) finally, the effects of the collection mats and then the collector cable, in addition to the mats, are investigated.

5.1.1 Case I

The first mode is when the train is at a location 100m from the start substation and is accelerating, while a current of 4400 amperes is fed to the train via the third rail. Such that, a current of 4060 amperes is fed to the train from the nearest substation. Fig. 11 shows the leakage current from rail and fig.12 shows potential of running rail along the way for three different grounding systems.

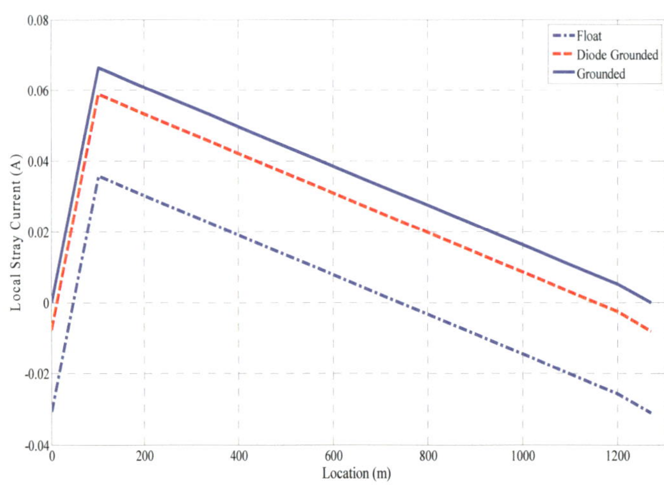

Fig. 11. Leakage current from rail for three different grounding system when the train is 100 meters away from the initial substation

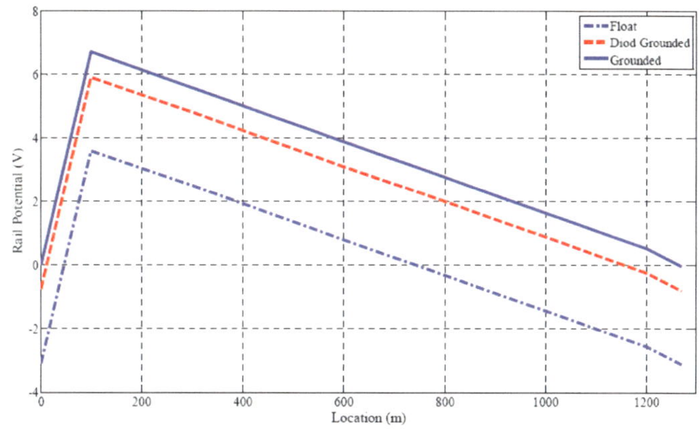

Fig. 12. The rail touch potential for three different grounding systems when the train is 100 meters away from the initial substation

It can be concluded from the presented graphs that in all three grounding system, the current is leaking from the rail to the ground in locations where rail to ground voltage is positive, and the amount of current flow depends on the rail voltage with respect to ground. In solidly grounded systems, due to the zero voltage at the substation and the positive rail to ground voltage, stray current leakage is observed in all locations of the line. In diode grounded system less stray current leakage is observed, since the rail voltage is negative at the substation. The reason for this is that diodes turn on only when the voltage of their cathode is -0.8v less than their anode voltage. In ungrounded system, the stray currents case is much improved. In the first half of the line, from the train to the substation, stray currents leak and flow from the rail to the ground (since the rail to ground voltage is positive,) and in the second half of the line, the leaked stray current flow back to the rail. The cumulative stray current in the floating system is 0.136 amperes, while it is 0.33 amperes in the diode grounded and 0.42 amperes in direct grounded systems.

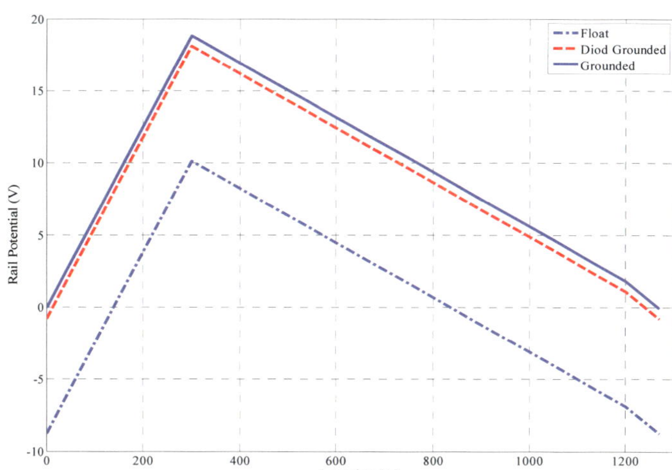

Fig. 13. The rail touch potential in three different grounding systems when the train is 300 meters away from the initial substation

When the train is 300 meters away from the initial substation, its current amount reaches a maximum of 4960 amperes. From this current, 3814 amperes are supplied by the first and 1146 amperes are supplied by the second substation. Because of the increase in traction current in this case, the stray current and touch potential are also amplified for the three grounding systems (Fig. 13 -14). As it is shown, touch potential in grounded system in nearly zero in substations and it's positive in other locations. Also it's approximately twice that of the floating system in train location. In this case, the amount of rails stray current increases to 1.1 amperes in the grounded system, 1.0 amperes in the diode grounded system and 0.3 amperes in the floating system.

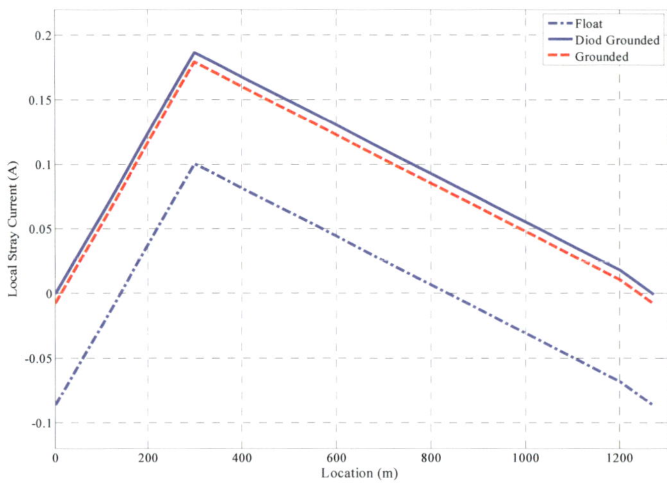

Fig. 14. Leakage current from rail for the three different grounding systems when the train is 300 meters away from the initial station

5.1.2 Case II

At first the investigation is done for a location that is 100m away from the initial first substation. Fig. 15 presents the resulting stray current for the grounding systems. As shown, the floating system has the lowest amount of stray current among all the existing grounding systems, and when the train is more than 200 meters away from the initial substation, current return from the ground to the rail is also observed. In the grounded system of Fig. 15, at all points of the rail, current leakage, which is more than the other systems, is observed. The highest amount of stray current is observed when the train is near the 100 meter point.

In solidly and diode grounded systems, in which the rail voltage about the substation is zero, even after the train has passed location 100m, the rail voltage near the substation remains positive and the current leakage continues, although at lower magnitudes. The traction current, however, increases up to location 330 m. In the floating system when the train passed location 200m, the rail voltage at location 100m point become zero and the flow of stray current stops. However, further train running makes the voltage of this location negative and therefore the stray current flows back to the rail.

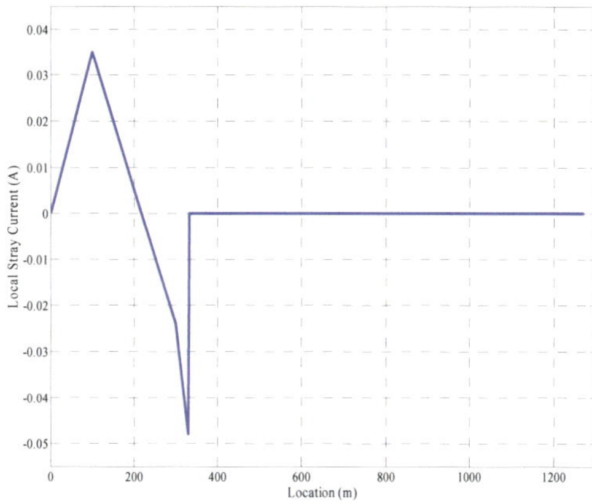

(a)

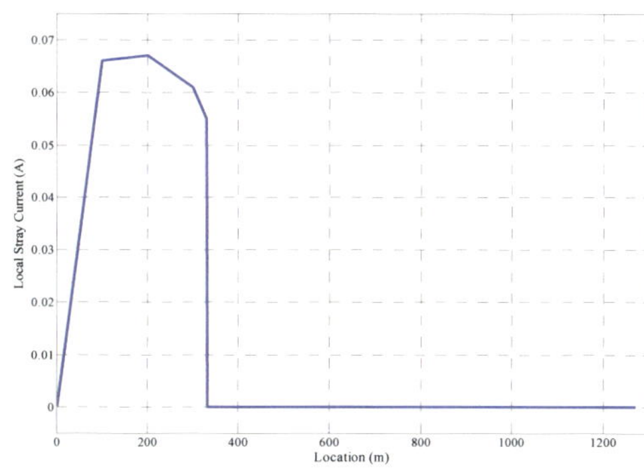

(b)

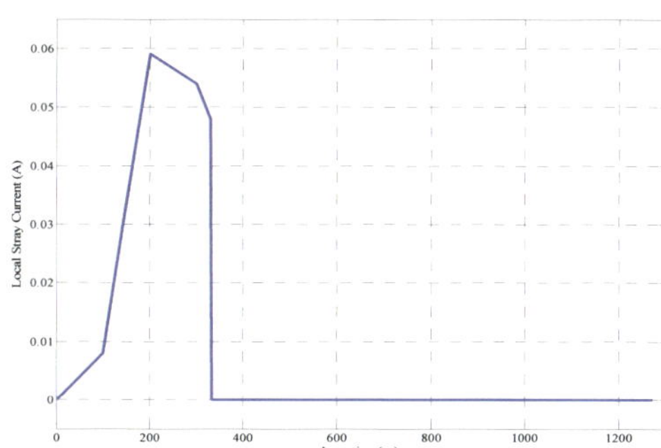

(c)

Fig. 15. Stray current at location 100 m on the rail in case II for a) The floating system b) Solidly grounded system c) Diode grounded system

Fig. 16 shows the stray current at location 300 m in the line. Like the previous instances, the stray current is at its peak when the train is also at this location. At location 300m, due to sufficient distance from the substation, unlike location 100 m, always in all grounding system, when there is current flow, the touch potential is positive and stray current exists.

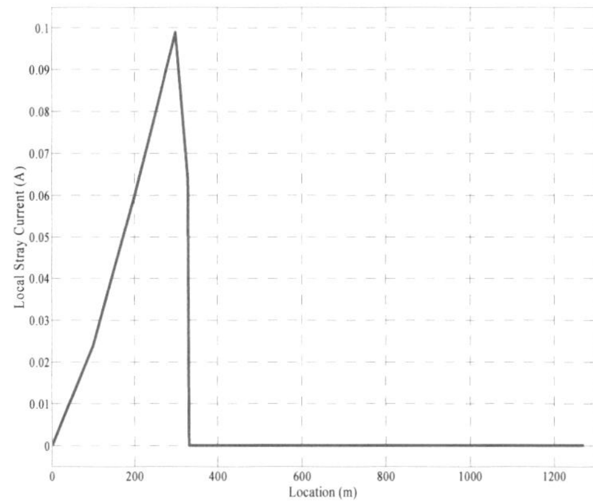

(a)

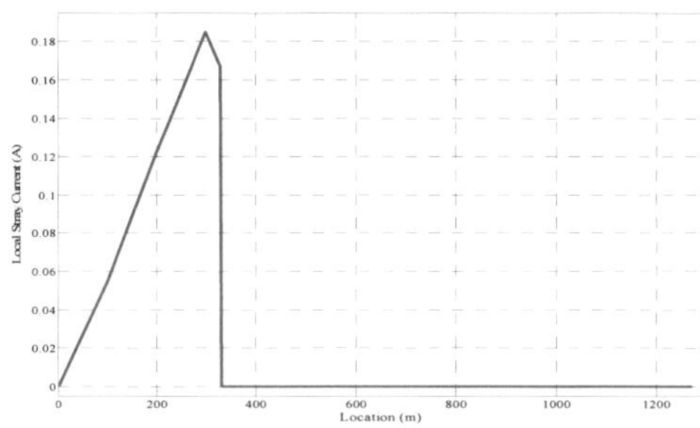

(b)

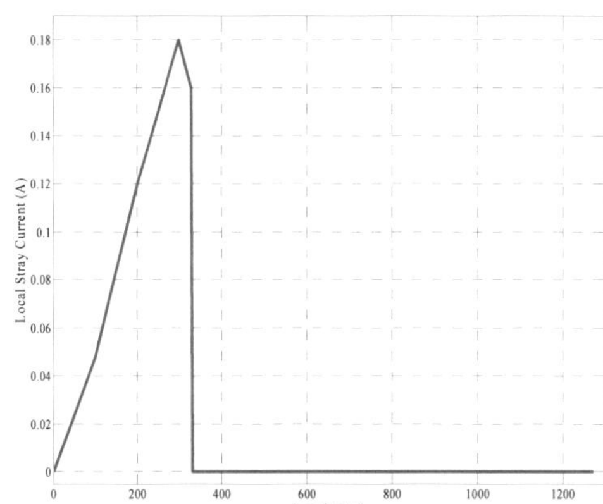

(c)

Fig. 16. The stray current at position 300 m, on the rail in case II for A) Floating system b) solidly grounded system c) Diode grounded system

Fig.17 shows the stray current at location 1000 m from the initial substation. The voltage at this location is just similar to the voltage at location 100 m, however the stray current is lower and the current amount that returns from the rail to this point is higher. In the floating system, since this location is closer to the first substation, its voltage remains negative and the current keeps flowing back to the rail.

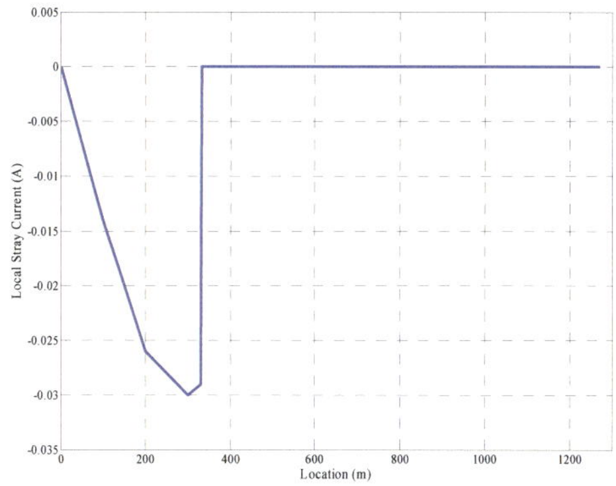

(a)

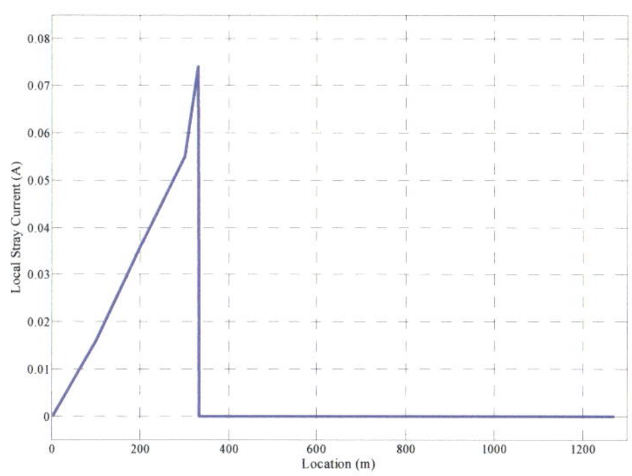

(b)

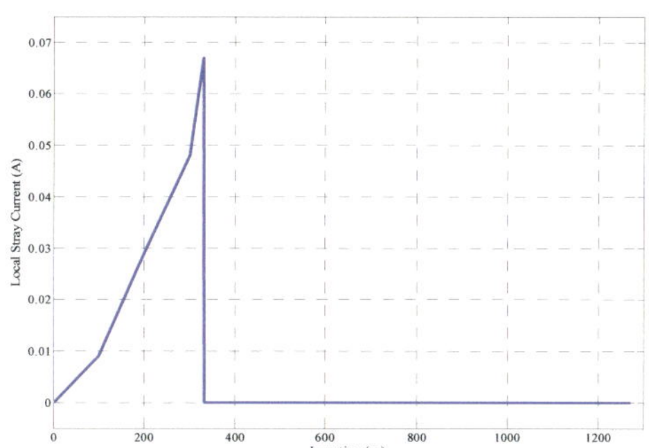

(c)

Fig. 17. Stray current at location 1000 m when the train is running for a) Floating system b) Solidly grounded c) Diode grounded system

5.2 Using stray current collection mat

The simulations in this section are performed assuming the presence of a reinforcement bars. If the metal bars in the concrete under the rail intersect with each other, they collect main portions of stray currents due to creating a low resistance path for conducting these currents from rail to traction substation. So this collection mat result increasing leakage current from rail because of making return path to traction substation. Some portions of stray currents leak from the rail to metal bars of the reinforced concrete and continue to flow through the underlying concrete structure. If this current is not returned to the substation through a specific current path, the current leak from the concrete structure to the ground would create corrosions in the metal bars of the concrete. In fact for executing reasons, the mats are installed in sections with length of 100 m such that there is a gap of nearly 100 mm between sections. If there is no electrical connections (by wire or cable) between separate sections of mat, entering current to this structure cause severe damages to them.

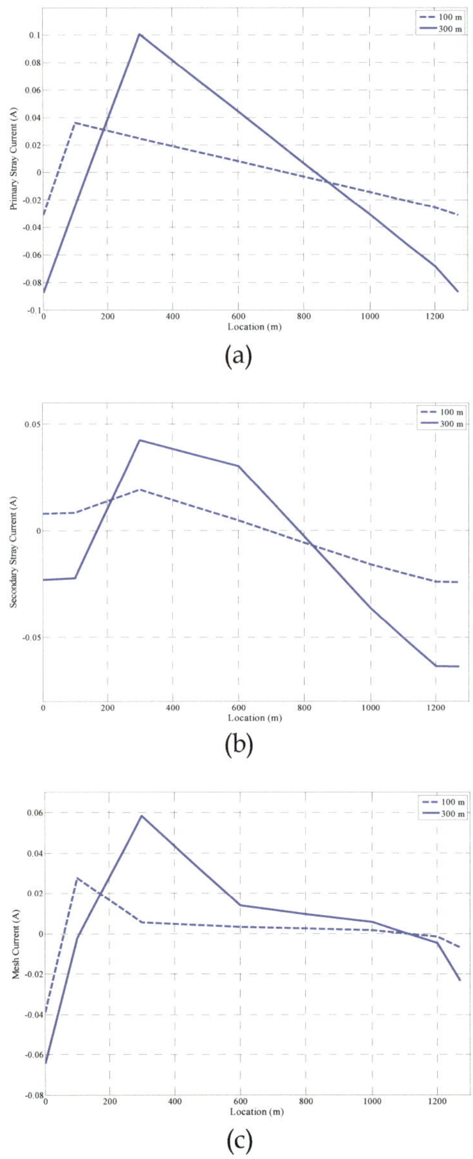

Fig. 18. Stray current when the train is at locations 100 m and 300 m from the initial station and the metallic concrete structure is unprotected a) Stray current leak from the rail b) Stray current leak to the ground c) Current entering the concrete collection mat

In the first part, this metallic structure is consumed disconnected and there is no direct return path to substation for current that enters which results are shown is fig. 18.

In this case, in the floating system, stray current leak from the rail when the train is at location 100m from the first substation is 1.37 amperes, of which 0.06 amperes flows through the concrete's metallic structure and causes severe damages to this structure. When the train is at location 300m, the stray current becomes 0.3 amperes, of which 0.145 amperes flows through the collection mat. Also about 50% of stray current from rail doesn't enter to this mat. To overcome this flaw, in addition to interconnecting all parts of the mat to each other, a path for connecting the collection mat to the negative busbar of the substation should be provided. This connection is done by means of a diode that helps having cathodic protection and making the current flow one directional (Fig. 19).

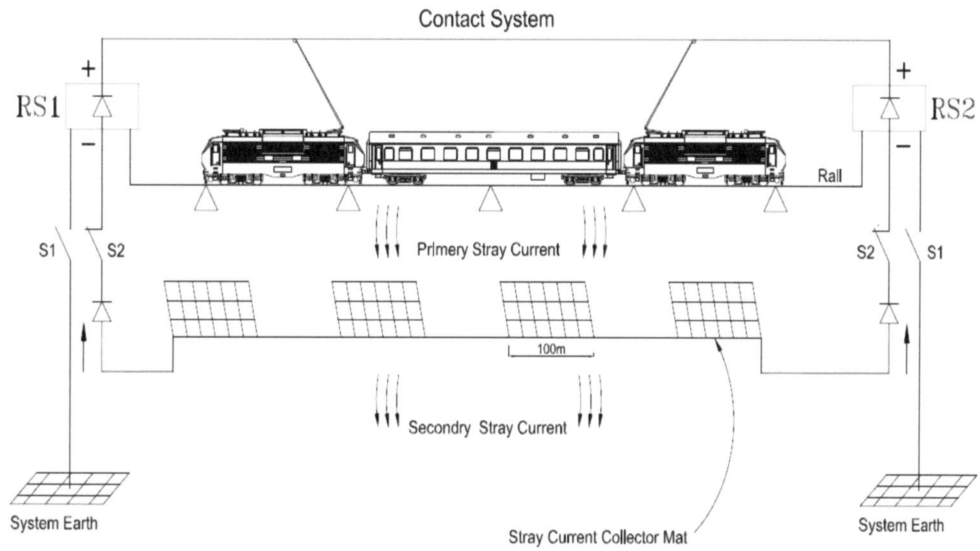

Fig. 19. Using the concrete metallic structure as the stray current collection mat

Fig. 20 shows the stray current from rail when the train is at locations 100m and 300m from the first substation and the stray current collection mat is used. Because of the mat to ground resistance and connection of the mat to the negative busbar by the diode, the situation here is like that of the diode grounded system. In this case, although the substation ground is considered floating, the substation voltage remains around zero (which is the diode on voltage) and the rail touch potential at the train's location becomes more than the floating system's voltage and, as a result, the stray current increases. Fig.21 shows the current which captured by the collection mat and as shown this amount has increased compared to previous part. However, as shown in Fig. 22, a large portion of the stray current is collected by the collection mat and only a small portion of it leaks to the ground. In this case, the total rail output current at locations 100m and 300m are 0.33 and 1.01 amperes, respectively, of which 0.065 and 0.099 amperes leak to the ground, respectively, and the rest is collected by the mats. Using equation (5) for evaluating the efficiency of the stray current collection system, the system performance becomes 81% at position 100m and 90% at position 300m.

The equation is

$$\eta = (I_{Collected} / I_{st}) \times 100 \quad (5)$$

in which $I_{Collected}$ is the amount of stray current collected by the mats and I_{st} is the total stray current that has leaked from the rail. As shown in Fig. 22, the highest amount of stray current leakage occurs in the middle point of the line. The reason for this is the long distance of this position from the substations. Although the highest stray current is observed at location 300m, the stray current at location 600m is also high and is 70% of the stray current amount at location 300m. Besides, since location 600m has the highest distance from the line terminating substations, the resistance remains high at this location for stray currents that enter the mats, and this makes this middle position to have the highest rate of stray current leakage to ground in the entire line.

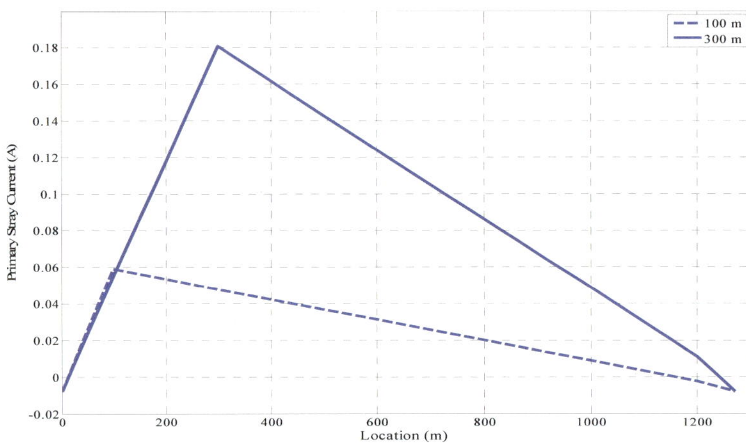

Fig. 20. Rail stray current when the train is at locations 100 m and 300 m and collection mats are used

In Fig. 23, the efficiency of the collection system, based on changes of the cross sectional area of the mat is presented. As mentioned before, the mats used in Tehran Metro line 4 have cross sectional areas of 1800 mm^2. The higher the cross sectional area of the mat, the lower its resistance per unit length and the higher its stray current collection amount would be. R_{mm} is 27.4 mΩ for a cross sectional area of 700 mm^2 and 8 mΩ for a cross sectional area of 2400mm^2. Fig. 23 shows the graph for the time the highest traction current supply, when the train is at position 300 m.

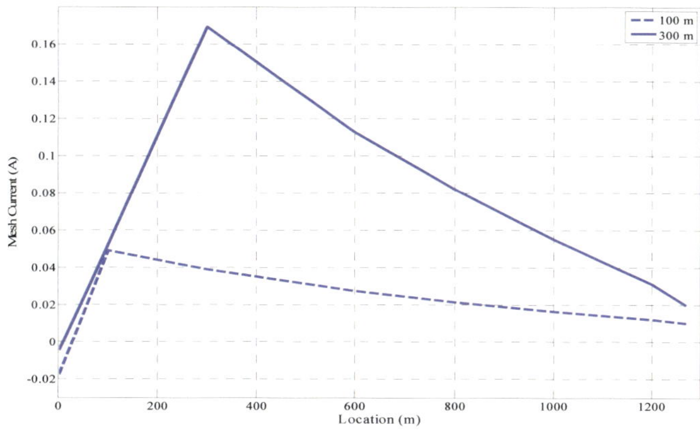

Fig. 21. The current collected by the mat when the train is at locations 100 m and 300 m from the initial substation and stray current collection mat is also used

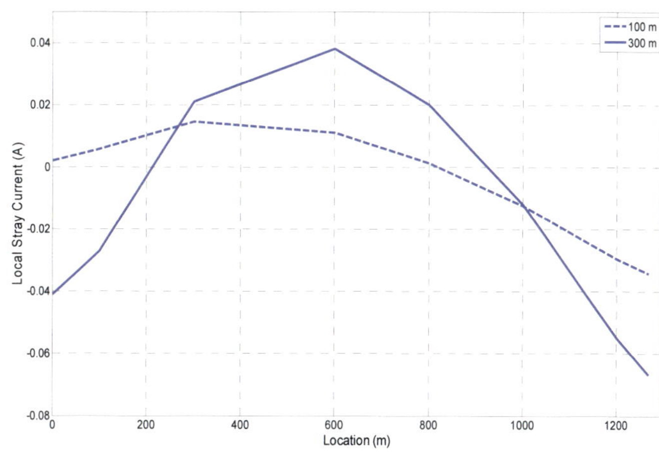

Fig. 22. Stray current leakage to ground when the train is at locations 100 m and 300 m from the first substation and collection mat is used

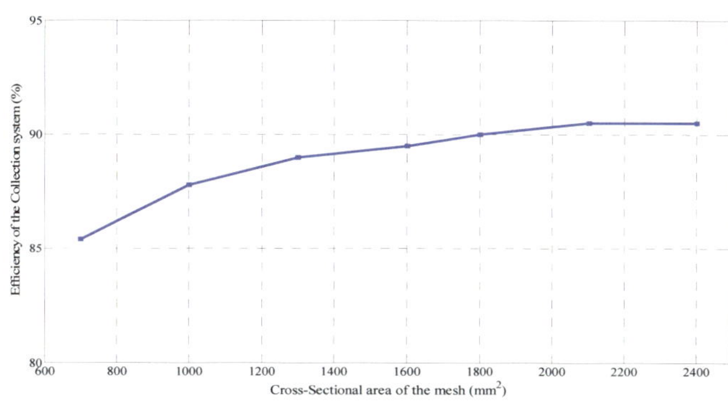

Fig. 23. Efficiency of Stray current collection system based on changes of the cross sectional area of the mat at location 300 m (the worst scenario)

5.3 Using stray current collector cable

Due to the problems in building a continuous collection mat system, existence of current in the mat system increases the chance for stray current leakage from the mat system itself (especially at connection points). For protecting and retaining high efficiency of the mat system, stray current collector cables are used. Stray current collector cables are installed alongside the rail and they are, at specific locations (e.g., connection points,) connected to the underlying stray current collection mat. In this way these cables provide a low resistance and insulated parallel current path that canalizes and directs the main part of the mat currents to the negative busbar of the substation (Fig. 24).

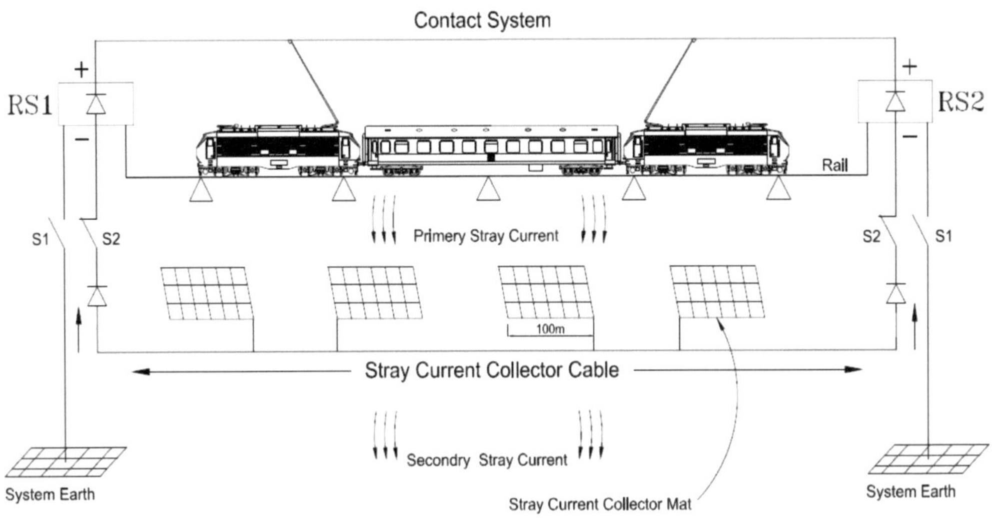

Fig. 24. Using stray current collection mat and cable

The collector cable not only creates a suitable path for the currents collected by the collection mat, but also avoids current leakage to ground due to its insulation. It directs all the collected currents to the negative substation busbar.

Fig. 25 shows the rail stray current for a line with a collector cable and when the train is at locations 100m and 300m from the initial substation. Since stray current collection mat resistance changes in comparison to resistance between rail and mat are so small, the leakage current from rail is not that much different from the previous scenario (fig.20) in floating system. But in comparison, the current captured by collection mat has increased (fig.26) and the current leakage to the ground has significantly decreased (Fig. 27). When the train is at location 300m, the total rail output current is 1.01 amperes, of which 0.049 amperes leak to the ground. In this case the efficiency of the collector system is 95%. When the train is at location 100m from the initial substation, the total rail current output is 0.33 amperes, of which 0.043 amperes enter the ground. The collector system's efficiency is 87% in this case.

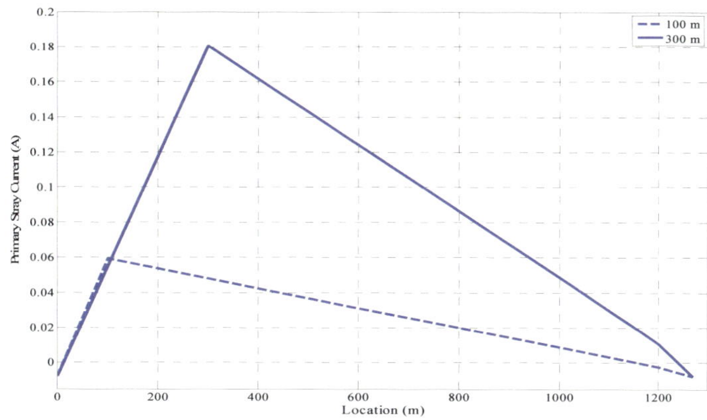

Fig. 25. The rail stray current when the train is 100 m and 300 m away from the initial substation and collector cable is used

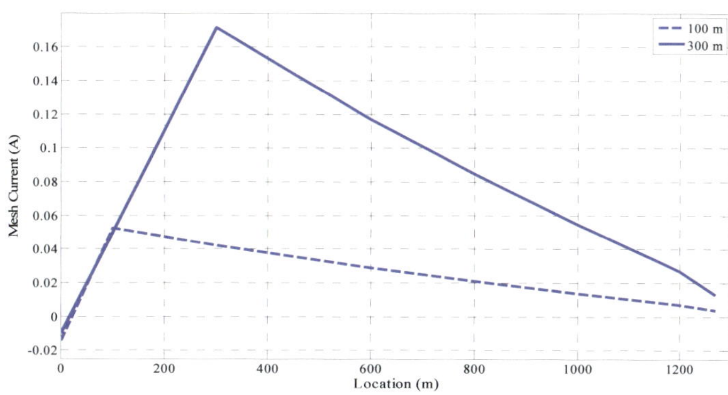

Fig. 26. The current collected by the collection system when the train is 100 m and 300 m away from the initial substation and collector cable is used

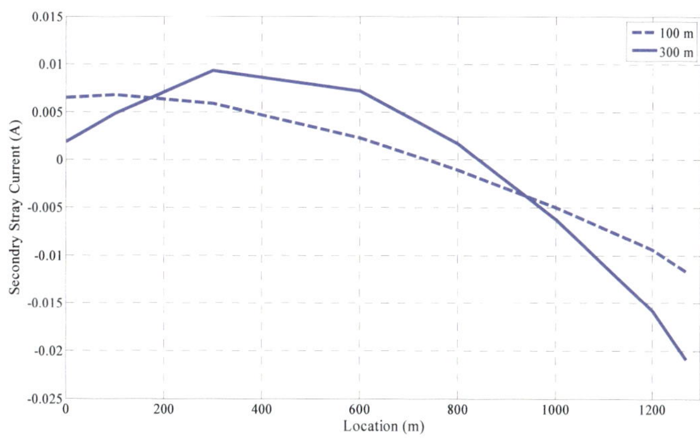

Fig. 27. Stray current leakage to ground when the train is 100 m and 300 m away from the initial substation and collector cable is used

The reason for the small effect of the collector cable on system efficiency at location 100m is the closeness of this location to the initial substation. At this short distance of the train from the substation where the rails have high current output amounts, the substation voltage cannot decrease enough so as to turn the current collector diode on. As a result, the current leakage to the ground remains high and the collector cable shows no significant effect on system efficiency. Fig. 28 shows the collector system efficiency based on the cross sectional area of the utilized cable, when the cross sectional area of mat is constant. Creation of a low resistance current path and insulation from the ground (via the collector cable) significantly decreases the stray current leakage to the ground. Changes in the cross sectional areas of the cable have effects around 1~2 % on the efficiency of the collector system. Resistances of cables with cross-sectional areas of 90 mm^2 and 270 mm^2 are 36 mΩ and 12 mΩ, respectively.

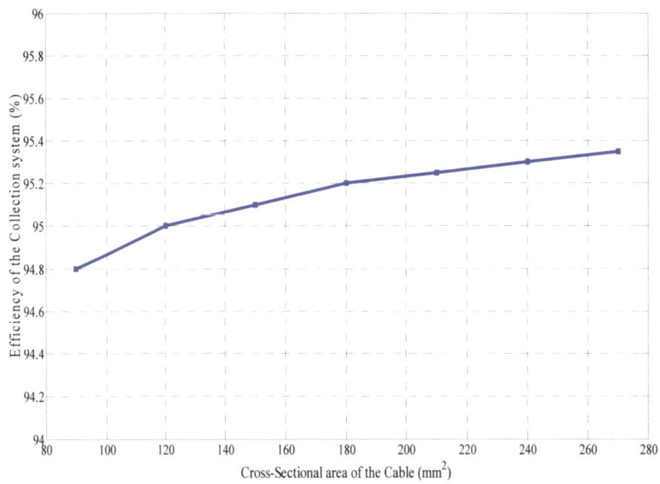

Fig. 28. Efficiency of the collector system based on the cross sectional area of the cable

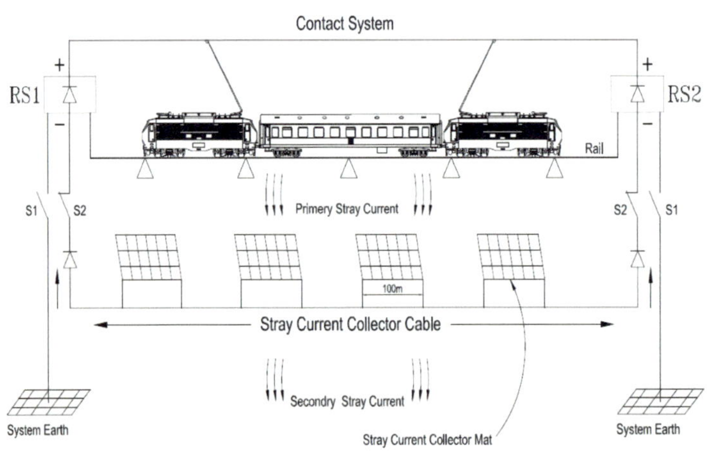

Fig. 29. Using the stray current collection mat and cable with two connection points between the mat and the cable.

The type of collector cable connections to the mats can be also changed. The above simulations were for scenarios where the mats were connected to the cables only at one point (Fig. 24). In Fig. 29, the mats are connected to the collector cables at two points. Figs 30-32 show stray current leakage from the rail, current leakage to the ground and the stray current collected by the collector system. In this case, the rail stray current becomes 1.01 amperes and the total stray current becomes 0.04, which creates an increase in system efficiency to 96.1%.

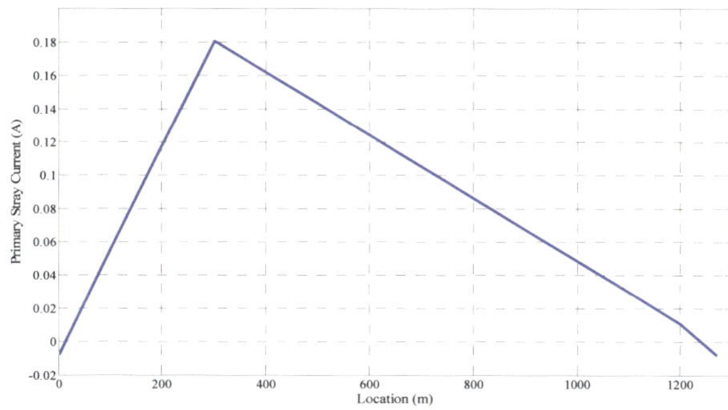

Fig. 30. The rail stray current when the train is 300 m away from the initial substation and mat and cable are connected at two points.

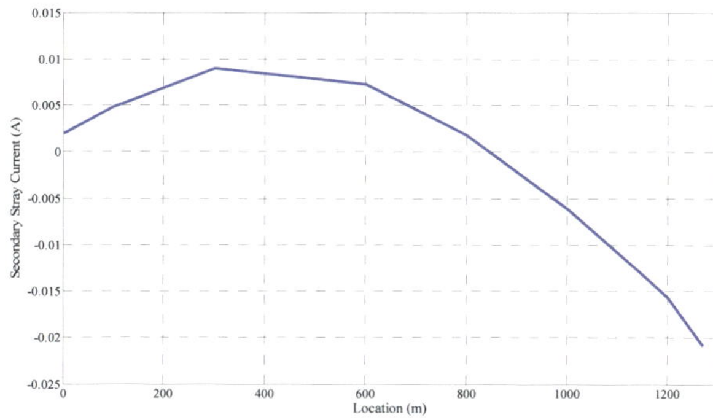

Fig. 31. Current leak to the ground when the train is at location 300 m from the first substation and the cable and mat are connected at two points.

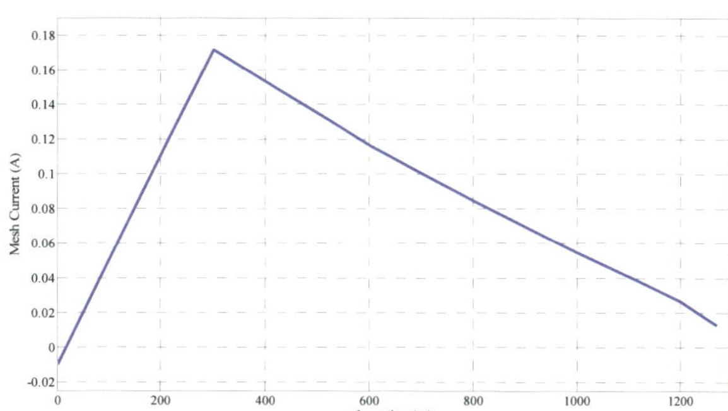

Fig. 32. The stray current collected by the mats when the train is at location 300 m from the initial substation and the cable and mat are connected at two points.

5.4 Simulation results in the case of usage passenger stations without traction substation

Some stations in line 4 of Tehran Metro are equipped with electrical substations. The traction power in these stations is supplied form substations of the neighboring stations. The train traction current should return to these substations via the running rails. Since the current path, compared to the previous cases, is increased, different stray current amounts and rail potentials are observed in these stations. The shorter headways between trains can result in presence of multiple trains in neighboring stations. The minimum headway in line 4 is planned to be two minutes. In this part of the research, in order to study one of the worst scenarios, the effect of presence of four trains at the following section is investigated. For this purpose, P4 station (which has no substation) and its neighboring O4 and Q4 stations (which have substation) are discussed.

In order to analyze the stray current and rail potential, four trains are assumed to be in the following locations:

Train A, in the southern line of O4 station (travel direction towards P4 station)

Train B, in the southern line of P4 station (travel direction towards Q4 station)

Train C, in the northern line of P4 station (travel direction towards O4 station)

Train D, in the northern line of O4 station (travel direction towards P4 station)

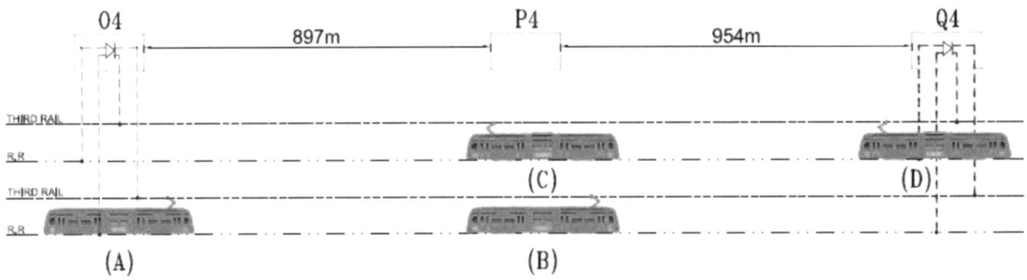

Fig. 33. Trains in the 4 train system between Q4 and O4 stations.

The order of the trains is shown in Fig. 33. It is assumed that all of the trains here start their trips simultaneously and follow the same trip profile. The mentioned parameters are investigated when the trains have traveled 300 m from their initial stations and require the highest amount of traction current. The order of trains, in this case, is shown in Fig. 34.

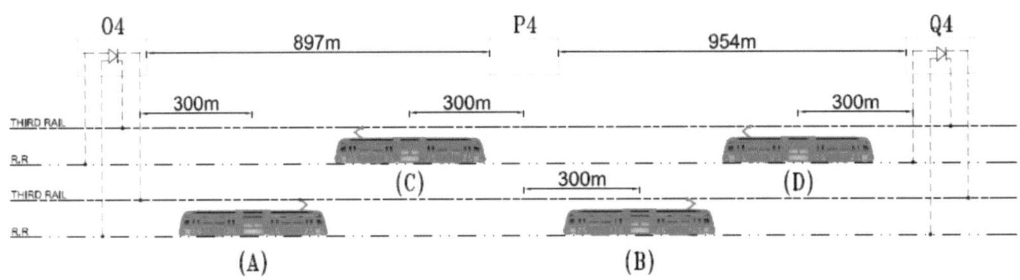

Fig. 34. Location of trains when they are 300m away from their initial stations.

Fig. 35 shows the rail stray current for the southern and northern lines when no collection mats is used. It is obvious that current leakage in P4 station is more than any other location in the line. These numbers also indicate that the rail potential is high in the neighborhood of these stations.

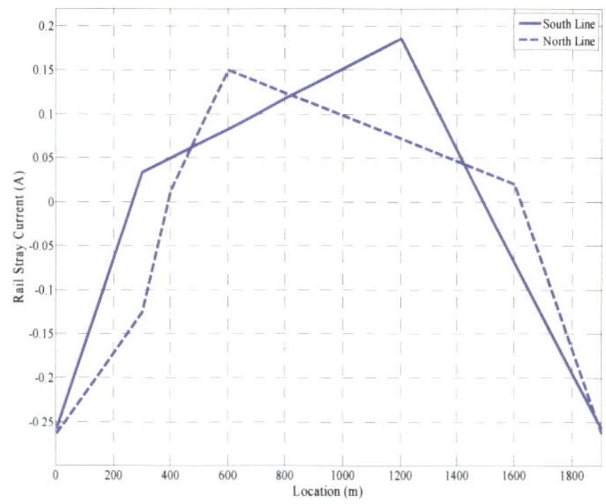

(a)

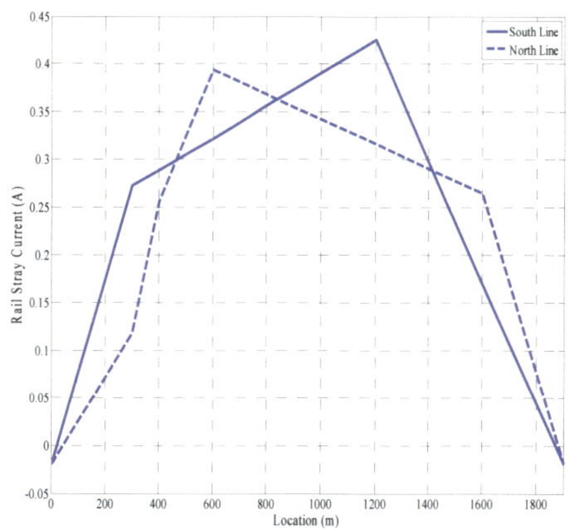

(b)

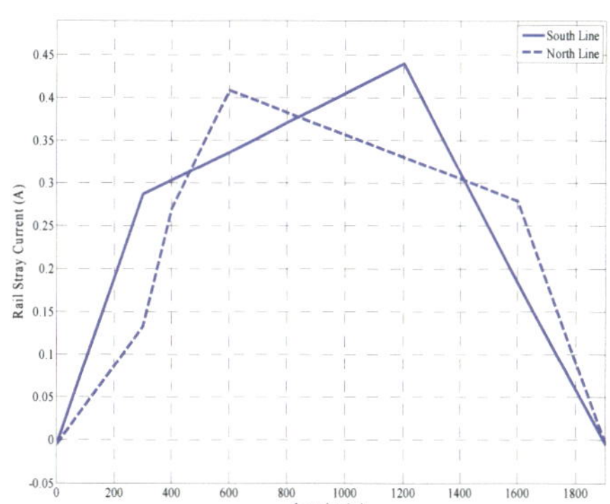

(c)

Fig. 35. The rail stray current in the four-train system in the southern and northern lines a) Floating system b) Diode grounded system c) Solidly grounded system.

The total rail stray current in the southern and northern line is 1.6 and 1.26 amperes for the floating system, 5.87 and 5.55 amperes for the diode grounded system, and 6.17 and 5.84 amperes for the solidly grounded system, respectively. These numbers are for the times when all 4 trains are consuming their maximum traction supply current from the network.

Utilizing stray current collection mats under the rails, with the previously mentioned characteristics, would highly minimize the amount of stray current leakage to the ground. Fig. 36 presents the rail voltage, rail stray current, stray current leakage to the ground and the stray current collected by the mats in the current scenario.

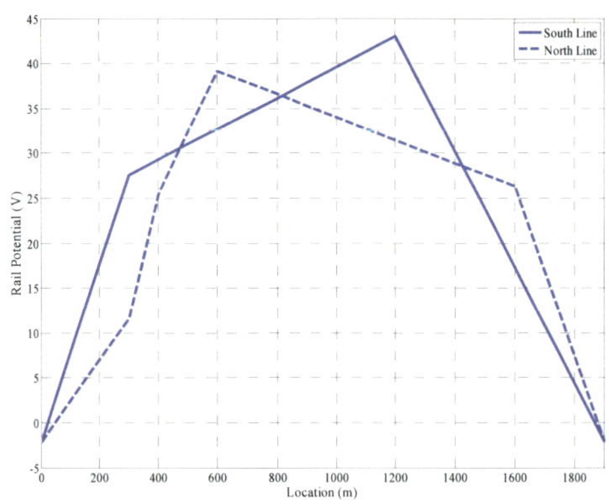

(a)

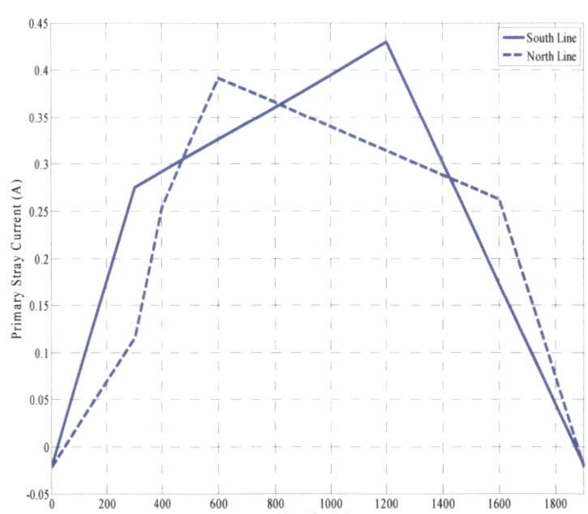

(b)

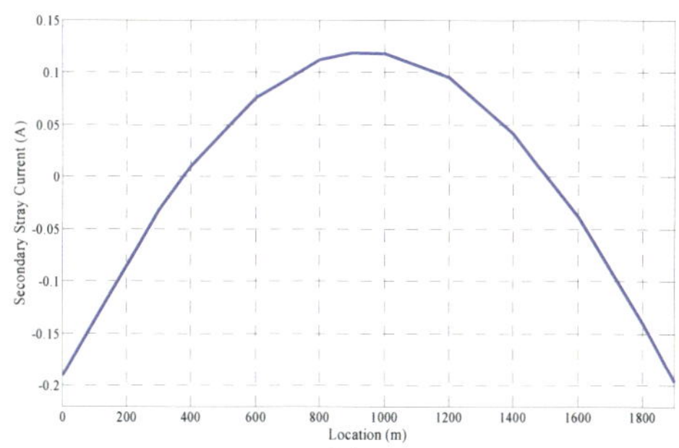

(c)

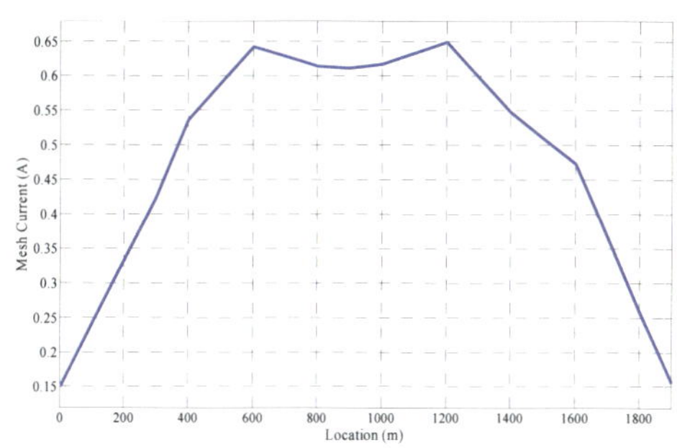

(d)

Fig. 36. The rail potential and stray current of the two lines when collection mats are used a) Rail potential b) Rail stray current c) Current leakage to the ground d) Current entering the metallic structure of the concrete.

In this case, the stray current leakage from the southern and northern lines is 5.95 and 5.51 amperes, respectively; however, only a current of 1.14 amperes leaks to the ground. In fact, the efficiency of the collector system is 90%. The system efficiency can be further improved by using a collector cable alongside the underlying collection mat, which results in even lower stray current leakage to the ground.

Fig. 37 shows the rail potential and currents in a system that has both the collection mat and the collector cable. In this system, stray current leakage from the southern and northern line is 5.96 and 5.52 amperes, respectively. Also, the current leakage to the ground is 0.71 amperes, which results in system efficiency of 94%.

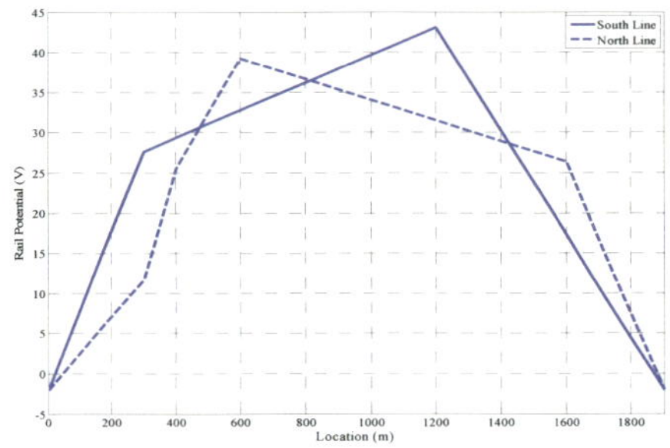

a)

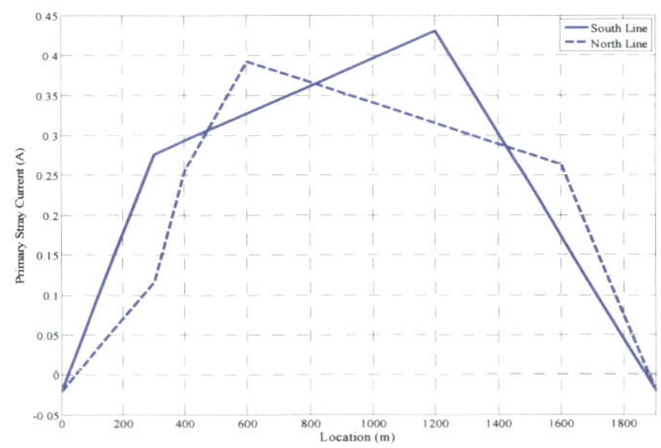

b)

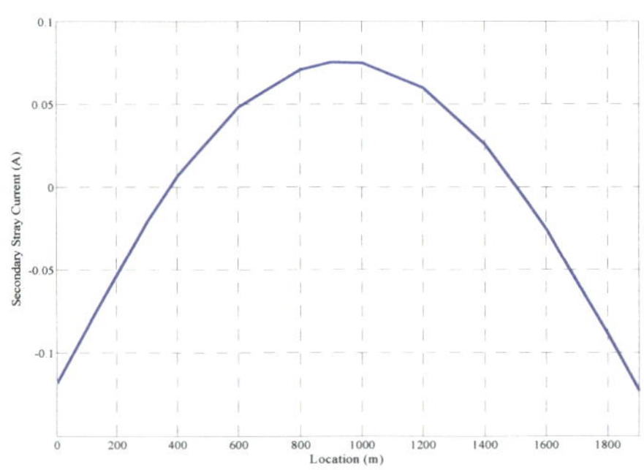

c)

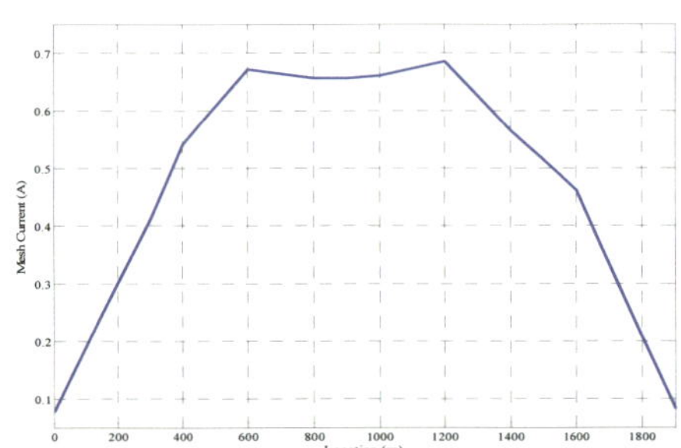

d)

Fig. 37. Rail potential and stray currents of the two lines when collection mats and collector cable are used a) Rail potential b) Rail stray current c) Current leakage to the ground d) Current entering the current collector mat.

6. Conclusions

Based on the performed simulations, it can be concluded that among the existing grounding schemes, the solidly grounded system creates the highest amounts of stray current. Using floating ground systems reduces the stray currents leakage. The systems that create direct rail to ground connections increase the corrosion rates significantly. The higher the connection current and the longer the time, the higher the magnitude of the resulting corrosions would be.

The metallic mats of track-beds and foundations increase the stray current leakage to the ground. The stray current travels through the mat and creates metallic corrosion in system terminations. If the mats are used as low resistance paths for absorbing stray currents and directing them to the substations, not only the corrosions of the mats, but also damages to the neighboring structures are avoided. For this purpose, the detached structures are bonded to each other and finally connected to the negative busbar of the substation. Employing stray current collection mats greatly reduces current leakages to ground. Increased cross sectional areas (of cables) and unified connectivity would further improve system efficiency and protection of the mats against corrosions. Stray current collector cables are also used for increasing system efficiency and protecting against corrosions. These cables are insulated from the ground and by collecting the stray currents from the mats, greatly diminish current leakages to soil. The cables can be connected to the collection mats at specific locations. The number of connections depends on the magnitude of stray current at a specific location.

Using collection systems as mats can help collect more than 0.85 of rails stray currents. Also addition of cables to these systems further boosts the system lifetime and stray current collection up to 0.94.

However, usage of stray current collection system causes floating system acts like diode grounded system, but as shown in the last scenario, use of voltage control device is necessary in stations without traction substations.

At locations where there is negative current leakage to the ground, the health of the existing metallic structures is threatened. Therefore it is recommended to use stray current and metallic structures voltage changes monitoring systems, at these locations, that conform to EN50122-2 standard [12].

7. Acknowledgements

The authors would like to thank Tehran Urban and Suburban Railway Company for assistance to perform the testing, study and evaluation. They express their appreciations to Mr. Hamed Zafari for his effort to revise this paper in English language.

8. References

[1] Y. C. Liu and J. F. Chen, Control scheme for reducing rail potential and stray current in MRT systems, IEE Proc. Electr. Power Appl.,2005, vol. 152issue 3, pp 612-618.

[2] D. Paul, DC traction power system grounding, IEEE Trans. Ind. Appl., 2002, vol. 38, pp 818–824.

[3] G. Yu and C. J. Goodman, Modeling of rail potential rise and leakage current in DC rail transit systems, Presented at IEE colloquium on Stray current effects of DC railways and tramways, 1990, pp 221-226.

[4] S. Case, DC Traction Stray Current Control. So What is the Problem?, Inst. Elect. Eng. Seminar (1999).

[5] C. H. Lee, and Wang, H.M., Effects of earthing schemes on rail potential and stray current in Taipei rail transit systems, IEE Proc. Electr. Power Appl., vol. 148(2001), 148–154.

[6] C. H. Lee and C. J. Lu, Assessment of Grounding Schemes on Rail Potential and Stray Currents in a DC Transit System, IEEE Trans. on Power Delivery,2006, vol. 21, pp.1941-1947

[7] C. Charalambous and I. Cotton, Influence of soil structures on corrosion performance of floating-DC transit systems, IET Electr. Power Appl., 2007, vol. 1, pp .9-16

[8] I. Cotton, P. Aylott and P. Ernst, Stray Current Control in DC Mass Transit Systems, IEEE Transaction on vehicular technology,2005, vol. 54, no.2. pp 722-730

[9] C. Lee, Evaluation of the Maximum Potential Rise in Taipei Rail Transit Systems, IEEE Transactions on Power Delivery, 2005, vol. 20, no. 2, pp. 1379-1384.

[10] C. H. Lee and Y. S. Tzeng, Assessment of grounding, bonding, and insulation on rail potential and stray currents in a direct current transit system, JRRT206, 2009, vol. 223, pp. 229-240

[11] W. M. Sim, C. F. Chan, Stray current monitoring and control on Singapore MRT system, IEEE international Conf. on power system tech., Powercon (2004), pp 1898-1903.

[12] European Standard EN 50122 -2, Railway applications - Protection again leaked currents, CENELEC, Bruxelles(1999).

10

Cellular Automaton Modeling of Passenger Transport Systems

Akiyasu Tomoeda
Meiji Institute for Advanced Study of Mathematical Sciences, JST CREST, Meiji University, 1-1-1 Higashi Mita, Tama-ku, Kawasaki, Kanagawa, Japan

1. Introduction

Jamming phenomena are observed everywhere in our daily live. These stagnations in flow occur not only on highways, but also in stadiums, in public transportation like buses and trains, in the world of the Internet, and even in our bodies. Almost everyone will have a negative image of a "jam", which means the clogging of the flow of traffic, however we do have positive reactions to some kinds of jams. For instance, it is gratifying to interrupt the transmission of infectious disease or prevent the spreading of fire.

The important point is that all these kinds of phenomena have the commonality of being a congestion in a transporting process which comes about through a universal jamming formation process. Especially from the point of statistical physics, these jamming phenomena are also interesting as a system of interacting particles, such as vehicles, pedestrians, ants, Internet packets, and so on, driven far from equilibrium. By considering all the above particles in various transporting processes as "Self-Driven Particles" (SDPs), we are allowed to treat various transportation phenomena universally under the physics of complex systems [1–3] . This interdisciplinary research on jamming phenomena of SDPs in various fields has been recently termed as "Jamology" [4, 5] .

Now, let us consider the state of "Jam", i.e., what is a "Jamming flow"? It is difficult for anyone to answer this question exactly. Japanese expressway companies incorporate specific threshold velocities to define the jamming flow. That is, if the average traffic velocity becomes less than the defined threshold velocity, the state of traffic is considered to have transitioned to a jamming flow. Whereas, if the average traffic velocity is above the threshold, the state of traffic flow corresponds to "Free flow". For example, one company defines the threshold velocity to be 40km per hour. In this case, if the average traffic velocity becomes less than 40km per hour, the state of traffic flow is called jamming flow. Whereas, another company identifies jamming flow by defining the threshold as 30km per hour. There is a lack of uniformity, since the definition of jamming flow depends on the company. Moreover, when it comes to considering the jamming phenomena in the dynamics of non-vehicles, such as pedestrians and ants, it becomes difficult to properly translate these definitions based on threshold velocity to another definition of the jamming state in the dynamics of non-vehicles. Thus, we should begin to provide a clear definition of jamming flow in the next section to study jamming phenomena as a mathematical science.

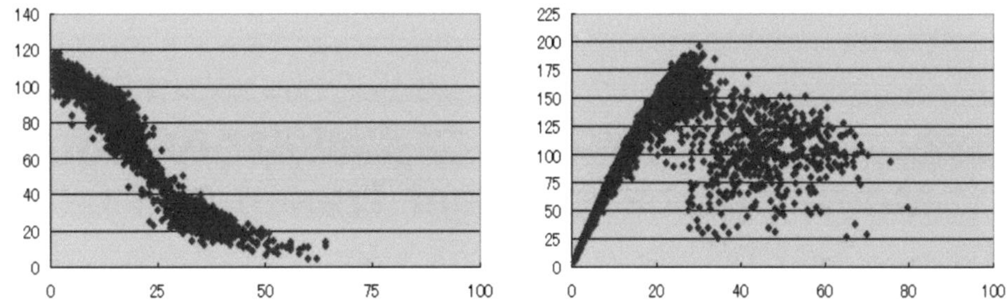

(a) density vs. velocity (b) density vs. flow

Fig. 1. Fundamental diagrams plotted by real data from a Japanese expressway (one-lane data). The horizontal axis indicates the density of vehicles (vehicles/km) and the vertical axis indicates (a) the velocity (km/hour) and (b) flow (vehicles/hour), respectively.

First, in Sec. 2 of this chapter, we introduce the *fundamental diagram* to provide a clear definition of jamming flow and explain two rule-based models for describing the dynamics of SDPs, the so-called *Asymmetric Simple Exclusion Process* (ASEP) and *Zero Range Process* (ZRP), which can capture fundamental features of jamming phenomena in various collective dynamical systems. These models have the important property of being exactly solvable, that is, their steady states are given by a form [6–11] . Therefore, we treat the behavior of particles in complex systems by not only numerical simulations but also analytical calculations in the steady state. In Sec. 3, as an extension of the above stochastic cellular automata, we explain in detail the *Public Conveyance Model* (PCM) [12] , which is a fundamental mathematical model for the passenger transport system by introducing a second field (passengers field) which tracks the number of waiting passengers. In addition, by introducing the route choice model of passengers explicitly into PCM, we have built a real-time railway network simulation tool "KUTTY", which has been applied to the *Tokyo Metro Railway Network* [13, 14] , as described in Sec. 4. Finally, Sec. 5 is devoted to concluding discussions.

The aim of this chapter is to understand the mathematical model for the passenger transport system built on analytical rule-based models and to introduce the real-time railway network simulation tool "KUTTY" as an application of our proposed model.

2. Fundamental diagrams and stochastic cellular automata

Now we introduce the *fundamental diagram* to provide a clear definition of jamming flow. The fundamental diagram is a basic tool in understanding the behavior of the flow in transportation systems: it relates the flow $Q(x,t)$ in the system and the density of vehicles $\rho(x,t)$ [1] . The fundamental is sometimes drawn to indicate the relation between velocity $v(x,t)$ and density $\rho(x,t)$, however, we can easily translate velocity into flow by the relation $Q = \rho v$.

Fig. 1 is an example of the fundamental diagrams of a Japanese expressway. From these diagrams, it is obvious that the state of traffic flow and velocity strongly depend on the density

[1] In the following, we use *"particle"* in mathematical models to represent a vehicle, a bus, or a train, to keep this discussion as general as possible.

of vehicles on the road. So long as the density is sufficiently small, the average velocity is practically independent of the density as the vehicles are too far apart to interact. Therefore, at sufficiently low density of vehicles, the system effectively acts in a state of "free flow". However, in practice, vehicles have to move more slowly with increasing density. This reality is correctly described in the fundamental diagrams.

As mentioned before, each expressway company in Japan defines the jamming state by a threshold velocity. Hence, there is no universal definition of the jamming state that can be treated in a mathematical sense. In order to provide a clear definition of "free flow" and "jamming flow", one transforms the vertically plotted value from velocity to flow, i.e., from Fig. 1 (a) to Fig. 1 (b). Surprisingly, this type of fundamental diagram (Fig. 1 (b)) shows the universal features not only in the dynamics of traffic vehicles but also for other general SDPs as follows (also see Fig. 2):

(a) At low density, there is almost a linear relation between the flow and the density, which intersects at zero. The slope at low density corresponds to the average velocity without congestion.

(b) If the density exceeds some critical value, the so-called *critical density* ρ^c, the flow decreases monotonically and it vanishes together with the velocity at some maximum density. [2]

(c) The flow has one maximum value at medium density.

The critical density indicates the changing point of the flow state from free flow to jamming flow. Therefore, free flow and jamming flow can be defined as the lower density region and higher density region which are separated by the critical density as shown in Fig. 2. Once free flow and jamming flow are defined in terms of the fundamental diagram, one can determine whether the flow is really in a state of "Jam" or not in a rigorous, mathematical sense, even in the dynamics of various other kinds of SDPs, where the jamming state has been considered undefinable or unclear. Thus, the fundamental diagrams are essential to treat the jamming phenomena as a mathematical science.

Now let us introduce two simple stochastic cellular automaton models, i.e., ASEP and ZRP, which are the simple models for non-equilibrium systems of interacting self-driven particles. In these cellular automaton models, the path (a road or rail) is partitioned into L identical cells such that each cell can accommodate at most one particle at a time, enforcing the so-called *exclusion principle*, that is, the excluded-volume effect is not supposed to be ignored unlike in the flow of water. [3] Generally, the dynamics of these models are described by a rule [4]. The rule for dynamics of particles in case of ASEP is very simple, i.e., "*If the front cell is empty, a particle can move forward with hopping probability p.*" as shown in Fig. 3. Note that, in general,

[2] In several situations involving vehicle dynamics, it has been observed that flow does not depend uniquely on density in an intermediate regime of density. This indicates the existence of a *hysteresis effect* and *metastable states*. The critical density in vehicle dynamics is about 25 (vehicles/km).

[3] In traditional queuing theory, this excluded-volume effect and spatial structure have never been introduced into the queuing model. An extension of the $M/M/1$ queuing process with a spatial structure and excluded-volume effect is introduced in [15, 16] , as the TASEP on a semi-infinite chain with open boundary.

[4] Some of them can be described in the form of equations, such as a "master equation" or "max-plus equation / tropical-polynomial".

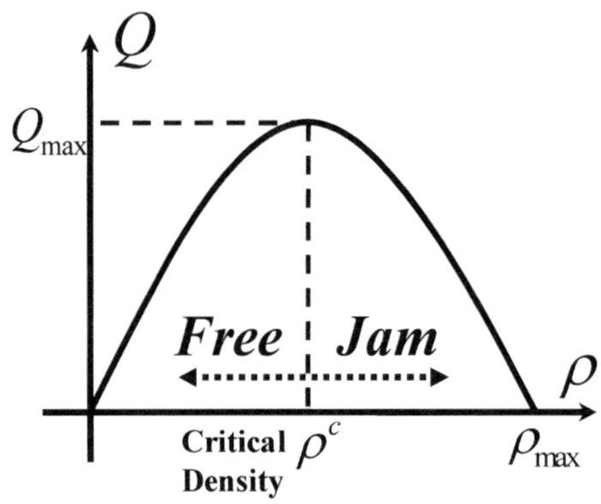

Fig. 2. Features and definition of the jamming state in the simplified fundamental diagram (density vs. flow).

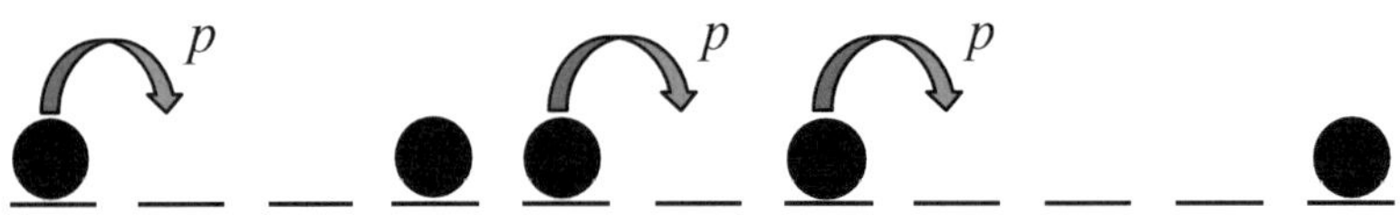

Fig. 3. Dynamics of the Asymmetric Simple Exclusion Process: if the next cell is empty, a particle can move forward with hopping probability p.

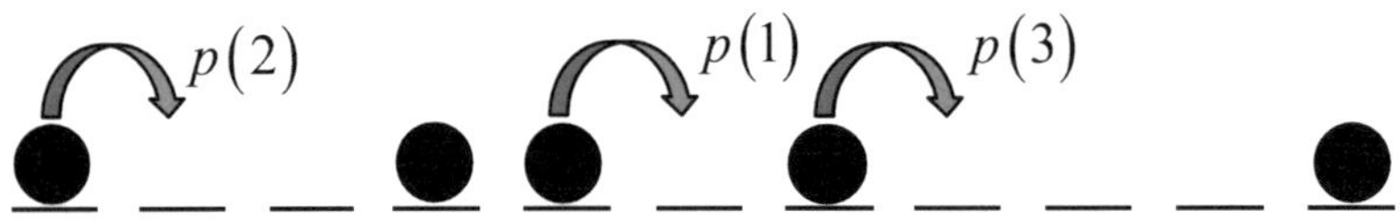

Fig. 4. Dynamics of the Zero Range Process after the mapping to the asymmetric exclusion process: if the next cell is empty, a particle can move forward with hopping probability $p(h)$, which depends on the distance to the next particle in front.

ASEP is characterized by the hopping rate

$$P_{\{10\}\to\{01\}} = p, \tag{1}$$

$$P_{\{01\}\to\{10\}} = q, \tag{2}$$

which is considered as a model of interacting random walks.

The most important special case, theoretically as well as in the application to transport systems, is known by the full name of *Totally Asymmetric Simple Exclusion Process* (TASEP), which has $q = 0$ so that its motion is allowed only in one direction. Here, we use the general name ASEP to mean TASEP for simplicity' sake. Moreover, the hopping probability p is usually called the hopping rate in ASEP, since time is continuous in ASEP. However, we call the rate p the hopping probability p, since we treat only the discrete time case in this chapter.

In the case of ZRP, after a suitably precise mapping, the rule is considered as "*If the front cell is empty, a particle can move forward with hopping probability $p(h)$, which depends on the distance to* *the next particle in front.*", as illustrated in Fig. 4. Indeed, ASEP is considered as a special case of ZRP.

Here, we impose periodic boundary conditions [5], that is, we consider that the particles to move on a circuit so that the transport system is operated as a loop. This means that the number of particles N on the circuit is conserved at each discrete time step. Moreover, we distinguish three basic types of dynamics: the dynamical variables may be updated one after the other in a certain order (*sequential update*), one after another in random order (*random-sequential update*), or in parallel for all sites (*parallel update*). [6] Now let us employ the parallel update for all sites in the system as the updating procedures.

As mentioned before, ASEP and ZRP have the important property of being exactly solvable, that is, their steady states are given by a form [6–11] . In the case of ASEP with periodic boundary condition and in parallel dynamics, one obtains the form (see [7] for details)

$$Q(\rho) = \frac{1}{2}\left[1 - \sqrt{1 - 4p\rho(1-\rho)}\right], \tag{3}$$

where $Q(\rho)$ is flow of particles, p is the hopping probability and $\rho = N/L$ is the density of particles.

On the other hand, in the case of ZRP, as a simple example, we now assume that the hopping probability is

$$p(1) = p, \quad p(h \geq 2) = q. \tag{4}$$

If $q = p$, this model is reduced to the ASEP with hopping probability p, as denoted above. Moreover, if $p = q = 1$ in ZRP or $p = 1$ in ASEP, this model is reduced to the deterministic version of ASEP, which is called the rule-184 cellular automaton. This rule-184 CA model is one of the elementary cellular automata, which are defined by S. Wolfram (1959 $\sim$) [17, 18] . Various extensions of this rule-184 CA are also proposed as a powerful model for realistically describing one-dimensional traffic flow.

The fundamental diagram of ZRP defined by (4) with periodic boundary conditions and parallel update in a parametric representation, where the density $\rho(w) = 1/(1+h)$ and the flow $Q(\rho) = w\rho(w)$ are calculated for the parameter $0 \leq w \leq 1$, is given by the following equations (see [11] for details)

$$F(w) = \Big(1 - p(1)\Big)(1+w) \sum_{n=0}^{\infty} \Big(w^n \prod_{j=1}^{n} \frac{1-p(j)}{p(j)}\Big), \tag{5}$$

$$h(w) = w\frac{\partial}{\partial w}\Big(\log F(w)\Big). \tag{6}$$

Fig. 5 shows the numerical simulation results (dots) and analytical calculated results (line) for ASEP and ZRP, respectively. In both cases, the analytical results show good agreements with the numerical results. Moreover, both figures captures the universal features $(a) - (c)$ of general SDPs in transport systems, which were introduced earlier in this section. Therefore,

[5] ASEP and ZRP with open boundary conditions are also well investigated.

[6] Sometimes, updating the dynamics in parallel for all sites of a given sub-lattice (*sub-lattice update*) is distinguished from this parallel update.

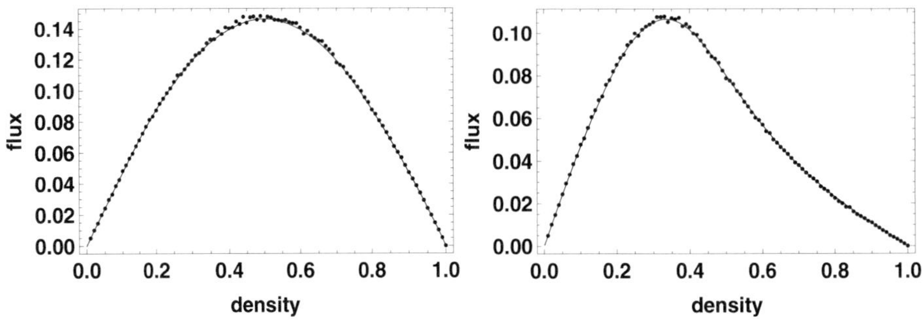

(a) ASEP (b) ZRP

Fig. 5. Fundamental diagram of (a) Asymmetric Simple Exclusion Process and (b) Zero Range Process with the parallel update. The dots and the line correspond to the simulation data and analytical results, respectively. The numerical simulations are done with $L = 100$ sites. In the case of ASEP, the hopping probability p is 0.5. In the case of ZRP, the hopping probability $p(h)$ is $p(0) = 0, p(1) = 0.1, p(2 \geq h) = 0.5$.

various extensions of these models have been reported in the last few years for capturing the essential features of the collective spatio-temporal organizations in wide varieties of systems, including those in vehicular traffic [1, 2, 19] .

Until now, public conveyance traffic systems such as buses, bicycles and trains have also been modeled by an extension of ASEP using similar approaches [12–14, 20, 21]. A simple bus route model [21] exhibits clustering of the buses along the route. The quantitative features of the coarsening of the clusters have strong similarities with coarsening phenomena in many other physical systems. Under normal circumstances, such clustering of buses is undesirable in any real bus route as the efficiency of the transportation system is adversely affected by clustering.

In the next section, a new public conveyance model (PCM) will be explained which is applicable to buses and trains in a transport system by introducing realistic effects encountered in the field (the number of stops (stations) and the behavior of passengers getting on a vehicle at stops) into the stochastic cellular automaton models.

3. Public conveyance model for a bus-route system

Now we will explain the PCM in detail. Although we refer to each of the public vehicles as a "bus", which is a one-dimensional example of a transport system, the model is equally applicable to train traffic on a given route. We impose periodic boundary conditions as well as the stochastic cellular automata described in the previous section and partition the road into L identical cells. Moreover, a total of S $(0 \leq S \leq L)$ *equispaced* cells are identified in the beginning as bus stops. Note that, the special case $S = L$ corresponds to the *hail-and-ride* system, in which the passengers could board the bus whenever and wherever they stopped a bus by raising their hand. In contrast to most of the earlier bus route models built on the stochastic cellular automaton, we assume that the maximum number of passengers that can get into one bus at a bus stop is $N_{\max}$, which indicates the *maximum boarding capacity* at each bus stop rather than the *maximum carrying capacity* of each bus.

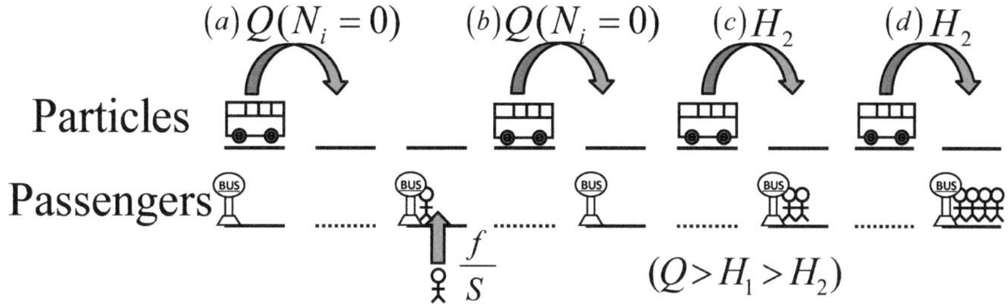

Fig. 6. Schematic illustration of the PCM. The hopping probability to the bus stop depends on the number of waiting passengers. Accordingly, if the waiting passengers increase, the hopping probability to the bus stop decreases. Although case (a) and (b) is the same hopping probability, the situation is different: (a) next cell is without bus stop, (b) next cell is with bus stop but without passengers. Case (d) has smaller probability than the case (c), since the hopping probability depend on the number of waiting passengers. Of course, case (c) is smaller than the case (b) (also case (a))).

The symbol H is used to denote the hopping probability of a bus entering into a cell that has been designated as a bus stop. We assume H has the form

$$H = \frac{Q}{\min(N_i, N_{\max}) + 1} \tag{7}$$

where $\min(N_i, N_{\max})$ is the number of passengers who can get into a bus which arrives at the bus stop i at the instant of time when the number of passengers waiting at the bus stop i $(i = 1, \cdots, S)$ is N_i. The form (7) is motivated by the common expectation that the time needed for the passengers to board a bus is proportional to their number. Fig. 6 depicts the hopping probabilities schematically. The hopping probability of a bus to the cells that are not designated as bus stops is Q; this is already captured by the expression (7) since no passenger ever waits at those locations.

If the form H does not depend on the number of waiting passengers but depends on the presence of passengers in the case $S = L$, i.e.,

$$H = \begin{cases} Q & \text{no waiting passengers,} \\ q & \text{waiting passengers exist,} \end{cases} \tag{8}$$

where both Q and q $(Q > q)$ are constants independent of the number of waiting passengers, this model corresponds to the *Ant-Trail-Model*, which also shows quite similar clustering phenomena to those of vehicles in the collective movement of ants and obtains the results through approximate analysis [22, 23] . Moreover, if the form H is always constant, this model is reduced to the ASEP.

In principle, the hopping probability H for a real bus would depend also on the number of passengers who get off at the bus stop; in the extreme situations where no passenger is waiting at a bus stop, the hopping probability H would be solely decided by the disembarking passengers. However, in order to keep the model theoretically simple and tractable, we ignore the latter situation and assume that passengers get off only at those stops where waiting passengers get into the bus and that the time taken by the waiting passengers to get into the bus is always adequate for the disembarking passengers to get off the bus.

The PCM model reported here can be easily extended to incorporate an additional dynamical variable associated with each bus to account for the instantaneous number of passengers in it. But, for the sake of simplicity, such an extension of PCM is not reported here [7]. Instead, we focus on the simple version of PCM. As shown in Fig. 7, the model is updated according to the following rules:

1. *Arrival of a passenger*
   A bus stop i ($i = 1, \cdots, S$) is picked up randomly, with probability $1/S$, and then the corresponding number of waiting passengers is increased by unity, i.e., $N_i \to N_i + 1$, with probability f to account for the arrival of a passenger at the selected bus stop. Thus, the average number of passengers that arrive at each bus stop per unit time is given by f/S.
2. *Bus motion*
   If the cell in front of a bus is not occupied by another bus, each bus hops to the next cell with probability H. Specifically, if passengers do not exist in the next cell the hopping probability equals Q because N_i is equal to 0. Otherwise, if passengers exist in the next cell, the hopping probability is $Q/(\min(N_i, N_{\max}) + 1)$. Note that, when a bus is loaded with passengers to its maximum boarding capacity $N_{\max}$, the hopping probability is $Q/(N_{\max} + 1)$, the smallest allowed hopping probability.
3. *Boarding a bus*
   When a bus arrives at the i-th ($i = 1, \cdots, S$) bus stop cell, the corresponding number N_i of waiting passengers is updated to $\max(N_i - N_{\max}, 0)$ to account for the passengers boarding the bus. Once the door is closed, no more waiting passengers can get into the bus at the same bus stop although the bus may remain stranded at the same stop for a longer period of time either because of the unavailability of the next bus stop or because of the traffic control rule explained next.

We introduce a traffic control system that exploits the information on the number of buses in each *segment* between successive bus stops, as well as a block section of the railway system. Every bus stop has information I_i ($i = 1, \cdots, S$) which is the number of buses in the i-th segment of the route between the i-th and next $(i+1)$-th bus stops at that instant of time. If I_i is larger than the average value $I_0 = m/S$, where m indicates the total number of buses, a bus remains stranded at a stop i as long as I_i exceeds I_0. In steps 2, 3, and the information-based control system (step 4), these rules are applied in parallel to all buses and passengers, respectively.

We use the average speed $\langle V \rangle$ of the buses, the number of waiting passengers $\langle N \rangle$ at a bus stop and the transportation volume R, which is defined by the product of velocity of the i-th bus $V_i \in \{0, 1\}$ and the number of on-board passengers M_i ($0 \le M_i \le N_{\max}$), i.e.,

$$R = \sum_{i=1}^{m} M_i V_i, \tag{9}$$

as three quantitative measures of the efficiency of the public conveyance system under consideration; a higher $\langle V \rangle$, a higher R and smaller $\langle N \rangle$ correspond to a more efficient transportation system.

[7] We have reported the extended PCM by incorporating the disembarking passengers explicitly for the case of the elevator system in [24] .

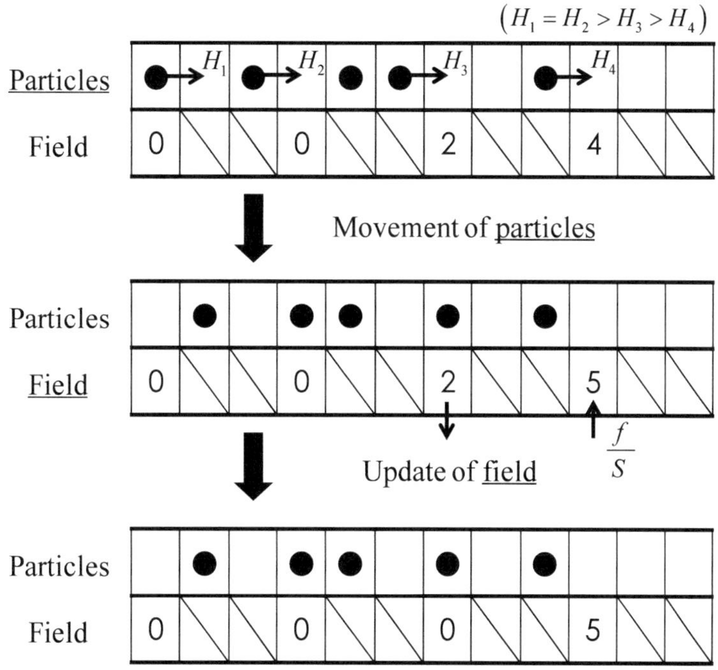

Fig. 7. Time development of public conveyance model. The upper path of cells indicates the state of particles and the lower one indicates the state of passengers. The number in each cell represents the number of waiting passengers. The cells with diagonal lines indicate cells without bus stops.

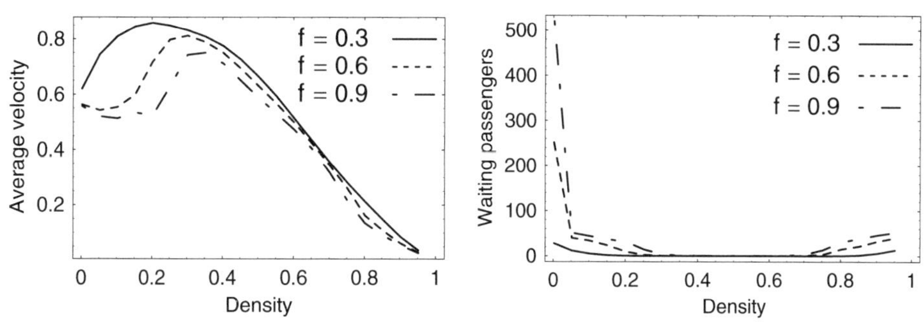

(a) Average velocity $\langle V \rangle$

(b) Waiting passengers $\langle N \rangle$

Fig. 8. The plot of $\langle V \rangle$ and $\langle N \rangle$ without information for $S = 5$ and $f = 0.3, 0.6$ and 0.9.

In the simulations presented here, we set $L = 500, S = 5, Q = 0.9, q = 0.5$ and $N_{\max} = 60$. The main parameters of this model, which we varied, are the number of buses m and the probability f of the arrival of passengers. The density of buses is defined by $\rho = m/L$ in the same way. We study not only the efficiency of the system but also the effects of our control system by comparing the characteristics of two traffic systems one of which includes the information-based control system while the other does not.

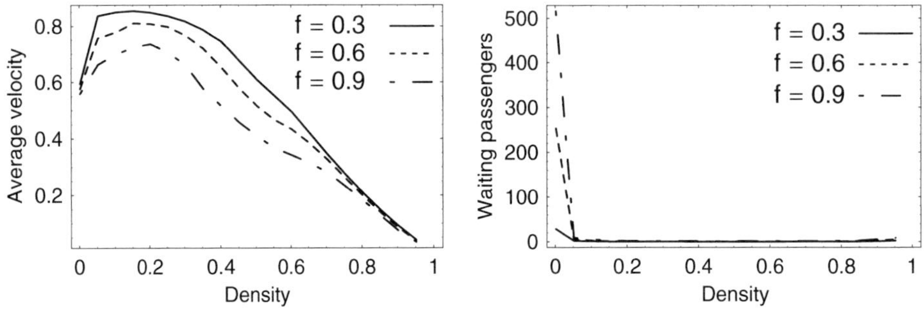

(a) Average velocity $\langle V \rangle$ (b) Waiting passengers $\langle N \rangle$

Fig. 9. The plot of $\langle V \rangle$ and $\langle N \rangle$ with information for $S = 5$ and $f = 0.3$, 0.6 and 0.9

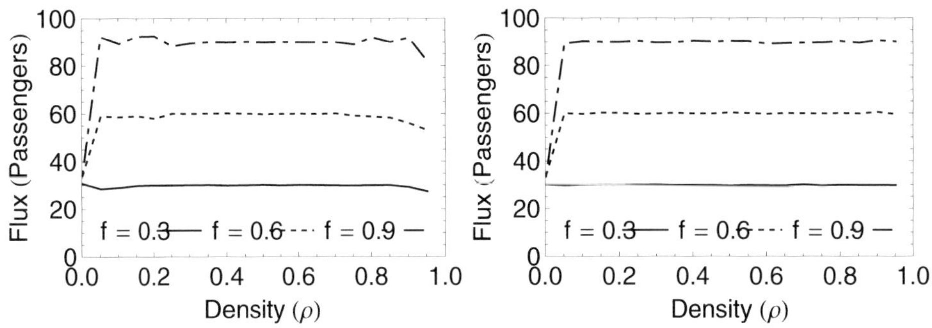

(a) without information (b) with information

Fig. 10. Fundamental diagrams for the transportation volume. (a) The case without the information-based control system, (b) The case with the information-based control system.

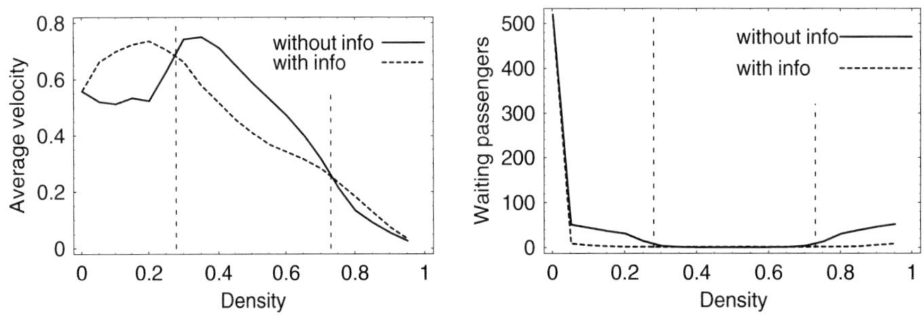

(a) Average velocity $\langle V \rangle$ (b) Waiting passengers $\langle N \rangle$

Fig. 11. Two efficiency plots $\langle V \rangle$ and $\langle N \rangle$ for the parameters $S = 5, Q = 0.9, q = 0.5$, and $f = 0.9$.

Some of the significant results of the numerical simulations of the PCM are as follows. In Fig. 8 and Fig. 9, we plot $\langle V \rangle$ and $\langle N \rangle$ against the density of buses for several different values of f. Fig. 8 (a) demonstrates that the average speed $\langle V \rangle$, which is a measure of the efficiency of the bus traffic system, exhibits a *maximum* at around $\rho = 0.2 \sim 0.3$, which reflects the bus bunching especially at large f. As shown in Fig. 8 (b), The average number of waiting

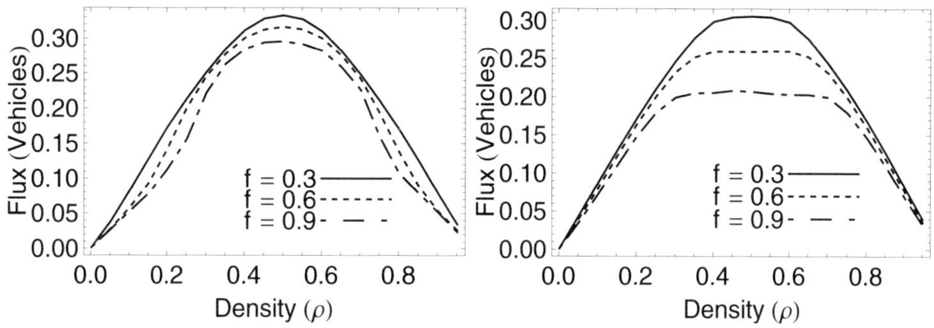

(a) without information

(b) with information

Fig. 12. Fundamental diagrams for the flow of vehicles. (a) The case without the information-based control system, (b) The case with the information-based control system.

passengers $\langle N \rangle$, whose inverse is another measure of the efficiency of the bus traffic system, is vanishingly small in the region $0.3 < \rho < 0.7$; $\langle N \rangle$ increases with decreasing (increasing) ρ in the regime $\rho < 0.3$ ($\rho > 0.7$). The results for the PCM with information-based traffic control system are shown in Fig. 9. The density corresponding to the peak of the average velocity shifts to lower values when the information-based traffic control system is switched on. The average number of waiting passengers $\langle N \rangle$ decreases between Fig. 8 (b) and Fig. 9 (b) in the regions $0.1 < \rho < 0.3$ and $0.7 < \rho < 0.9$.

The other measurement of the efficiency, the transportation volume R, is shown in Fig. 10. The optimal density which shows higher $\langle V \rangle$ and lower $\langle N \rangle$ does not always correspond to the most efficient operation for the transportation volume, since that is maintained substantially constant except at low density, even though the density of vehicles increases. This is because the number of buses with small transportation volume increases even though the average velocity decreases. Thus, we have found that the excess buses result in unneeded buses which has no passengers since the transportation volume is the same.

The data shown in Fig. 11 establishes that implementation of the information-based traffic control system does not necessarily always improve the efficiency of the public conveyance system. In fact, in the region $0.3 < \rho < 0.7$, the average velocity of the buses is higher if the information-based control system is switched off. Comparing $\langle V \rangle$ and $\langle N \rangle$ in Fig. 11, we find that the information-based traffic control system can improves the efficiency by reducing the crowd of waiting passengers. However, in the absence of waiting passengers, introduction of the information-based control system adversely affects the efficiency of the public conveyance system by holding up the buses at bus stops when the number of buses in the next segment of the route exceeds I_0. Therefore, we have found the information-based traffic control system can improve the efficiency in a certain density region, but not in all possible situations.

The typical fundamental diagrams for the flow of vehicles in the PCM are given in Fig. 12. The flow of vehicles without the information-based control system gradually decreases as the arrival rate of passengers increases. In contrast, the flow with the information-based control system drastically decreases for intermediate densities, where there are no waiting passengers in Fig. 9, showing a trapezoidal shape. This trapezoidal shape is similar to the fundamental diagram in [25] and [26] , where a blockage effect is artificially introduced into the rule-184 cellular automaton to take a flow bottleneck into account. Thus, in the absence of waiting

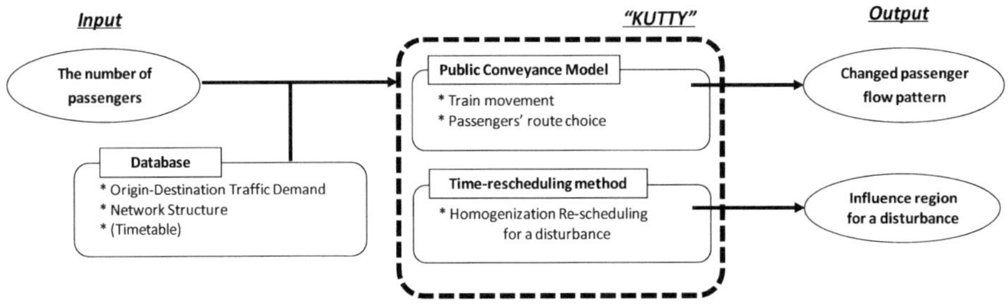

Fig. 13. The design flow diagram of our simulator "KUTTY".

passengers, implementation of the information-based control system corresponds to creating a bottleneck effect and decreases the flow of vehicles.

As a theoretical approach, mean-field analysis is also applied to the model to estimate the efficiency. The details of mean-field analysis can be found in [12–14] .

Next, let us explain an expanded PCM for the railway transportation system in the Tokyo metropolitan area by introducing the route choice model of passengers.

4. Public conveyance model for Tokyo metro subway network

Disturbances of well scheduled trains occur everyday. The disturbances may be caused by, for example, a slight delay of trains, traffic accidents, or other factors. Under such circumstances, how does the train company adjust and operate the train schedule? The conventional method of rescheduling the timetable is that the scheduling specialists adjust the timetable by their own empirical rules. This conventional method usually only considers the performance of a single (metro) line and does not take into account the performance of the whole network. Furthermore, this method tries to adjust the trains so that they are equally-spaced in time without considering the flow of passengers. Recently, various kinds of models of the train network, which treat the network as a complex system, have been proposed and studied [27–30]. However, the behaviors of passengers have not been considered in these models either.

A public conveyance model (PCM) which could reproduce the bus clustering phenomenon and estimate the efficiency of the bus system on a one-dimensional route was proposed in the previous section, where passengers arrive at a stop randomly and their destinations have not been considered explicitly. Furthermore, since buses run on a single route, passengers do not need to choose the route. In this section, we have extended the PCM by introducing realistic passengers' behaviors explicitly and proposed a real-time railway network simulation tool "KUTTY", which has been applied to the *Tokyo Metro Railway Network*. It is shown that the passenger flow pattern is well simulated. Moreover, we have presented a homogenization re-scheduling method to alleviate congestion of crowded trains, and it is found that our method is more efficient than the conventional one.

4.1 Database

Our real-time simulator "KUTTY" is operated not only based on the input data but also by extracting data from a database (see Fig. 13). In this section, we present the assembly of the database and explain its components: *OD (origin-destination) traffic demand* and *network structure*.

4.1.1 Origin-destination traffic demand estimation

In order to apply our PCM to the railway transportation system in a realistic way, real data of the OD traffic demand of passengers is indispensable on the network. In general, it is very difficult to obtain this sort of data. Fortunately, the Tokyo Metro Company posts one-day rider-ship of all stations on its web-site. Moreover, at the mutual entry stations [8], the rider-ship also includes the influx and outflux of passengers. Nevertheless, the data only record the sum of passengers getting through the station gate and do not distinguish whether the passengers enter or leave the station. Therefore, we assume that half of the passengers enter the station and the other half leave the station. This is reasonable because most people go to work or school in the morning and go back home in the evening. Thus they return to the station where they originally entered .

Under these reasonable assumptions, we can estimate the OD demand from one-day rider-ship data of all stations and have verified that the estimated number of passengers is suitable for the data.

When a passenger enters a station i, the probability that the passenger's destination is j is

$$P_{i \to j} = \frac{N_j}{\sum_j N_j}, \tag{10}$$

where N_j indicates the number of passengers who leave from station j.

4.1.2 Network structure

In 2007, the Tokyo Metro Railway Network consisted of 8 lines and 138 stations [9] . Some of these stations are transfer stations, in which several lines intersect. In our model, the stations are mapped to nodes. Under our scheme, each node corresponds one-to-one to the ID number of each station on each line. That is, a transfer station is mapped to more than one nodes, depending on how many lines intersect (e.g., "Otemachi" of the Tozai Line (T09) and "Otemachi" of the Marunouchi Line (M18) are mapped to different nodes). The connection between neighboring stations is referred to as a "segment" and the connection inside a transfer station is referred to as a "link". Using this system, the Tokyo Metro Railway Network has 169 nodes, 170 segments and 50 links. Passengers can travel from any station to any other station in this network with at most two transfers. Therefore, the calculation time in searching for a possible route notably is decreased by restricting transfers to more than two. We have made the database of all paths by using *Dijkstra's Algorithm* [31].

4.2 Models and homogenization method

In this section, we first present the route choice model of the passengers. Then, the train movement model, which is built on PCM, is introduced. After that, we propose a homogenization re-scheduling method to alleviate congestion.

---

[8] The trains still move on from these stations, beyond which the transportation system is operated by other companies, such as Japan Railway. The mutual entry stations are shared by several companies.

[9] Since 2008, a new line has opened to traffic, but we have not included the new line here.

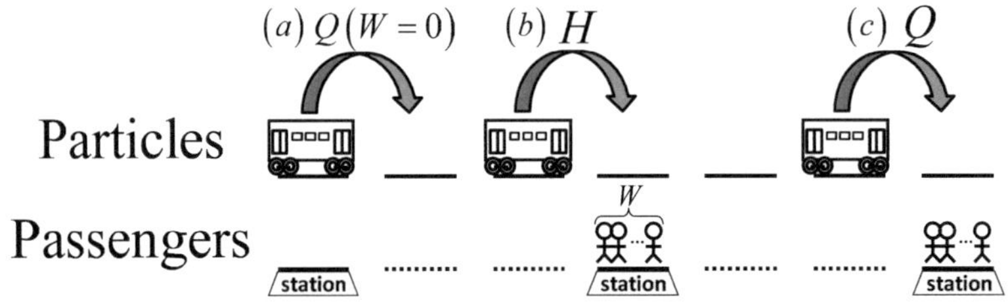

Fig. 14. Illustration of all cases of the hopping probability from a train: (a) Hopping into a non-platform cell. (b) Hopping into a platform cell (first time). (c) Hopping into a platform cell after one stop in the case of (b).

4.2.1 Route choice model

When a passenger p arrives at a station, his/her destination is determined by (10). Suppose $S(p)$ is the set of all routes from the origin to the destination. The cost E of each route $s \in S(p)$ at time t is calculated by

$$E\Big(T(p,s), C(p,s), D(p,s,t)\Big) = aT(p,s)^{\alpha} + bC(p,s)^{\beta} - cD(p,s,t)^{-\gamma}, \tag{11}$$

where $a, b, c, \alpha, \beta, \gamma$ are parameters with positive value. In this formula, $T(p,s)$ is the expected travel time on route s, $0 \leq C(p,s) \leq 2$ is the transfer number on route s. $T(p,s)$ is calculated by $T(p,s) = nT_t + C(p,s)T_c$. Here n is the number of stations on the route (excluding the origin) and T_t is set to 2 minutes, which denotes the average travel time between two neighboring stations. T_c is the average transfer time, which is set to 1 minute. The term $bC(p,s)^{\beta}$ is adopted to reflect how passengers are reluctant to transfer. Sometimes, passengers would rather choose a route with smaller $C(p,s)$, despite the fact that the travel time on this route is longer. $D(p,s,t)$ is defined as $D(p,s,t) = \max(D_{tr}, D_{pl})$, where $D_{tr} = \max_{i\in I}(D_{tr,i})$ and $D_{pl} = \max_{j\in J}(D_{pl,j})$. Here $D_{tr,i}$ is passenger density on train i at time t, and $D_{pl,j}$ is passenger density on platform j at time t; I denotes the set of all trains that move on route s at time t; J denotes the set of all platforms on route s.

The normalized probability $P(p,s,t)$, for the passenger p to select route s, is described as follows,

$$P(p,s,t) = \frac{\exp\Big[-E(p,s,t)\Big]}{\sum_{s_i \in S(p)} \exp\Big[-E(p,s_i,t)\Big]}. \tag{12}$$

This means that the smaller the cost $E(p,s,t)$ is, the larger is the probability that the route will be selected.

4.2.2 Train movement model

For simplicity, we assume that each *segment* of all lines is partitioned into four identical cells and each *platform* is designated as one cell such that each cell can also accommodate at most one train at a time. Fig. 14 depicts the hopping probabilities in the train model schematically. Let us denote the probability that trains hop into a non-platform cell as Q, the hopping probability of a train to a platform cell as H, the number of passengers waiting at the platform

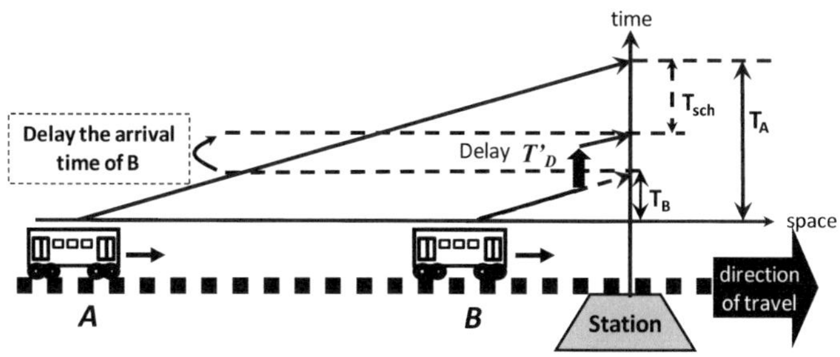

(a) One of the traditional methods

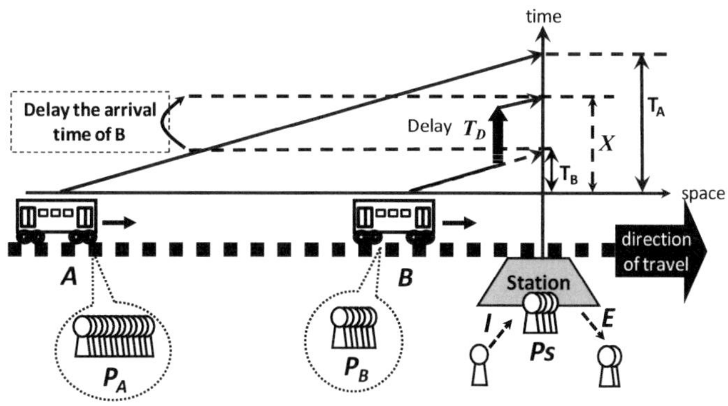

(b) Our homogenization method

Fig. 15. Two types of the re-scheduling methods. The traditional rescheduling method (a) is to adjust the time gap to T_{sch}. However, our homogenization method (b) is to adjust the time gap to a value which depends on the distribution of the number of passengers.

at the instant of time when a train arrives at the upstream neighboring cell as W and the maximum carrying capacity of trains as $W_{\max}$, we assume that

$$H = \min\left(\frac{1}{a' \min(W, W_{\max}) + 1}, Q\right), \tag{13}$$

where a' is a parameter. However, if the train fails to hop into the platform cell once, then it will hop with probability Q in the next time step. This is because trains are controlled in units of very small segment by the railway signaling system, and it is not realistic that trains stop stochastically again and again. In this section, the parameters are set to $a' = 0.2$, $Q = 1$.

4.2.3 Homogenization re-scheduling

Now let us propose a homogenization re-scheduling method. Suppose there is a sudden increase in passengers at a station. Since there are a lot of passengers waiting on the platform, the nearest upstream train from the station will become congested and thus delayed. If there is no re-scheduling, this train will become more and more delayed at the next stations.

In order to solve this problem, some re-scheduling method could be adopted by delaying the preceding trains of the delayed train, so that the congestion could be shared by preceding trains. In the conventional re-scheduling method, this is fulfilled by equalizing the time gaps between neighboring trains. For instance (see Fig. 15), suppose that train B is expected to arrive at the next station at time $t = T_B$ and train A is expected to arrive at the station at time $t = T_A$. If $T_A - T_B > T_{sch}$, then the delay time of train B is determined by $T'_D = T_A - T_B - T_{sch}$ (see Fig. 15 (a)). Here T_{sch} is the scheduled time gap.

The delay times of trains located further downstream can be calculated similarly.

In our homogenization re-scheduling method, the number of passengers in each train is homogeneously-distributed by adjusting the distribution of trains as shown in Fig. 15 (b). Here P_B and P_A correspond to the number of boarding passengers on trains B and A respectively. P_s denotes the number of passengers waiting at the platform. E_B and E_A denote the number of passengers that will get off at the station i. Let symbol I be the number of passengers who arrive at the station per unit time. The expected arrival times of B and A at the station i are still denoted as T_B and T_A. Our objective is not equalizing the time gap but homogenizing the number of passengers, that is, the number of passengers on B and A is homogenized by extending T_B to X. We calculate X from the equation [10]

$$P_B + (P_s + IX) - E_B = P_A + I(T_A - X) - E_A. \tag{14}$$

The left-hand side (right-hand side) of (14) is the number of passengers on B (A) after departing from the station. The delay time T_D of B is thus decided by

$$T_D = X - T_B. \tag{15}$$

Having obtained the delay time of train B, the delay times of trains located further downstream can be calculated similarly.

4.3 Simulations and results

Fig. 16 shows two snapshots of our simulator "KUTTY" which display the direct simulation model and the flow pattern of the passengers of each segment as a visualization on the route map.

By simulating the flow of passengers quantitatively in all segments all over the network, we have found that the most congested area in the Tokyo Metro Railway Network is Otemachi Station on the Tozai Line (T09). Based on this result, we have simulated the case where a virtual accident occurs at Otemachi Station on the Tozai Line so that the trains of Tozai Line could only be operated on two sides of the station. Under this circumstance, the flow pattern of passengers will be changed significantly. This simulation, therefore, provides a very important clue for train scheduling with respect to the potential needs of users for alternative routes in case of accidents.

Fig. 17 shows the quantitative results of several simulations. Fig. 17 (a) shows that the number of passengers who take the Tozai Line decreasing by about 25 percent from normal operation due to the accident at T09. In contrast, the number of passengers who take the Ginza Line and the Hibiya Line increases by about 15 percent and 10 percent respectively. We believe this is

[10] Here we would like to mention that Eq.(5) is not always valid. For example, when $P_B + P_s - E_B > W_{\max}$, Eq.(5) will be invalid. In order for Eq.(5) to be valid, the conditions $T_B < X < T_A$ and $0 < P_B + (P_s + IX) - E_B < W_{\max}$ should be met. Fortunately, in our simulations, this is always the case.

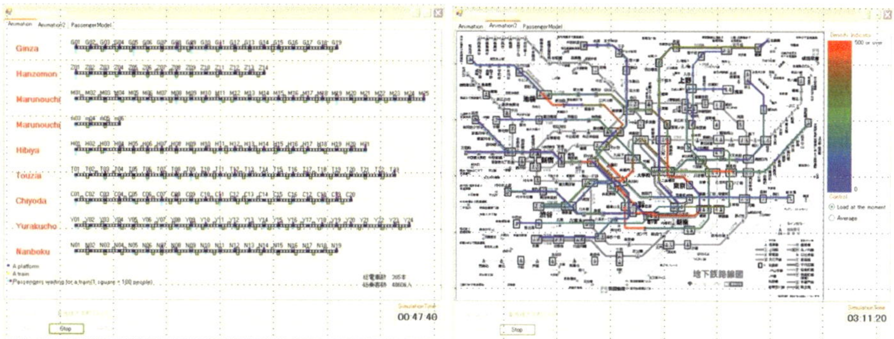

(a) Dynamical animation window (b) Visualization window

Fig. 16. Snapshots of our simulator "KUTTY". (a) The dynamical animation window of our model. (b) The entire map view, used to visualize the changing-flow in the route map. In KUTTY, the high flow regions (low-flow regions) are colored red (blue). This flow dynamically changes with time.

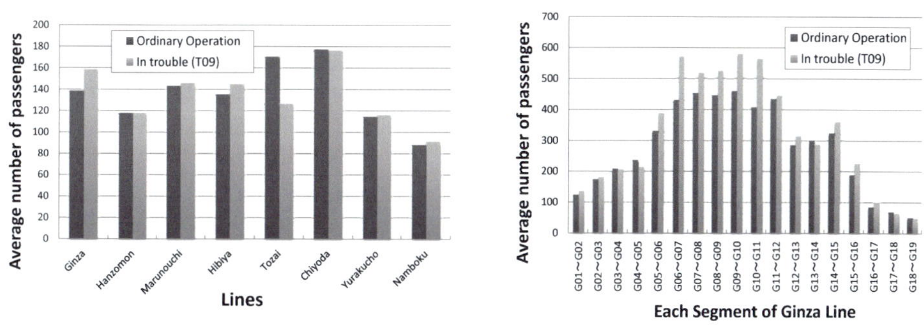

(a) per train in each line (b) per segment of the Ginza Line

Fig. 17. Comparison plot of the number of passengers between normal operation and operation with an accident at T09. (a)The number of passengers in each line comparing ordinary operation and congested conditions. (b)The number of passengers in each segment of the Ginza Line comparing ordinary operation and congested conditions.

because the Ginza Line intersects with the Tozai Line at T10, and the Hibiya Line intersects with the Tozai Line at T11, both of which are important transfer stations. In contrast, the Hanzomon Line intersects with the Tozai Line at T07, which is not such an important transfer station as T10 and T11. As a result, the passenger flow of the Hanzomon Line is essentially unaffected. Moreover, this result also implies that Otemachi Station is not the destination of most passengers on the Tozai Line, because otherwise the flow rate of the Hanzomon Line would increase (from T10 to Z09 to Z08, and from T07 to Z07 to Z08). Fig. 17 (b) shows the number of passengers on all segments of the Ginza Line. It can be seen that in the area from G05 to G11, the number of passengers increases remarkably. By transferring at these stations, passengers could change to the Marunouchi Line at G09, the Chiyoda Line at G06, and the Namboku Line at G05 and G06. Note that under normal operation, passengers change to the

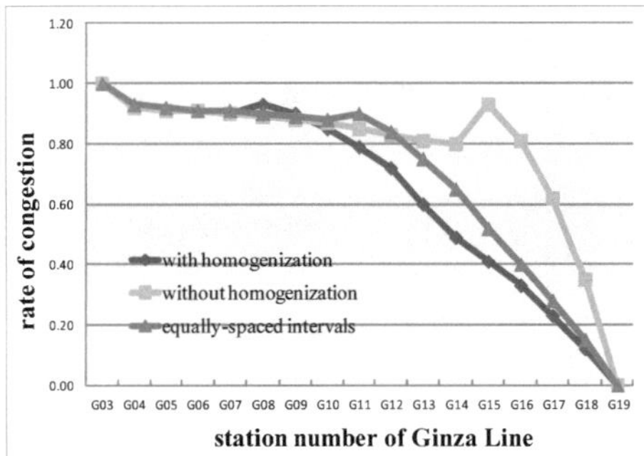

Fig. 18. Comparison plot of the passenger density of train A among the three systems. The congestion rate among the uncontrolled system, ordinary controlled system, and our homogenization system.

Marunouchi Line and to the Chiyoda Line at Station T09 and to the Namboku Line at Station T06.

Next we investigate the effect of our homogenization re-scheduling method. We suppose that the number of passengers waiting at G03 on the Ginza Line increases suddenly at $t = t_0$, so that the passenger density of the nearest upstream train A becomes 1 when it arrives at G03 at $t = t_1$. We have compared the evolution of the passenger density on train A in three systems: the system without re-scheduling, the system with conventional re-scheduling, and the system with homogenization re-scheduling. In Fig. 18, it can be seen that systems with re-scheduling decrease the passenger density of the train when the train is between G13 and G18. Moreover, we have found that our homogenization method is better than the conventional re-scheduling method.

5. Concluding discussions

In this chapter, we have proposed a new mathematical model for passenger transport systems, the so-called *public conveyance model*, built on the stochastic cellular automaton, which is exactly solved in the steady state. First, we defined the jamming state as a mathematical science and introduce the fundamental diagram to discuss the flow of particles. As two examples of analytical rule-based models, ASEP and ZRP, the fundamental diagrams obtained from numerical simulations and analytical calculations have been demonstrated.

As a one-dimensional case of public conveyance model, we investigated the bus route system and its efficiency by introducing three measurements: average velocity, the number of waiting passengers and transportation volume. Moreover, the effectiveness of an information-based control system, in which the number of particles between successive stops is adjusted, was discussed by comparing the case without control and with control in terms of these three measurements. As we found that implementation of the information-based traffic control system does not necessarily always improve the efficiency of the public conveyance system.

As an application of the public conveyance model, we have proposed a network simulator "KUTTY", which is based on the route choice behaviors of passengers. "KUTTY" takes into account the complex topology of the Tokyo Metro Railway Network and the OD demand estimated from the rider-ship data provided by the Tokyo Metro company. "KUTTY" can immediately provide an estimation of the passenger flow pattern required when an accident occurs in the Tokyo Metro Railway Network. Furthermore, we have also presented a homogenization re-scheduling method to alleviate congestion of a crowded train. It is based on the idea that the number of passengers in each train should be homogeneously-distributed. We found that our method is more efficient than the conventional approach.

Finally, we hope that this mathematical model and this simulator as an application of the mathematical model can be applied to other transportation systems and help in demonstrating the dynamical patterns and estimating the efficiency, traffic volume, and the other measurements to optimize their operation.

6. Acknowledgements

This chapter is owed to colaborated works with my colleagues. I would like to give a huge thanks to Debashish Chowdhury, Andreas Schadschneider, Katsuhiro Nishinari, Rui Jiang, Daichi Yanagisawa, Ryosuke Nishi, Mitsuhito Komatsu, Il Yun Yoo, Makoto Uchida, and Ryo Takayama for enjoyable collaborations.

7. References

[1] D. Chowdhury, L. Santen and A. Schadschneider, Phys. Rep. 329 (2000), 199-329.
[2] D. Helbing, Rev. Mod. Phys. 73 (2001), 1067-1141.
[3] A. Schadschneider, D. Chowdhury and K. Nishinari, "STOCHASTIC TRANSPORT IN COMPLEX SYSTEMS FROM MOLECULES TO VEHICLES", Elsevier, (2010).
[4] K. Nishinari, “Distributed Autonomous Robotic Systems 8”, p.175 (2009).
[5] K. Nishinari, Y. Suma, D. Yanagisawa, A. Tomoeda, A. Kimura and R. Nishi, “Pedestrian and Evacuation Dynamics 2008”, pp.293-308 (2010).
[6] B Derrida, M R Evans, V Hakim and V Pasquier, J. Phys. A: Math. Gen. 26 1493 (1993).
[7] M. Schreckenberg, A. Schadschneider, K. Nagel and N. Ito, Phys. Rev. E 51 2939 (1995).
[8] F. Spitzer, Adv. Math. 5, 246 (1970).
[9] M. R. Evans, J. Phys. A: Math. Gen. 30, 5669 (1997).
[10] M. R. Evans and T. Hanney, J. Phys. A: Math. Gen. 38, R195 (2005).
[11] M. Kanai, J. Phys. A: Math. Gen. 40 pp.7127-7138 (2007).
[12] A. Tomoeda, D. Chowdhury, A. Schadschneider and K. Nishinari, Physica A, 384, 600-612 (2007).
[13] A. Tomoeda, M. Komatsu, I. Y. Yoo, M. Uchida, R. Takayama and K. Nishinari, Cellular Automata (Lecture Notes in Computer Science, Springer), 5191 (2008), 433.
[14] A. Tomoeda, M. Komatsu, I. Y. Yoo, M. Uchida, R. Takayama, R. Jiang and K. Nishinari, GESTS International Transaction on Computer Science and Engineering, 54, 81, (2009).
[15] C. Arita, Phys. Rev. E, 80, 051119, (2009).
[16] C. Arita and D. Yanagisawa, J. Stat. Phys., 141, 829, (2010).
[17] S. Wolfram, *Theory and Applications of Cellular Automata* (1986) (Singapore: World Scientific).
[18] S. Wolfram, *Cellular Automata and Complexity* (1994) (Reading, MA: Addison-Wesley).
[19] A. Schadschneider, Physica A 313, 153 (2002).
[20] R. Jiang, B. Jia and Q.S. Wu, J. Phys. A 37, 2063 (2004).
[21] O. J. O’Loan, M. R. Evans, M. E. Cates, Europhys. Lett. 42, 137 (1998); Phys. Rev. E 58, 1404 (1998).

[22] D. Chowdhury, et al., J. Phys. A, 35, L573 (2002).
[23] A. Kunwar, et al., J. Phys. Soc. Jpn., 73, 2979 (2004).
[24] A. Tomoeda and K. Nishinari, in the proceedings of "SICE Annual Conference, 2008" (2008) 549.
[25] K. Nishinari and D. Takahashi, J. Phys. A: Math. Gen. 32, 93 (1999).
[26] S. Yukawa, M. Kikuchi and S. Tadaki, J. Phys. Soc. Japan 63, 3609 (1994).
[27] M. Pursula, Simulation of traffic systems - an overview, Journal of Geographic Information and Decision Analysis 3 (1999) 1.
[28] V. Latora and M. Marchiori, Physica A, 314 (2002) 109.
[29] D. J. Watts and S.H. Strogatz, Nature, 393 (1998) 440.
[30] D. Meignan, O. Simonin and A. Koukam, Simulation Modeling Practice and Theory, 15 (2007) 659.
[31] E. W. Dijkstra, Numerische Mathematik, 1 (1959), 269.

11

Gaming Simulations for Railways: Lessons Learned from Modeling Six Games for the Dutch Infrastructure Management

Sebastiaan Meijer
[1]*Delft University of Technology*
[2]*Royal Institute of Technology, Stockholm*
[1]*The Netherlands*
[2]*Sweden*

1. Introduction

The Dutch railway system is a highly complex and heavily utilized network (Goverde, 2005; CBS, 2009). Worldwide it is one of the most densely driven networks, yet its capacity has to increase further. Improvements in the domain of capacity management and traffic control are increasingly difficult to implement because of the large interconnectedness of all processes. The de-bundling of rail infra management (ProRail) and train services (predominantly NS, and some smaller regional lines by Syntus, Veolia, Arriva a.o. plus freight train operators) has created an operational process in which multiple offices and platform/line operations need to synchronize to control the daily train flow.

ProRail, the Dutch railway infrastructure manager, has stated a goal to increase the capacity by 50% as a challenge till the year 2020. This cannot be done the 'old way' through increased amounts of physical infrastructure, as both money and geographical space are insufficient. Furthermore, the complexity and interconnectedness of the network is yet at such a level that more of this will lead to less resilience and becoming (even more) prone to disturbances. Because of the 50% growth challenge till the year 2020, new and smarter ways of managing capacity and traffic are key for the success of the Dutch rail infrastructure for society. The ProRail organization has taken up gaming simulation as a key method to improve the innovation process (Meijer, forthcoming).

Unique for gaming simulation is the highly detailed simulation of both technical and process variables of rail infrastructures and the decision and communication function of real people in their real roles. The method does not assume models of decision-making but draws upon the real-world knowledge of professionals in the operation. Over the course of the projects that ran in 2009, 2010 and 2011, the specific setting of the ProRail organization proved to be both a complex and fruitful environment for gaming simulation. The complexity was found in the large number of stakeholders both in and outside the organization and in the interconnectedness of every aspect of train traffic control on the performance of passenger and freight train service providers.

In the year 2009, the gaming group of Delft University of Technology was asked to facilitate three projects using gaming simulation methodology. These projects ran so successful that the organization asked the Delft researchers to identify where in the organization large-scale implementation of gaming simulation methodology would be most promising. Based upon a series of interviews through the organization, ProRail and TU Delft jointly formulated a four-year research and implementation proposal that is now in operation. The first gaming sessions in this new collaboration have been held and results have led to methodological lessons-learned on how to model. This chapter reports on three modeling issues crucial to gaming simulation for railway and similar systems. How to abstract from the nitty-gritty details while keeping the simulation real and valid enough for real-world operators to participate and do their job is the focus of this chapter.

2. Problem description

Innovation in the Dutch railways is on one hand much needed, while on the other hand very complex to achieve. The 1995 politically instigated de-bundling of rail infra management (ProRail) and train services (predominantly NS, and some smaller regional lines by Syntens, Veolia, a.o.) has created an operational process in which multiple offices and platform/line operations need to synchronize to control the daily train flow. The increasing importance of rail services for individual provinces in the Netherlands has led to multi-party tendering (Van de Velde et al, 2008). In this complex multi-actor and multi-level environment the strategic safeguarding of public values in managing operations proofs often impossible (Steenhuisen et al, 2009). The combination of these events and trends leads to a challenge to innovate on two aspects, being quality in operations and ways to increase the capacity.

2.1 Quality in operations – Robustness and resilience

Over the past decade, the railways in The Netherlands have received major criticism for the quality of its operations. From a policy perspective this has led to performance contracts for both the main train service operator (NS) and the publicly owned infrastructure manager ProRail (Van de Velde et al, 2009). Over the past decade the performance has seen improvements on the critical performance indicators, but still it is not regarded to be a high quality service due to many small delays, overly crowded trains and non- or mal-informed passengers. The rail system often suffers from small defects, leading to bigger delays when the problems spread like an oil spill over the regions and lines. If we define robustness as the degree to which a system is capable to withstand problems within the limits of the designed system, then the robustness of the railways is questionable.

A lower score on robustness would not have been so detrimental is the railways were more resilient. Hollnagel et al (2006) define resilience as the ability of a system or an organization to react to and recover from disturbances at an early stage, with minimal effect on the dynamic stability. The challenges to system safety come from instability, and resilience engineering is an expression of the methods and principles that prevent this from taking place. Furthermore the recent years have shown that snow, storms, national festivities and other outliers in the situation for which the system is not specifically designed cause total or at best partial collapse of the national system, as soon as small problems start to occur. This has led to Parliamentary Investigation (Rekenkamer, 2011). According to Hale and Heijer

(2006), railways, from their assessment of safety operations at the Dutch Railways, would seem to be examples of poor, or at best mixed, resilience, which can, however, still achieve high levels of safety, at least in certain areas of their operations. Hence safety is achieved by sacrificing goals, traffic volume and punctuality. The system does not achieve all its goals simultaneously and flexibly and is not resilient.

2.2 Capacity increases

The Dutch railway sector will face a massive growth of transport demand in the forthcoming decade. This growth is both expected in passenger and in freight transport. Currently, the Dutch railway network is one of the most densely used networks in the world, approaching its maximum capacity given the current infrastructure and control mechanisms. The projected increase in demand requires a step-change in both the physical and control aspects of the railways. ProRail formulated an ambitious program, called 'Room on the Railways" (Ruimte op de Rails, in Dutch) to increase the number of trains on the network by 50% before the year 2020. One of the major components of this program is the plan for high-frequency passenger trains on the major corridors. Currently there are (on average) 4 intercity, 2 to 4 local and 1 or 2 freight trains per hour on the major corridors. This should increase to 6 intercity, 6 local and 2 freight trains before 2013. This new frequency of trains is often called 'untimetabled travelling" as the passenger can just go to a station without checking departure times: the next train will be there soon. The official title of the schedule is High Frequency Train Transport.

The projected increase of capacity cannot be achieved by building new infrastructure alone: the costs for the complete program would be around 9 billion euro, and the time for procedures and construction would frustrate the transport demand for years. ProRail has taken up the challenge to achieve the goals with only half of this budget by combining strategic choices for new infrastructure with new control and management solutions.

3. Gaming simulation for process innovations

Gaming simulation, here defined as 'simulating a system through gaming methods' is one of the terms in a loosely demarcated field of interactive participatory activities, aiming to involve participants, who may be the real stakeholders, in an activity. Other terms used are simulation game, policy exercise and serious gaming. The word gaming will be used here as the short term for gaming simulation. Different authors have different preferences, but in in general the terms depend on the intended use of the method.

Game theory and gaming simulation are often intertwined. Game theory is the mathematical approach of analyzing calculated circumstances where a person's success is based upon the choices of others. In games, these situations often occur, and therefore is game theory a popular method of modeling artificial intelligence in games. This chapter does not use game theory per se, however a more prominent and explicit use is foreseen in future games that incorporate automated agents.

Given the number of gaming titles and scientific publications, the use of gaming methods for learning is the most popular by far, typically occupying 'serious gaming' and 'simulation game' for usually computer-supported games that place the player in a simulated world

(Bekebrede and Mayer, 2005; De Freitas and Martin, 2006; Kriz and Hense, 2003). Learning about innovation in games is a popular topic for MBA-style versions, typically related to markets and supply chains (Meijer et al, 2006; Meijer, 2009). Learning and communicating complex issues are in this stream highly related. An important aspect for ProRail is the opportunity to communicate ideas. While a slideshow can communicate a message, a gaming simulation enables you to experience it for yourself (Bekebrede, 2010). The aspects about which it is sometimes difficult to communicate at present include: the impact of new timetables (on all categories of employees), the need for precision in carrying out tasks (employees), the influence of disruptions on the network as a whole (general public) and to experience the key aspects of traffic control / capacity management (general public). At present, visualizations of train flow models such as FRISO and SIMONE (Middelkoop and Loeve, 2006) are available, but it is not possible to experience these aspects by sitting at the controls. The opportunity for communication gives employees the chance to play a role that they do not have in reality. This can help clarify different points of view.

In the world of policymaking, there is half a century of history in using gaming as an intervention to bring together policy makers and other stakeholders in participatory events. Games provide a way to collectively decide firstly on the system boundaries and secondly on the dynamics of the system that will be played. Then, policies can be formulated in this simulated environment (Duke, 1974; Duke and Geurts, 2004; Mayer, 2010). This approach relies on Duke and Geurts' (2004) 5-C's of gaming simulation for improving policy making, namely by understanding the Complexity, enhancing Creativity, enabling Communication, reaching Consensus and Commitment to action. Within ProRail this role of gaming simulation is particularly relevant for management questions.

Increasingly popular is the possibility to try out the effect of policies on a simulated system, and see whether innovation in roles, rules, objectives and constraints can be made. This approach, although very relevant for policy-making, is actually a third use of gaming, for testing hypotheses (Peters et al, 1998). This application is less common and puts great emphasis on the verification and validation of the gaming simulation (Klabbers, 2003, 2006; Noy et al, 2006; Meijer, 2009). For innovation at ProRail, this use is at the core of the reasoning behind choosing gaming simulation as a new method in reducing uncertainty in more complex, system level changes.

A fourth use that is emerging is linked to the gamification of society (Hiltbrand and Burke, 2011). Innovation can take place through game play if the incentives are such that the crowd can generate and implement their ideas in a system. Few scientific literature on this exists as of yet, but examples are UK innovation in pensions (Gartner, 2011), crowd sourcing of ideas in an insurance company (Bekebrede and Meijer, Forthcoming)

4. Modeling challenges in gaming simulation for railways

In the world of gaming simulation several design guides and principles exist on how to capture real world problems in a gaming simulation. In the field of policy making the most important method is the one that Duke and Geurts (2004) describe, where for learning, sense-making and related issues the more recent work of Harteveld (2011) is gaining footage among some others. However, the problem is that these methods are so generic that the specific issues for technical domains like the railways have to be addressed specifically for that domain. A second issue, especially with the more policy-oriented approaches and the popularity for learning in higher education is the focus on participants with a relatively large capability in thinking abstract, as policy makers and students tend to have more of this skill than the average operator. Peters et al (1998) describe the process from real world to simulated game world as a process of abstraction and reduction. The big question is how far can you abstract and reduce from reality before operators loose their grip on the simulated reality?

The operational skill training is recently getting more and more addressed in the gaming literature. Druckman (1994) proved already the need for more 'fidelity' (that could be translated as 'detailed realism') when training less abstract skills. Applications for operational skill training is getting common in the domains of image-based medical procedures (like laparoscopy, gastrointestinal flexible endoscopy, image-guided neurosurgery, and endovascular surgery) (Gaba, 2001; Botden et al, 2008; Hamdorf and Hall, 2000), aviation (Proctor et al, 2007), and safety training for dike inspection (Harteveld, 2011) and the oil and gas industry (Meijer and Poelman, 2011). Each of these domains finds a solution in 3D-based computer games that model an environment, either geographical or the organs in a body, through which the player has to navigate and perform a coherent set of actions. There is an overlap between the fields of virtual reality, simulation and gaming here.

Involving operators in games for policymaking or for testing hypotheses is almost undocumented, with some notable exceptions like the work at CIRAD and Cemagref (Barreteau, 2003). Traditionally the questions in policymaking and the hypotheses tested are at a higher level of abstraction. In Meijer (2009), the author argued that involving the real operators in a gaming simulation has the benefit of avoiding models and assumptions about their behavior, and thus can increase the validity of the behavior of the entire socio-technical system simulated. This has been proven in the domain of supply chain management research, studying the organization of transactions.

In our work we focus on the behavior of the people in the daily operations in railway systems, with a focal point at the train traffic controllers. Within the scope of the infrastructure management ProRail, their behavior has the most direct influence on the robustness and resilience of the network. To base decisions upon their behavior in gaming simulations it is essential to consider the validity of this behavior.

The most common critique for behavior observed in a session is "it is only a game.....". In the literal meaning the statement is true. A gaming simulation is a model of reality, and the roles, rules, objectives and constraints are necessarily different from the real world. The insinuation of the statement is, however, that behavior observed in a session is unlike behavior in the real world and is no valid representation of real-world behavior. Peters et al (1998) discuss the validity of games (gaming simulation) based upon the work of Raser (1969) who defined validity of models in the following way: "A model can be said to be valid to the extent that investigation of that model provides the same outcomes as would investigation in the reference system." Raser (1969) suggests four aspects of validity that apply to gaming simulation:

- Psychological reality: To what degree does the gaming simulation provide an environment that seems realistic to the participants?
- Structural validity: To what degree is the structure of the gaming simulation (the theory and assumptions on which it is built) isomorphic to that of the reference system?

- Process validity: To what degree are the processes observed in the gaming simulation isomorphic to those observed in the reference system?
- Predictive validity: To what degree can the gaming simulation produce outcomes of the historical or future reference system?

The psychological reality demands that sessions are conducted in such a way that participants are emotionally involved and really play their role. The situation of the session in the life of the participants, the consequences of participation or non-participation and the location and atmosphere of a session and its moderation are important factors. This requires craftsmanship of the game leader that is hard to operationalize in a scientific context. Various authors have made attempts at determining the quality of conducting sessions. Kriz and Hense (2006) offer an elaborate and theory-based evaluation methodology, that according to Klabbers (2008) does a good job in (temporarily) bridging the gap between analytical and design sciences. Kriz and Hense's approach is an adapted version of the theory-based evaluation method by Reynolds (1998). They distinguish between concept, design and application that can be evaluated.

The psychological reality and process and structural validity of Raser (1969) come together in the concept of situation awareness (Endsley, 2004) for operators. Operators should get involved psychologically when they can recognize sufficient components of the processes and structures they are used to in their real work. In the medical world this has led to consensus guidelines for validation of virtual reality surgical simulators (Carter et al, 2005), but in railways this work is only done for train drivers (Hamilton and Clarke, 2005)

When we take the concept of situation awareness as the central concept for considering the validity of railway operator behavior in gaming simulations, the list of items in the situation awareness still becomes vast. The modeling issue could include nearly any technical aspect of the railways, interface and representation items as well as the cognitive state of operators during their normal workdays. This chapter focuses on three important issues, posed as how-to questions:

1. How to immerse train traffic controllers in a gaming simulation?
   Immersion is one of the important indicators of presence and therefore psychological reality in simulated environments (Witmer and Singer, 1998). Therefore an indication of how to model a game so that railway operators get immersed in a first important step towards validity.
2. How to model time?
   Where real-world train flow is a continuous time process, this does not necessarily translate one-to-one to a gaming simulation, as the research question may ask for another solution than continuous, like step-wise, round-based, or asynchronous time.
3. How to present operational data?
   In the real operations the data flow to operators is bundled in machine interfaces or is fairly constant as time tables typically change only once a year, and infrastructure doesn't change fast either. In a game the standard tools may not be available and timetables and infrastructure may be new to the participants. How to present the information so that operators can still use it?

These questions will be answered in the remainder of this chapter. The next section discusses six projects from which the experiences are gained, then Section 6 translates this into lessons learned.

5. Modeling gaming simulations for rail innovation projects

As of 2011, the gaming collaboration resulted in six different gaming simulations, specifically built for innovation projects within ProRail. Each of these projects used gaming simulation to investigate various solution strategies with the aim of increasing capacity utilization, resilience and robustness on the rail network. The initial pilot project covered three projects while the long-term collaboration yielded three so far.

From the launch of the initial project, ProRail formulated three preliminary cases to study using gaming simulation. TU Delft was to developed unique approaches for each of these cases, after which the initial success of gaming simulation for the Dutch Railways would be re-evaluated. The cases differed in nature. The first was about the potential value of market mechanisms for management of demand of cargo capacity. This game could be seen as a management game on the tactical level. The second case was about studying a control concept for high-frequency train transport at the Bijlmer junction. This game was at the operational level of train dispatching and network control. The third case was about the opening regimes of the bridge over the river Vecht. This game was purely about train dispatching at the operational level.

During the course of these three cases, the success became very apparent to the senior management involved at ProRail. This led to an Intermezzo phase after the third game to reflect upon the results so far and to identify the value from interviews with ProRail internal stakeholder held by Delft researchers. The launch of a large four-year project was marked by a kick-off case that convinced the last skeptics. In the following sub-sections each of the cases and the intermezzo phase are described.

5.1 Rail cargo market game (RCM)

The first and kick-off subproject called Goederenmarktplaats (Freight Market) introduced ProRail to a paper-based and partly computer-supported game with a high degree of abstraction. This game type was referred to as a management game, due to the focus on more abstract policy-related aspects. Most of the participants were managers, with one session including a small number of network controllers.

Table 1 lists the core description of this game, more information can be found in Meijer et al (2009).

The game sessions delivered results timely, and in a positive and active manner. This game is still referred to two years later in the organization. Important to note for the introduction of gaming is that this project happened to have many people on board in senior staff functions from two different divisions (Traffic Control and Capacity Management) who appeared to be key people in later problems that called for gaming simulation methodology. The foundation in terms of exposure to key personnel therefore couldn't be better.

The research team then conducted interviews within ProRail to evaluate the pilot project and identify the opportunities it presented. In these interviews, the management game was repeatedly described positively. However, this generated few new ideas as regards applicability. Many of the issues encountered within the ProRail organization are operational and thus call for less abstract forms of gaming simulation.

Core aspect	Description
Purpose	Studying the potential value of various market mechanisms for better capacity allocation of cargo paths.
Roles	Clients with demand for transport, Rail Cargo Transporters, Passenger Transport, Rail Capacity Planning, Rail Asset Management
# of players	15 - 25 depending on step
Own/real/fictitious role	Real role, but selected for knowledge for instance from previous job position.
Scenarios	3 - 4 scenarios per session. First scenarios that explored the more fundamental market mechanisms. Then scenarios to validate the successful configurations.
Intervention range	Facilitator could start and stop the scenario and dissolve disputes only on the process steps.
Simulated world	Stylized train path market, stylized transport demand
Immersion	Fast, once roles were clear and adopted. Lively play including some conflicts. Capacity planners in second session had issues getting insight in their track system.
Time model	Continuous
Data presentation	Simplified to stylized network, simplified timetable and simple contracts. Big jump between session 2 and 3 when replacing capacity planners with computer reservation system that was similar conceptually.
# of sessions	3 subsequent games each with 1 session during 1 full day.
Type of data generated	Quantitative and qualitative, testing hypotheses about mechanisms that are assumed to have a certain effect on capacity allocation.
Consequences	Policy formulated but put out of scope for 2010/2011, possible application in 2012. Politically very sensitive.

Table 1. Core description of Rail Cargo Market Game

With respect to the three modeling issues we learnt that it is very hard to have operational level people in an abstract simulation when they have to work on infrastructure and timetabling they do not recognize. The usual flexibility that is commonly found in gaming with groups of higher education is not working here. To overcome this, we automated some of their tasks in the game into a computer model for train path reservations. This worked flawless for the more management-like question of this game.

5.2 Bijlmer junction game

This subproject introduced ProRail to a computer-based gaming simulation developed on ProRail's own MATRICS simulator (Van Luipen and Meijer, 2009). This simulation pushed the envelope in terms of utilizing the technical specifications of MATRICS. This type of game was described as a multi-player process simulation due to its detailed reflection of real-life operational processes. The participants play a pre-defined role that is 100% identical to their job description, to carry out their real-life duties in a simulated game environment. Table 2 lists the core description of this game. For a full description we refer to Meijer et al (2009).

Core aspect	Description
Purpose	Testing and validating a control concept for high frequency train transport.
Roles	Train driver (2), Train traffic controller (3), Network controller (5)
# of players	10 plus 2 facilitators and 2 experts.
Own/real/fictitious role	Own role, participant selected by their team leaders
Scenarios	3 Scenarios, gradually testing more complexity.
Intervention range	Facilitators could start, stop and pause scenarios and interfere with train driver behavior.
Simulated world	Detailed infrastructure between Amsterdam and Utrecht, detailed timetable.
Immersion	Very fast and deep for train drivers and network controllers. Difficult for train traffic controllers; see discussion in Meijer et al (2009).
Time model	Continuous
Data presentation	Highly detailed through computer interface. Interface different from real-world abstraction.
# of sessions	1 full day session
Type of data generated	Quantitative (failed) and qualitative.
Consequences	Data generated in the game yielded insights in key materials and resources needed for implementation of the control concept, and high-frequency planning in general.

Table 2. Core description of Bijlmer Junction Game

ProRail had assigned a project team to come to new control and steering procedures that suite the future reality of high-frequency passenger trains. The challenge of this project team was to come up with new concepts that would both be supported by train traffic controllers and network controllers, and would yield a stable, controllable control and routing operation when put into place. The question was raised: how to test new control and steering concepts when there is no option to test in real life? The Bijlmer Junction Game was targeted at this. In the game the interaction of train drivers, traffic controllers and network controllers was crucial, as studied earlier by Albrecht (2009).

The gaming simulation session yielded insights in key materials and resources needed for implementation of the control concept, and high frequency planning in general. The importance of buffer areas with sufficient space to side-track a train without disturbing other services, platforms asides the entire train for passenger exit, and alternative departure options for all passengers within reasonable time is a clear outcome for ProRail. Furthermore, train traffic controllers do not yet seem to realize what the projected high-frequency planning will mean in practice for their tracks.

As described in Meijer et al (2009), this game was not a break-through success. We learned that involving the operational people in the organization in a game that modeled the infrastructure and timetabling as detailed as they are used to, requires interfaces that connect to the situation awareness capabilities of these operators. Simple said: even though we checked our approach upfront with the operators, they were not able to do what they though were capable of due to different visualization. Luckily, the debriefing and discussions still yielded sufficient data of sufficient quality for ProRail to be able to contribute to the problem solving. For the gaming team, this experience led to the development of the following game.

5.3 Railway bridge game

The subproject Railway Bridge Game (for a bridge over the river Vecht) introduced ProRail to the process management game, a computer-based gaming simulation for which new software was developed. Over the course of one week, various train traffic controllers played this game in a single-player environment using a series of scenarios. The type of game was described as a single-player process simulation. Table 3 gives the core description of this game. More information can be found in Kortmann and Sehic (2010).

Core aspect	Description
Purpose	Studying a new regime for bridge openings on the busy Amsterdam - Amersfoort corridor.
Roles	Train traffic controller. Bridge operator (simulated)
# of players	1
Own/real/fictitious role	Own role.
Scenarios	5, each subsequent day the same train traffic controllers played one scenario of increasing complexity
Intervention range	Facilitator played other roles
Simulated world	Detailed infrastructure, detailed timetable
Immersion	Good to very good. More experienced train traffic controllers had more hesitancy towards the computer system, but once used to it scored better with more situation awareness.
Time model	Continuous
Data presentation	Detailed through near-familiar computer interface.
# of sessions	1 session, full week
Type of data generated	Mainly quantitative (measured actions and train throughput, questionnaires) and qualitative from interviews
Consequences	None as of 2011, new game with improved interfacing planned for winter 2012 testing more details.

Table 3. Core description of Railway Bridge Game

The Railway Bridge Game was positively received. It learned that the drawbacks of the interface problem signaled in the Bijlmer Junction Game could be overcome by making special gaming modules. In these modules the representation of the infrastructure and the control options can be made closer to the real world systems. Given the differences between experienced and less experienced controllers we conclude that more resemblance is better for immediate immersion, but not necessarily related to the quality of the decision once a certain threshold of realism is reached.

Playing the game showed its potential to help solve the bottleneck of the Vecht Bridge on the OV-SAAL rail corridor. Under increasing loads of timetabling the experienced operators scored significantly better than operators in training. Under light loads this was the other way around, showing the difficulty of experienced people to overcome differences in the user interface, but as soon as craftsmanship was required to minimize delays the experience helped keeping control.

Both single-player and multi-player gaming simulation were readily welcomed by almost all of the stakeholders in the organization as a valuable new resource for ProRail as an organization. The aspect of the multi-player gaming simulation that prompted a particularly positive response was the opportunity to test the feasibility of timetables, control concepts and exceptional situations in a setting that includes several layers of management and/or control areas. The aspect of the single-player gaming simulation that prompted a particularly positive response was the opportunity to train and practice in relation to exceptional situations and future timetables and infrastructures in an offline setting, using simulated trains.

5.4 ETMET 2010

One of the two strategic innovation trajectories to come to the desired capacity increase is the program to come to a metro-like timetable on the major corridors. On the Amsterdam - Eindhoven corridor this program is titled 'Every Ten Minutes A Train' (Elke Tien Minuten Een Trein - in Dutch), shortly ETMET. In the fall of 2010, the largest train operator National Railways (NS) and ProRail tested this concept for a full month in the real operation. This program required substantial preparation, and gaming simulation was selected through the senior staff involved in earlier games to answer questions about two ways of handling a major disruption under the new timetable. This resulted in the ETMET 2010 Game, described in Table 4.

In the ETMET 2010 Game we simulated the train flow and all processes and interactions in the train control, personnel and rolling stock processes. The wish was to have the train traffic controllers working on gaming modules similar to the one in the Railway Bridge Game. Soon during the development we found out that the underlying rail traffic simulators available did not support the required actions of turning around, skipping a service or renumbering rolling stock to different train services. Therefore the decision was made to create a complete manual, analog simulator, observed with cameras overhead the infrastructure maps, distributing views similar to the regular computer visualizations to three rooms with operators.

The session delivered the data required to answer the question on the differences between two methods of handling a major disruption. The project management assumed the new method to be beneficial for resilience, however they proved wrong. The new method essentially provided a pre-defined pattern for guiding trains over a double track where one track is blocked. The network and service controller had to makes their choices out of the set of trains currently running on the tracks, approaching the blocked track. Remaining trains have to be cancelled or coupled. This was assumed to be a better solution than the old solution in which there is a separate document for every possible interaction between two trains. It appeared however, that the choices for assigning trains to the pattern were impossible to make, given the interactions that all the trains available have with other parts of the system. While working on a solution the situation changed too fast to make a single decision in time, while overseeing all of the complexity.

Core aspect	Description
Purpose	Testing the differences between two mechanisms of handling a major disruption under High Frequency Transport scheduling
Roles	Train traffic controllers, Passenger information, Driver rescheduling, Rolling Stock rescheduling, Platform coordinator, Decentralized network controller Network controller, Service controller.
# of players	14 in role, 9 in support roles in analog simulator center, 6 observers, 1 host, 1 game leader
Own/real/fictitious role	Own roles, invited on personal title however with support of management.
Scenarios	2 scenarios: first the 'old' way and then a new mechanism
Intervention range	Facilitators could start, stop and pause the scenarios.
Simulated world	Detailed infrastructure Utrecht - Geldermalsen, detailed high-frequency timetabling, essentials of communication lines between different offices involved. Stylized passenger flow.
Immersion	Full immersion in a few minutes. Conflicts arose, leading to a time-out by the game leader to settle the issues and go back to a state all could agree on. Extremely involved and lively game play.
Time model	Continuous
Data presentation	Infrastructure representation in familiar schematics, detailed timetabling on paper, time and delays through simple interfaces.
# of sessions	1 session, full day
Type of data generated	Quantitative and qualitative, testing hypotheses about differences between 2 mechanisms.
Consequences	Proposed solution abandoned based on data generated in the gaming session.

Table 4. Core description of ETMET 2010 Game

In this game all data was completely detailed available to all participants, on paper, and mostly in a format they recognize, using all real-world abbreviations and notations. Contrary to many games in which the designer abstracts and reduces to a level where it is not about managing large amounts of data, it proves to be very well possible to give operators this data. It even helps to give them situation awareness, as in the debriefing all were confident that their behavior and reactions were similar to what they would to in practice, and all could work with the data supplied.

5.5 NAU

Utrecht Central station is the heart of the Dutch rail network: here come trains from all directions together in a versatile, but consequently complex knot. The complexity and interdependence of the many train movements and other activities makes Utrecht very vulnerable to chain reactions of delays. For large disturbances, history has repeatedly shown that the risk of flooding and even completely crashing the traffic flow is high. Resilience and robustness of the Dutch rail network therefore has to consider Utrecht as a key parameter.

In previous years the timetable has been 'disentangled', meaning that trains are assigned to a corridor and that these corridors are planned to have as little interference with each other on the physical infrastructure as possible. Now, after ProRail disentangled the schedule the aim is to reduce the interference further by matching the control concept to the corridors. For this the NAU (New Action plan Utrecht) program was launched.

Within the NAU program five goals have been identified:

1. ensure that the basic plan remains within the corridors;
2. limiting defects through maintenance / inspection;
3. Limiting deviations from shunting;
4. limit abnormalities in major disruptions;
5. adjust the division of labor in the Traffic Control Post.

Gaming simulation has been selected as a tool to try out the concept first in a simulated environment before it will be brought to the control post. During the game the effect of the new concept on goals 1,4 and 5 had to be researched. Table 5 describes the game details.

The main result of the NAU game was empirically based insight in the fundamental consequences of reducing the number of switches used and corridor control on capacity, resilience and robustness. In the old situation, the capacity reduces rapidly as the disturbance level increases. Due to the many switches, many options remain in heavily disturbed situations to continue driving, requiring a mastery level of the train controller. These options cause 'infection' of problems of one corridor to others. When using corridor control in its strict sense, the process remains more manageable with mild to moderate disturbances. But because the control options are limited to the corridors, there comes a moment in the corridor that all capacity is lost, still not affecting the other corridors. There's a tipping point where the limits for a disturbed corridor become unacceptably high. At this tipping point it can help to deviate from the corridor principle. The ideal situation is to control & isolate the corridor as long as possible and use other parts of the infrastructure only when the critical level of disturbance is reached. In other words: you want to cash the potential of corridor control and avoid potential losses. Where the disturbance level is critical, how often this situation occurs and what specific deviations must be allowed, is still open for further research.

In the NAU game modeling we re-used the infrastructure schematic layout and timetabling information that was so successful in the ETMET game. Yet again this proved to immerse the participants in the simulation within a few minutes, and to make them enact their role perfectly. In this post-game evaluation the participants rated their behavior as highly realistic. The only exceptions to this were the network and service controller who both work at the national level. For them there was no game material to play with apart from information derived from the simulated area. This resulted in less emersion and a bit grumpy atmosphere in which they were mocking about the new concept. For the project this proved functional as their comments in the discussion raised important points for the improvement of the concept, but the game play from them was not optimal. Therefore no direct conclusions could be drawn from the interaction between the national and regional level.

Core aspect	Description
Purpose	Testing the improvements in resilience and robustness when introducing a new control concept for Utrecht Central station.
Roles	Train traffic controllers, Decentralized network controller, Driver rescheduling, Rolling Stock rescheduling, Platform coordinator, Network controller, Service controller.
# of players	9 in role, 4 in support roles for analog simulator, 3 observers, 1 host, 1 game leader
Own/real/fictitious role	Own roles, invited on personal title however with support of management.
Scenarios	2 scenarios: first the 'old' way and then a new mechanism
Intervention range	Facilitators could start, stop and pause the scenarios.
Simulated world	Detailed infrastructure Utrecht Central, detailed current timetabling, face-to-face communication lines between different offices involved. Stylized planning tools
Immersion	Instant and very good for all players, except for network controller and service controller who were less immersed, showing in discussions about other topics.
Time model	Continuous
Data presentation	Infrastructure representation in familiar schematics, detailed timetabling on paper, time and delays through simple interfaces.
# of sessions	1 session, full day
Type of data generated	Quantitative and qualitative, testing hypotheses about improvements with new control concept. Numbers real enough to base decisions on.
Consequences	New control concept embraced, actions defined to counterbalance penalty for major disruptions. Invention of the concept of pre-defined handling scenarios for non-availability of small parts of the infrastructure. Heavy post-game discussions leading to high-level decisions on the project.

Table 5. Core description of NAU Game

5.6 Platform overnight parking (POP)

In the capacity planning process for 2012, two service areas have been declared out of capacity and ProRail is obliged, according to law, to find solutions to solve these capacity shortages. The goal of this game is to determine whether it is possible to orchestrate a 'carousel process' around Hoofddorp so that scarce capacity in the service area can be increased. This means that after servicing and technical controls at the service area the train is then drawn along the (platform) tracks of Hoofddorp station or Middle Hoofddorp.

Core aspect	Description
Purpose	Answering the question how many pieces of rolling stock could be parked along the platforms of stations during the night, given the processes of cleaning and maintenance that have to be performed at service areas. Question asked for two locations: Amsterdam-Watergraafsmeer and Hoofddorp.
Roles	Train traffic controller (2), Foreman of cleaning (2), Train driver, Service area supervisor.
# of players	6 in role, 2 support for illiterate cleaning foreman, 3 observers, 1 host, 1 game leader
Own/real/fictitious role	Own roles, invited on personal title however with support of management.
Scenarios	1 scenario per location (Plus 30 minutes 'warm up' scenario)
Intervention range	Facilitators could start the simulation and determine the time required for each 5 minutes of simulated time.
Simulated world	Detailed infrastructure Amsterdam Central - Amsterdam Watergraafsmeer and Hoofddorp, detailed timetabling for end-of service of trains. Detailed service demand,
Immersion	Immediate for train traffic controllers and service area supervisor. Foremen took some time, being illiterate and not used to any abstraction, but came in role in 15 minutes.
Time model	Step-wise.
Data presentation	Infrastructure representation in familiar schematics, detailed timetabling on paper, time through simple interface, cleaning capacity in simple game objects.
# of sessions	1 session, full day
Type of data generated	Quantitative and qualitative, delivering a range of rolling stock feasible to park
Consequences	Potential yield for Amsterdam-Watergraafsmeer too unsecure given additional complexity of extra night maintenance in the years 2012 - 2013. Solution considered for 2014. For Hoofddorp the yield found in the game was verified in the field and implemented for 2012.

Table 6. Core description of Platform Overnight Parking Game

By 'gaming' these processes it should become clear whether and to what extent the (platform) rail capacity can be used for the preparation of passenger rolling stock. If a carousel process is theoretically feasible, then follow-up actions are defined to carry out a practical test. Table 6 describes the core features of the game.

The game delivered the results requested in time, by which it became the first game in the row of six that not only drew conclusions based upon the mechanisms of the game play, but also on the numbers generated in the game. Halfway the first scenario in the session there was an intervention required because the service area supervisor felt that the game play was 'not realistic'. After a thorough joint review it appear there were two trains left in the wrong location. The game leader corrected this, and from this point all agreed the outcomes were valid and representative for a normal evening with no major disturbances.

In this game the modeling of infrastructure and information followed the infrastructure schematic layout and timetabling information that was previously successful in the ETMET and NAU games. The time was for the first time not continuous but step-based. The reason for this was the long time to simulate (6 hours) during which many moments are trivial, as trains stand still and some cleaning is done. As no more game time than 2 hours per scenario was available, a speed-up was required, but just faster time would not contribute to the game as some time periods need more attention than others. The solution was found in 5-minute time steps in the game time that could take anywhere between 30 seconds to 20 minutes to execute in clock-time. In this way the players (most of them operational practitioners, two of them near illiterate) could keep up with the more abstract representation of their real work. The two foremen of the cleaning teams had most issues getting involved. Their whole task consisted out of taking 1 toothpick out of a wagon once it was cleaned, and they could each take out one stick per 5-minute step. Once they got used to this task they could make choices for priority over the service area easily and got their behavior realistic according to both their own and others judgments.

6. Methodological challenges

In this section the lessons learned on methodological challenges are discussed on two levels. The first subsection answers the three modeling issues raised in Section 4. The second subsection discusses how to guarantee validity in gaming simulations for railways.

6.1 Modeling issues

In the six projects, the modeling issues appeared to have a large interaction. The question how to immerse train traffic controllers in a gaming simulation appears to be largely dependent on the display of information. For train traffic controllers we learned that a detailed representation of the infrastructure is key for their involvement. However, the geographical representation did not work, where the common abstracted versions as used in practice performed flawlessly, both in digital (RBG) and in analog game board format (ETMET, NAU, POP). More abstraction and reduction of complexity of the infrastructure does not work for operators (RCM). For the timetabling and similar information like personnel and rolling stock planning similar mechanisms worked: give the players the information on a detailed level but keep the format they are used to in practice, even tough the medium (computer or paper) is different. The same held true for delays and other process information. Once the delays are presented directly after a train number in the format of +3, +5, + 10 minutes everybody understood it immediately. Once the players could understand the information well they could concentrate on their task, which they automatically did fully immersed.

Regarding the question how to model time we learned that the logical model of continuous time for rail operations works well and puts pressure on the process. In the ETMET game the frustrations over problems under time pressure became so high that the game leader had to intervene, and other games showed real pressure on the players who are so aware of the real-time nature of their real-world process that this can be triggered immediately in the game. Care should be taken to give players sufficient situational awareness without all their real tools available. Based on the experiences with the Railway Bridge Game versus the

analog games to expect that continuous time will improve on stress and pressure level when computer models are more easily deployable and integrated in the games. Analog simulators are surprisingly good but require extensive and thus expensive expert support.

6.2 Validation

The sessions usually run only once. Drawing conclusions on just one session puts emphasis on the validity of the behavior observed and decisions made in the simulation. The number of people to validate a full game with is limited in terms of availability (they work in de 24/7 operation) and costs, validation approaches need to be done differently. By modularizing the toolkit of gaming into sub-models and software components, validation of the components can be done outside of the final game sessions in analogy with the recent insights in multi-agent simulation of social systems (Gilbert, 2011). Work on the validation requires deeper understanding of train traffic control and train driver behavior. This encompasses the knowledge base in the organization. Work on this gives methodological challenges that go beyond the literature on gaming methodology (Peters et al, 1998).

In the railway gaming simulation described above (but for the RCG) processes of self-validation were used to overcome the validation issue for now. During every session signs of discomfort of the players and comments on 'how real' something was were constantly monitored and discussed openly even if this led to time-outs or moments of difficult discussion during the game play. The game leader always stated that everything to make the session better would be welcome at any time. In the debriefing the explicit questions were asked: which part is realistic and which part would be different in the real world, and why? This gave often very valuable information, even when in case of the NAU game when the network and service controller were not very involved in the game play, but could comment and criticize the validity very well.

By ensuring immersive game play and having the self-validation during the games the Raser categories of psychological reality and process validity are addressed to an extent that is satisfying for the organization. The structural validity is a design issue and is difficult to improve when using analog simulators. You simply cannot model the exact train flow and safety and interlocking systems in an analog way. Computer simulators have a lot to offer here in interaction with the players during a game session. This is future work for integration. The predictive validity is currently under review as the project follows all game projects longitudinally to determine the extent to which the conclusions based upon game sessions hold true in the real operation. Future work will report on outcomes.

7. Conclusions

The series of six projects shows the purpose and usefulness of building gaming simulations to help the Dutch railway infrastructure manager ProRail innovate on its core processes. Over the projects methodological lessons on involving operators as game participants have been drawn, as well as for the abstraction and reduction of information and the modeling of time. These modeling challenges appear to be highly interrelated. The lessons learned show the need, contrary to the traditional modeling approaches in gaming simulation, for very little abstraction and reduction in modeling the game where it concerns items that the operators have to play with. The model for less operational aspects can be more abstract, in line with literature on the need for fidelity for learning in games. While this finding may be not surprising to experienced game developers, the value of using abstractions that are used in the real world for the game is new.

As the game projects reported in this chapter are not for learning but for testing of designs and hypotheses, the findings on how to make real operators show valid behavior in a game session contributes to the small but growing field of gaming simulation for testing. For the categories psychological reality and process validity the current approach has found ways to address given the limited time and capacity available for traditional validation. For structural validity and predictive validity future work is defined.

The sequence of gaming simulations led to a successful introduction in the ProRail organization of the gaming method. Full support has led to a four-year partnership between academics and the operation to make gaming suited for ProRail and ProRail suited for gaming. Once this project has been carried out, ProRail will have at its disposal a gaming suite that connects with existing rail traffic simulators. The gaming suite will make it possible to configure a game simulation session without the need to call in outside expertise by selecting timetables, locations, actors, duration and measurement variables. The key feature is the possibility to create 'what-if' scenarios. The outcomes of these scenarios support the decision-making process by providing an understanding of the problems and the pros and cons of the possible solutions.

8. Acknowledgements

This research has been funded by ProRail and the Next Generation Infrastructure Foundation (NGI). Special thanks go out to the team members both on the side of ProRail (Jelle van Luipen, Emdzad Sehic, the steering committee, amongst many others) and the side of TU Delft (Rens Kortmann, Igor Mayer, Alexander Verbraeck, Bas van Nuland, Gert Jan Stolk and the Game Lab a.o.). Research like this is teamwork.

9. References

Albrecht, T. (2009). The Influence of Anticipating Train Driving on the Dispatching Process in Railway Conflict Situations. In: Networks and Spatial Economics, Springer, vol. 9(1), 2009: pages 85-101.

Barreteau, O. (2003) The joint use of role-playing games and models regarding negotiation processes: characterization of associations. Journal of Artificial Societies and Social Simulation vol. 6, no. 2

Bekebrede, G. (2010). Experiencing Complexity: A Gaming Approach for Understanding Infrastructure Systems. Delft: Next Generation Infrastructures Foundation.

Bekebrede, G. and I.S. Mayer (2005). Build your seaport in a game and learn about complex systems. In: Journal of design research, 5(2): pp 273-298.

Bekebrede, G. and S.A. Meijer (Forthcoming). Gaming for innovation in the health insurance sector.

Botden S.M.B.I., Buzink S.N., Schijven M.P., Jakimowicz J.J. (2008) ProMIS Augmented Reality training of laparoscopic procedures: Face, expert and referent validity. Simulation in Healthcare 3 (2), 97-102

Carter, F.J., M. P. Schijven, R. Aggarwal, T. Grantcharov, N. K. Francis, G. B. Hanna and J. J. Jakimowicz (2005). Consensus guidelines for validation of virtual reality surgical simulators. Surgical Endoscopy. Vol. 19: 12, pp 1523-1532

CBS (2009) Hoe druk is het nu werkelijk op het Nederlandse spoor? Het Nederlandse spoorgebruik in vergelijking met de rest van de EU-2. Editors: P. Ramaekers, T. de Wit, M. Pouwels. Netherlands Statistics Bureau CBS - in Dutch

De Freitas, S. and M. Oliver (2006). How can exploratory learning with games and simulations within the curriculum be most effectively evaluated? Computers and Education 46 (3) 249-264

Druckman, D. (1994). The educational effectiveness of interactive games. In D. Crookall and K. Arai (Eds) Simulation and gaming across disciplines and cultures. SAGE publications, pp. 178-187.

Duke, R.D. (1974). Gaming: the future's language. Sage, Beverly Hills / London.

Duke, R.D. and J.L.A. Geurts (2004). Policy games for strategic management. Dutch University Press, Amsterdam, The Netherlands.

Endsley, M.R. (2004). Situation awareness: Progress and directions. In S. Banbury & S. Tremblay (Eds.), A cognitive approach to situation awareness: Theory, measurement and application (pp. 317–341). Aldershot, UK: Ashgate Publishing.

Gaba, D.M., S.K. Howard, K.J. Fish, B.E. Smith, YA. Sowb (2001). Simulation-Based Training in Anesthesia Crisis Resource Management (ACRM): A Decade of Experience. Simulation Gaming Vol. 32 no. 2 pp 175-193

Gartner (2011) Case Study: Innovation Squared: The Department for Work and Pensions Turns Innovation Into a Game. Accessed through the web 2011-09-30 http://www.gartner.com/DisplayDocument?id=1476216

Gilbert, N. (2011). Keynote address to Artificial Economics conference, The Hague, The Netherlands, 2011-09-01.

Goverde, R.M.P. (2005). Punctuality of Railway Operations and Timetable Stability Analysis. TRAIL Research School, Delft, ISBN 90-5584-068-8

Hale, A. and T. Heijer (2006) Is Resilience Really Necessary? The Case of Railway. In: Resilience engineering: concepts and precepts. Hollnagel, E., D.D. Woods & Nancy Leveson (eds), Ashgate Publishing, Hampshire, UK, pp 125 - 147

Hamilton, W.I. and T. Clarke (2005). Driver performance modelling and its practical application to railway safety. Applied Ergonomics Vol. 36: 6, pp 661-670

Hamdorf, J.M. and J.C. Hall (2000): Acquiring surgical skills. British Journal of Surgery. Vol: 87: 1- pp 28-37

Harteveld, C. (2011) Triadic Game Design: Balancing Reality, Meaning and Play. Springer

Hiltbrand, T. and M. Burke (2011). How Gamification will change Business Intelligence. Business Intelligence Journal, Vol 6: 2.

Hollnagel, E., D.D. Woods & Nancy Leveson (2006) Resilience engineering: concepts and precepts. Ashgate Publishing, Hampshire, UK.

IEEE (2010) IEEE Recommended Practice for Distributed Simulation Engineering and Execution Process (DSEEP), IEEE Std 1730 - 2010, IEEE Explore.

Klabbers, J.H.G. (2003). Simulation and gaming: introduction to the art and science of design. Simulation and Gaming, 34(4): 488-494.

Klabbers, J.H.G. (2006). Guest editorial. Artifact assessment vs. theory testing. Simulation & Gaming, 37(2): 148-154.

Kortmann, L.J., and E. Sehic (2010) The Railway Bridge Game - usability , usefulness , and potential usage for railways management. In proceedings of ISAGA 2010, Spokane.

Kriz, W.C. and J.U. Hense (2006). Theory-oriented evaluation for the design of and research in gaming and simulation. Simulation Gaming, 37: 268 - 285.

Mayer, I.S. (2010) The Gaming of Policy and the Politics of Gaming: A Review. Simulation & Gaming 40(6) pp. 825–862

Meijer, S.A. (2009) The organisation of transactions : studying supply networks using gaming simulation. Wageningen Academic. ISBN 9789085853343

Meijer, S.A. (forthcoming): Introducing gaming simulation in the Dutch railways. Accepted for TRA 2012, Greece.

Meijer, S.A. and R. Poelman (2011) Supervisor: a 3D serious game for hazard recognition training in the oil industry. Proceedings of ISAGA 2011, Poland.

Meijer, S.A.; Hofstede, G.J.; Beers, G.; Omta, S.W.F. (2006): Trust and Tracing game: learning about transactions and embeddedness in a trade network. Production Planning & Control 17: 6. pp. 569 - 583.

Meijer, S.A., Kracht, P. van der, Luipen, J. van & Schaafsma, A. (2009). Studying a control concept for high-frequency train transport. In E Subrahmanian & J Schueler (Eds.), Developing 21st Century Infrastructure Networks (pp. 1-6). Chennai, India: IEEE.

Meijer, S.A., I.S. Mayer, J. van Luipen & N. Weitenberg (2011): Gaming Rail Cargo Capacity Management: Exploring and Validating Alternative Modes of Organization. Simulation & Gaming 1046878110382161, first published on February 1, 2011 as doi:10.1177/1046878110382161

Middelkoop, A.D. and L. Loeve (2006). Simulation of traffic management with FRISO. Computers in Railways X: Computer System Design and Operation in the Railway and Other Transit Systems.

Noy, A, D.R. Raban and G. Ravid (2006). Testing social theories in computer-mediated communication through gaming and simulation. Simulation & Gaming, 37(2): 174-194.

Peters, V., G. Vissers and G. Heijne (1998). The validity of games. Simulation & Gaming, 29(1): 20 - 30.

Proctor, M.D, M. Bauer, T. Lucario (2007) Helicopter Flight Training Through Serious Aviation Gaming. The Journal of Defense Modeling and Simulation: Applications, Methodology, Technology. Vol. 4:3 pp 277-294

Raser, J.C. (1969). Simulations and society: an exploration of scientific gaming. Allyn & Bacon, Boston.

Rekenkamer (2011) Besteding Spoorbudgetten door ProRail. Report to Parliament by Algemene Rekenkamer, The Netherlands (in Dutch)

Reynolds, A. (1998). Confirmatory program evaluation: a method for strengthening causal inference. The American journal of evaluation, 19: 203 - 221.

Steenhuisen, B., W. Dicke, J.A. de Bruijn (2009). "Soft" Public Values in Jeopardy: Reflecting on the Institutionally Fragmented Situation in Utility Sectors. International Journal of Public Administration, Volume 32, Number 6, May 2009 , pp. 491-507(17)

Van de Velde, D.M., W.W. Veeneman, L. Lutje Schipholt (2008). Competitive tendering in The Netherlands: Central planning vs. functional specifications. Transportation Research Part A: Policy and Practice, Volume 42, 9: pp 1152-1162

Van de Velde, D.M., J. Jacobs, M. Stefanski (2009). Development of railway contracting for the national passenger rail services in the Netherlands. In: International Conference Series on Competition and Ownership in Land Passenger Transport - 2009 - Delft, The Netherlands - Thredbo 11

Van Luipen, J.J.W. and S.A. Meijer (2009). Uploading to the MATRICS: Combining simulation and serious gaming in railway simulators. In: Proceedings of 3rd Int. Conf. on Rail Human Factors, Lille, France, 2009

Witmer B.G., M.J. Singer (1998) Measuring Presence in Virtual Environments: A Presence Questionnaire. Presence Teleoperators Virtual Environments. Vol 7: 3, pp 225-240

12

Application of 3D Simulation Methods to the Process of Induction Heating of Rail Turnouts

Elżbieta Szychta, Leszek Szychta, Mirosław Luft and Kamil Kiraga
Technical University of Radom, Institute of Transport Systems and Electrical Engineering
Poland

1. Introduction

Keeping turnouts fully functional is necessary for safety of both train passengers and personnel operating railroads, particularly in winter. This is of particular importance since turnouts, key railroad elements, are exposed to adverse weather conditions. Proper function of turnouts is obstructed by snow falling between the point and the rail and between components of the setting lock which, combined with low ambient temperatures, causes the point to freeze to the sliding chairs and the components of the setting lock to freeze. This leads to blocking of a turnout[1]. Effective protection of turnouts against such weather conditions greatly reduces accident rates and improves efficiency of rail traffic.

The presented methods of heating railway turnouts all over the World are designed to prove operation efficiency turnouts during the winter. The authors set the goal of work to develop a new method of heating railway turnouts, using an inductive heating phenomenon. To do that it is necessary to know the properties of magnetic and electric rails. These properties are not explicitly specified by the manufacturers of rails, so the determination of their value by the laboratory tests is required. The chapter discusses the measurement methods used to determine the basic properties of electric and magnetic rails. Based on laboratory results a simulation model of inductive heating rails in 3D space is developed. In the final part of the chapter results of a simulation model tests are presented and discussed.

2. Contemporary methods of turnout heating

Methods of clearing snow and ice have evolved in line with the state of engineering art, weather conditions in a region as well as availability and costs of energy sources. The most common methods of heating turnouts in Europe include[2]:

- gas heating,
- water heating, used in smaller rail facilities, mainly in Germany. The first systems of this type in Poland were installed at Boguszów station,
- geothermal heating,

---

[1]Kiraga K., Szychta E., Andrulonis J. (2010). Wybrane metody ogrzewania rozjazdów kolejowych - artykuł przeglądowy
[2]Brodowski D., Andrulonis J. (2002). Ogrzewanie rozjazdów kolejowych, Problemy kolejnictwa

- electric heating by means of resistance heaters (most often supplied with 3X400 [V] systems, 230 [V] power supply, or with 15 [kV] $16\frac{2}{3}$ [Hz] traction networks via voltage-reducing transformers).

Gas heating is used in extremely hard weather conditions (heavy snowfall and low temperatures of down to - 30°C) on Austrian, Norwegian and Swiss railroads (the Alpine region) or in the Netherlands, where the weather is mild but very damp. Gas systems of turnout heating are characterised by high thermal efficiency. Gas burners reach power of up to 1000 [W/running metre of a rail]. Liquid mixtures of propane and butane or butane alone are most often employed as fuel. The gases display varied pressure depending on temperatures of condensed gas. The pressure reduces as the temperatures diminish. Mixtures of propane and butane are utilised where ambient temperatures are above - 17°C, replaced with butane itself below that temperature. Gas turnout heating is diagrammatically presented in Figure 1.

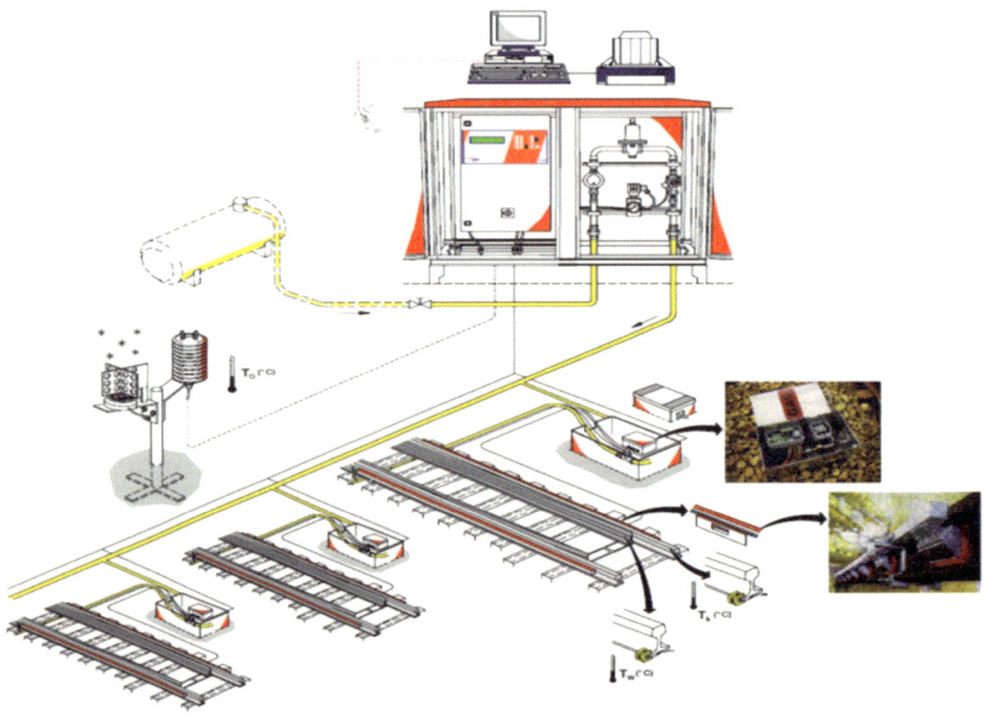

Fig. 1. Gas system of turnout heating

Water heating is another way of heating rail turnouts. An oil unit heats the working fluid, a mixture of water and anti-freeze agents. The heated fluid is supplied to pipe heat exchanges (heaters) which are fitted along a rail and turnout saddles. The fluid yields its heat and returns to the unit, cooled, to be re-heated. In a water heating system, the energy contained in the fuel (heating oil) is converted into heating power in the heating unit placed in the immediate vicinity of turnouts (normally in a building nearest to a signal cabin, which can itself be heated by the unit as well)[3].

Design of a MAS water heating system comprises the following elements[4]:

- a heating station - including a heating unit, fuel tank, and a control system, 220V~/12V= (or 24V =) power supply to the heating unit, 12V (24V) battery as a reserve power source for the unit, and plumbing (surge tanks, pumps, pipes, cut-off valves),
- heating elements, i.e. heat exchangers inside a turnout,
- installation feeding the operating fluid from the heating station to turnouts.

Layout of a MAS system is shown in Figure 2.

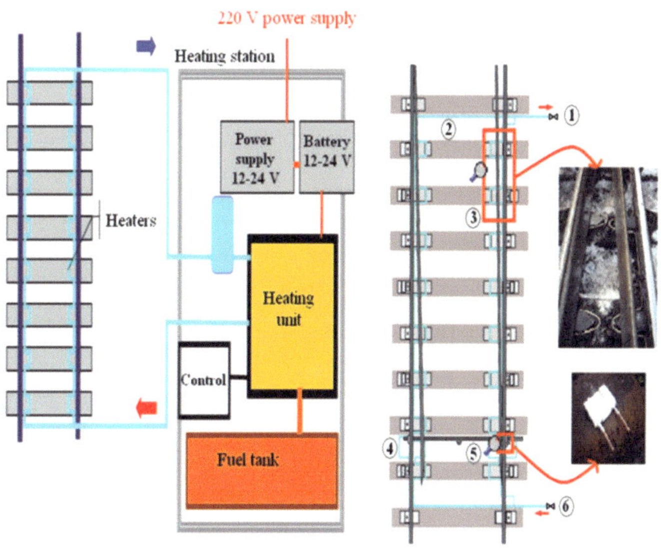

Fig. 2. Single-flow MAS water heating of a short turnout: 1 – return to unit, 2 – Ø 12mm steel pipe (heater), 3 – a heater located near a rail edge and sliding chair,
4 – heating of a setting lock, 5 – support heater, 6 - input

Geothermal systems are the third method of turnout heating discussed here. Geothermal heating, which uses the natural underground heat, is a new alternative to the systems presented above. A heat pump is its core element. Depending on the season and depth, soil temperatures may range from 4 to 8°C. At more than 15m below the ground level, season-related thermal motions cease and the temperature is constantly around 8-10°C. Still deeper, the soil is regenerated by flowing underground waters, heat from the earth's core and from above.

[3]Materiały seminaryjne CNTK. Wodne ogrzewanie rozjazdów kolejowych typu MAS
[4]Badania eksploatacyjne wodnego system ogrzewania rozjazdów typu MAS-Guben

Heat pumps operate like fridges which take advantage of the hot, not the cold section of the heat cycle. An appropriate working agent is compressed and decompressed producing a desired heating or cooling effect. To generate useful heat, for instance, soil or underground water heat (at low temperatures of approx. 10°C) is employed to evaporate the operating agent (harmless gas R497C) that boils at a low temperature. Thus, the originally liquid working agent leaves an evaporator (heat exchanger on the side of ground collector) as gas. The gas is then compressed and condensated in a liquefier (heat exchanger on the side of heating installation) at high temperatures (50-60°C), yielding the condensation and compression heat to the water contained in the heating installation. The still pressurised working agent is subsequently decompressed in a valve and enters the low-pressure section, thereby initiating the cycle once again.

A complete system of geothermal turnout heating is illustrated in Figure 3.

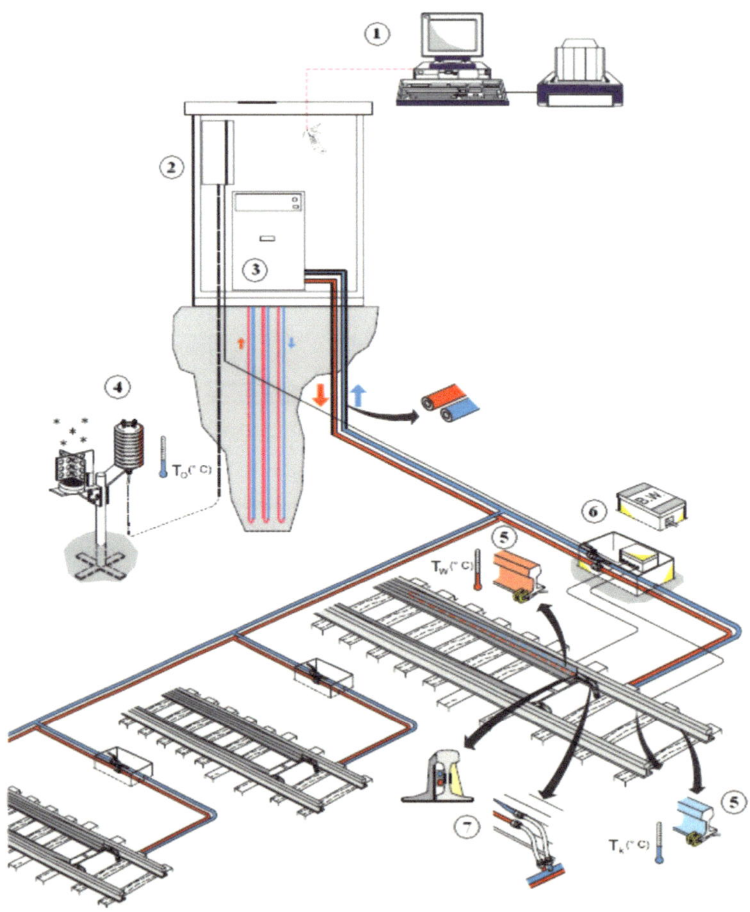

Fig. 3. Design of a geothermal turnout heating system including: 1 – a local control and monitoring system, 2 – an automatic control system containing an automatic weather unit, 3 – heat pump and heat cycle, 4 – snowfall and ambient temperature sensor, 5 – humidity and rail temperature sensors, 6 – junction box, 7 – heat exchangers (heaters) with heat-conducting insulation

Electric heating is the final method of turnout heating to be discussed here. The electric heating currently prevails in Poland among the equipment used to melt snow and ice on rail turnouts. It functions on more than 18 000 turnouts. Its combined installed power reaches approximately 110 [MW][5]. The electric systems heat for an average of 300 hrs in a season. Providing 330W per running metre of rail assures effective warming of railroad turnouts during the heating season. The electric heating employing 330W/m heaters provides for good functioning of railroad turnouts under normal weather conditions (i.e. temperatures above -20°C) and with average (other than catastrophic) snowfall. A diagram of electric turnout heating is shown in Figure 4.

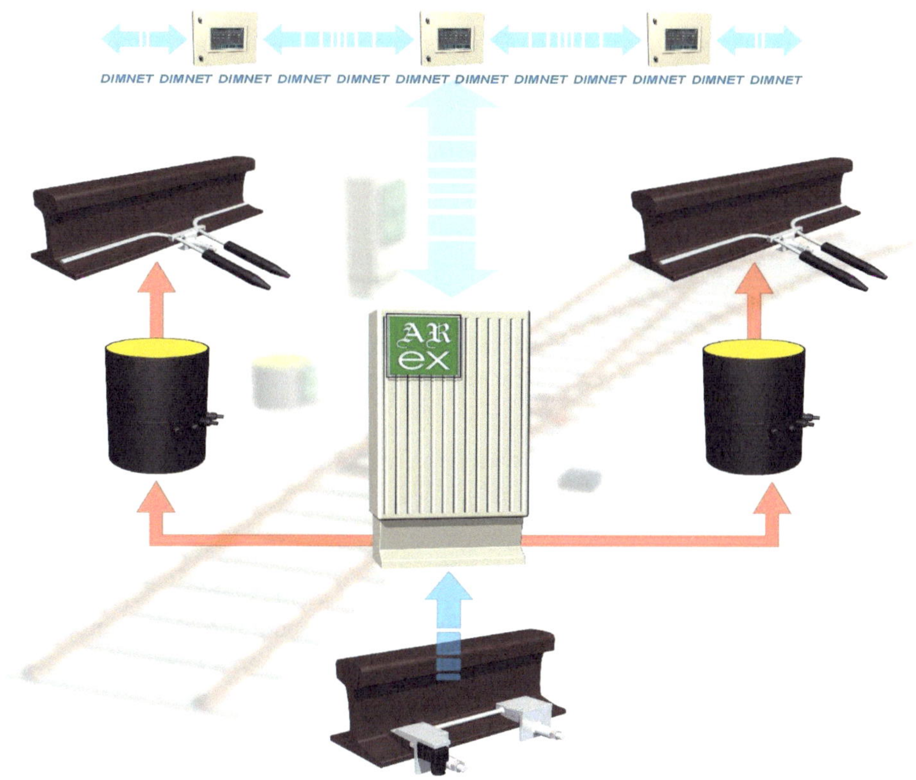

Fig. 4. Electric turnout heating

Recent years of cold winters and violent snowfalls have unfortunately shown that electric systems fail to provide effective heating of turnouts. The method of electric heating has therefore been modified to enhance its efficiency.

Performance trials are under way using heat insulation on Dutch, German, and Polish railroads. The rail foot is additionally insulated as part of these solutions. Arrangement of heat insulations on a rail is presented in Figure 5 below[6].

[5]Brodowski D., Andrulonis J.(2000). Efektywność ogrzewania rozjazdów kolejowych
[6]Prospekt informacyjny o otulinach firmy Haet Point, 2009.

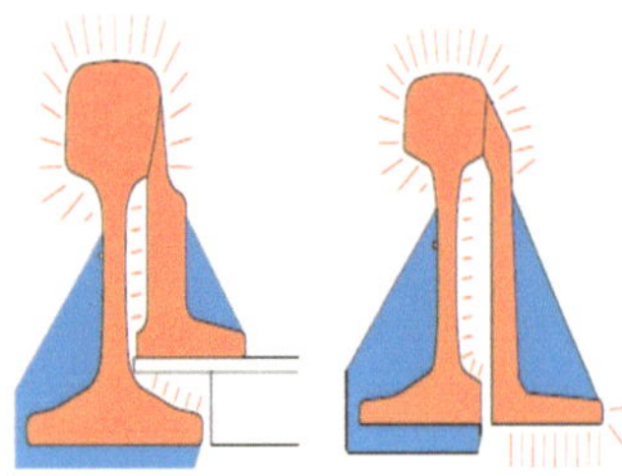

Fig. 5. Situation of insulations on the internal and external sides of a rail and on rail foot for every rail type

Such a solution is offered to PKP (Polish State Railroads) by Heat Point of the Netherlands, Research and testing will demonstrate potential advantages and drawbacks of this solution.

3. Induction heating of turnouts now and in the past

Induction turnout heating (an original Polish concept) was tested by PKP in 1978/1979 on selected turnouts and stations[7]. Insulated heating wires were used. They were not in galvanic contact with rails. Rails were heated with eddy currents induced inside the rails. The wires were heated to temperatures in the range +15°C to +20°C.

The heating wires were made of copper, wrapped with Tarflon tape, and placed inside a steel envelope. 3-3.3 [V] and 50 [Hz] were supplied to the wire, where the current was 350 [A].

The following types of heating wires were used:

- 2.55 [m], power 750 [W] for the rail UIC-49
- 3.00 [m], power 900 [W] for the rail UIC-60

A 2800 [VA] transformer was employed to supply power to each type of heaters. Given the mains frequency of 50 [Hz], the heating wires vibrated and produced human-audible acoustic waves whose frequency was twice greater than the frequency of the supplied voltage.

The inductive nature of power distribution system loading by ior equipment of the time required an additional capacitor to set off the reactive power and to improve the power factor cosφ from approx. 0.5 to 0.85 - 0.9. Capacitors capable of adjusting reactive power of 4 [kVA] or more were employed in a single given turnout. Work on the system and its application was abandoned due to insufficient technological resources at the time (1978/1979). The material on testing of inductive turnout heating discussed here is the only, scarce material still extant in the archives of the then COBiRTK (Centre for Rail Engineering Research and Development), currently named IK (Rail Engineering Institute).

These authors have decided to revive the idea of induction turnout heating and to use it, as part of a greater operating frequency system, to heat rail turnouts. A flow diagram of induction turnout heating as proposed by the authors is shown in Figure 6.

[7]Praca zbiorowe: Studium na temat wyboru optymalnego systemu ogrzewania rozjazdów

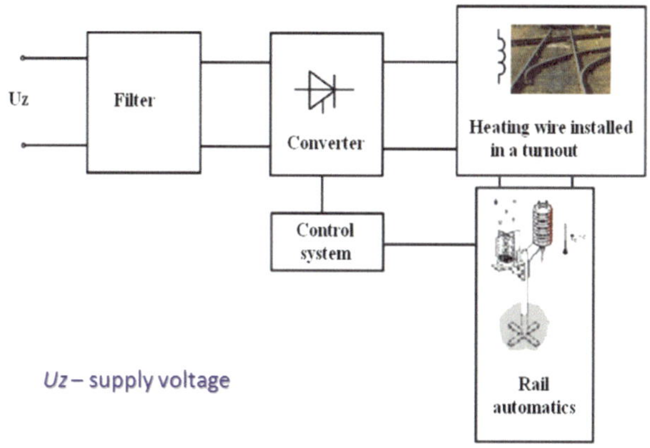

Fig. 6. System of induction turnout heating

Elementary theory concerning induction heating is not introduced as it is commonly known and easily available in engineering literature[8] [9]. Knowledge of fundamental electric and magnetic properties of rails forming parts of turnouts is necessary to construct a fully functional induction heating system. Since not all of those properties are easily accessible or clearly defined, their values had to be determined experimentally by one of the methods described in the following section.

Attention focused then on the mathematical apparatus based on Maxwell's equations and used to describe the electromagnetic field in space. Final element method (FEM) was subsequently discussed. Maxwell's equations and FEM provide the foundations for an analytical model of induction turnout heating as executed in Flux 3D software.

4. Electric and magnetic properties of e160 rail

Rails (stock) are fundamental design elements of a turnout, beside switch points, sliding chairs or switching closure assemblies. Rails are principally designed to set the proper travel direction of rolling stock wheel sets. Shape of a rail comprises three characteristic sections: head (the part along which rolling stock wheels move), web, and foot (the part supporting the whole and carrying the load on to sleepers).

Two main rail types are used on routes administered by Polish State Railways PKP PLK: 60E1 and 49E1. They differ in the weight of a running metre and cross-section dimensions. 49E1 (49.39 kg/ running metre and cross-section surface area of 62.92 cm^3) is used on routes with light rolling stock load. 60E1 (60.21 kg/ running metre and cross-section surface area of 76.70 cm^3) is operated on heavily loaded routes where trains travel at speeds over 100 km/h.

Table 1 presents steel grades to be used in rail manufacture[10] [11]. The grade references follow from applicable European and Polish standards (PN-EN 10027-1 and PN-EN10027-2). The

[8]Sajdak Cz., Samek E. (1985). Nagrzewanie indukcyjne. Podstawy teoretyczne i zastosowanie

[9]Gozdecki T., Hering M., Łobodziński W. (1979). Urządzenia elektroniczne. Elektroniczne urządzenia grzejne

[10]Wielgosz R. (2009). Łączenie bezstykowych szyn kolejowych

symbols of rail materials are based on rolling surface hardness, in Brinell degrees, with the added symbol of an element used to refine the rail steel or in reinforcement heat treatment. Table 1 also includes references to previously used steel grades, of chemical compositions similar to the new steels recommended by the EU in accordance with EN 13674-1:2003 (E).

Two steel grades most commonly used in Poland are R260 (hardness range 260÷300 HB) and R350HT (hardness range 350÷390 HB, heat treated head).

Steel marking	Description	Material number	Previous marking
R200	Carbon-manganese	1.0521	R0700
R220	Carbon-manganese	1.0524	R0800
R260	Carbon-manganese	1.0623	R0900; St90PA
R260Mn	Carbon-manganese	1.0624	R0900Mn; St90PB
R320Cr	Low alloy	1.0915	R1100Cr
R350HT	Heat treated carbon-manganese	1.0631	R1200
R350LHT	Heat treated low alloy	1.0632	

Table 1. Rail steel markings

The steels formerly used in rails contain 0.40 to 0.82% carbon, 0.60 to 1.70% manganese, 0.05 to 0.90% silica, and some additionally contain up to 1.30% chromium. The new steels recommended by the EU contain: 0.38 - 0.82% carbon, 0.65 - 1.70% manganese, 0.13 - 1.12% silica, and some additionally contain up to 1.25% chromium.

As steel is heated and cooled and temperatures change in rail manufacturing processes, structural transformations occur. In the final production process, a rail is hot rolled at temperatures of 700 - 900°C. The (heterogeneous) structure and material properties of the rail result from such processes.

60E1 is selected for purposes of testing electric and magnetic properties because it serves trans which travel at up to 100 km/h. Standardised dimensions of normal-gauge 60E1[12] [13] are illustrated in Figure 7.

The following elements of a heating circuit are analysed as part of testing aimed at developing a method of induction turnout heating: structure of rail material, resistivity and magnetic permeaability, skin effect and depth of magnetic field penetration into the rail structure, discharge of active power in the form of impact of the magnetic field on the rail. These parameters depend, inter alia, on magnetising current frequency and are not defined by rail manufacturers. It appears rails may come from different charges, may be produced by means of diverse rolling, straightening, and possibly hardening technologies. It was therefore important to determine locations from which rail samples would be removed. To generate a maximum quantity of data for further research, samples were withdrawn (cut) from key points of a rail. Sample removal locations and their geometric dimensions are shown in Figure 8.

---

[11]strona internetowa odnośnie szyn kolejowych: www. inzynieria-kolejowa.dl.pl

[12]Grobelny M. (2009) Budowa, modernizacja, naprawa i remonty nawierzchni kolejowej – urządzenia i elementy

[13]Instrukcja eksploatacji i utrzymania urządzeń elektrycznego ogrzewania rozjazdów

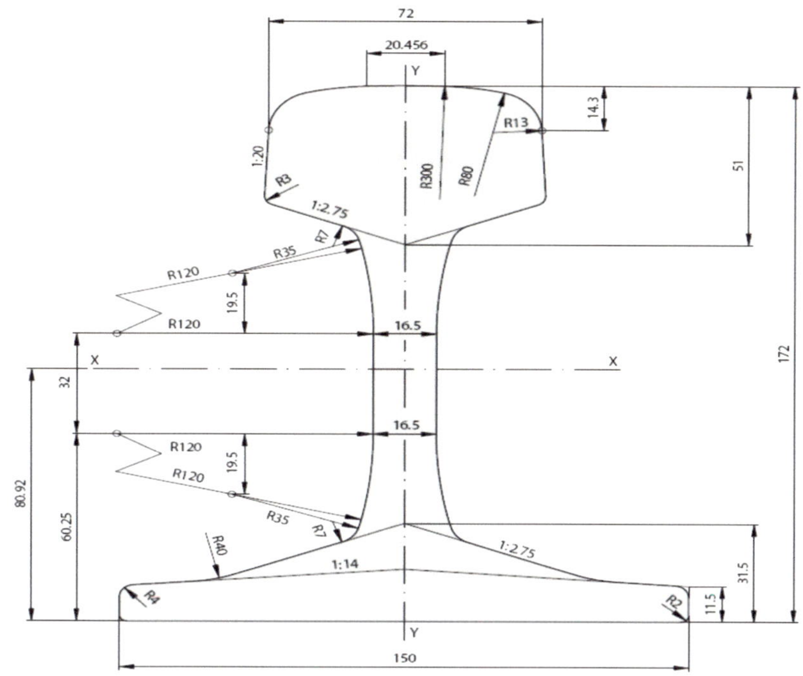

Fig. 7. Geometrical dimensions of the normal-gauge E160

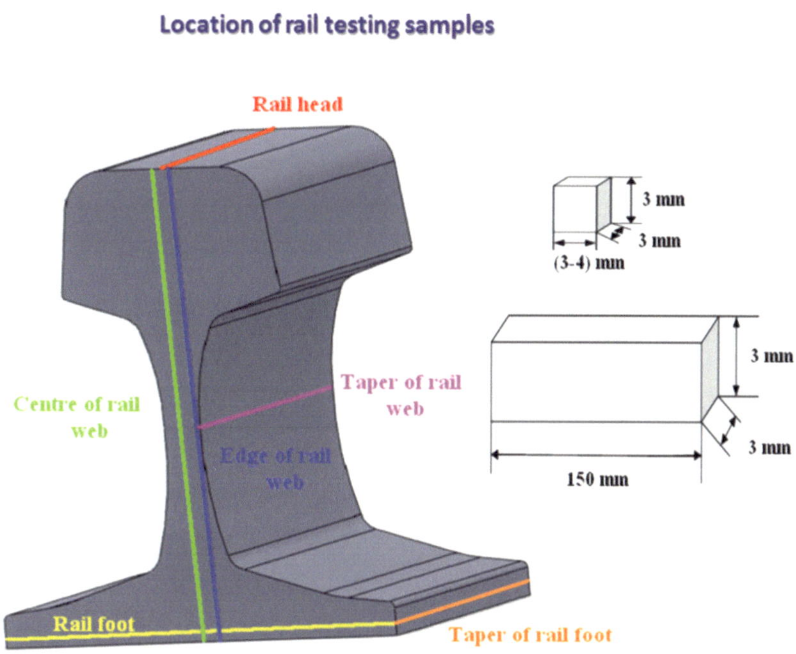

Fig. 8. 60E1 sample removal locations and their geometric dimensions

Determining key electric and magnetic parameters of construction materials for railroad turnouts was essential in designing a system of induction heating and employed a variety of testing methods. Electric and magnetic parameters were determined by means of the following methods:

- *four-point linear probe method* to determine electric resistivity (taken into consideration as rotary currents arise when the magnetic field penetrates the internal structure of a rail).

A flow diagram of electric resistivity measurements is illustrated in Figure 9.

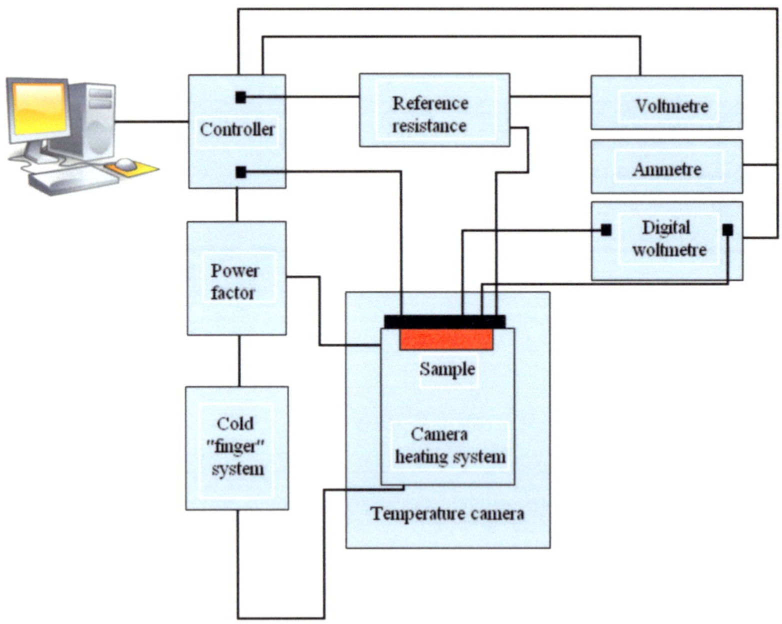

Fig. 9. Flow diagram of electric resistivity measurement system

Electric resistivity can be measured by means of a four-point linear probe and a measurement system which employs a PC to record the voltage drop across a tested sample and across a reference resistor R_w. Voltage is checked by running the current (10 times) in two directions. Once averaged, both the results are added and contact effects are eliminated in this way. The result is averaged and saved to the memory. The current's value can be calculated according to:

$$I_x = \frac{U_{Rw}}{R_w} \tag{1}$$

With known geometric dimensions of the sample, that is, its height *a* and width *b*, electric resistivity can be determined on the basis of:

$$\rho = \frac{U_x}{I_x} \cdot \frac{S}{l} = R_{mierz} \cdot \frac{S}{l} \tag{2}$$

where:

U_x - voltage drop across the sample,
I_x - measurement current,
S - cross-section of the sample,
l - distance between the measurement probes.

Table 2 summarises results of electric resistivity measurements for samples from characteristic rail locations.

Source of the sample	**Mean electric resistivity ρ [Ωm]**
Edge of rail web	$2.62 \cdot 10^{-7}$
Centre of rail web	$2.85 \cdot 10^{-7}$
Rail foot	$2.75 \cdot 10^{-7}$
Taper of rail foot	$2.57 \cdot 10^{-7}$
Taper of rail web	$2.7 \cdot 10^{-7}$
Rail head	$2.58 \cdot 10^{-7}$

Table 2. Values of electric resistivity for the tested samples

- *HP bridge:* active and passive magnetic hardness as well as the loss tangent are determined. Initial magnetic hardness can also be defined since the intensity of the magnetic field is low.

Hewlett Packard 4284A 20Hz - 1MHz Precision LCR Meter helps to conduct measurements in order to determine relative magnetic permeability μ of selected rail sections.

A connection link is replaced with a measurement coil of known length l and number of coils z. A sample is placed inside the coil. The bridge circuit is supplied with AC. A millivoltimetre of very high internal resistance is connected in parallel to the coils with the sample inside. It can measure the voltage drop across this element and, consequently, select the magnetising field as appropriate. The device also provides for de-magnetising of samples using a 50Hz field whose amplitude reduces towards zero, for regulation and stabilisation of temperature.

The system used to determine relative magnetic permeability μ of the tested rail sections consisted of:

- a PC and its software,
- Aligent 34401A 6 ½ Digital Multimeter,
- (hp) Hewlett Packard 4284A 20Hz - 1MHz Precision LCR Meter,
- and a measurement coil.

Figure 10 contains a diagram of the measurement stand serving to determine μ.

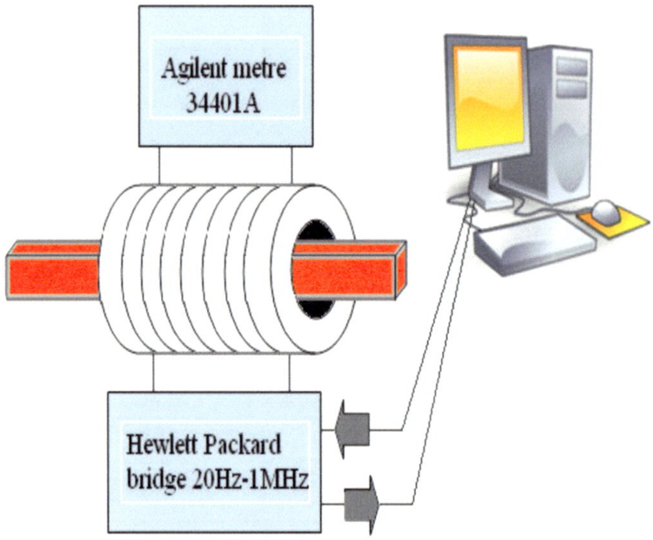

Fig. 10. Measurement apparatus serving to determine relative magnetic permeability μ

An AC bridge is the chief component of the system which provides for accurate measurement of combined magnetic permeability:

$$\underline{\mu} = \mu_{cz} + i\mu_b \tag{3}$$

The combined permeability comprises:

- the component active magnetic permeability μ_{cz} responsible for magnetising processes,
- the component passive magnetic permeability μ_b responsible for magnetic losses.

The active magnetic permeability can be formulated:

$$\mu_{cz} = \frac{(L_x - L_0) \cdot l}{\mu_0 z^2 S_p} + 1 \tag{4}$$

here:

L_x – inductivity of the coil containing the sample,
L_0 – inductivity of an empty coil,
μ_0 – magnetic permeability of the vacuum $\mu_0 = 4\pi \cdot 10^{-7}$H/m,
S_p – cross-sectional surface area of the tested sample,
z – number of coils,
l – length of the coil.

As $\mu_{cz} >> 1$, 1 is ignored in the calculations.

The passive magnetic permeability is computed:

$$\mu_b = \frac{R_{rdz} \cdot l}{\mu_0 \cdot Z^2 \cdot 2\pi \cdot f \cdot S_p} \tag{5}$$

R_{rdz}, or the core's resistance, is calculated:

$$R_{rdz} = R - R_0 \tag{6}$$

where:

R_0 – resistance of an empty coil,
R – resistance of the coil containing the sample.

Substituting (6) to (5) produces the following equation of passive magnetic permeability:

$$\mu_b = \frac{R_{rdz} \cdot l}{\mu_0 Z^2 2\pi f S} = \frac{(R - R_o) \cdot l}{\mu_0 Z^2 2\pi f S} \tag{7}$$

Measuring voltage drop across the resistance R helps to determine intensity of the electric current across the winding and then the intensity of the magnetic field at which magnetic permeability is measured using the dependence:

$$H = \frac{Z \cdot U}{l \cdot R} \tag{8}$$

where:

z – number of coils,
U – voltage drop across the resistor R,
l – length of the measurement coil.

Figure 11 illustrates the initial magnetic permeability μ_{initial}=f(f) of a sample from the rail web edge (red) and μ_{initial} = f (f) of a sample from the rail's head (black) as well from the rail's core (the green curve).

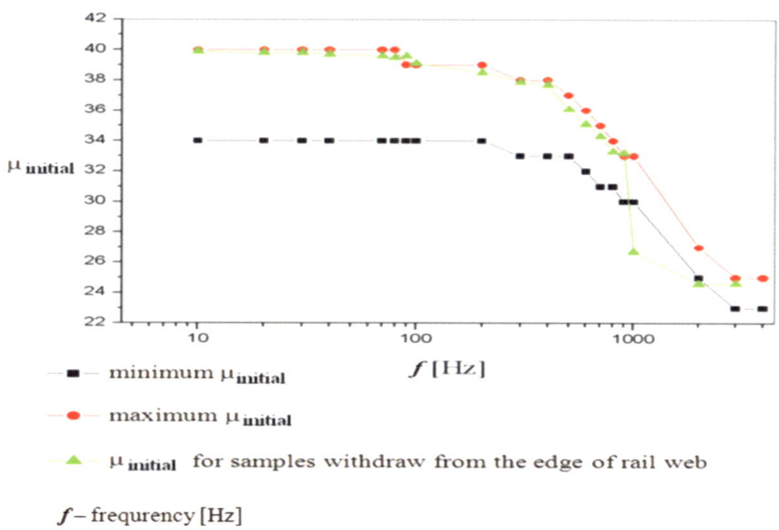

Fig. 11. Magnetic permeability μ_{inital}=f(f) for a sample from the edge of rail web and μ_{inital} = f (f) for the remaining samples.

- A *fluxmetre* serves to plot the curves $B = f(H)$ and $\mu = f(H)$[14 15],

Prime magnetising curves were determined by means of the measurement system shown in Figure 12.

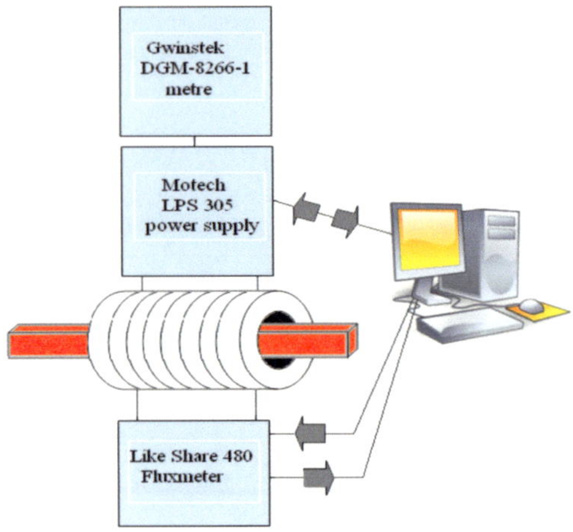

Fig. 12. Measurement system employing the fluxmetre and its components

The tested sample is placed inside the coil in magnetising and measurement winding. A fluxmetre to measure magnetic flow variations in the tested material sample is the element of the measurement system. Variations in the tested material sample result from commutations of the current across the magnetising winding z_m. The electric current pulse induced in the measurement winding over time dt is measured by the fluxmetre. The magnetic flow variations $d\Phi$ generate an electromotor force ε which can be expressed:

$$\varepsilon = -\frac{d\phi_c}{dt} = -z_p \cdot \frac{d\phi}{dt} \tag{9}$$

Fluxmetre readings are proportional to magnetic flow variations.

When a uniform external magnetic field is applied, magnetic induction B can be computed according to:

$$B = \frac{d\phi}{2z_p S} \tag{10}$$

where S is the cross-sectional surface area of the tested sample. The sample is demagnetised prior to each measurement by means of the system.

A prime magnetising curve for a sample from the rail web and maximum magnetic permeability in respect of the same sample determined using the fluxmetre are shown in Figure 13.

---

[14]Gignoux D., Schlenker M.(2005). *Magnetism Fundamentals*

[15]Jiles D. (1991). Introduction to Magnetism and Magnetic Materials

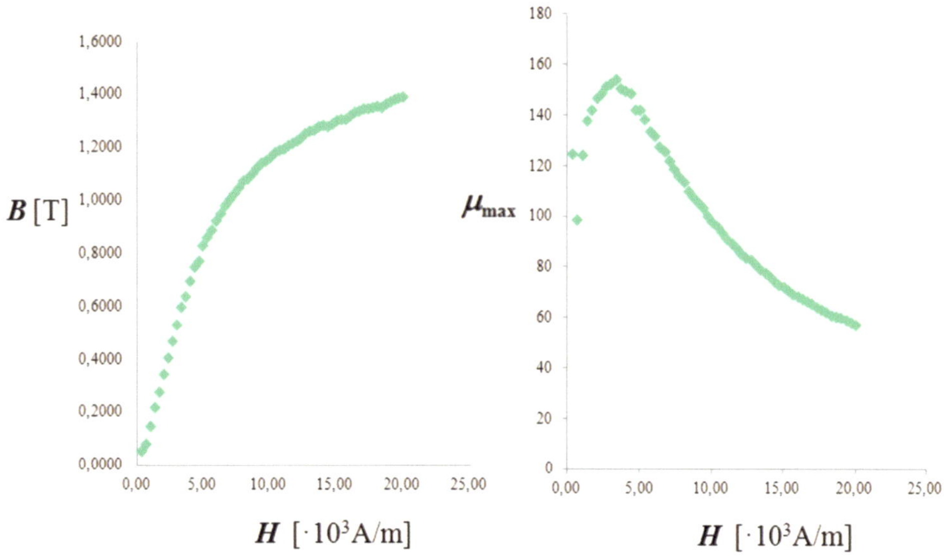

Fig. 13. Curves B = f(H) and μ = f(H) for a sample of the rail web

- *coercion metre* measures intensity of the coercion field,

Intensity of the coercion field of magnetically soft and hard materials can be measured by means of a coercion metre. The relevant measurement diagram is presented in Figure 14.

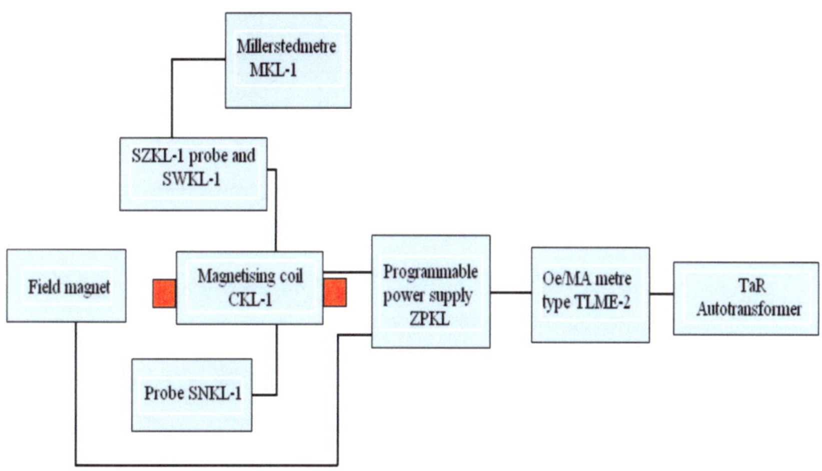

Fig. 14. Measurement diagram using a coercion metre and its components

Magnetic coercion[16] (also referred to as coercive force) is the value of an external magnetic field that must be applied to a material (e.g. a ferromagnetic material) to bring the magnetic residue down to zero. The magnetic residue (also remanence or residual magnetisation) is the magnetic induction remaining after an external magnetic field magnetising a given material is removed.

[16]Kuryłowicz J. (1962). Badania materiałów magnetycznych

Table 3 summarises measurement results of magnetic coercion intensity for samples removed from key rail locations.

Source of the sample	Sample number	Hc [A/m]	Mean Hc [A/m]
Edge of rail web	1	820	823
	2	820	
	3	828	
Centre of rail web	1	772	774
	2	776	
	3	772	
Rail foot	1	772	778
	2	772	
	3	788	
Taper of rail foot	1	876	912
	2	884	
	3	1003	
	4	884	
Taper of rail web	1	860	830
	2	860	
	3	804	
	4	796	
Rail head	1	812	850
	2	860	
	3	820	
	4	908	

Table 3. Coercion field intensity in respect of all samples

- *PPMS VSM* was utilised to plot the curve $J = f(H)$,

A Physical Property Measurement System (PPMS) by Quantum Design (San Diego, USA) is a unique, state-of-the-art concept of a laboratory facility.

The PPMS platform comprises the following elements:

- superconducting magnet of up to 7 Tesla (and, more recently, even 16 Tesla),
- specific heat measurement system (Heat Capacity 4He) in the temperature range 2K - 400K and magnetic fields of up to 7 Tesla,
- specific heat measurement system (Helium-3) in the temperature range 350mK - 350K and magnetic fields of up to 7 Tesla,
- AC/DC magnetisation measurement system for magnetic fields of up to 7 Tesla,
- vibration magnetometer VSM for precise magnetising measurements In a broad temperature range of 2K to 1000K. It is additionally fitted with an oven (P527 Sample Magnetometer Oven) for measurements of up to 1000K.
- heat conductivity and thermal force measurement system,

- measurement capability of electric resistance in the range of 350mK,
- vertical rotator to regulate sample position in relation to the magnetic field,
- and an easyLab Pcell 30kbar pressure chamber to measure resistance at high pressure of up to 30kbar.

The overall flow diagram of PPMS platform is shown in Figure 15.

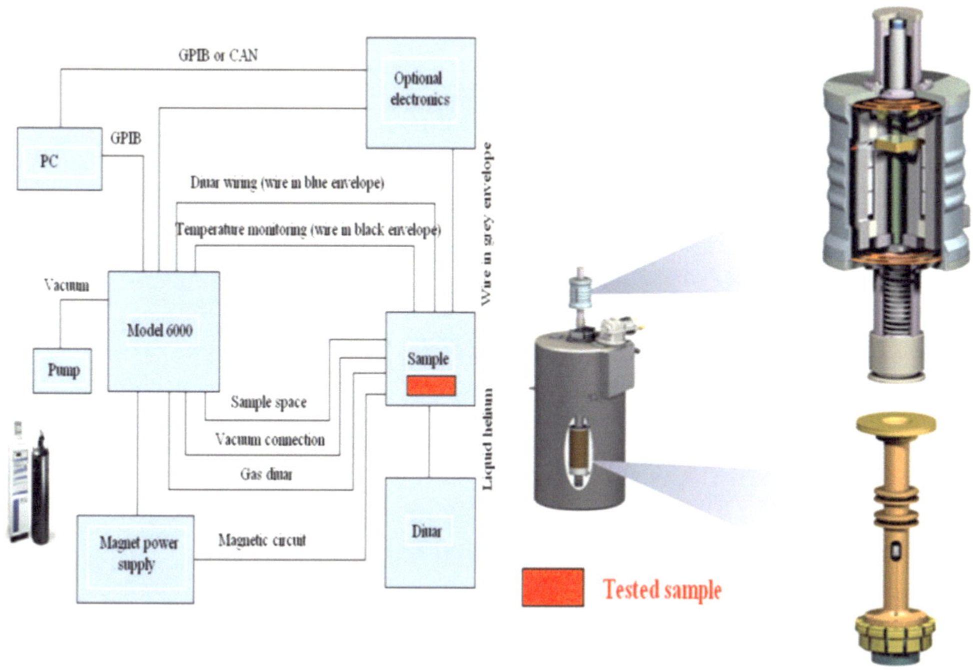

Fig. 15. Physical Property Measurement System to test magnetic properties of rail samples

A magnetometer suction cup with the vibrating sample was employed to plot prime magnetising curves and to determine saturation of the tested rail samples.

A sample is positioned on a non-magnetic, mobile bar and vibrates vertically at a set frequency. The sample's oscillations generate (induce) a variable voltage signal in the measurement coil system under impact of the magnetic field. The signal is proportional to the magnetic moment of the sample and to parameters characterising its motion, i.e. to the amplitude and vibration frequency. It can be described as follows:

$$V_{cewki} = -\frac{d\phi}{dt} = -\left(\frac{d\phi}{dz}\right)\left(\frac{dz}{dt}\right) = C\ m\ A\ \omega\ \sin\ (\omega\, t) \tag{11}$$

where:

C -proportionality constant,
m - a known moment of the sample,
A - vibration (oscillation)amplitude,
ω - frequency.

A flow diagram of vibrating sample magnetometer (VSM) is illustrated in Figure 16.

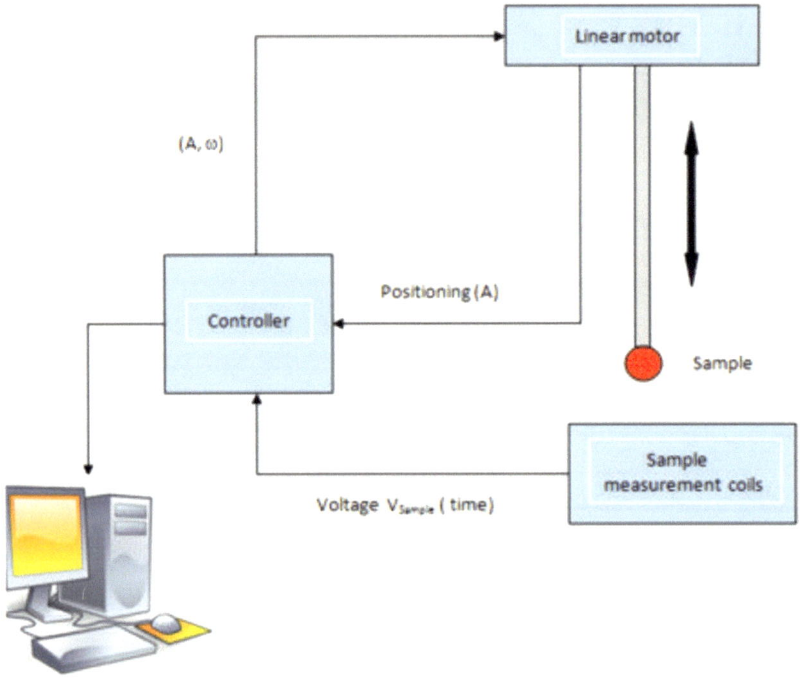

Fig. 16. Flow diagram of VSM measurement platform

Figure 17 presents a magnetising curve J = f(H) and magnetic saturation J_s for rail head samples.

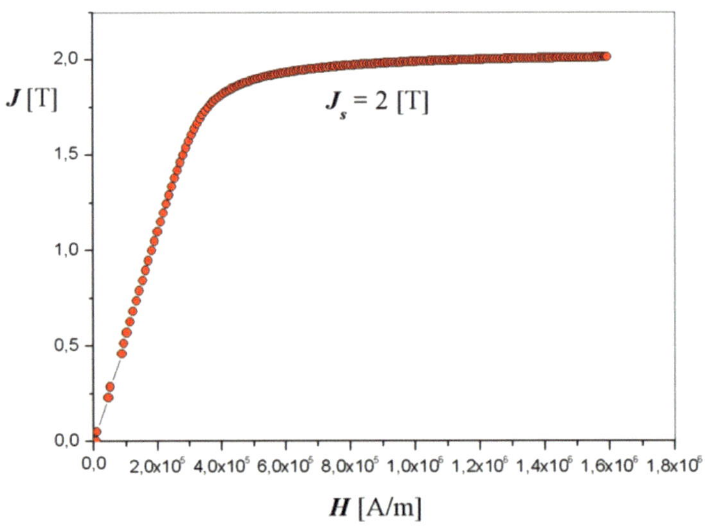

Fig. 17. Magnetising curve J = f(H) and magnetic saturation J_s for rail head samples

- *ACMS* determines combined magnetic susceptibility, including its real and imaginary components, the loss tangent for varied intensities of the magnetic field and for different temperatures.

ACMS (AC Measurement System) provides for a single sequence of measurements of AC and DC susceptibility in a broad temperature range from 1.9K - 350K. AC measurements are thrice as sensitive (2×10^{-8}emu (2×10^{-11} Am^2) at 10kHz) as DC measurements. Better results may be occasionally obtained by means of the DC method. Application of a powerful external field can increase the sample's magnetic moment and generate signals above the noise (the range of DC magnetising DC: 2.5×10^{-8}Am^2 - 5×10^{-3}Am^2 (2.5×10^{-5}emu - 5emu). DC measurement uses a constant field in the measurement area where the sample moves by means of detection coils and induces voltage there according to Faraday's law. Amplitude of this signal depends on the sample's magnetic moment and the rate of its removal. As part of this system, the sample is extracted at an approximate speed of 100cm/s. The signal's force grows considerably as a result compared to other systems. The measurements can be executed as a function of both temperature and of the magnetic field.

Flow diagram of ACMS and its key elements are shown in Figure 18.

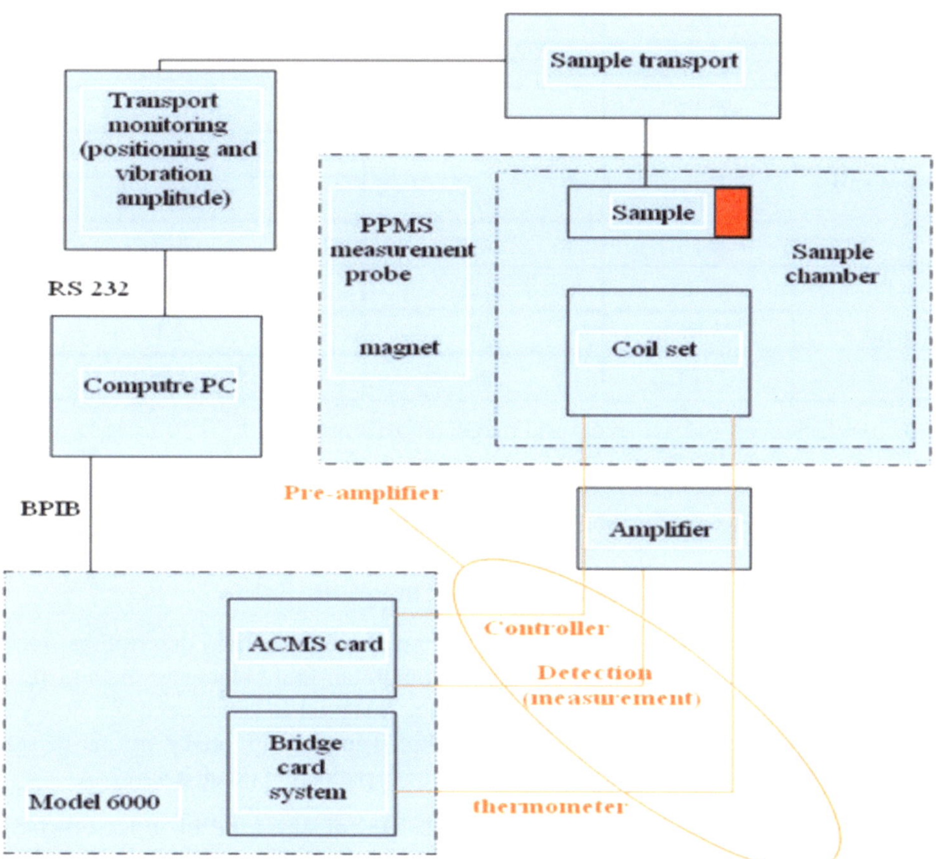

Fig. 18. Flow diagram of ACMS

Table 4 summarises loss factors on eddy currents and magnetic hysteresis for samples collected in key rail locations determined at varied temperatures by means of ACMS platform.

Source of the sample	Temperature [°C]	Loss factor on eddy currents w [10^{-4} s]	Loss factor on magnetic hysteresis loop [10^{-5} m/A]
Edge of rail web	0 [°C]	2.49	2.31
	25 [°C]	2.33	2.62
	-25 [°C]	2.7	2.44
Centre of rail web	0 [°C]	2.53	2.42
	25 [°C]	2.33	2.8
	-25 [°C]	2.72	2.78
Rail foot	0 [°C]	2.31	1.42
	25 [°C]	2.13	1.45
	-25 [°C]	2.49	1.63
Taper of rail foot	0 [°C]	2.39	1.94
	25 [°C]	2.2	1.97
	-25 [°C]	2.6	1.77
Taper of rail web	0 [°C]	1.83	1.55
	25 [°C]	1.69	1.76
	-25 [°C]	2.56	1.83
Rail head	0 [°C]	2.62	2.49
	25 [°C]	2.42	2.4
	-25 [°C]	2.83	2.68

Table 4. Loss factors on eddy currents and magnetic hysteresis

A 3D simulation model of induction turnout heating was constructed in FLUX 3D software on the basis of 60E1 rail's electric and magnetic parameters discussed above.

5. Issues relating to spatial descriptions of magnetic field

Beside the foregoing electric and magnetic parameters, a magnetic field description in three-dimensional space by means of scalar magnetic potential, finite elements method (FEM), and normalised geometric dimensions of 60E1 (Fig. 7) served to construct a 3D simulation model of induction turnout heating. The magnetic field description by means of scalar magnetic potential and finite element method will be explained in detail below

In a non-linear environment, the electromagnetic field at any type of input function is described with Maxwell's equations which relate intensities of the magnetic field, electric field, and charge as parts of the following differential dependencies (12 – 15):

$$\nabla \times \underline{H} = \underline{J} + \frac{\partial \underline{D}}{\partial t} \tag{12}$$

$$\nabla \times \underline{E} = -\frac{\partial \underline{B}}{\partial t} \tag{13}$$

$$\nabla \cdot \underline{B} = 0 \tag{14}$$

$$\nabla \cdot \underline{D} = \rho \tag{15}$$

Maxwell's equations are integrated into:

$$\oint_l \underline{H} \cdot d\underline{l} = \int_S \underline{J} \cdot d\underline{S} + \frac{\partial}{\partial t} \int_S \underline{D} \cdot d\underline{S} \tag{16}$$

$$\oint_l \underline{E} \cdot d\underline{l} = -\frac{\partial}{\partial t} \int_S \underline{B} \cdot d\underline{S} \tag{17}$$

$$\int_S \underline{B} \cdot d\underline{S} = 0 \tag{18}$$

$$\int_S \underline{D} \cdot d\underline{S} = \int_V \rho dV \tag{19}$$

Maxwell's equations are complimented with the material dependences:

$$\underline{B} \equiv \underline{B}(\underline{H}) \tag{20}$$

$$\underline{D} \equiv D(\underline{E}) \tag{21}$$

The vector quantities present in Maxwell's equations meet the following dependencies at environment boundaries:

$$n \cdot (\underline{J}_1 - \underline{J}_2) = 0 \tag{22}$$

$$n \cdot (\underline{B}_1 - \underline{B}_2) = 0 \tag{23}$$

$$n \times (\underline{H}_1 - \underline{H}_2) = \underline{J}_S \tag{24}$$

$$n \cdot (\underline{D}_1 - \underline{D}_2) = \rho_S \tag{25}$$

$$n \times (\underline{E}_1 - \underline{E}_2) = 0 \tag{26}$$

In the case of numerical electromagnetic field calculations for low frequencies, Maxwell's equations are solved indirectly with the aid of a couple of potentials and boundary conditions. Where the potentials are given, the field vectors E, D, H, B, J can be determined.

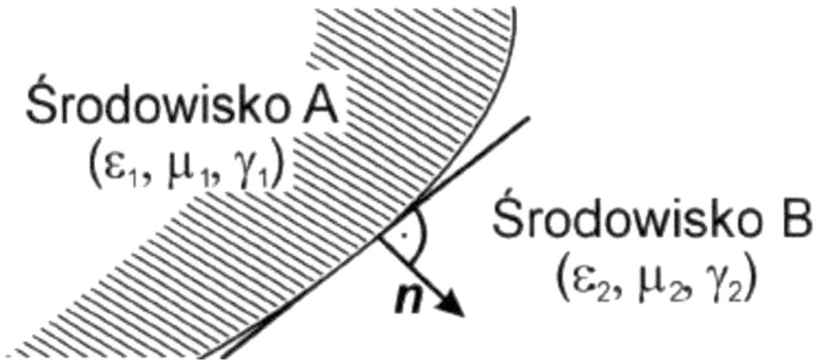

Fig. 19. Boundary of environments of diverse material properties

Maxwell's equations are most often solved by two types of potentials:

- Magnetic vector potential A ($\underline{B} = \nabla \cdot \underline{A}, \nabla \cdot \underline{A} = 0$),
- Total magnetic scalar potential Ψ ($\underline{H} = -\nabla\Psi$).

In the literature describe the vector of magnetic field intensity using a scalar potential function Ψ, expressed as:

$$\underline{H} = -\nabla \psi \tag{27}$$

When the material formula $\underline{B} = \mu(\underline{H})\underline{H}$ and the condition of source-free magnetic field are taken into account, a differential equation of total magnetic scalar potential results:

$$\nabla \cdot (\mu \nabla \psi) = 0 \tag{28}$$

Where conduction currents appear, the vector of magnetic field intensity H includes two components:

$$\underline{H} = \underline{H}_s + \underline{H}_m \tag{29}$$

where: H_s – magnetic field intensity component enforced by flow of currents in a uniform environment, H_m – magnetic field intensity component arising from magnetisation of the environment material.

H_s is computed according to Biot-Savart law as:

$$\underline{H}_s = \int \frac{\underline{J} \times 1_r}{4\pi\, r^2} dV \tag{30}$$

where: r is the distance from the observation point $O=(x, y, z)$ where H_s is calculated to the source point $Z=(x', y', z')$. 1_r is a unit vector oriented from Z to O.

In addition, H_s in the area V fulfils the condition:

$$\nabla \times \underline{H}_s = \underline{J} \tag{31}$$

The following obtains for H_m:

$$\nabla \times \underline{H}_m = 0 \tag{32}$$

Expressing H_m as a function of magnetically reduced scalar potential Φ results in:

$$\underline{H}_m = -\nabla\Phi \tag{33}$$

Based on (30) and (31), the magnetic intensity can be expressed:

$$\underline{H} = \frac{1}{\mu(\underline{B})}\underline{B} = \int \frac{\underline{J} \times 1_r}{4\pi\, r^2} dV - \nabla\Phi \tag{34}$$

Considering (14), (34) becomes:

$$\nabla \cdot (\mu\nabla\Phi) = \nabla \cdot (\mu\underline{H}_s) \tag{35}$$

The following material dependency is applied to models comprising a permanent magnet:

$$\underline{B} = \mu\ (\underline{H})(\underline{H} - \underline{H}_c) \tag{36}$$

Consequently, (35) describing the magnetically reduced scalar potential becomes:

$$\nabla \cdot (\mu\nabla\Phi) = \nabla \cdot (\mu\underline{H}_s) - \nabla \cdot (\mu\underline{H}_c) \tag{37}$$

In current-free areas, the resultant vector of magnetic field intensity H can only be described by means of the total magnetic potential.

$$\nabla \cdot (\mu\nabla\Psi) = \nabla \cdot (\mu\underline{H}_c) \tag{38}$$

Low accuracy of determining the magnetic field in the area of magnetic materials is the fundamental drawback of the reduced scalar potential method. This results from the fact that components of the magnetic field intensity vector H_s and H_m have similar values but opposite orientations. In effect, the resultant vector of intensity H reduces.

Where magnetic permeability μ is high, great errors arise. To avoid the error in effect of the diminishing resultant vector of magnetic field intensity in numerical calculations, the area is divided into a sub-area V_Φ of permeability μ_0 including current sources and a sub-area V_Ψ which includes the remaining area under consideration. Field distribution across V_Ψ is described by global scalar magnetic potential whereas V_Φ is described by reduced scalar magnetic potential. To obtain a unique solution to (37) and (38), conditions present on boundaries of the different sub-areas described with the different types of scalar magnetic potential must be defined. This relates to the need to provide continuity of the normal component of magnetic induction vector and the tangential component of the magnetic field intensity vector on the boundary of V_Ψ and V_Φ, which can be expressed:

$$\left(-\frac{\partial\Psi}{\partial t}\right)_{S_{\Phi-\Psi}} = \left(-\frac{\partial\Phi}{\partial t} + H_{st}\right)_{S_{\Phi-\Psi}} \tag{39}$$

where:

$S_{\Phi\text{-}\Psi}$ – boundary area between V_Ψ and V_Φ;
H_{st} – active component of H_s on the area $S_{\Phi\text{-}\Psi}$.

One of the most commonly used numerical methods of solving field (boundary) problems is the Finite Element Method. Its idea is to divide an area under consideration into some discrete sub-areas of any shape (finite elements). The smaller the elements into which an analysed area is split, the more precise results of the calculations.

Triangular, quadrangular, six- or four-sided elements are those used most frequently. A four-sided element can serve to interpolate a given area (Figure 20) to produce the equations below (38):

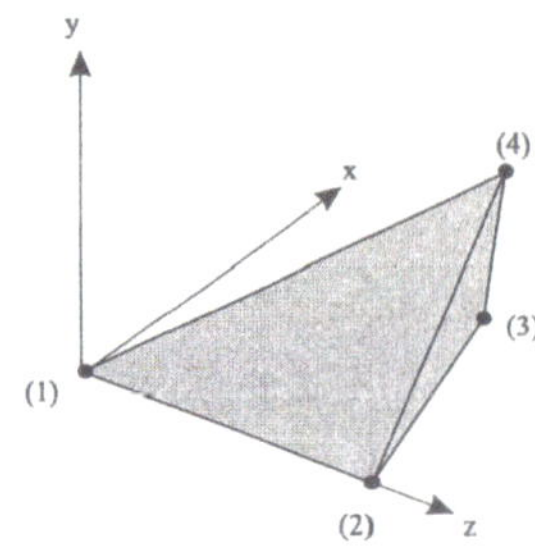

Fig. 20. Nodes of a sample four-sided element

$$\begin{bmatrix} u_1^e \\ u_2^e \\ u_3^e \\ u_4^e \end{bmatrix} = \begin{bmatrix} 1 & x_{11} & x_{21} & x_{31} \\ 1 & x_{12} & x_{22} & x_{32} \\ 1 & x_{13} & x_{23} & x_{33} \\ 1 & x_{14} & x_{24} & x_{34} \end{bmatrix} \begin{bmatrix} \beta_1 \\ \beta_2 \\ \beta_3 \\ \beta_4 \end{bmatrix} \tag{40}$$

where: e –element number, x_1, x_2, x_3 – coordinates of a point inside the element, β_1, β_2, β_3, β_4 - constants of an approximating function.

Solving this equation produces field values in the individual nodes. Figure 21 shows the most common discretisation shapes applied to 3D problems as part of FEM.

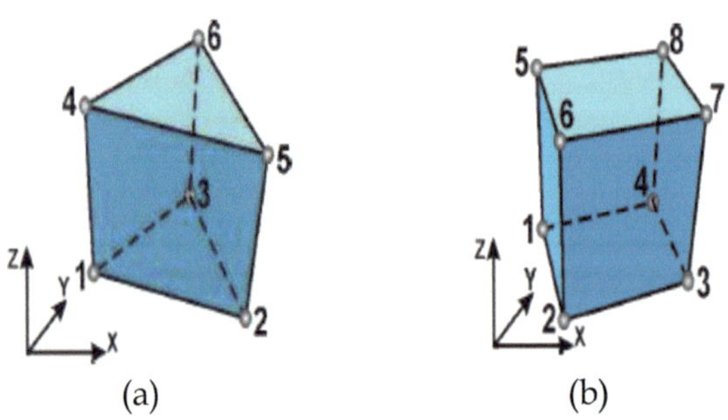

Fig. 21. Examples of three-dimensional elements utilised in FEM: a) five-sided element, b) six-sided element

Triangular three-node elements are most often used to discretise 2D areas. A sample division of Ω of boundary Γ into triangular elements is illustrated in Figure 22.

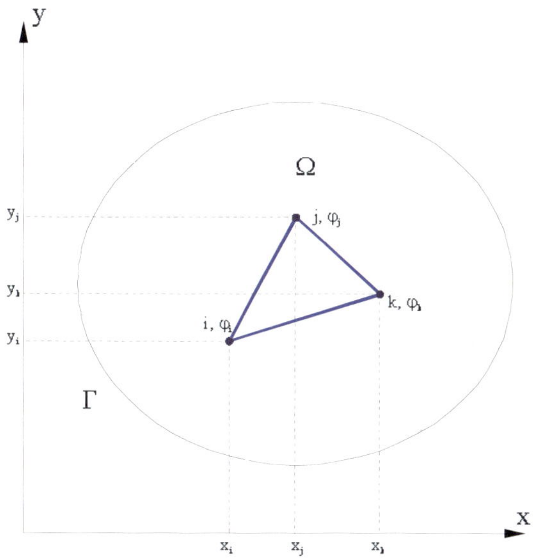

Fig. 22. Division of a two-dimensional area Ω into triangular elements

Dependence of the sought quantity $\varphi(x,y)$ inside the triangular element (e) is approximated by means of a first-degree polynomial:

$$\varphi^{(e)} = \alpha_1 + \alpha_2 \cdot x + \alpha_3 \cdot y \tag{41}$$

Values of $\varphi(x,y)$ at nodes of eth element are described by a system of equations:

$$\begin{cases} \varphi_i^{(e)} = \alpha_1 + \alpha_2 \cdot x_i + \alpha_3 \cdot y_i \\ \varphi_j^{(e)} = \alpha_1 + \alpha_2 \cdot x_j + \alpha_3 \cdot y_j \\ \varphi_k^{(e)} = \alpha_1 + \alpha_2 \cdot x_k + \alpha_3 \cdot y_k \end{cases} \tag{42}$$

The system will solve for factors α_1, α_2, α_3. Substituting them in (41) produces:

$$\varphi^{(e)} = \frac{(a_1 + b_1 x + c_1 y)\varphi_i + (a_j + b_j x + c_j y)\varphi_j + (a_k + b_k x + c_k y)\varphi_k}{2\Delta} \tag{43}$$

where:

$$a_i = x_j y_k - x_k y_j\,,\quad b_i = y_j - y_k\,,\quad c_i = x_k - x_j \tag{44}$$

$$2\Delta = \begin{vmatrix} 1 & x_i & y_i \\ 1 & x_j & y_j \\ 1 & x_k & y_k \end{vmatrix} \tag{45}$$

The remaining factors are obtained by cyclical shifting of the indices i, j, k. (43) can be expressed as a matrix:

$$\varphi^{(e)} = [N_i \quad N_j \quad N_k] \begin{bmatrix} \varphi_i \\ \varphi_j \\ \varphi_k \end{bmatrix} = \mathbf{N}\varphi \tag{46}$$

where:

$$N_{i,j,k} = \frac{a_{i,j,k} + b_{i,j,k}x + c_{i,j,k}y}{2\Delta} \tag{47}$$

N_i, N_j, N_k are functions of variables x and y, referred to as *shape* or *base functions* while **N** is a *shape function matrix*. (46) describes the value of $\varphi^{(e)}$ above the surface of a single element (e) (three values of $\varphi^{(e)}$ for nodes i, j and k). Solving with the aid of the finite element method consists in finding values of φ in respect of all the nodes of Ω.

$$\varphi = \begin{bmatrix} \varphi_1 \\ \varphi_2 \\ \vdots \\ \varphi_r \end{bmatrix} \tag{48}$$

where r is the number of discretisation network nodes. To this end, the functional $J(\varphi)$ is minimised in relation to φ:

$$\frac{\partial J}{\partial \varphi} = \begin{bmatrix} \dfrac{\partial J}{\partial \varphi_1} \\ \dfrac{\partial J}{\partial \varphi_2} \\ \vdots \\ \dfrac{\partial J}{\partial \varphi_r} \end{bmatrix} = 0 \tag{49}$$

The energetic functional with regard to the fields described by Laplace equation including Dirichlet's or Neuman's boundary conditions becomes:

$$J(\varphi) = \frac{1}{2}\iint_{\Omega} [k_x(\frac{\partial \varphi}{\partial x})^2 + k_y(\frac{\partial \varphi}{\partial y})^2]dxdy + \int_{\Gamma} \varphi dl \tag{50}$$

where k_x and k_y are the material constants in the direction of x and y (e.g. magnetic or electric permeability) while l is the length of the arc along the boundary Γ. The principle of summing all elements of Ω applied to the functional, therefore (49) can be formulated:

$$\frac{\partial J}{\partial \varphi} = \sum_{i=1}^{s} \frac{\partial J_i^{(e)}}{\partial \varphi} = 0 \tag{51}$$

where s is the number of all elements in the calculation area. Differentiating (50) with respect to the sought $\varphi(x,y)$ results in:

$$\frac{\partial J^{(e)}}{\partial \varphi_m} = \iint_{\Omega^{(e)}} [k_x \frac{\partial \varphi}{\partial x} \frac{\partial}{\partial \varphi_m}(\frac{\partial \varphi}{\partial x}) + k_y \frac{\partial \varphi}{\partial y} \frac{\partial}{\partial \varphi_m}(\frac{\partial \varphi}{\partial y})] dxdy \,. \tag{52}$$

for values i, j and k of m. Taking (43) and (47) into consideration, the differential expressions in (52) can be found:

$$\begin{cases} \frac{\partial \varphi}{\partial x} = \begin{bmatrix} \frac{\partial N_i}{\partial x} & \frac{\partial N_j}{\partial x} & \frac{\partial N_k}{\partial x} \end{bmatrix} \cdot \begin{bmatrix} \varphi_i \\ \varphi_j \\ \varphi_k \end{bmatrix} \\ \frac{\partial \varphi}{\partial y} = \begin{bmatrix} \frac{\partial N_i}{\partial y} & \frac{\partial N_j}{\partial y} & \frac{\partial N_k}{\partial y} \end{bmatrix} \cdot \begin{bmatrix} \varphi_i \\ \varphi_j \\ \varphi_k \end{bmatrix} \\ \frac{\partial}{\partial \varphi_m}(\frac{\partial \varphi}{\partial x}) = \frac{\partial N_m}{\partial x} \\ \frac{\partial}{\partial \varphi_m}(\frac{\partial \varphi}{\partial y}) = \frac{\partial N_m}{\partial y} \end{cases} \tag{53}$$

Substituting (52) in (51) results in:

$$\frac{\partial J^{(e)}}{\partial \varphi} = \begin{bmatrix} \frac{\partial J^{(e)}}{\partial \varphi_i} \\ \frac{\partial J^{(e)}}{\partial \varphi_j} \\ \frac{\partial J^{(e)}}{\partial \varphi_k} \end{bmatrix} = \begin{bmatrix} h_{ii}^{(e)} & h_{ij}^{(e)} & h_{ik}^{(e)} \\ h_{ji}^{(e)} & h_{jj}^{(e)} & h_{jk}^{(e)} \\ h_{ki}^{(e)} & h_{kj}^{(e)} & h_{kk}^{(e)} \end{bmatrix} \cdot \begin{bmatrix} \varphi_i \\ \varphi_j \\ \varphi_k \end{bmatrix} = h^{(e)} \varphi^{(e)} \tag{54}$$

where:

$$h_{pq} = \iint_{\Omega^{(e)}} [k_x \frac{\partial N_i}{\partial x} \frac{\partial N_j}{\partial x} + k_y \frac{\partial N_i}{\partial y} \frac{\partial N_j}{\partial y}] dxdy \tag{55}$$

for values i, j, k of p and q. The matrix $h^{(e)}$ in (53) is called *element rigidity matrix* (the upper index '(e)' is a reference to an element) and specifies material properties. On appropriate transformations of (44), (45), and (48), the following can be said in respect of a triangular element:

$$h^{(e)} = \begin{bmatrix} k_x b_i b_i + k_y c_i c_i & k_x b_i b_j + k_y c_i c_j & k_x b_i b_k + k_y c_i c_k \\ k_x b_j b_i + k_y c_j c_i & k_x b_j b_j + k_y c_j c_j & k_x b_j b_k + k_y c_j c_k \\ k_x b_k b_i + k_y c_k c_i & k_x b_k b_j + k_y c_k c_j & k_x b_k b_k + k_y c_k c_k \end{bmatrix} \cdot \frac{1}{4\Delta}. \tag{56}$$

(51) describes a differential of functional J with respect to the variable $\varphi(x,y)$ sought for (e). When components of (54) in s elements are summed according to:

$$H_{ij} = \sum_{k=1}^{S} h_{ij} \tag{57}$$

a differential of the functional can be expressed for the entire area:

$$\frac{\partial J_i}{\partial \varphi} = H\varphi = 0 \tag{58}$$

H in (58) is known as a *condition* or *rigidity matrix.* This is a band square matrix of the dimension r and a band width lower than the matrix's dimension.

To introduce the node variables defined by means of Dirichlet's boundary conditions to (58), the equations describing nodes of known φ can be eliminated. This procedure can be troublesome, however, when computer calculation algorithms are created as it requires appropriate lines and columns to be removed from the rigidity matrix. Another method of introducing Dirichlet's conditions has been proposed by Payne and Irons. Elements of H diagonals relating to a specific boundary node must be multiplied by a great number (e.g. 10^{15}) and the resultant product must be entered in an appropriate position of the zero vector which forms the right-handed side of (58). This procedure is widely used as it is easy to programme and does not require many operations, thereby minimising the time and cost of the calculations.

To find an approximate solution to a problem using FEM, the objective function needs to be defined, most often as a minimum error of the solution, Galerkin's method is of use in solving non-linear problems. The best solution for an area V delimited with certain boundary conditions is zeroing of the weighted average residuum $R = \overline{\zeta} - \zeta$, where ζ is the precise solution and $\overline{\zeta}$ an approximate solution. A general solution according to Galerkin's method can be presented as:

$$\int_V w_i R dV = 0 \quad i = 1,\ 2,\ 3, \ldots,\ m \tag{59}$$

where: w_i - tapering functions

According to the weighted residuum method, the tapering functions are those which interpolate an approximate distribution of the solution across the area under analysis V. *Poisson's* equation is the most commonly applied differential equation:

$$\frac{\partial}{\partial x}\left(v_x \frac{\partial \zeta}{\partial x}\right) + \frac{\partial}{\partial y}\left(v_y \frac{\partial \zeta}{\partial y}\right) + \frac{\partial}{\partial z}\left(v_z \frac{\partial \zeta}{\partial z}\right) = -f(x,\ y,\ z) \in V \tag{60}$$

where ζ may represent both scalar and vectoral magnetic potential.

Vectoral magnetic potential A is the sought quantity in 2D problems involving magnetic field. Scalar potential Ψ-Φ is the sought quantity in 3D problems (A can also be determined in FLUX 3D). These solutions overcome the need to apply analytical methods to solving of complicated differential equations, which is occasionally time-consuming and complex. At the present stage of computer technology, introducing several or a dozen thousand unknowns instead of a mere few makes little difference. FEM is additionally popular when compared to other numerical methods since:

- choice of shapes (and dimensions) which will serve to discretise is unlimited,
- any boundaries of discretised shapes can be very precisely approximated with straight- or curving-line elements,
- any desirable accuracy of calculations (restrictions on analysis time) can be attained by varying element sizes,
- diverse boundary conditions can be defined,
- the method is universal and can help to solve electromagnetic, electrostatic, magnetic and heat-flow problems.

The need to analyse complex physical models requires application of numerical methods that provide approximate results. Such results are subject to errors, however, due to a range of factors, such as:

- differential equation errors,
- errors due to simplifying assumptions (e.g. ignoring certain properties of some phenomena).

In the literature[17] [18] have discussed errors at the stage of problem solving (a central element of numerical calculations). These errors depend on the calculation method and are sources of distortions that may lead to misrepresentations of phenomena. Sources of errors as part of FEM include:

- interpolation errors;
- approximation errors generated when solutions are sought within a limited area;
- errors relating to discontinuity of an environment's physical parameters;
- errors in representations of structure (geometry);
- errors relating to rounding of node values.

Additionally, 3D problems are exposed to:

- errors arising at magnetic field boundaries described by means of different potentials;
- errors arising in ferromagnet areas described by means of reduced scalar potential.

A local error in a point of a model generated by FEM is inherently connected to the size of its elements surrounding a given point whereas it is only loosely related to average size of

---

[17]Leśniewska E. (1997). Zastosowanie symulacji pól elektromagnetycznych w projektowaniu przekładników

[18]Wincenciak S. (1998). Metody i algorytmy optymalizacji kształtu obiektów w polu elektromagnetycznym

elements in a space under consideration[19 20]. The latter type of errors are more significant and more difficult to eliminate in non-linear problems.

In the literature[21 22 23 24 25] have presented various examples of applying FEM to magnetic field calculations, explained problems of discretising models, and discussed detailed requirements of shape functions.

6. Simulation model of induction heating in flux 3D software

Based on laboratory research which determined electric and magnetic properties of a 60E1 rail, the internal rail structure was divided into areas of different properties. A fundamental magnetic model of a rail was created in this manner. Rail web and its edges are characterised by diverse material and magnetic properties. Magnetic properties of the remaining rail portions (e.g. centre) are identical. A 3D model on the basis of geometric dimensions given in Figure 7 is extended with areas in respect of which separate prime magnetising curves were obtained, evidence of structural differences generated by the rolling process. Figure 23 shows a rail divided into areas of varying magnetic properties, associated with appropriate characteristics (prime magnetising curves in the respective colours).

Figure 24 illustrates the way material properties of the individual rail areas are defined. 'Magnetic property' tab is selected, a curve type is chosen by means of 'Isotropic spline saturation', then specific values frem the characteristic **B**=f (**H**) are input. The more points are defined, the more accurate description of a given area.

Figure 25 presents a 3D rail model including divisions into areas of varying magnetic properties. The Figure distinguishes head, web edges, and central part of a rail.

Construction of a grid for further field calculations was the next step. Flux 3D automatically creates and generates a calculation grid from among three sizes available in the programme. The grid may also be more or less dense. The denser a calculation grid, the more precise the field calculations. Too dense a grid considerably prolongs the calculation process, unfortunately. A selected grid type is then generated and imposed upon the model. Figures 26 and 27 show a calculation grid for a constructed rail model.

A numerical model of the rail, serving to simulate induction heating, is shown in Figure 18 including a complete calculation grid imposed and a heating wire near a web edge.

---

[19]Mendrela E., Łukaniszyn M., Macek-Kamińska K. (2002). *Tarczowe silniki prądu stałego z komutacją elektroniczną*

[20]Wróbel R. (2000). *Analiza wpływu parametrów obwodu magnetycznego i elektrycznego na pracę silnika tarczowego prądu stałego z magnesami trwałymi i elektronicznym komutatorem*

[21]Binns K.J., Lawrenson P.J., Trowbridge C.W. (1995). *The Analytical and Numercial Solution of Electric and Magnetic Fields*

[22]Bolkowski S., Stabrowski M., Skoczylas J., Sroka J., Sikora J., Wincenciak S. (1993). *Komputerowe metody analizy pola elektromagnetycznego*

[23]Gawrylczyk K.M. (2007). *Analiza wrażliwościowa pola elektromagnetycznego z użyciem metody elementów skończonych*

[24]Jianming J. (1993). *The finite element method in electromagnetic*

[25]Sikora R. (1997). *Teoria pola elektromagnetycznego*

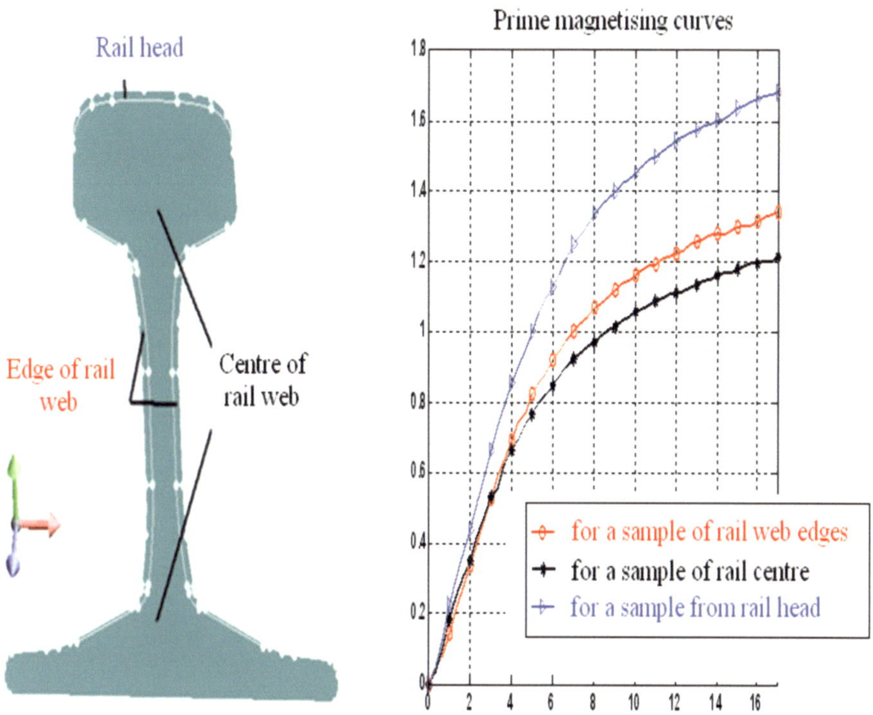

Fig. 23. Division of a rail into areas of varying magnetic properties, associated with appropriate characteristics (prime magnetising curves)

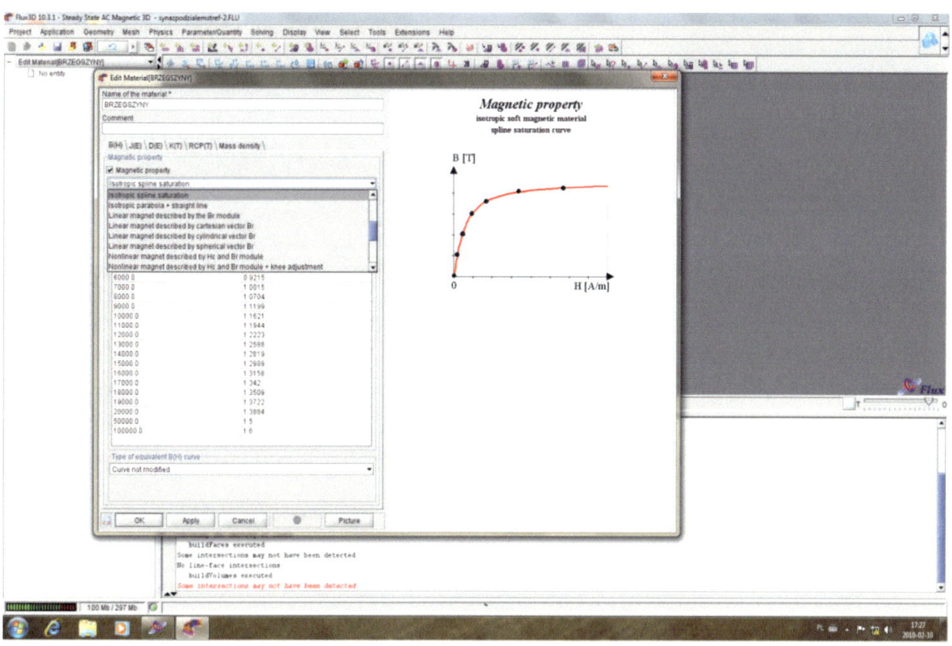

Fig. 24. Process of defining new materials and their magnetic properties

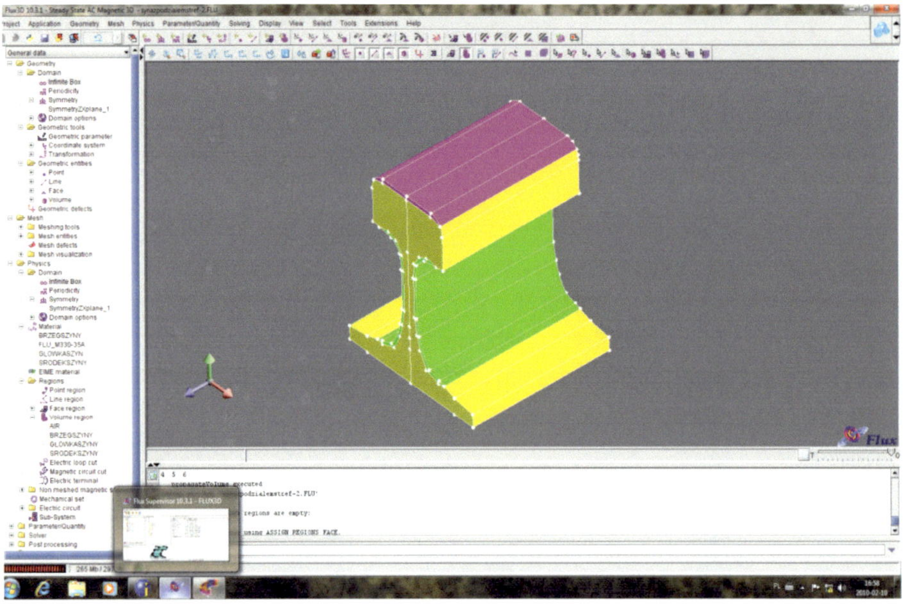

Fig. 25. Magnetic model of 60E1 rail in FLUX 3D software.

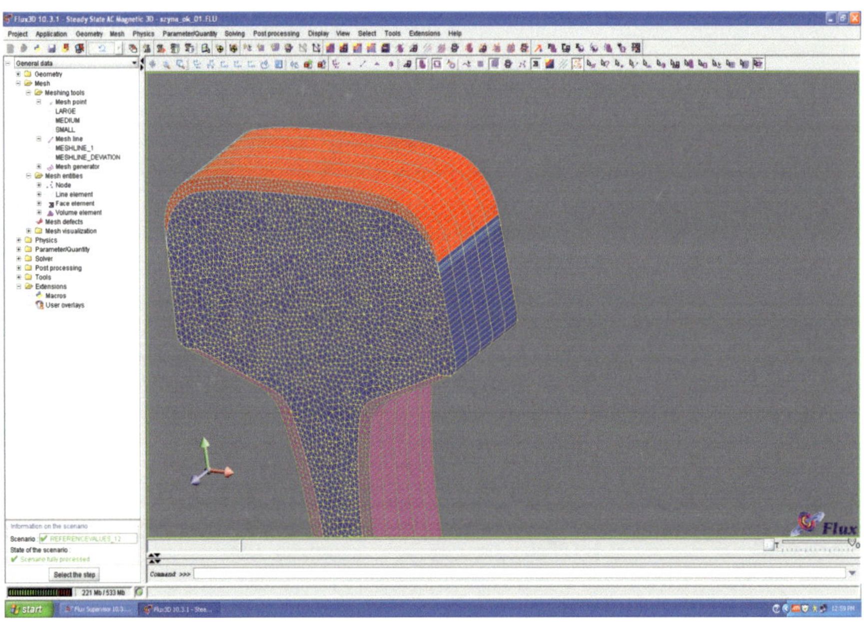

Fig. 26. Calculation grid for the rail head – red

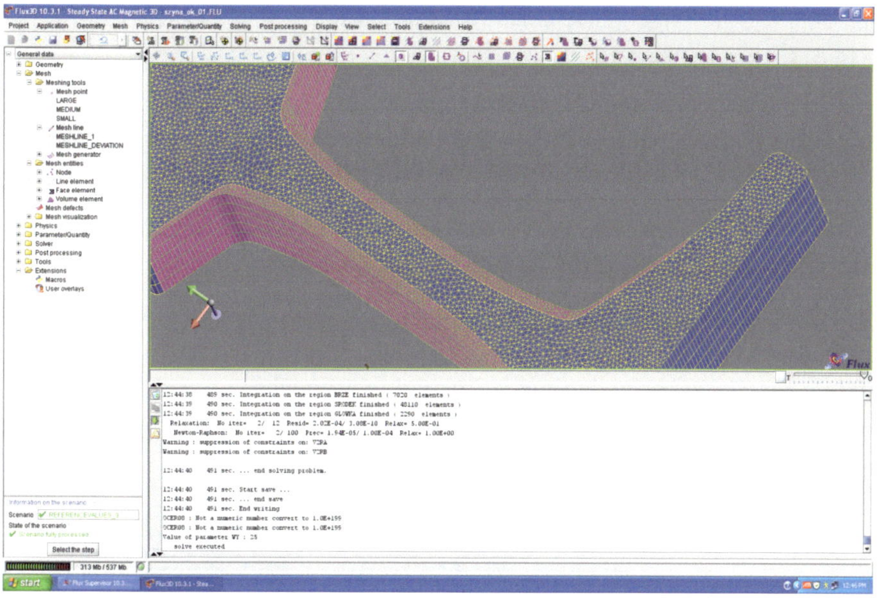

Fig. 27. A view of calculation grid executed in Flux 3D for rail web edges (pink) and rail centre (blue).

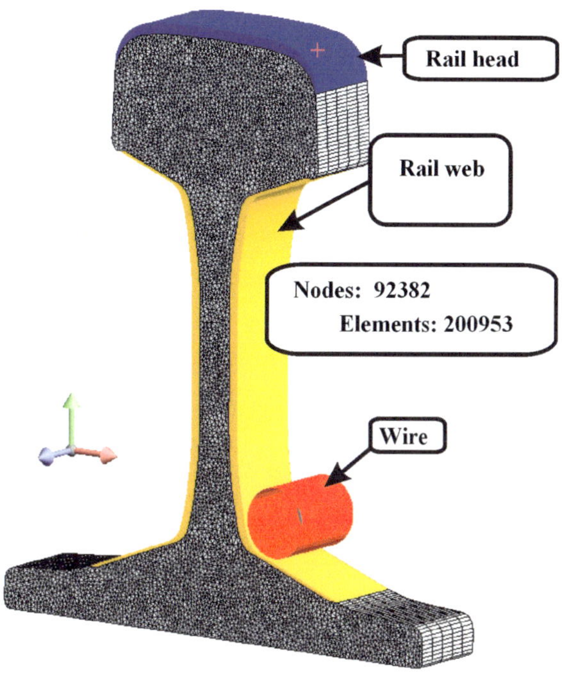

Fig. 28. Numerical model of the rail including a heating wire

This model is then subjected to the simulation process in order to determine behaviour of a rail when heated with eddy currents as part of induction heating.

The following assumptions underlie simulatiton testing of a rail at the time of induction heating in FLUX 3D software[26 27]:

- constant density of current across the heating wire,
- magnetic field description in 3D space with the aid of scalar magnetic potential,
- electric and magnetic parameters of a rail: electrical resistivity **ρ**, prime magnetising curve **B** = f (**H**) or **J**=f (**H**), intensity of the coercion field $\mathbf{H_c}$, loss factors of eddy currents **w** and of magnetic hysteresis loop **h**,
- zero boundary conditions: **n** ·**H** = 0.

The sample results were obtained in the simulation process at 650H frequency and 11.14 MA/mm^2 density of the current across the wire. Set ranges of frequency and rms values of the current across the wire can be varied as part of this model.

Figure 29 presents a sample distribution of the absolute values of magnetic induction as obtained in the simulation process.

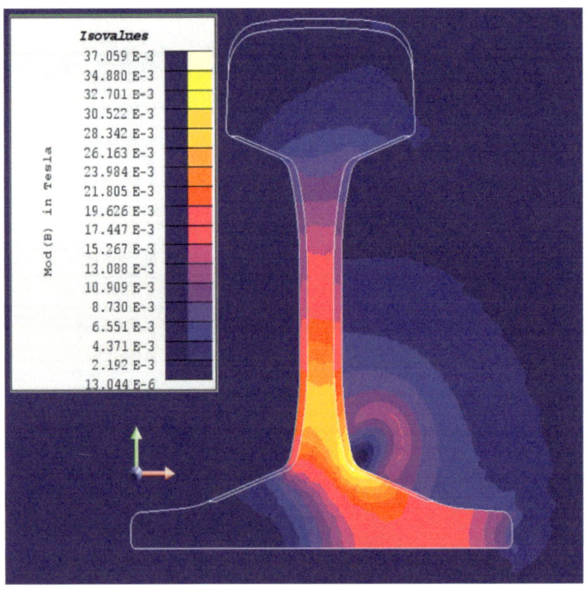

Fig. 29. Distribution of magnetic induction in the rail and its environment

Magnetic induction on the lateral web surface and on its foot reached a maximum of 0.037 T. This low value is due to the fact that an air gap appeared between the rail and the heating wire, impeding and dispersing the magnetic field. The wire itself was circular and its entire surface did not adhere to the rail, therefore magnetic field penetration into the rail was not effective.

Figure 30 shows the direction of magnetic field lines with regard to a rail placed inside the magnetic field of the heating wire.

[26]Femm, User's Manual, 2009

[27]Flux3d, *User's guide*, vol. 1-4, 2009

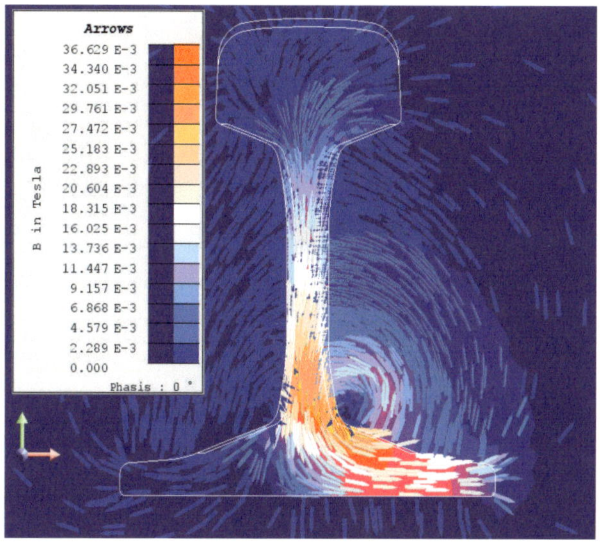

Fig. 30. Direction of magnetic field lines generated by the heating wire

This is the first model of induction heating of a rail forming part of a turnout to be developed by these authors. It was designed to demonstrate what phenomena occur when the rail is under the impact of a magnetic field. The rail and the wire do not form a full, closed magnetic loop. The magnetic field around the wire partly escapes into the air. As a result, penetration of the magnetic field into the rail structure is weaker and the magnetic induction on the rail surface is low. This weakened magnetic field may be insufficient to produce high eddy current densities in the rail and, in effect, it will be impossible to use induction heating for turnouts. Greater intensity of the magnetic field may improve the value of induction yet power losses associated with dispersion effects would increase.

Intensity of the magnetic field generated by the current across the coil is shown in Figure 31.

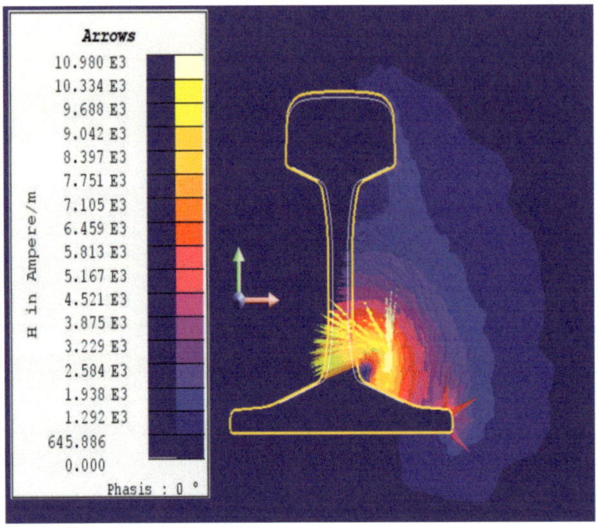

Fig. 31. Intensity of the magnetic field generated by the current across the heating wire

It is easy to read the value of magnetic field intensity arising from the heating wire: it approaches 11000A/m in the centre of the wire and reduces to 3000A/m at the contact of the rail and the wire after covering a distance of circa 1 centimetre (drag).

The air gap must be fully eliminated in continuing research and, should it prove impossible, the gap needs to be minimised in order to reduce magnetic field dispersion as much as practicable.

7. Conclusion

This chapter has presented elementary knowledge concerning 3D model illustrations of induction heating as applied to rail turnouts. It should be borne in mind, however, that a simulation model can differ widely from reality due, for instance, to simplifications discussed by the authors. The magnetic model itself must be modified in order to solve the issue of the air gap, for example.

It is also necessary to verify simulation results against those obtained in an actual model of turnout induction heating. Work on developing an actual model is in progress.

At the present stage, the magnetic model developed by the authors in Flux3D provides for observation of electric and magnetic effects in the rail's internal structure triggered by flow of eddy currents. The model will be utilised to determine the depth of magnetic field penetration into the rail structure as dependent on variations of magnetising current frequency and will serve to determine a temperature distribution along the rail in the process of heating. Knowledge of this temperature distribution or, to be more exact, of maximum temperature values attained by the individual rail sections is the key to success of this research.

8. References

Badania eksploatacyjne wodnego system ogrzewania rozjazdów typu MAS-Guben, CNTK Warszawa, Styczeń 2004

Binns K.J., Lawrenson P.J., Trowbridge C.W. (1995). *The Analytical and Numercial Solution of Electric and Magnetic Fields,* A Wiley-Interscience Publication, John Wiley & Sons, INC., New York

Bolkowski S., Stabrowski M., Skoczylas J., Sroka J., Sikora J., Wincenciak S. (1993). *Komputerowe metody analizy pola elektromagnetycznego,* Wydawnictwo Naukowo - Techniczne, Warszawa

Brodowski D., Andrulonis J.(2000). *Efektywność ogrzewania rozjazdów kolejowych,* CNTK, Warszawa

Brodowski D., Andrulonis J. (2002). *Ogrzewanie rozjazdów kolejowych, Problemy kolejnictwa,* zeszyt 135, CNTK

Femm, *User's Manual, 2009*

Flux3d, *User's guide,* vol. 1-4, 2009

Gawrylczyk K.M. (2007). *Analiza wrażliwościowa pola elektromagnetycznego z użyciem metody elementów skończonych,* Instytut Naukowo-Badawczy ZTUREK, Warszawa

Gignoux D., Schlenker M.(2005). *Magnetism Fundamentals,* Springer, Grenoble

Gozdecki T., Hering M., Łobodziński W. (1979). *Urządzenia elektroniczne. Elektroniczne urządzenia grzejne,* Wydawnictwa Szkolne i Pedagogiczne, Warszawa

Grobelny M. (2009) *Budowa, modernizacja, naprawa i remonty nawierzchni kolejowej – urządzenia i elementy,* RYNEK KOLEJOWY, 2009-03-09

Instrukcja eksploatacji i utrzymania urządzeń elektrycznego ogrzewania rozjazdów, PKP Polskie Linie Kolejowe S.A., Warszawa, 2007

Jianming J. (1993). *The finite element method in electromagnetic,* A Wiley-Interscience Publication, John Wiley & Sons, INC., New York

Jiles D. (1991). *Introduction to Magnetism and Magnetic Materials,* Chapman & Hall, ISBN 0-412-386-30-5,New York

Kiraga K., Szychta E., Andrulonis J. (2010). *Wybrane metody ogrzewania rozjazdów kolejowych – artykuł przeglądowy,* PRZEGLĄD ELEKTROTECHNICZNY, ISSN 0033-2097, R. 86 NR 2/2010

Kuryłowicz J. (1962). *Badania materiałów magnetycznych,* Wydawnictwo Naukowo-Techniczne, Warszawa

Leśniewska E. (1997). *Zastosowanie symulacji pól elektromagnetycznych w projektowaniu przekładników,* Zeszyty Naukowe Politechniki Łódzkiej, Nr 766, Łódź

Materiały seminaryjne CNTK. (2004). *Wodne ogrzewanie rozjazdów kolejowych typu MAS,* Warszawa, 21-22 Kwiecień

Mendrela E., Łukaniszyn M., Macek-Kamińska K. (2002). *Tarczowe silniki prądu stałego z komutacją elektroniczną,* Wydawnictwo Gnome, Katowice

Praca zbiorowe: *Studium na temat wyboru optymalnego systemu ogrzewania rozjazdów,* COBiRTK, 1971

Prospekt informacyjny o otulinach firmy Haet Point, 2009.

Sajdak Cz., Samek E. (1985). *Nagrzewanie indukcyjne. Podstawy teoretyczne i zastosowanie,* Wydawnictwo „Śląsk", Katowice

Sikora R. (1997). *Teoria pola elektromagnetycznego,* Wydawnictwo Naukowo-Techniczne, Warszawa

strona internetowa szyn: www. inzynieria-kolejowa.dl.pl.

Wielgosz R. (2009). *Łączenie bezstykowych szyn kolejowych,* MECHANIKA CZASOPISMO TECHNICZNE, Wydawnictwo Politechniki Krakowskiej, 2-M/2009, Zeszyt 6, Rok 106.

Wincenciak S. (1998). *Metody i algorytmy optymalizacji kształtu obiektów w polu elektromagnetycznym,* Oficyna wydawnicza Politechniki Warszawskiej, Warszawa

Wróbel R. (2000). *Analiza wpływu parametrów obwodu magnetycznego i elektrycznego na pracę silnika tarczowego prądu stałego z magnesami trwałymi i elektronicznym komutatorem,* Rozprawa doktorska, Łódź

13

EMC Analysis of Railway Power Substation Modeling and Measurements Aspects

S. Baranowski[1], H. Ouaddi[1], L. Kone[1] and N. Idir[2]
[1]*Université Lille 1 Sciences et Technologies, USTL, IEMN/TELICE Laboratory*
[2]*Université Lille 1 Sciences et Technologies, USTL, L2EP Laboratory, F-59650 Villeneuve d'Ascq, France*

1. Introduction

The first part of the chapter will present the global aspect of the railway power infrastructures and specially the power supply substation. The goals of this study consist in proposing a high frequency model of the railway systems and verifying by simulation the conformity with the EMC standards. Thus, each component of the railway power infrastructure (transformer, power rectifier) is modeled and the simulation results of the conducted emissions are compared to measurements on a reduced scale of the power supply substation.

2. EMI sources in railway system

A railway system might pollute, in an electromagnetic sense, the surrounding environment, disturbing radio and communication systems which can be not related to the railway itself. In this case the whole railway can be considered as a source of Electromagnetic Interference (EMI). In order to avoid the disruption of the electronic equipments near to the railway, the overall field generated by railway system must be kept below certain safety values given by standards [1]. Obviously, there are many sources of the electromagnetic field and their contributions can come from:

- Each part of the railway: for example, a locomotive or the power supply substation can induce EMI that exceed the EMC standards limits set for the electromagnetic field, because of power circuits that might become a source of emissions.
- The whole system: the power supply line (cable) can behave as an antenna, which radiates an electromagnetic field proportional to the current. The spectrum and the intensity of the current depend not only on the power absorbed by a train, but also from the structure (geometrical dimensions ...) of the line, that may cause resonances as shown in Fig.2.

However, a train might be or be not compliant to field standards depending on the line characteristics and on its position on the railway. Nowadays, trains are designed to meet EMC rules, but the non-compliances can be remedied thanks to identification of the different resonances frequency, which can occur mainly in a frequency range from some kilohertz up to one gigahertz.

2.1 Description of standards EMI measurement methods

The EMC standard EN 50121 [1] is used to characterize the EM environment, in the railway systems; it notably aim to limit the EMI levels from the railway infrastructures to the external environment. This standard EN 50121 describes the methodologies and the limits to apply, relating to the EM radiations and immunity of railway equipments, vehicles and infrastructures. The emissions of the whole railway system, including vehicles and infrastructure, are dealt with the section 2 of the EN 50121. The objective of the tests specified in this standard is to verify that the EM emissions produced by the whole railway systems do not disturb the neighboring equipments and systems.

The methodology then consists in measuring the radiated EM emissions at a distance of 10 m from the middle of the tracks and at about 1.5 m from the floor (Fig.1) and in comparing them with the maximum levels (limit curve in red in Fig.2). The measurements protocol and the limits are specified for the frequencies varying from 9 kHz to 1 GHz.

For each frequency band (9kHz-150kHz; 150kHz-30MHz; 30MHz-1GHz), the standards define among others:

- which component of field, magnetic (up to 30MHz) or electric (above 30 MHz), has to be measured,
- the resolution bandwidth (RBW) of the spectrum analyzer that must be used for the measurements. Then the limit curve depends on these parameters and is not a constant as shown until 1MHz in fig.2

During measurements of the electromagnetic field radiated at 10 m from the railway track, we can observe that, sometimes, for some frequency, the standards limit can be exceeded. An example of measurement from 10 kHz to 1 MHz is given in fig. 2. The exceeding is characterized by resonance phenomena which appeared for some frequencies of the power supply current.

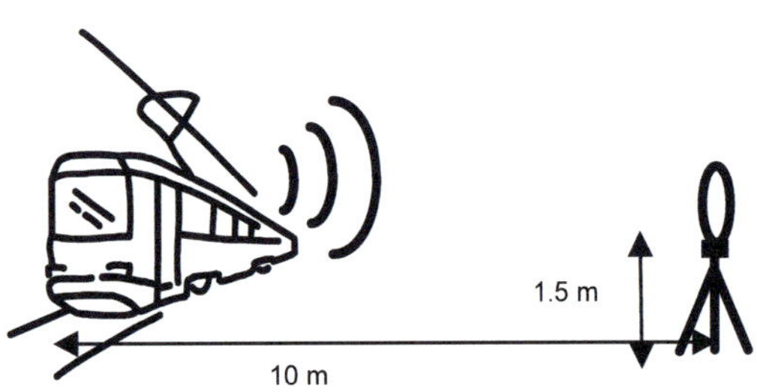

Fig. 1. On site measurement of the electromagnetic radiation of the train (stationary and moving)

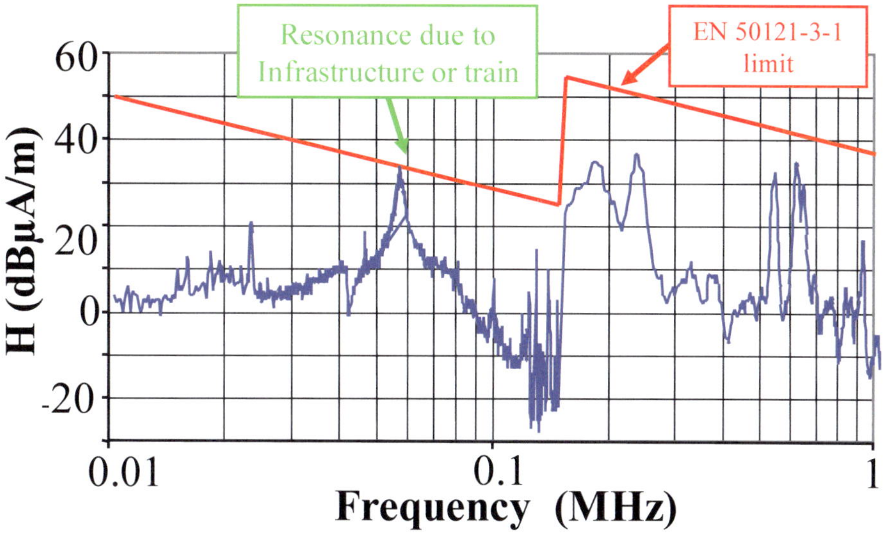

Fig. 2. Example of the measured magnetic field

2.2 Effect of the resonances

As presented previously, the aim of the standards is to limit the disturbance of the external environment produced by the railway infrastructure but also on the railways system itself. Indeed, the new generation of rolling stock is equipped with safety communication systems with on board antennas which must not be disturbed neither. These elements of safety system operate in frequency range from 10 kHz to 30MHz, which constitute the frequency range considered in this study but also above, for example, the GSMR signal frequency is around 900MHz. Thus, it becomes important to know which source can introduce the disturbances.

Besides, sensitivity of these communication systems must also be evaluated for the resonant frequencies.

3. Railway power infrastructures

A schematic representation of the railway infrastructure is shown in Fig. 3. In this figure, we can see that the power supply system constituted by the substation, the rails, the catenaries and the train which constitutes the load of the power line. This figure also shows a location system based on telecommunication between on board antenna and antenna along the track.

The railway power supply system is generally composed by two main systems:

- The first one is the "lines system" constituted by the catenaries and the rails; this transmission line has already been modeled in a previous work [2], taken into account all important parameters in an EMC point of view (non uniformity and camber of the lines, multi-conductors structures, conductivity of the ballast, ...)
- The second main part is defined by the power supply substation which contains often a power transformer, sometimes static converters, and cables, bus bars...etc.

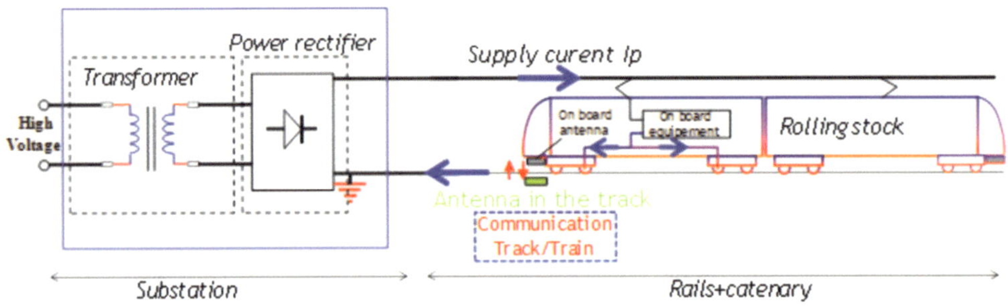

Fig. 3. Simplified diagram of railway infrastructure

4. Problem of EMC in railway

In order to avoid EMI with high levels of electromagnetic disturbances, measurements are usually performed. Test results can lead to different comments and from an industrial point of view, it is important to distinguish the disturbances due to the railway power supply infrastructure (substation - line catenaries/rail) and others due to the train itself. Indeed, it is important to know which kind mitigations solutions have to be done to reduce the disturbances and who should do that: the train manufacturer or the power supplier.

In a previous research project [2], the lines system has been modeled. The proposed model enables to estimate the field radiated by the power line taken into account various parameters like the geometry of the line; the conductivity of the soil … The model is based on line theory, the train is the load (Z_{train}) and a generator feeds the lines.

An example of the simulation conditions and the obtained results is given in figures 4 and 5

This model gives good results and many details on how each parameter has been taken into account can be found in [2]. But it is only a model of the lines and the generator used in this simulation is a sinusoidal one with constant amplitude.

Then, it was necessary to model the power substation to obtain a more realistic generator and this has constituted the goal of the CEMRAIL project of which some results are presented here.

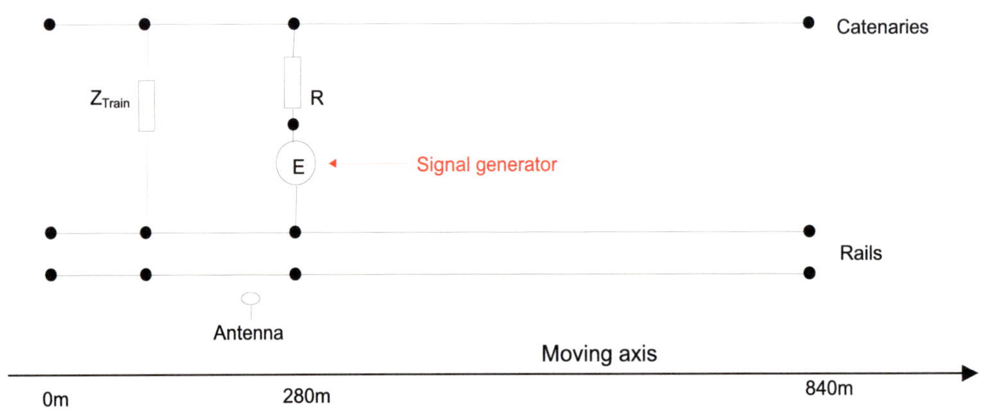

Fig. 4. Simplified model of the power supplier line

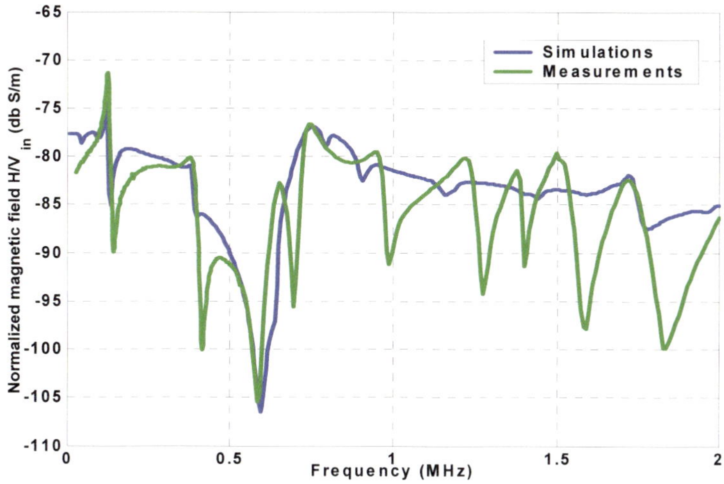

Fig. 5. Example of obtained results (corrsponding to the Fig. 4 configuration). The generator give a sinusoidal signal of 1V amplitude.

5. Power supply substation

In the world, many kind of railway power supply exist: alternative (AC) or direct current (DC), with various amplitudes: 1500kV, 3kV, 25kV… Then, we have worked on a general power substation constituted by a transformer and a power converter as presented in fig. 6.

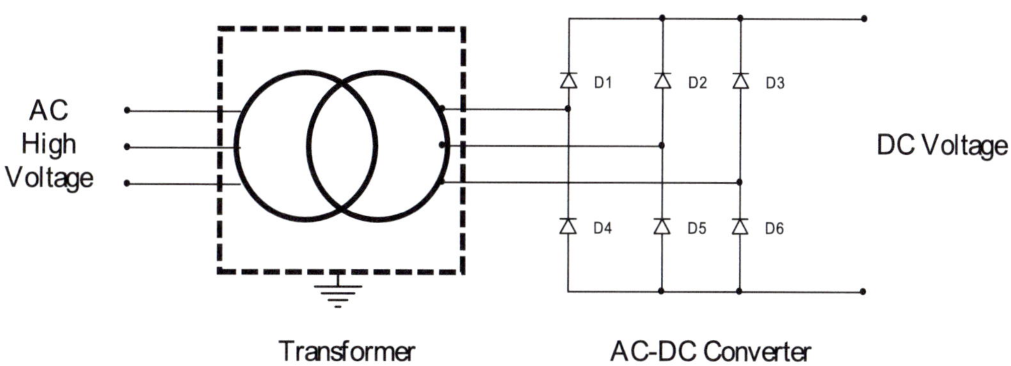

Fig. 6. Schematic circuit of a railway substation

In case of AC power supply substation, there is no AC/DC converter. The power transformer constitutes the main device of the substation.

In order to propose a high frequency model of the whole system, it's necessary to determine a model for each element, especially of power transformer, which accuracy will be highly dependent of the accuracy of each primary model. The low frequency model of the transformer, defined at 50Hz, is not valuable when the frequency increases. Indeed, some phenomena can be predominant in high frequencies as the Eddy current, which depends on the variation of the magnetizing impedance and the winding resistance, but also the parasitic capacitances which appear between windings due to the insulated parts of the transformer. The analysis of the transformer behavior shows that it's possible to use an equivalent electrical circuit model which is valid in the considered frequency band.

During the past decades, many models of power transformer have been studied for several applications. Most of those models, often in a reduced frequency band, are based on the representation of the transformer by an arrangement of resistive, inductive and capacitive elements which can take into account the physical behavior of power transformer [3]; some others are wide band models established using the black box principle [7]. Obviously, it's also possible to apply the FEM (Finite Element Method) method to have a precise model, but in this case, the exact constitution of the transformer must be known. Some additional difficulties are found in the studied problem: how to found the appropriate data sheet when the transformer is operated in the railway system since many years? In the following sections, we will present the proposed equivalent model and two techniques used to identify the transformer parameters.

6. High frequency model of transformer

6.1 Preliminary study

Making measurements in railways power substation is not easy, then preliminary investigations have been done on various power transformers. Primary and secondary winding impedances measurements in various configurations have been done on two different transformers:

- 15 kVA three phase laboratory transformer,
- 2.38 MVA train power transformer.

Figure 7-a (15kVA transformer) and Fig. 7-b (2.38MVA transformer) show the secondary winding impedance measured when the primary is short circuited for these two transformers.

These two impedances have been measured in the same frequency band: 100 kHz to 40MHz. We can note that these two responses are globally identical. Thus, we will model, in the first time, a low power transformer before to apply this model to a real substation transformer.

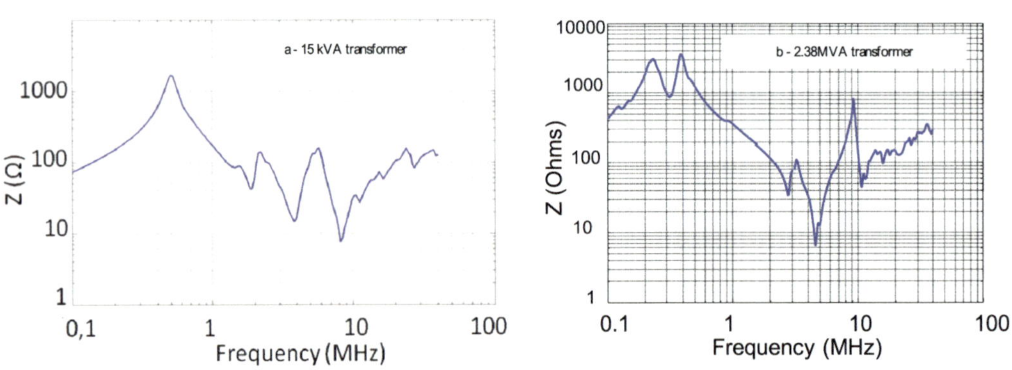

Fig. 7. Secondary winding impedance measurements when the primary is short circuited for two tested transformers (a- 15kVA and b- 2.38MVA)

6.2 Laboratory test bench

In order to analyze the behavior or the railway substation we have built a test bench which reproduces the substation in small scale.

The experimental setup is composed with a three phase power transformer, a diodes rectifier and loads. The transformer is a 15 kVA - 220 V/110 V with Δ connected high voltage (HV) winding and Δ connected low voltage (LV) winding. The power rectifier is designed with twelve power diodes and is connected to 47 Ω resistor. All these elements are placed over a common ground plane (Fig. 8). The different devices are connected using 1m long cable.

6.3 Transformer model

A high frequency model of the power transformer, which takes into account the various physical phenomena has been studied [4][8]. Figure 9 presents this model applied to the 15kVA three phase power transformer used in the laboratory test bench (fig.8).

Starting from an ideal transformer model with transformation ratio η, the proposed model takes into account, for each phase, the leakage inductances, the skin effects, the magnetizing impedance and stray capacitances of which become significant at high frequency. Moreover the iron core has been considered linear at frequency above 10 kHz [12] [13] consequently its effect can be neglected at high frequencies.

This model is valid in the frequency band varying from 40 Hz to 30 MHz.

The subject of the work is to identify the important physical parameters which contribute to induce resonance phenomena. We have chosen to propose a model as physical as possible and after, if necessary, to add black box to model more intricate phenomena.

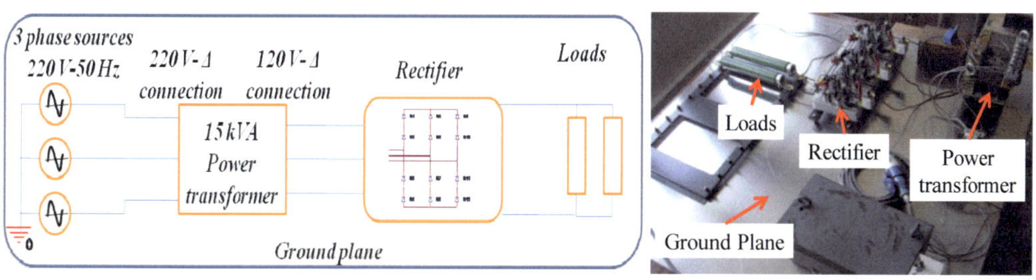

Fig. 8. Circuit diagram and photo of the test bench

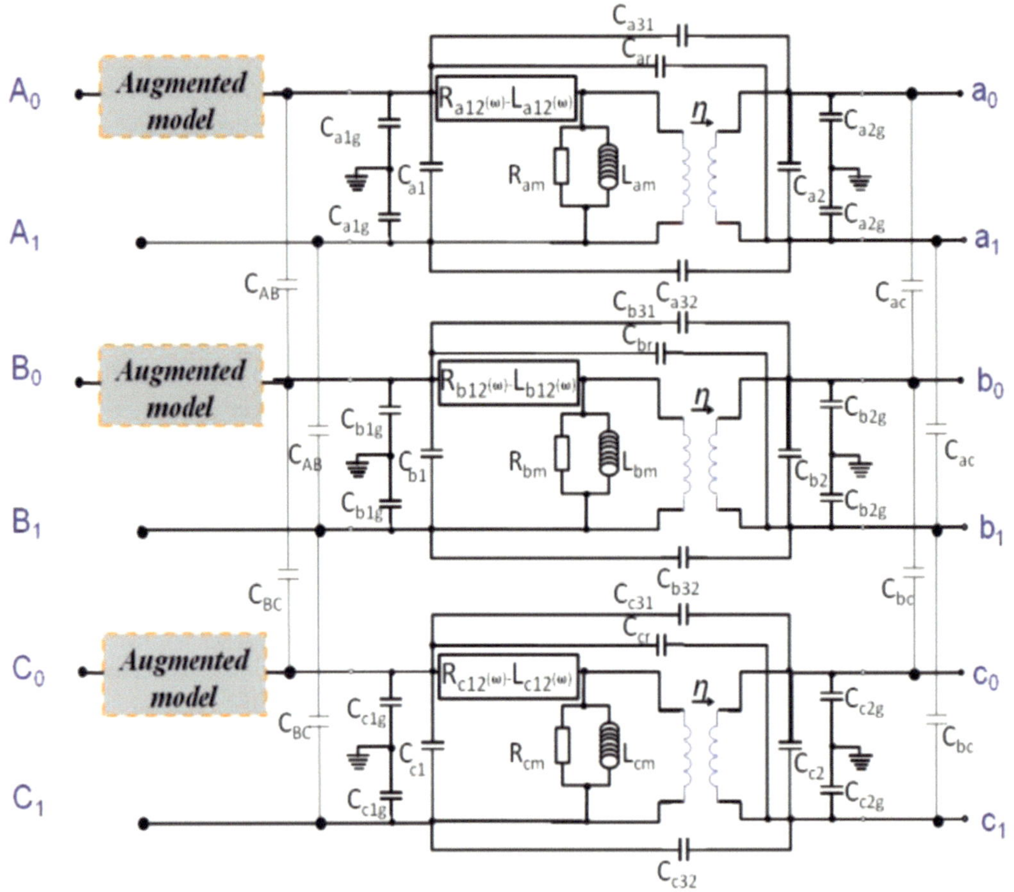

Fig. 9. High frequency model of three-phase power transformer

Each element of the equivalent circuit will be detailed and the method used to determine these parameters will be presented in the next section.

- The block R_{a12} - L_{a12} (R_{b12} - L_{b12} and R_{c12} - L_{c12} resp.) shown in the Fig. 9 represents the leakage inductance and wire resistance (due to the skin effects) of the phase A (B and C resp.) [14].
- The magnetizing impedance is modeled thanks to a resistance (R_{am}) with an inductance in parallel (L_{am}) for phase A (resp. R_{bm}, L_{bm} and R_{cm}, L_{cm} for phase B and C resp.). Of course, the magnetizing impedance changes with the frequency [7] [15].
- In the proposed model, the considered capacitances, presented in fig 9, are listed below [16], [17] :
  - Turn-to-turn capacitance of the primary and secondary windings: C_{a1}, C_{a2}, C_{b1}, C_{b2}, C_{c1}, C_{c2},
  - Capacitances between windings (divided in two capacitances): C_{a31}, C_{a32}, C_{b31}, C_{b32}, C_{c31}, C_{c32},
  - Capacitance between the input of the primary winding and the output of the secondary: C_{ar}, C_{br}, C_{cr},

- Capacitances between the winding and the ground: C_{a1g}, C_{a2g}, C_{b1g}, C_{b2g}, C_{c1g}, C_{c2g},
- Capacitances between the phases (C_{AB}, C_{BC}, C_{ab}, C_{bc}).

- Impedance called "augmented model": the impedance measurements show fast fluctuations above 10 MHz (Fig.10) and their modeling using electric circuits can be time consuming. However, the use of macro model can be a good solution to take into account these fast variations at high frequency. These impedances are modeled by blocks named "augmented model" as shown in Fig.9 and defined by using vector fitting method; details are given in [4-5-6].

Note: The proposed model can be used to model a single phase transformer by using only one circuit by phase.

6.4 Validation results

The experimental results, in time or in frequency domain presented in the next sections, allow determining the various parameters of the proposed model. Figure 10 shows an example of modeling results compared to experimental data.

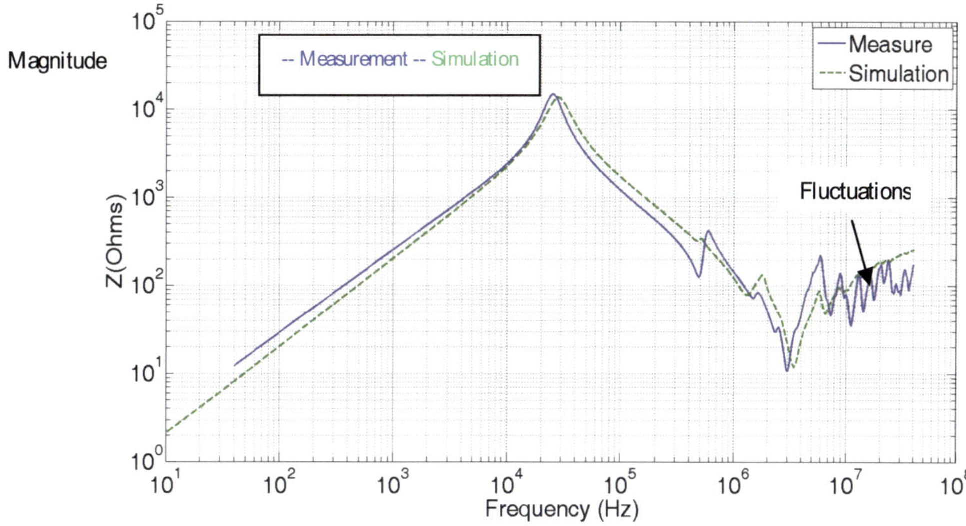

Fig. 10. Comparison within measurements and simulation results (impedance of the primary winding, the secondary being open)

7. Determination of the various elements of the transformer model

The determination of each parameter of the equivalent circuit (Fig.9) of the transformer model is realized from impedance measurements versus frequency, in different test configurations [7] [8].

7.1 The magnetizing impedance

The magnetizing impedance of each phase is modeled by a resistance in parallel with an inductance; their values are deduced from the impedance measured on the primary winding, at low frequency, when the secondary winding is open as shown in Fig 11. The values of these parameters for the phase A and B are given as following:

R_{am}= 13.44 kΩ, and L_{am}=31.19 mH,

R_{bm}= 15.87 kΩ and L_{bm}=32.14 mH

However, if we take into account the geometry of the transformer, phases A and C are considered, to have the same behavior, then the corresponding parameters are supposed to have the same value.

7.2 The leakage impedance

The leakage inductance and the wire resistance (the skin effect) of the winding can be determined by measuring the primary impedance when the secondary winding is short-circuited. Winding losses can be estimated from the impedance measured in the low frequency band as shown in Fig. 12.

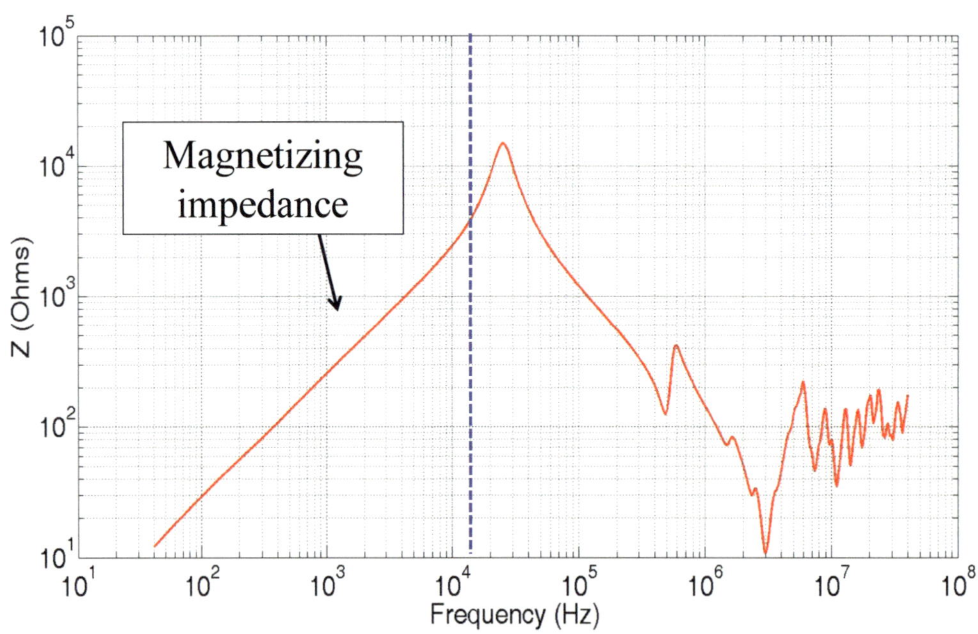

Fig. 11. Primary winding impedance when the secondary is opened (measuremens on the Phase A of the test bench transformer)

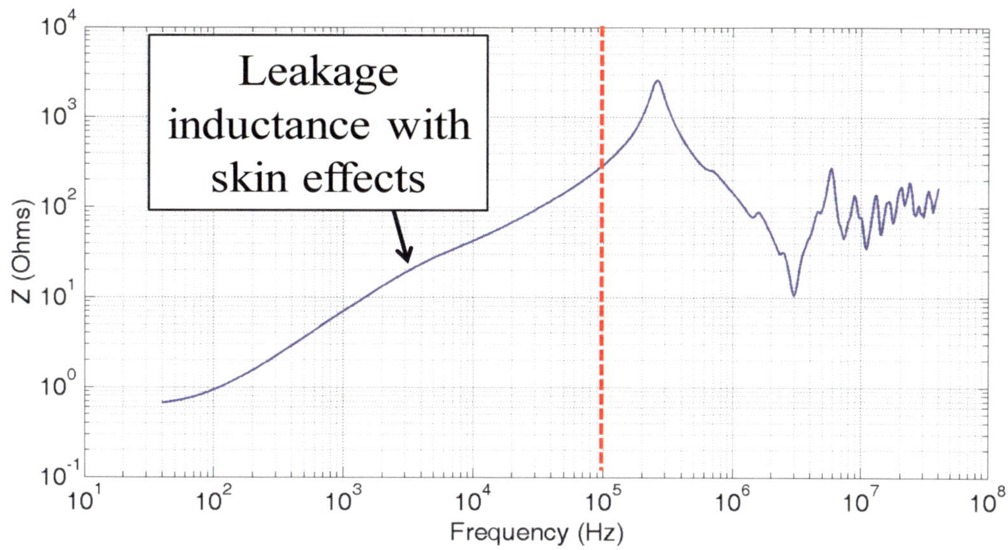

Fig. 12. Primary impedance with secondary short circuited (phase A)

The leakage impedance can be modeled using a R-L ladder network as shown in Fig. 13

In order to have an efficient approximation, the parameters of the equivalent circuit are computed with a less square algorithm; the identification being performed with the results of measurements.

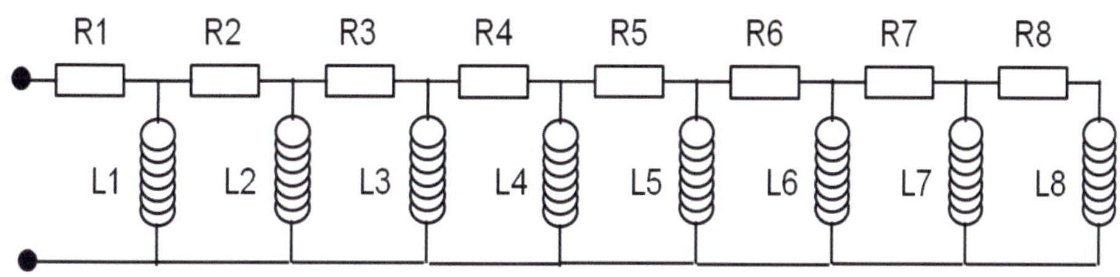

Fig. 13. Equivalent circuit of the leakage impedance

Table1 presents the parameters of the equivalent circuit of the leakage impedance obtained for the studied transformer.

R_i in Ω	R_1	R_2	R_3	R_4	R_5	R_6	R_7	R_8
Phase A	0.6	58	49	132	45	472	746	679
Phase B	0.6	58	0.001	33	37	1082	15	39
L_i in mH	L_1	L_2	L_3	L_4	L_5	L_6	L_7	L_8
Phase A	1.1	1.08	3.6	100	13	11	54	14
Phase B	1.1	2.3	2.4	2.5	8.3	11	36	69

Table 1. Parameters of the leakage impedance model of the laboratory transformer (phase C is supposed equivalent to phase A)

7.3 Stray Capacitances

The stray capacitances of each phase of the transformer are estimated separately. As an example, here, the phase A (A0A1-a0a1) is studied. The transformer is modeled as a "black box", and it's necessary to determine the capacitance Ca1, Ca2, Ca31, Ca32, Car, Ca1g and Ca2g. Their effects are located in high frequencies so they will be evaluated in this frequency range.

For each phase, seven measurement configurations are necessary to determine these capacitances. Thus, they provide seven equations which unknowns are the capacitances of the equivalent circuit. Following, the equation system is solved using mathematical tools. Figure 14 presents the seven measurements configurations with the associated equations, in the case of the phase A. The arrows show the measurements points, the short circuits are presented by the lines and Cme represents the measured capacitance. As well as the magnitude, the phase of the impedance is measured in order the make an efficient characterization.

Moreover some precautions are taken for example it is proper to short-circuit the two other phases when the third is characterized in order to inhibit their influences.

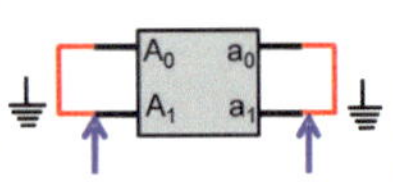	$C_{me} = C_{ar} + C_{a32} + C_{a31} + \frac{2C_{a1g}C_{a2g}}{C_{a1g} + C_{a2g}}$	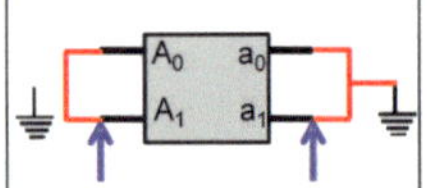	$C_{me} = C_{ar} + C_{a32} + C_{a31} + 2C_{a1g}$
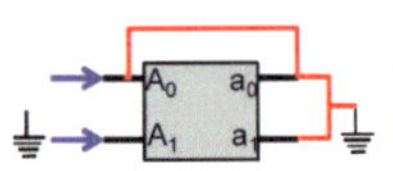	$C_{me} = C_{a1} + C_{a32} + \frac{C_{a1g}}{2}$	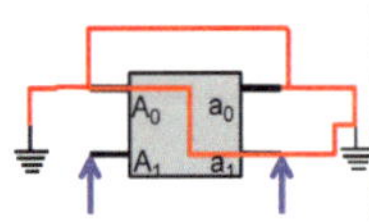	$C_{me} = C_{a32} + \frac{(C_{a1g} + C_{a1}}{C_{a1g} + C_{a1} +}$
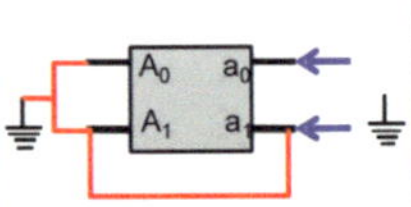	$C_{me} = C_{a2} + C_{a31} + \frac{C_{a2g}}{2}$	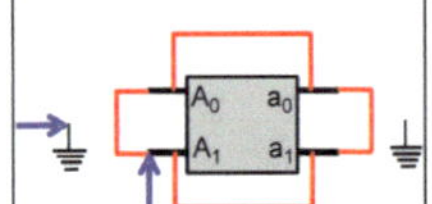	$C_{me} = 2C_{a2g} + 2C_{a1g}$
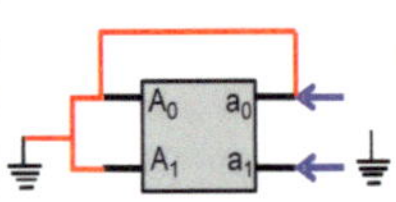	$C_{me} = C_{a2} + C_{a32} + \frac{C_{a2g}}{2}$		

Fig. 14. Measurement configuration used for determining capacitances (example for phase A)

7.4 The augmented model impedance

The impedance characteristics presented in Figures 11 and 12 show the same evolution at high frequency (above 3 MHz) with fast variations and consequently it costs time to model. Thanks to "augmented method" [5], it is possible to consider this part by using an inductance in parallel with a "black box". The "black box" is elaborated by using the principle of the "vector fitting" [13]. This algorithm consists in approximating a frequency response with a rational function, expressed by a sum of partial fractions. This principle is implemented in the IdEM® software [6] [18] used for our study. This code generates a macro-model which could be integrated in software like Pspice. Figure 9 presents the main equivalent circuit of the transformer with the added impedance issued from "augmented model" (Fig.15) which takes into account the high frequency effects. Figure 16 shows the effect of this 'augmented model impedance' on the HF accuracy of the model, by comparison of the measurements results with the model ones.

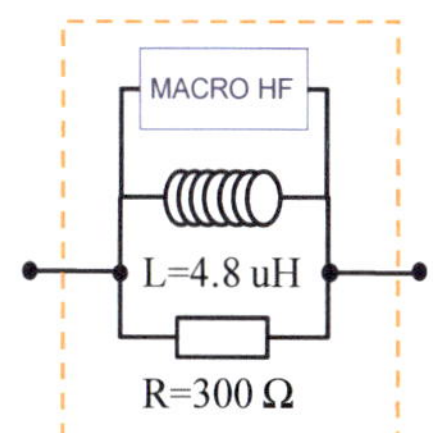

Fig. 15. Equivalent circuit for high frequency behavior of the transformer.

This circuit is repesented by a component named "augmented model" in figure 9.

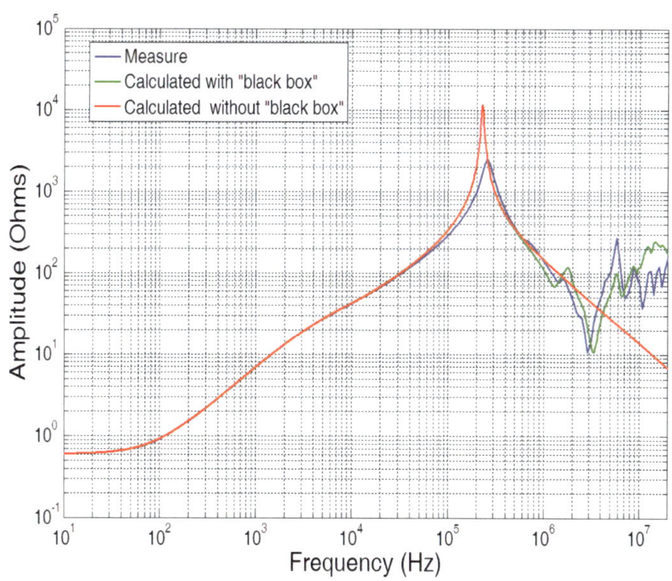

Fig. 16. Primary impedance with secondary short circuited (Phase A). Comparison between measurement and simulation result with and without the augmented model (named black box in the figure).

8. Characterization of the transformer in frequency domain

As detailed in the previous section, the determination of the proposed model parameters is based on impedance measurements which can easily be done in frequency domain with impedance or network analyzer.

This method gives good results for low power transformer but there are some problems to make this kind of measurement with very high power transformer as the ones used in railways substation, which are the followings:

- The connection between transformer and measurement apparatus (network or impedance analyzer) is not easy and some effects due to non ideal connection can be seen, especially at high frequency.
- The windings are often constituted with 'bus-bar' because the nominal power is high. Then, measurements with the previous apparatus are done with very low level signal compared to nominal power and can induce many errors (signal to noise ratio, non linearity effect).

We propose a measurement method, carried out in time domain and based on the injection of higher level signals in the transformer. The principle of the proposed method and the obtained results are presented in the next section. [9]

9. Characterisation in time domain

9.1 Measurements principle

A square voltage waveform is applied at terminals of the transformer in the various configuration tests. The input current and input voltage are measured in time domain and determined in frequency domain via Fourier transforms and then the corresponding impedance can be deduced.

This experimental method has been applied in reduced scale on the laboratory test bench, to characterize the 15kVA three phases transformer. For this characterization, a square signal of magnitude 30 V with a period equal to 30µs, duty cycle equal to 0.5 and the rise time and fall time equal to 10ns. A differential voltage probe (bandwidth: 100MHz) is used to measure the voltage at the generator terminals. The current is measured with a current probe (bandwidth: 100MHz). The connection wires between the generator and the transformer are chosen as short as possible. Figure 17 shows a diagram of the measurement bench; we can see the generator, one phase of the transformer and the position of measurements probes.

Figure 18 and 19 show respectively the waveforms of the primary voltage and current of one phase of the transformer, in the configuration described in figure 17 when the voltage signal is applied at the primary winding with the secondary being open. These measurements allow determining the magnetizing impedance. The sample frequency used for these measurements is 125 Msample/s. All impedances can then be measured with this technique, but now the data have to be processed.

9.2 Impedance calculation

The measurements results obtained in time domain will be converted in frequency domain. The impedance can be expressed as the ratio between the Fourier calculation of the voltage and the Fourier calculation of the current which can be calculated through a FFT algorithm for all measurements. The various impedances of the equivalent circuit shown in Fig. 9 can then be determined through various configurations of the transformer.

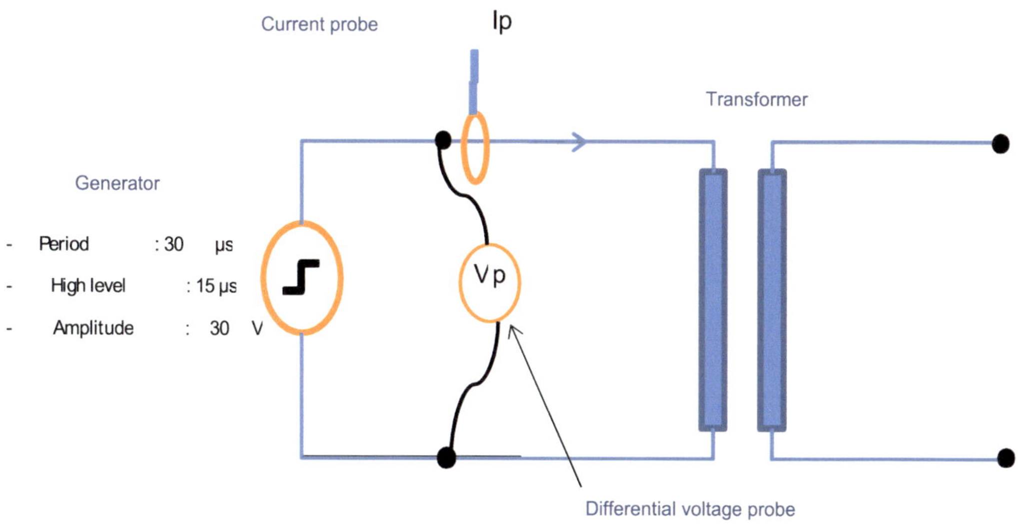

Fig. 17. Measurement bench in time domain

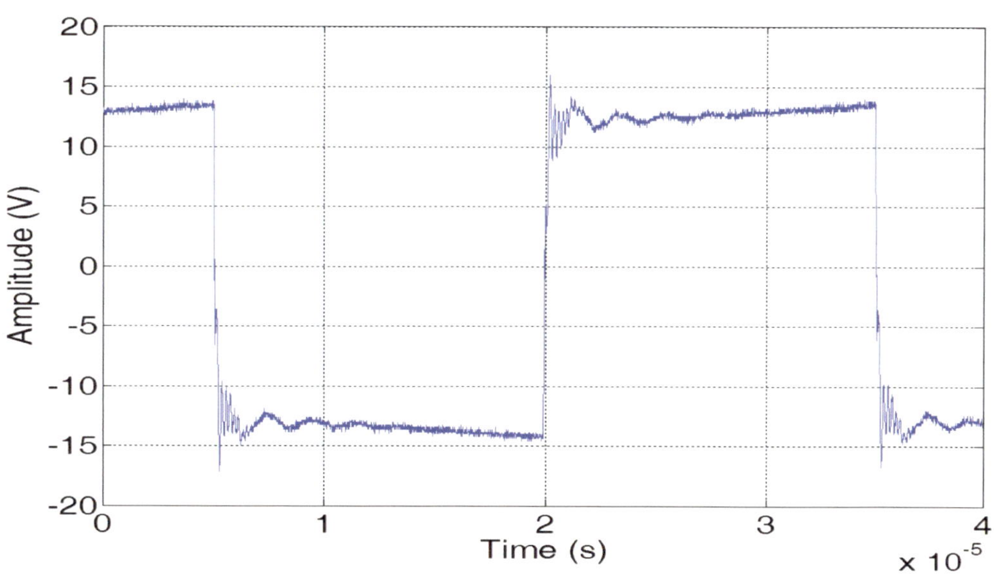

Fig. 18. Voltage waveform in primary winding when the secondary is open

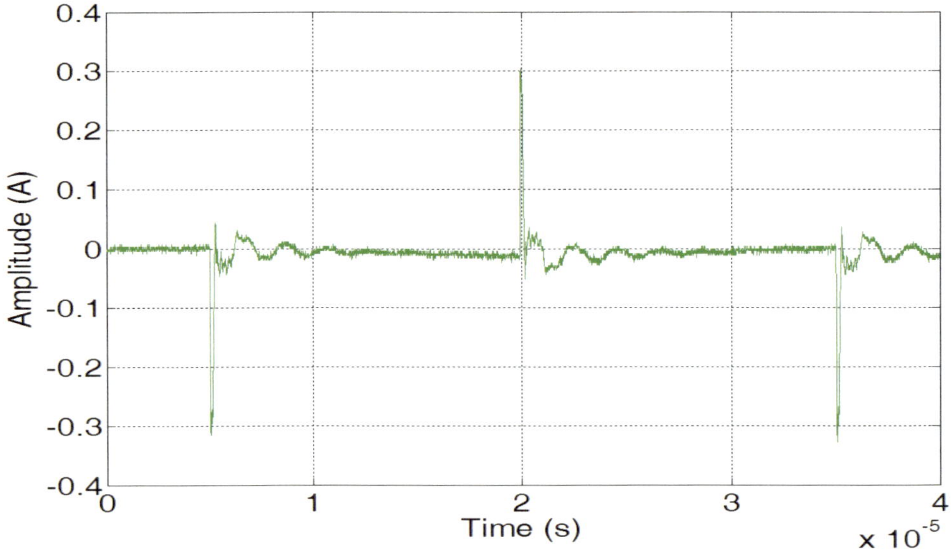

Fig. 19. Current waveform in primary winding when the secondary is open

Figure 20 shows the comparison between the impedance measured at the primary winding when the secondary is short circuited measured directly in frequency domain with impedance analyzer and the impedance issued from the data processing applied to measurements in time domain.

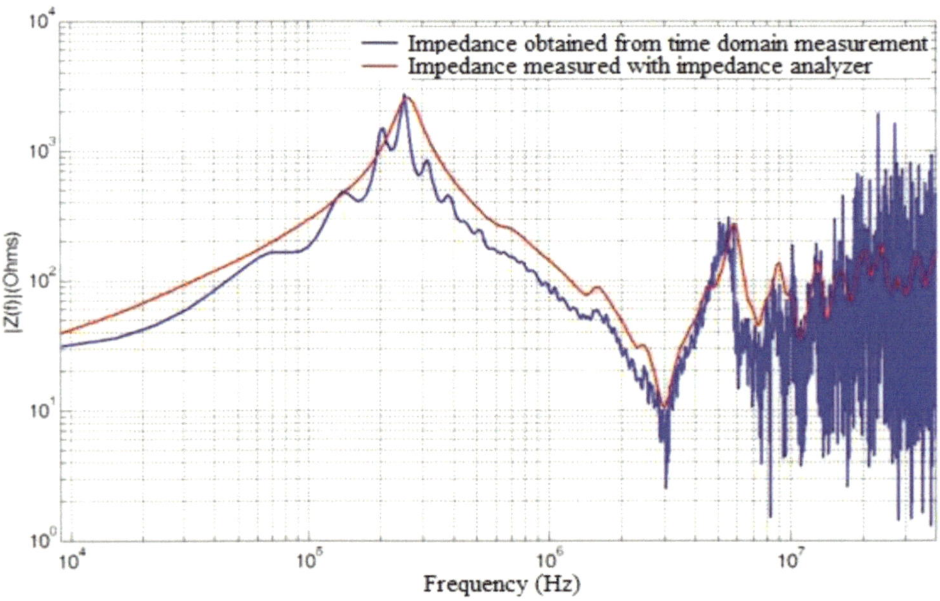

Fig. 20. Comparison between measurement of impedance in frequency domain and with temporal method applied to impedance of the primary winding, the secondary being short circuited.

The obtained results show a good agreement between these methods. However, we note a small difference, in the impedance variation between these methods in low frequency, which may be due to two reasons: the low accuracy of measurement equipments in the time domain (oscilloscope, probes ...) and the behavior of the magnetic material at low frequency.

We can say that these results validate the proposed temporal method. This method is based on measurement of the voltage and current waveforms in the study system by reference to the method of Frequency Response Analysis FRA [10] which is based on the determination of the transfer function and voltage measurements. In the proposed method, the injection of higher current values allows us to obtain the nominal operating of the transformer.

10. Application to the test bench

10.1 The laboratory setup without the line

The test bench (Fig. 8) is composed by the 15 kVA power transformer, a rectifier (twelve diodes) and a load constituted by four power resistances. The high frequency model of the power transformer is used in this simulation.

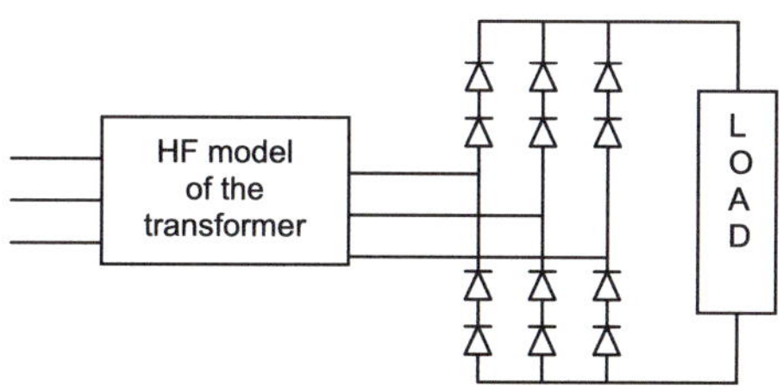

Fig. 21. Circuit diagram of the test bench

The output current of the power transformer is measured for each phase using an oscilloscope and current probe.

Figure 22 shows an example of current measured for phase A and obtained by the simulation. The test bench is simulated by using the high frequency equivalent circuit of the power transformer, presented previously, loaded with twelve diodes which model is available in the components library of the software simulator (SPICE) and four resistors.

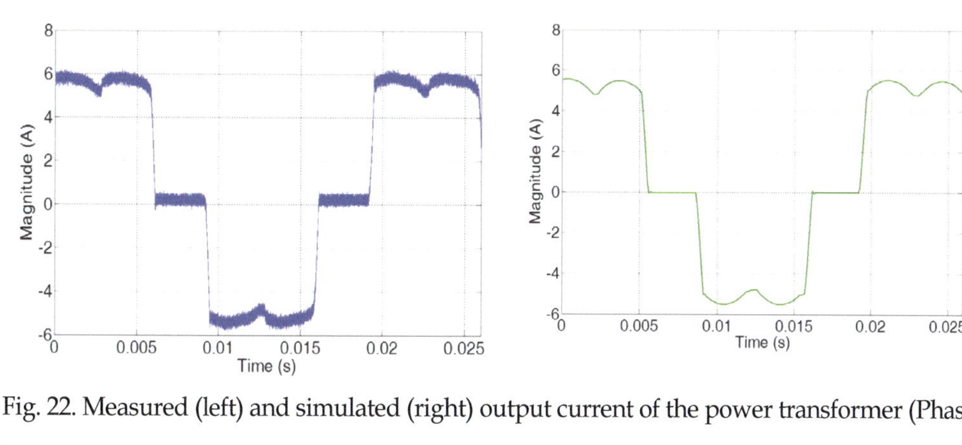

Fig. 22. Measured (left) and simulated (right) output current of the power transformer (Phase A)

As shown in Figure 22, we note a good agreement between the measured and simulated currents. Nevertheless, the waveform of the measured current is disturbed by digital noise of the oscilloscope. Moreover in our simulation we used sources without noise which does not exist in real case.

10.2 The laboratory setup with the line

As a conclusion of the laboratory study, we have realized a complete simulation of a railway power system by adding a power line and a power load to the transformer. Fig. 23 shows the experimental setup.

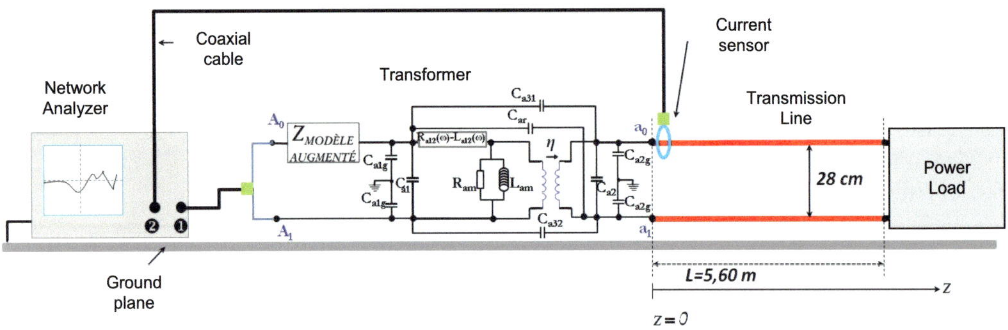

Fig. 23. Laboratory test bench simulating a railway power infrastucture

The test bench is constituted by the previous laboratory transformer (15kVA) used in one phase transformer configuration (the two others phases are short circuited), connected to a 5.6m power line and load by power resistances.

This system is fed by a network analyzer (connected to port 1) which allows measuring the current in the power line via the S_{21} parameter as presented in figure 23, thru a current sensor (connected to the port 2 of the analyzer) .

Figure 24 shows a comparison between measurements and modeling for two extreme positions of the current sensor: near the transformer (z = 0) and near the load (z = 5.6m).

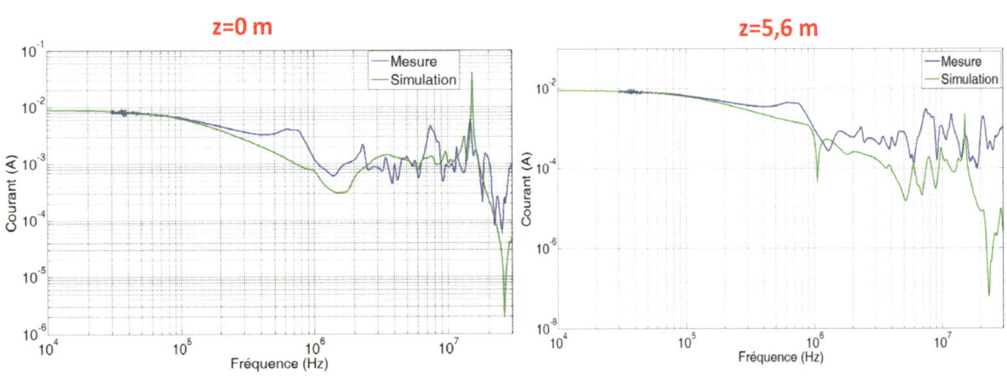

Fig. 24. Comparison between measurement and modeling : frequency variation of the current in the power line for two positions of the sensors

These curves show that the model seems very good until 10MHz. These results on a reduced scale power system are very encouraging and this study has to be continued on a real railway power substation

11. Conclusion

The results obtained with a laboratory power transformer show a good agreement between simulations and experimental results in time and in frequency domain.

Depending on the application, the measurements approach, in frequency domain, is often used and gives good results. However, for the high power system it is not possible to use to impedance analyzer (specific apparatus) thus other methods can be used as the Frequency Response Analysis (FRA). Nevertheless, the power necessary for these experimental determinations is low and can be a problem when the goal of the measurements is to define a model functioning at high power level for a large frequency band. The proposed experimental method, in time domain, allows making measurements with high injected power. The preliminary results, obtained on a laboratory transformer are very interesting.

12. Acknowledgment

This work has been done with the help of V. Deniau and J. Rioult from IFSTTAR LEOST laboratory and G Nottet from Alstom Transport within CEMRAIL project, with competitiveness cluster I-TRANS

13. References

[1] European Standards EN 50121: 2006 Railway applications - Electromagnetic Compatibility.

[2] A Cozza, "Railways EMC : assessment of infrastructure impact" PhD thesis Lille and Torino , 2005

[3] C. Andrieu, E. Daupahant and D. Boss, "A frequency- Dependant Model for a MV/LV Transformer", IPST'99 - International Conference on Power Systems Transient, June 20-24,1999, Budapest

[4] H. Ouaddi, S. Baranowski, N. Idir "High frequency modelling of power transformer : application to railway substation in scale model", XIV International Symposium on Electromagnetic Fields in Mechatronics, Electrical and Electronic Engineering, Arras, France, September 2009.

[5] J. Kolstad, C. Blevins, J. Dunn, A. Weisshaar, " NewCircuit Augmentation Method for Modeling ofInterconnects and Passive Components", IEEE Trans, Advanced packaging, Vol. 29, no.1, February 2006

[6] www.emc.polito.it section IDEM®.

[7] S. Chimklai, J.R Marti, "Simplified Three-Phase Transformer Model for Electromagnetic Transient Studies", IEEE Trans, Power Delivery, vol. 10, no. 3, July 1995, pp. 1316-1325.

[8] H. Ouaddi, S. Baranowski, N. Idir " High frequency modelling of power transformer: Application to railway substation in scale model" Przegląd Elektrotechniczny (Electrical Review), ISSN 0033-2097, R.86 May 2010. pp 165-169.

[9] H. Ouaddi, S. Baranowski, G. Nottet, B. Demoulin, L. Koné, "Study of HF Behaviour of Railway Power Substation in Physical and Geometrical Scale" European Phys. J., Appl. Phys., 53, 3 (2011) 33603-1-7 February , 2011

[10] WG A2.26 report. "Mechanical condition assessment of transformer windings using frequency response analysis (FRA)" Electra n°228 October 2006.

[11] H. Ouaddi, G. Nottet S. Baranowski, L. Koné, N. Idir, 'Determination of the high frequency parameters of the power transformer used in the railway substation', IEEE VPPC 2010, Lille, 1-3 September 2010. proceeding sur clef USB

[12] A. Oguz Soysal, "A method for wide frequency range modelling of power transformers and rotating machines", IEEE Transactions on Power Delivery, vol. 8, No. 4,pp. 1802–1810, October 1993.

[13] B. Gustavsen and A. Semlyen, "Rational approximation of frequency domain response by Vector Fitting", IEEE Trans, Power Delivery, vol. 14, no. 3, July 1999, pp. 1052-1061

[14] B.Cogitore, J. P. Kéradec, "The two-winding transformer: an experimental method to obtain a wide frequency range equivalent circuit", IEEE transactions on instrumentation and measurement, Vol. 43, No. 2, April 1994.

[15] T. Noda, H. Nakamoto, S. Yokoyama, "Accurate modeling of core-type distribution transformers for electromagnetic transient studies", IEEE transactions on power delivery, Vol.17, October2002.

[16] F. Blache, J. P. Kéradec, B. Cogitore, "Stray capacitance of two winding transformers: equivalent circuit, measurements, calculation and lowering", IEEE tr IEEE transactions on Powers Electronics Specialists Conference,, 1994.

[17] H.Y. Lu, J.G. Zhu,V.S. Ramsden, S.Y.R. Hui, "Measurement and modeling of stray capacitance in high frequency transformers", IEEE transactions on Powers Electronics Specialists Conference, Vol. 2, Page(s): 763 - 768, 1999.

[18] I. A. Maio, P. Savin, I. S. Stievano, F. Canavero, " Augmented models of high frequency transformees for SMPS", Proceeding, 20th Int. Zurich Symposium on EMC, Zurich 2009.

Part 3

Signalling, Security and Infrastructure Construction in Railway

14

Criteria for Improving the Embankment-Structure Transition Design in Railway Lines

Inmculada Gallego, Santos Sánchez-Cambronero and Ana Rivas
University of Castilla-La Mancha
Spain

1. Introduction

In the design of a railroad track there are some situations in which the introduction of a structure in the track is needed, for example a bridge, a viaduct or a pontoon. This circumstance is even more frequent in the High Speed lines, since the design criteria, fundamentally slopes and radius, are stricter than those for conventional lines. The introduction of a structure determines the appearance of a point with an abrupt change in the vertical stiffness from a track cross section to another.

The experience has shown that these transition zones between embankment and structure are the source of many problems (related to safety, passengers' comfort, maintenance expenses, etc.), causing differential settlements among adjacent track cross sections and originating which is known as "dip" (European Rail Research Institute, ERRI., 1999)

In order to diminish this unfavorable effect, the well known "technical blocks" are designed in a length determined between the structure and the embankment of access to this one. However, in spite of this structural disposition, it has not been found yet any design solution that notably reduces the track geometrical quality defects that have been observed in the mentioned zones. This is an important issue, because they produce a relevant increase in the maintenance expenses of the High Speed lines and they affect the availability of the track (Gallego, López, Ubalde, & Texeira, 2005).

2. Theoretical foundation of embankment-structure transition behavior

The wheel load transmitted by a train to the track does not correspond to the static load; instead, random dynamic overloads appear due to the sprung and un-sprung masses. Among the great amount of existing formulations relative to these overloads, there is an outstanding contribution made by Prud 'Homme (Prud`Homme, 1970) according to the expression (1)

$$\sigma(\Delta Q_{NS}) = 0.45\frac{V}{100}b\sqrt{mK\varphi(\varepsilon)}\ , \qquad (1)$$

where: $\sigma(\Delta Q_{NS})$ is the standard deviation of the dynamic overloads due to the un-sprung masses of the material; V is the running speed of the vehicle; b is a variable related to the track defects and to the vehicle defects; m is the un-sprung mass of the vehicle; k is the vertical track stiffness; $\varphi(\varepsilon)$ is Damping of the track.

Expression (1) introduces a new criterion to reduce the mutual aggressiveness between track and vehicle. From that it is deduced the importance of having a low value of the vertical stiffness of the track *(K)* and of the un-sprung mass of the vehicle *(m)* to avoid increasing the dynamic overloads due to the un-sprung masses. This influence is a more relevant fact in high speed trains.

In addition, it is not only the stiffness value which determines the dynamic overloads. They are also influenced by the variation of the stiffness value that might exist from one sleeper to another.

The first studies carried out to deal with this problem were carried out by Amielin, 1974, later on in the eighties they stand out those by Lopez, 1983, Hettle, 1986, Hunt, 1997, Esveld, 2001, López A. , 2001 or Teixeira, 2003 shown that as the difference between the stiffness values of two consecutive sleepers increases, the reaction on the sleepers increases, thereby increasing the load on the sleeper. On the other hand, next to these increments of stress, the experience has proved that some differential settlements are originated. As a result of these two factors hanging sleepers can be developed that in turn increase the stress on the ballast. In order to avoid this deterioration experimented in the transitions, these sections are built the well known "technical blocks", whose aim is achieving a gradual increase in the stiffness from one sleeper to the following one, as we reach the structure.

Now, it is interesting to know: How are these designs? What criteria are used to define them? To answer these questions a revision of the designs used by the different European Railway Administrations has been made. Five types of the most used measures have been identified. They are enumerated next, being the first one the most frequently employed:

- Backfill behind the abutment either with materials of a high compression level or granular material treated with cement.
- Use of a transition slab built with reinforced concrete or another material.
- Introduction of horizontal layers on a track formation of different materials.
- Use of geosynthetics to achieve an abutment reinforced backfill.
- Treatment of the track bed and sub-ballast with cement.

Along with these measures, they have been also identified a variety of track formation materials behind the abutment. There are three types that stand out, just as it is schematized in Fig 1. The first two types are the more frequently used, and with regard to that work they will be called slope type PA and slope type PB.

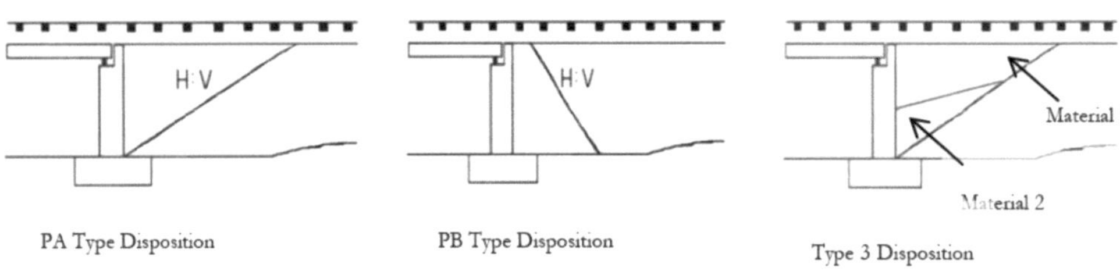

Fig. 1. Types of dispositions of the backfill behind the abutment in embankment-structure transitions.

This revision of the solutions employed, shows a lack of homogeneity of the design criteria: each Railway Administration uses different designs for the longitudinal sections of the embankment-structure transitions.

3. Motivation of the chapter

The previous analysis shows a lack of homogeneity in the design criteria. Besides, it must be added what the experience has pointed out, a remarkable deterioration of the quality in the transitions. These two facts, lead to think about two reflections: a) it does not exist a precise knowledge of the behavior of transitions, and b), the current designs have certain limitations since they are unable to reduce in a remarkable way the deterioration experimented in these areas. The first reflection invites us to deepen in the knowledge of the deterioration process and the second one induces us to consider two more aims: 1) introducing new design criteria and 2) giving an analytical basis to the already existent, to overcome some of those limitations. In order to achieve those aims a numerical modelization of the embankment-structure transition is shown in this chapter (Gallego & López, , 2009)

Therefore, within the scope of this work it is sought to carry out a model of finite elements which simulates the behavior of embankment-structures transitions. This model will enable to quantify the vertical stiffness of the track according to the type of disposition of the materials which the transition is carried out with (PA and PB, Fig 1) as well as the geotechnical characteristics of these materials (See Tab 1 and Tab 2).

Material	**E (N/m^2)**	**υ**	**c (N/m^2)**	**ϕ (º)**	**ρ (N/m^3)**
Rail steel	$2.1x10^{11}$	0.3	-	-	7.5 10^4
Elastic bearing	$2.952x10^8$	-	-	-	-
Sleeper element 1	$8.01x10^{10}$	0.25	-	-	-
Sleeper element 2	$5.02x10^{10}$	0.25	-	-	-
Sleeper element 3	$3.68x10^{10}$	0.25	-	-	-
Ballast	$1.3x10^8$	0.2	0	45	$1.9x10^4$
Sub-ballast	$1.2x10^8$	0.3	0	45	$1.9x10^4$
Track bed	8.10^7	0.3	0	35	$2x10^4$
Material QS1	$1.25x10^7$	0.4	15000	10	$2x10^4$
Material QS2	$2.50x10^7$	0.3	10000	20	$2x10^4$
Material QS3	$8x10^7$	0.3	0	35	$2x10^4$
Cement-treated granular material	$1.6x10^8$	0.25			$2.3x10^4$
Rock	$3x10^9$	0.2			$2.7x10^4$

Table 1. Values of geotechnical parameters considered in the models

GEOMETRIC PARAMETERS		GEOTECHNICAL PARAMETERS		Name of case studies H embankment (H7)-Disposition – value of the slope- material 1-material2-original ground
Type of design (Fig 1)	Slope value (H:V)	Type material of transition (type1/type2)	Original ground	
PA	1:1	QS2/QS3	QS1	H7PA11QS2QS3QS1
PB	3:1	QS2/MGT* QS3/MGT*	QS2 QS3 ROCA	

*MGT = Cement-treated granular material

Table 2. Geometric and geotechnical parameter considered in the model

4. Description of the numerical model and the work assumptions

In order to quantify the vertical track stiffness value and the incidence that the disposition and type of material of the track formation have on it, it is considered more accurate to analyze the track as a whole system. So as to carry out this analysis, the most appropriate approach is to apply the finite elements method. This method enables the numerical simulation of different materials and diverse boundary conditions, facilitating the study of the interaction among the different elements that compose the railway superstructure and infrastructure.

The employment of the finite elements method is accurate to evaluate the global behavior of the track structure, but it is very limited to quantify the efforts in the ballast: the inter-granular stress is very different from stress and strain assumed in a continuous medium. However, the use of a discrete elements model, or a mixed one, finite and discrete elements, implies a tremendous computational cost and an enormous complexity.

In the Railway field, the finite elements method has been used by authors as López A. , 1977, Sauvage & Larible, 1982, Profillidis, 1983, Sahu, Rao, & Yudhbir., 1999 Mira, Férnández, Pastor, Nasarre, & Carrillo, 2000, among others. Some studies developed in the eighties stand out, such as Profillidis, 1983, since the results of those works were integrated by the Committee D-117 of the ORE in the Record (Comité D-117 (ORE), 1983). The sizing graphics of the track bearing structure are collected in this work. Together with the previous works, it is also worth mentioning those carried out by the "Ministerio de Fomento Español", which have been the basis for making some recommendations for the Railways track construction.

In order to generate the model proposed in this chapter, the contributions collected in the different models of Railways track formations carried out so far and enunciated previously have been taken into consideration.

4.1 Description of the analyzed domain

The length of an embankment-structure transition depends on the type of structure and the height of the access embankment. In the case of embankment heights around 15 meters, the technical block can reach lengths of up to 85 meters (Fig 2). A 3D Model for such that length require an extremely powerful software and hardware, implying a huge computational cost due to the long time calculation that would be needed.

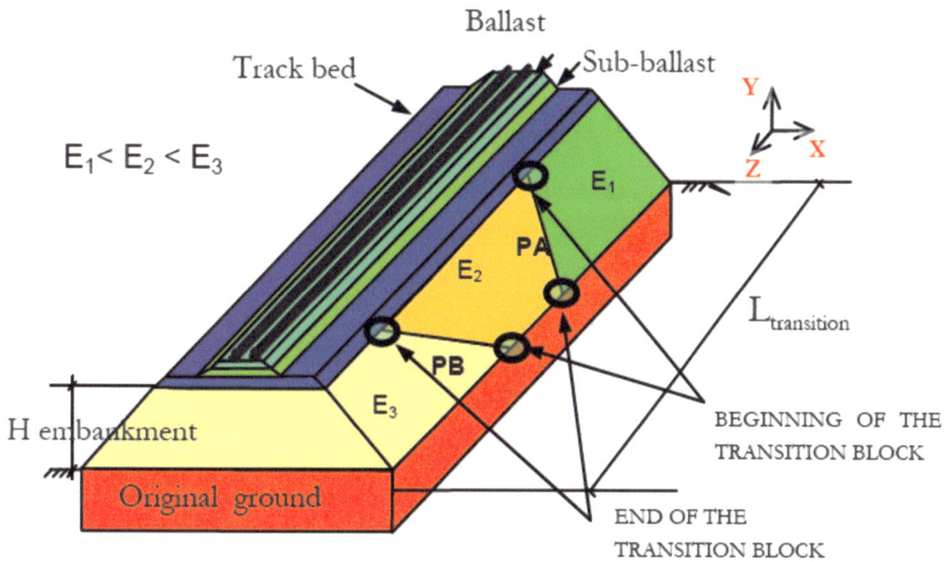

Fig. 2. General schematic of an embankment-structure. The transition is the zone adjacent to the structure (see Gallego I.,2006)

Suitable accuracy can be achieved by modeling only a transition section. However, the track section chosen to be modeled should be such that possesses the fundamental characteristics of a transition. However, the track section chosen for modeling should have the fundamental characteristics of a transition. For both slope types considered (PA and PB), these characteristics occur when the material changes from one material to another, either at the beginning or end of the slope (see Fig 2). Thus, the sections to be modeled are those shown within the rectangles in Fig 3.

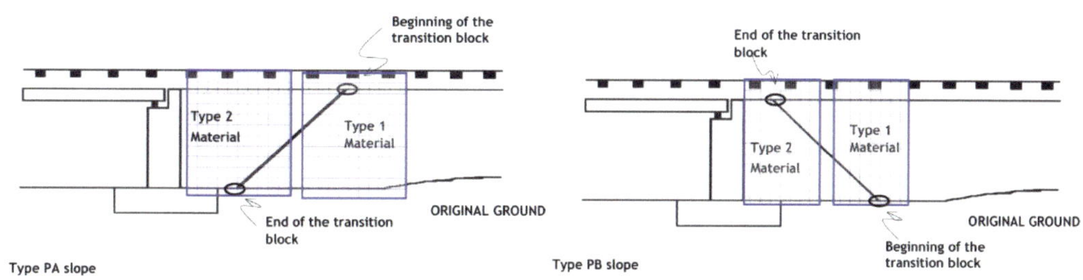

Fig. 3. Schematic of the beginning and end of the technical block in the slope of type PB and PA

4.2 Geometry of the analyzed domain

The directions considered for the model are the following ones: the axis x indicates the sleeper direction, the axis y the vertical and the axis z the rail direction (see Fig 2).

4.2.1 Transverse section

The measures adopted for the transversal section are the ones used in the High Speed line Madrid-Sevilla, but considering the track as simple and applying symmetry with respect to the plane $x = 0$.

The slope of the embankment varies with the type of material from which it is made; between H/V=3/2 and 1/2. The value 3/2 has been adopted in the model, because it was employed in the Madrid–Seville line.

4.2.2 Vertical direction

In the vertical direction (axis y) not only all the elements that compose the super structure are considered but also the, sub-ballast , the formation layer, 7m for the embankment and 3m for the original ground.

Values of 30, 25, and 60 cm, respectively were adopted for the thicknesses of the ballast under a sleeper, the sub-ballast the formation layer. These values are the same values as those applied in the transversal section of the High Speed line between Madrid and Seville.

4.2.3 Longitudinal direction

For domain analysis, the load applied on a sleeper is transmitted to the adjacent sleepers through some transmission coefficients. These coefficients clearly decrease as the distance from the point of load application increases; the coefficient is only 7 per cent in the third sleeper, when the first sleeper is defined as the one on which the load is applied (Comité D-117 (ORE), 1983)

To determine the real value of the settlement of the head of the rail when the load acts on it, one must not only solve for that load but also consider the history of the previous loads that have affected that sleeper. The load applied on a sleeper transmits it to the two adjacent sleepers. In order to observe the behavior of two consecutives sleepers with different stiffness under a load, while taking into account their load history also, four successively loaded sleepers (T5, T6, T7, and T8) were considered in order to analyze the behavior of sleepers T7 and T8.

In 1983, a test in Derby showed that important phenomena are apparent up to the fourth sleeper from the one loaded (Comité D-117 (ORE), 1983). Therefore, four unloaded sleepers were introduced at both ends of the model. This set-up avoided artefacts and yet included a substantial number of sleepers onto which the load could be applied. This approach required consideration of a transition sector comprising 12 sleepers, leading to a model system with a total length of 7.20 m.

4.3 Modelling rail track, elastic bearing, and sleeper section

In order to model the rail track, its resistance to bending was simulated in the most accurate way possible (Fig 4), which is why the inertia of the modelled rail must be equal to that of the real rail.

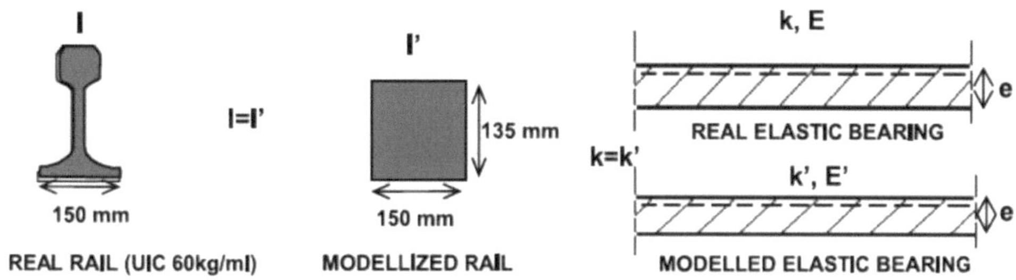

Fig. 4. Model of the rail and elastic bearings carried out in this study

The model also sought to make the vertical stiffness equal for all of the elastic bearings (see Fig 4). The vertical dimension and the modulus of elasticity were fixed so that the vertical stiffness of the element coincided with the stiffness of the elastic bearing provided by the manufacturer. For the high-speed Madrid–Seville line, the elastic bearings have a stiffness of nearly 500kN/mm (López A. , 2001)

Because the sleeper section is not constant along its entire length, the dimensions of its most representative section were used for the sleeper model elements. For each element (See Fig 5), the modeled flexural stiffness must be equal to the real flexural stiffness, as follows

$$E_{model} I_{model} = E_{real} I_{real} \tag{2}$$

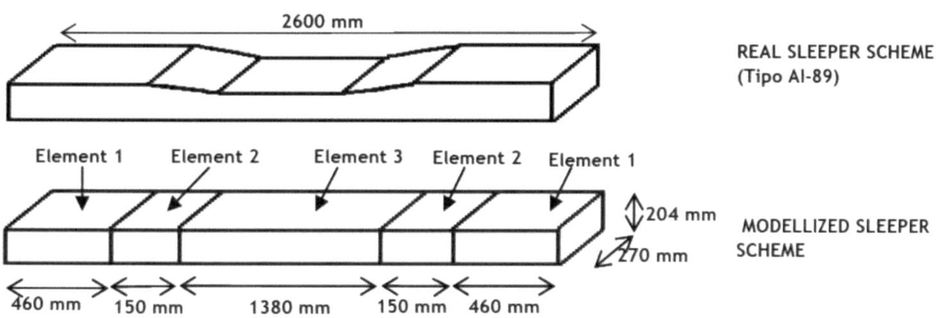

Fig. 5. Schematic of a real sleeper and a modeled sleeper

To obtain homogeneity in the calculations, the elements of the model had to have a constant width, which is not the case in the real sleeper. Thus, the model width must be considered to be an average width. This average width must be such that the load bearing surface in the model is equal to that in reality.

4.4 Sleeper–ballast contact

The sleeper–ballast contact zones contain a high concentration of strains. This local phenomenon requires refining of the mesh used to model these zones. However, applying this procedure is sometimes impossible because of the computational resources and model complexity needed. The most common alternative to modeling the contact zones is to use bounded degrees of freedom. In fact, this solution was adopted by ORE Committee D-117 and was used by the Railway Track Formations Project in its recommendations on railway track construction (Ministerio de fomento, 1999).

The use of bounded degrees of freedom requires the introduction of different nodes for each material at the contact surface. These nodes must move equivalently in the direction perpendicular to the contact plane (see Fig. 6). However, these nodes can move at different values in the directions parallel to the contact plane.

This solution is effective because it solves the tensional discontinuities that appear at the interface between two materials that differ significantly in their stiffness. In this model, bounded degrees of freedom were used at the sleeper–ballast contacts.

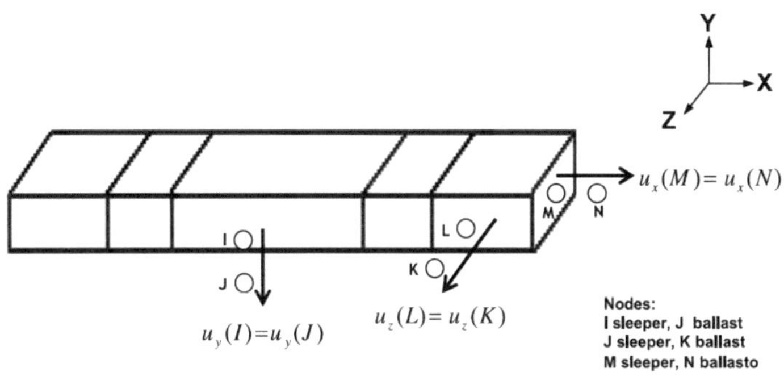

Fig. 6. Schematic of the ballast-sleeper contact

4.5 Boundary conditions

The model in this study differs from many existing models of railway track construction (Gallego, López, Ubalde, & Texeira, 2005), in which all vertical planes are constrained in all directions. In the Supertrack project (European comunity, 2005) and in this work, the planes that shape the slopes of embankments are left completely free, with no restrictions. In particular, the boundary conditions used here are as follows (see Fig. 7):

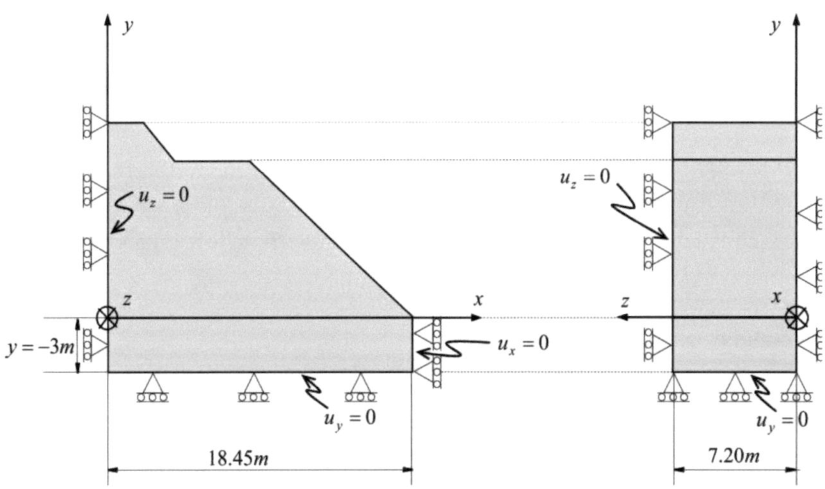

Fig. 7. Boundary conditions

- In the vertical plans limits of the model, z=0 and z=7.20m, the boundary condition adopted is to impose the nullity of movement in the perpendicular direction to these plans (u_z=0).
- In the vertical plans limits of the model, x=0 and x=18.45m, the boundary condition is, like the previous one, to impose the nullity of movement in the perpendicular direction to these plans (u_x=0).
- In the horizontal inferior plan of the model y=-3m, the condition to be imposed is the null vertical displacement (u_y=0).

4.6 Material constitutive model

An elastic, isotropic, and linear model was used to develop a mechanic model of the rail tracks, elastic bearings, sleepers, and the granular material processed with cement. In contrast, a perfect plastic model, i.e. the Drucker–Prager model (Oliver & Arlet, 2000), was used to model the rest of the materials, including the ballast, sub-ballast, track bed, embankment fill, and original ground. Finite strain was used to simulate the kinematics of the continuous medium.

Granular material treated with cement is considered to behave elastically, at least until it reaches a substantial percentage of its stress limit; one can assume its modulus of elasticity to remain essentially constant under normal stress.

To model the embankment material on which the track is laid, a perfect elasto-plastic behavior was assumed. This assumption implies that reloading occurs in the same way as downloading, and that the material experiences no hardening (hardening parameter $H=0$).

For modeling of the yield surfaces, the most accurate approach is to use a model dependent on hydrostatic pressure. These models are the Drucker–Prager and Mohr-Coulomb models, which limit the material behavior for states of hydrostatic stress (in traction). For the present study, the Drucker–Prager model was selected because it has been used in several elasto-plastic models used to design railway projects, and it has been validated by ORE Committee D-171.

The principle formulated by Drucker and Prager in 1952 includes the influence of pressure through the first invariant of the stress tensor I_1 and the internal friction angle φ. It also depends on the second invariant J_2 of the deviatoric stress tensor, as well as on two parameters: the friction angle among particles φ and the cohesion c. This criterion is expressed by means of the principal stress invariant and J_2 (the second invariant of deviatoric stress), as follows:

$$\mathrm{F}(I_1, J_2, c; \phi) = \bar{\alpha}(\phi) I_1 + \sqrt{J_2} - \bar{\mathrm{K}}(\kappa, \phi) = 0 , \qquad (3)$$

where

$$\bar{\mathrm{K}}(\kappa, \phi) = 6\, c(\kappa)\, \cos\phi \,/\, (3\sqrt{3} + \sqrt{3}\,\mathrm{sen}\,\phi) \qquad (4)$$

and

$$\bar{\alpha}(\phi) = 2\,\mathrm{sen}\,\phi \,/\, (3\sqrt{3} + \sqrt{3}\,\mathrm{sen}\,\phi) \qquad (5)$$

When the function that defines the yield surface (3) is represented in the main stress space, a cone is obtained, the axis of which is the hydrostatic axis (see Fig 8).

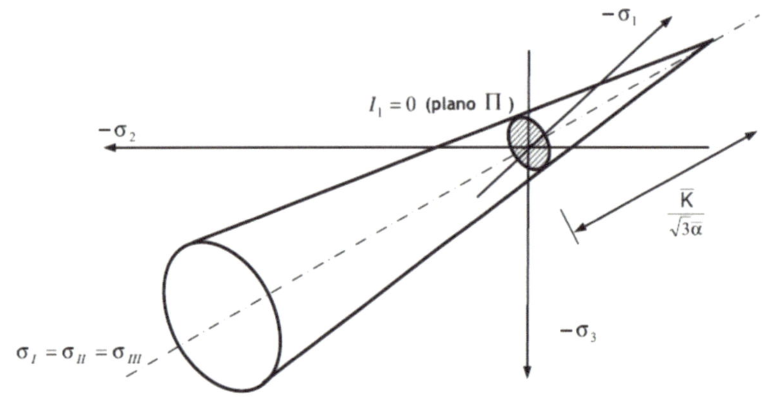

Fig. 8. Schematic of yield surfaces Drucker–Prager

Of the elements used, a quadratic 'brick' element of 20 nodes was selected; this is the most common type of element in three-dimensional models used to design railway tracks. The final meshes are shown in Fig 9.

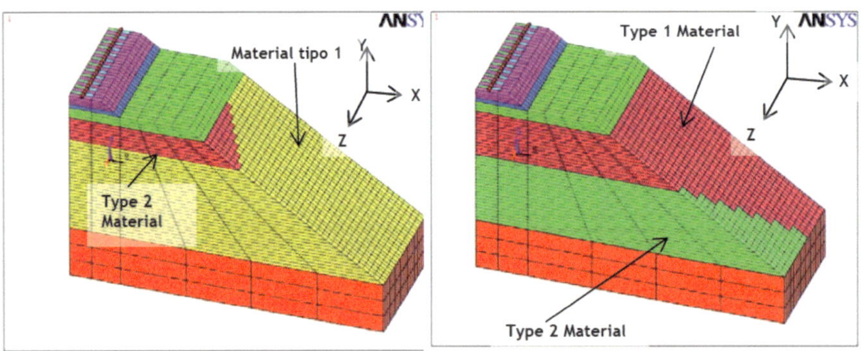

Fig. 9. Finite-element model for transitions with slopes of type PA31 and PB11

4.7 Hypothesis adopted

The load application must be carried out in several stages. In the first stage, only the material's own weight is considered until reaching the stress balance, while at later stages the loads due to the train are also taken into account. The stresses and displacements of interest are the ones that correspond to the application of the train loads; therefore, they can be calculated from the difference between the totals obtained after applying the train loads to the first stage.

Here, it was convenient to apply four load states due to the train passage, matching each state to the application of the static load per wheel in the four central sleepers of the model: T5, T6, T7, and T8 (See Fig 10).

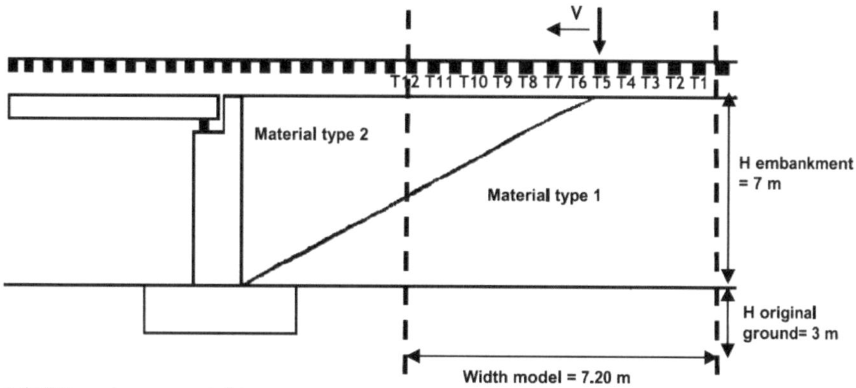

Load state 1 (LS1): only own weight

Load state 2 (LS2): static load on sleeper T5

Load state 3 (LS3): T5 sleeper download and T6 load

Load state 4 (LS4): T6 sleeper download T6 and T7 load

Load state 5 (LS5): T7 sleeper download and T8 load

Fig. 10. Schematic of the different load states

To simulate the constructive process of an embankment and to ensure convergence of the solution, the first load state (only the material weight) was divided into 250 substeps of the gravitational load application and 50 balance iterations for each one. For railway loads, 15 sub-steps proved to be sufficient to achieve convergence. The program used is ANSYS Structural, which enables nonlinear analysis.

To carry out the calculations, the load of only one axle was considered while assuming the effects of the remaining axles to be negligible.

The increased value of the dynamic overloads was calculated using Prud'Homme's formulation, with 1mm used as the value for *b* and the speed set to be 300 km/h. With this approach, the value of a dynamic overload was obtained according to the dynamic stiffness, which is unknown because its calculation requires knowing the value of the point dynamic load on the track. It is customary to carry out railway calculations assuming that the dynamic stiffness has a value similar to that of the static stiffness; this assumption has been validated by experience and confirmed by calculations (Ministerio de fomento, 1999). Thus, this assumption was also made in the present study, and numerical tests were used to justify this choice.

5. Results from the model: Presentation and critical analysis

This modeling study was carried out in two phases with different aims. In the first phase, different case studies were analyzed numerically, and the results were studied. These results then determined whether to carry out more case studies or proceed with the second phase. The main objective of the first phase was to observe how changing from a less rigid material to a more rigid one would affect the vertical stiffness of the track.

In the second phase, variations caused by the different stiffness values were analyzed for all the case studies from the first phase.

5.1 First phase

In the first phase, the main differences among the designs of the embankment–structure transitions are the types of materials used and the construction design. This justifies defining two types of case studies based on different geotechnical and geometrical parameters.

The geometrical parameters include the type and grade of slope; the latter is defined, e.g., as $1:1$ and $3:2$ ($H:V$, horizontal and vertical). The geotechnical parameters are the modulus of elasticity of the materials that compose the embankment fill and the original ground. For the modulus of elasticity, a range of values was used, matching those adopted in the numerical model presented by ORE Committee D-117. The values coincide with the lower limit values corresponding to the material types QS1, QS2, QS3, and rock (see Tab 1).

To fill the embankment, granular material processed with cement (MGT) was added to the model, since this material is so frequently used. Combining all these values with the technical block and the four load states already described, yielded a total of 48 case studies (see Tab 2). The modeling results for the case studies are shown in Tab 3.

To make additional comparisons, the stiffness was calculated for cases in which the fill corresponded to conventional embankments made with the same type of material (see Tab 4).

1. It is useful to apply stiffness values not only at the beginning but also at the end of the technical block; therefore, the cases corresponding to the ends of the technical blocks were calculated. Since the calculated stiffness values in the first 48 cases were similar for original ground QS1 and QS2, and for QS3 and rock, it was sufficient to calculate the cases corresponding to QS2 and QS3, thereby reducing the number of cases from 48 to 24 (see Tab 5).

5.2 Second phase

The second phase involved analyzing the results of the first phase. The criteria were to limit the following:

- The upper value of the vertical stiffness
- The lower value of the vertical stiffness
- The value of the longitudinal variation

TRANSITION TYPE	ORIGINAL GROUND QS1	ORIGINAL GROUND QS2	ORIGINAL GROUND QS3	ORIGINAL GROUND ROCK
H7PA11QS2QS3	11.242	17.227	37.206	42.752
H7PB11QS2QS3	18.847	19.957	36.054	39.325
H7PA31QS2QS3	10.800	16.380	34.139	39.744
H7PB31QS2QS3	14.015	19.296	34.519	38.512
H7PA11QS2MGT	15.380	25.499	43.859	47.795
H7PB11QS2MGT	28.721	32.644	37.657	40.000
H7PA31QS2MGT	13.182	18.275	36.549	42.077
H7PB31QS2MGT	21.524	25.961	35.777	38.592
H7PA11QS3MGT	32.193	53.563	71.692	75.374
H7PB11QS3MGT	54.186	59.936	69.294	70.473
H7PA31QS3MGT	29.354	51.068	70.875	74.650
H7PB31QS3MGT	48.668	58.069	70.340	73.444

Table 3. Values of vertical static track stiffness at the beginning of transition K_I (kN/mm) for all cases studied

EMBANKMENT MATERIAL	ORIGINAL GROUND QS1	ORIGINAL GROUND QS2	ORIGINAL GROUND QS3	ORIGINAL GROUND ROCK
FULL QS2 ($K(QS2)$)	10.186	15.717	33.226	38.433
FULL QS3 ($K(QS3)$)	28.899	50.929	70.606	74.027
FULL MGT ($K(MGT)$)	67.782	73.676	83.214	87.800

Table 4. Values of vertical static track stiffness K for conventional embankments (kN/mm)

TRANSITION TYPE	ORIGINAL GROUND QS2		ORIGINAL GROUND QS3	
	BEGINNING (K_I)	END (K_F)	BEGINNING (K_I)	END (K_F)
H7PA11QS2QS3	17.227	37.581	37.206	61.927
H7PB11QS2QS3	19.957	47.071	36.054	64.276
H7PA31QS2QS3	16.380	41.982	34.139	65.175
H7PB31QS2QS3	19.296	49.312	34.519	67.536
H7PA11QS2MGT	25.499	62.550	43.859	72.812
H7PB11QS2MGT	32.644	66.571	37.657	80.345
H7PA31QS2MGT	18.275	68.029	36.549	76.708
H7PB31QS2MGT	25.961	69.294	35.777	79.658
H7PA11QS3MGT	53.563	68.278	71.692	79.658
H7PB11QS3MGT	59.936	71.145	69.294	78.319
H7PA31QS3MGT	51.068	71.418	70.875	82.844
H7PB31QS3MGT	58.069	71.009	70.340	82.115

Table 5. Values of vertical static track stiffness at the beginning K_I and end of transition K_F *(kN/mm)* for all cases studied

These criteria were applied in two steps. Initially, the first two criteria were applied, and solutions were discarded if they resulted in either very high stiffness values, which generate elevated dynamic overloads, or very low stiffness values, which generate excessive rail deformations of the rail. To determine whether the stiffness values were high or low, they were compared to the values for designs using the same original ground material and the same materials as the simulated transition. Tab 6 shows the solutions remaining after this elimination process.

Materials of transition: Type1/Type 2	ORIGINAL GROUND QS1	ORIGINAL GROUND QS2	ORIGINAL GROUND QS3	ORIGINAL GROUND ROCK
QS2/QS3	PB31 PA11	PB31 PA11	PB11 PA31	PB11 PA31
QS2/MGT	PB31 PA11	PB31 PA11	PB11 PA31	PB11 PA31
QS3/MGT	PB31 PA11	PB31 PA11	PB31 PA31	PB31 PA31

Table 6. Transition types obtained after eliminating the transition types that with extreme stiffness values

The second step consisted of applying the third criterion, limiting the longitudinal variation value. From among the case studies, the cases selected were those with the smallest increase in the K value at the beginning and end of the technical blocks. In this way, the most appropriate solutions were obtained for each type of original ground material (QS2 and QS3). This approach yields certain design recommendations, which are described in the following section.

6. Proposed design recommendations

The analysis identified some problems related to the type of material. Based on these, the most relevant recommendations include:

- Excessive deformations were observed in the rail when material of type QS1 exists in the original ground (see Fig 11). These deformations reached 14mm under the rail when transitions QS2 and QS3 were used (Tab 7). In these cases, the deformations were large, as were the deformation values between adjacent sleepers. For this reason, it is appropriate to substitute the original ground material of type QS1 with another material or to treat the existing QS1 material in such a way as to obtain a modulus of elasticity corresponding to that of a material of at least type QS2.

In transitions from material of type QS2 to treated granular material, which occurs commonly in buried structures, the stiffness increased significantly at the beginning of the technical block when there was a relatively compressible material in the original ground. The solution with the smallest increase in the stiffness value was PA31. Fig 12 shows this value to be 16.4% which is too large. For this reason, transition-type QS2/MGT is not appropriate for use in the surface structure.

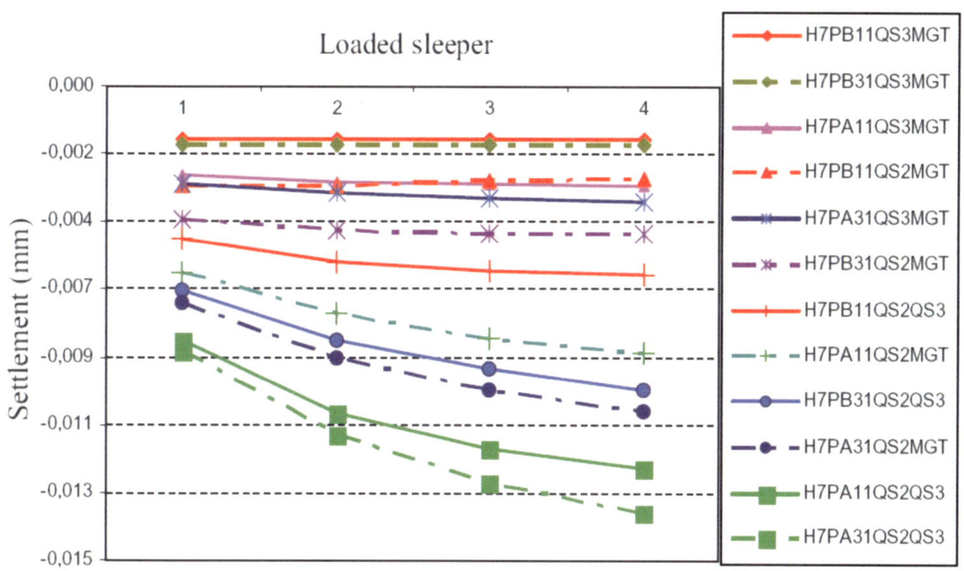

Fig. 11. Deflection of the head of loaded rails

	SLEEPER 5	SLEEPER 6	SLEEPER 7	SLEEPER 8
H7PA11QS2QS3				
LOADED SLEEPER 5	**-0.0083**	-0.0077	-0.0069	-0.0063
LOADED SLEEPER 6	-0.0104	**-0.0106**	-0.0100	-0.0091
LOADED SLEEPER 7	-0.0113	-0.0117	**-0.0118**	-0.0112
LOADED SLEEPER 8	-0.0117	-0.0120	-0.0123	**-0.0125**
H7PB11QS2QS3				
LOADED SLEEPER 5	**-0.0050**	-0.0045	-0.0038	-0.0032
LOADED SLEEPER 6	-0.0054	**-0.0057**	-0.0052	-0.0044
LOADED SLEEPER 7	-0.0051	-0.0057	**-0.0060**	-0.0055
LOADED SLEEPER 8	-0.0049	-0.0054	-0.0059	**-0.0062**
H7PA31QS2QS3				
LOADED SLEEPER 5	**-0.0086**	-0.0081	-0.0074	-0.0069
LOADED SLEEPER 6	-0.0111	**-0.0114**	-0.0109	-0.0101
LOADED SLEEPER 7	-0.0121	-0.0126	**-0.0129**	-0.0124
LOADED SLEEPER 8	-0.0127	-0.0131	-0.0136	**-0.0139**
H7PB31QS2QS3				
LOADED SLEEPER 5	**-0.0067**	-0.0062	-0.0055	-0.0049
LOADED SLEEPER 6	-0.0079	**-0.0083**	-0.0078	-0.0071
LOADED SLEEPER 7	-0.0083	-0.0089	**-0.0092**	-0.0087
LOADED SLEEPER 8	-0.0085	-0.0090	-0.0096	**-0.0099**
H7PA11QS2MGT				
LOADED SLEEPER 5	**-0.0061**	-0.0056	-0.0048	-0.0043
LOADED SLEEPER 6	-0.0072	**-0.0074**	-0.0069	-0.0062
LOADED SLEEPER 7	-0.0074	-0.0079	**-0.0082**	-0.0077
LOADED SLEEPER 8	-0.0076	-0.0080	-0.0084	**-0.0087**
H7PB11QS2MGT				
LOADED SLEEPER 5	**-0.0032**	-0.0027	-0.0020	-0.0015
LOADED SLEEPER 6	-0.0030	**-0.0032**	-0.0027	-0.0020
LOADED SLEEPER 7	-0.0026	-0.0029	**-0.0031**	-0.0026
LOADED SLEEPER 8	-0.0022	-0.0025	-0.0028	**-0.0030**
H7PA31QS2MGT				
LOADED SLEEPER 5	**-0.0071**	-0.0066	-0.0059	-0.0053
LOADED SLEEPER 6	-0.0086	**-0.0089**	-0.0084	-0.0077
LOADED SLEEPER 7	-0.0091	-0.0096	**-0.0099**	-0.0094
LOADED SLEEPER 8	-0.0094	-0.0098	-0.0103	**-0.0106**
H7PB31QS2MGT				
LOADED SLEEPER 5	**-0.0043**	-0.0038	-0.0031	-0.0025
LOADED SLEEPER 6	-0.0044	**-0.0047**	-0.0042	-0.0034
LOADED SLEEPER 7	-0.0040	-0.0045	**-0.0048**	-0.0043
LOADED SLEEPER 8	-0.0036	-0.0040	-0.0045	**-0.0048**
H7PA11QS3MGT				
LOADED SLEEPER 5	**-0.0029**	-0.0025	-0.0020	-0.0016
LOADED SLEEPER 6	-0.0028	**-0.0031**	-0.0027	-0.0022
LOADED SLEEPER 7	-0.0024	-0.0029	**-0.0032**	-0.0028
LOADED SLEEPER 8	-0.0021	-0.0025	-0.0029	**-0.0032**
H7PB11QS3MGT				
LOADED SLEEPER 5	**-0.0017**	-0.0014	-0.0010	-0.0007
LOADED SLEEPER 6	-0.0015	**-0.0017**	-0.0014	-0.0010
LOADED SLEEPER 7	-0.0011	-0.0015	**-0.0017**	-0.0014
LOADED SLEEPER 8	-0.0008	-0.0011	-0.0015	**-0.0017**
H7PA31QS3MGT				
LOADED SLEEPER 5	**-0.0032**	-0.0028	-0.0022	-0.0018
LOADED SLEEPER 6	-0.0031	**-0.0035**	-0.0031	-0.0025
LOADED SLEEPER 7	-0.0028	-0.0033	**-0.0036**	-0.0033
LOADED SLEEPER 8	-0.0025	-0.0029	-0.0034	**-0.0037**
H7PB31QS3MGT				
LOADED SLEEPER 5	**-0.0019**	-0.0016	-0.0012	-0.0009
LOADED SLEEPER 6	-0.0017	**-0.0019**	-0.0016	-0.0012
LOADED SLEEPER 7	-0.0013	-0.0017	**-0.0019**	-0.0016
LOADED SLEEPER 8	-0.0010	-0.0013	-0.0016	**-0.0019**

Table 7. Settlements of rail for the different load steps. Case study: original ground QS1

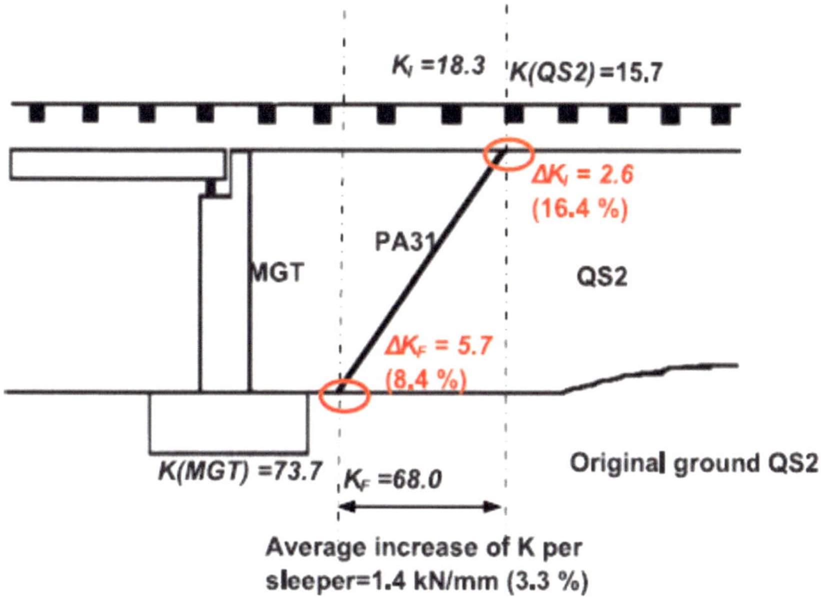

Fig. 12. Analysis of the stiffness variations assuming material transition from type QS2 to granulartreated material

- Near the abutment, the stiffness increased abruptly when using a granular material treated with cement as fill (MGT, 3% by weight). In fact, the stiffness value of the abutment could oscillate between 200KN/mm and 300KN/mm. The maximum of 87.8 KN/mm obtained for the case of treated granular material (Tab 4) is outside of that range. In this case, it is necessary to put a material with greater stiffness than MGT next to the abutment.
- A direct transition from QS2 or QS3 material to the structure is not recommended. At best, the stiffness is 38.43 KN/mm when the embankment material is QS2 and 74.027 KN/mm if it is QS3 (see Tab 4), as compared to200-300 KN/mm of vertical stiffness on the abutment.
- It is sometimes necessary to build the embankment adjacent to the abutment before building the structure. For example, this occurs when the original ground adjacent to the abutment is preloaded, and the load is removed immediately before building the transition. In that case, a slope of type PA (right column) is used instead of a slope of type PB (left column) in the transition from embankment material QS2 to QS3. In this research, a suitable value is obtained for the slope of the transition for each type (PA, PB). For that reason, it is necessary to distinguish between both slope types when considering whether a better approach would be to build the embankment before or after fabricating the structure.

In this way, based on these analyses, this study can make some recommendations about construction designs. Moreover, this study has proposed some ideas relating to the geometric designs of the different materials and has classified the designs according to the original ground and according to when the adjacent embankment is built, either before or after the construction of the structure (Fig 13).

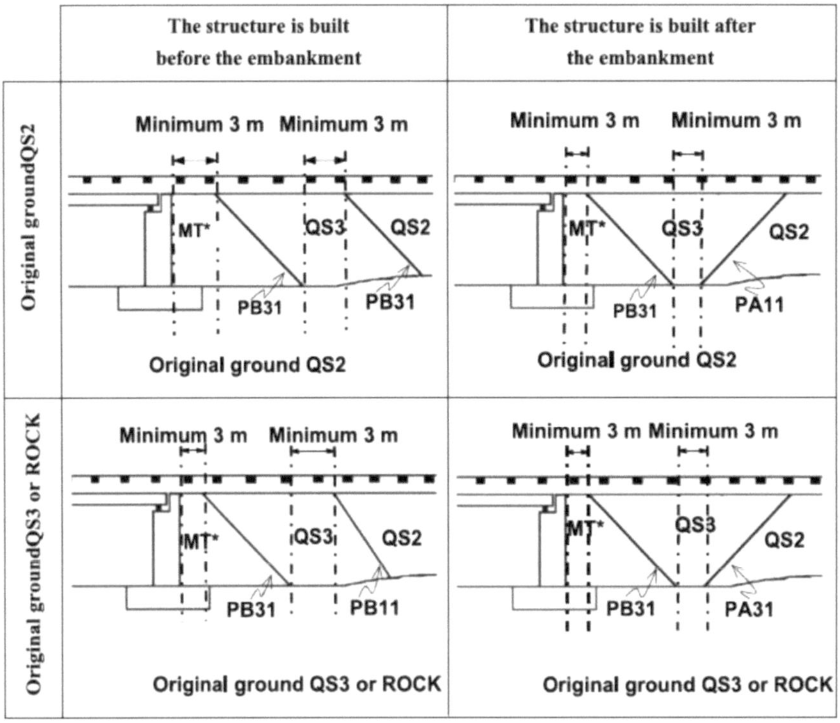

Fig. 13. Proposed design schemes

7. Conclusions

The designs currently used in embankment-structure transitions lead to zones that suffer significant deterioration. In addition, the various European railway systems have adopted many different design specifications when constructing these zones.

Increased railway speeds enhance the deterioration problems in the embankment-structure transitions, which has important implications for operating and maintenance costs, as well as for passenger safety and comfort.

From the numerical analysis carried out in this study, the following conclusions are drawn:

- The embankment must not be built on excessively compressible material, as shown for material of type QS1 on original ground. Such material must be replaced with another material or treated to obtain a modulus of elasticity corresponding to that of a material of at least type QS2.
- Transitions from material of type QS2 to material of type MGT should be carried out when there the original ground is of type QS3, or when the buried structure is treated.
- For each disposition type (PA, PB), there exists an optimal value for the slope. This value is not always the lowest one (3:1), as expected. The most suitable slope value (3:1 or 1:1) depends on the of original ground material, on the disposition type, and on the materials used in the transition.

- An abrupt increase in the stiffness takes place when a granular material treated with cement (MGT at 3% by weight) is used near the abutment. This problem remains unresolved, but future solutions should focus on improving the modulus of elasticity of the material without producing excessive stiffness increases at the extremes of the transition from the material of type QS3 to the treated material.

The application of all of those conclusions leads to a succession of recommendations for the most suitable building designs. When developing such designs, those factors which have an influence on the transition behavior (original ground, transition materials, and slope type) must be considered.

8. References

Amielin, S. (1974). La vía y el servicio de vía. (Spanish translation from the original in Russian, by Fundación de los Ferrocarriles Españoles).

Comité D-117 (ORE). (1983). *Adaptation optimale de la voie classique au trafic de l'avenir.Rapport no. 27. Comportement des structures d'assise de la voiesous charges répétées.* Office de Recherches et d'Essais del'Union Internationale des Chemins de Fer.

Esveld, C. (2001). *Modern railway track.* The Netherlands: MRTProductions.

European comunity. (2005). *Sustained Performance of Railway Tracks-SUPERTRACK Project Numerical simulation of train-track dynamics.*

European Rail Research Institute, ERRI. (1999). *Code D-230.1/RP3, Bridge ends embankment. Structure transitions.*

Gallego, I. (2006). *Heterogeneidad resistente de las vías de Alta Velocidad: Transición terraplén-estructura.6.* Ciudad Real: Doctoral Thesis, University of Castilla-La Mancha.

Gallego, I., & López, A. (2009). Numerical simulation of embankment-structure transition design. *Journal of Rail and Rapid Transit , 223,* 331-342.

Gallego, I., López, A., Ubalde, L., & Texeira, P. (2005). Track deterioration in high-speed railways: the influence of longitudinal variatioical stiffrness in the embankment-structure transitions. *Congress on Railway Engineering.* London.

Hettle, A. (1986). Modelluntersuchungen der Gleissetzungen am Ubergang Brücke (Studies in a reduced scale model on the track sinking/downfall/collapse in a section of a bridge/embankment transition). *ETR (Spanish translation from the original in German) .*

Hunt, H. (1997). Settlement of railway track near bridge abutment. *Proc. Instn. Civil Engrs. Transp,* (págs. 68-73).

López, A. (1977). Análisis de la deformabilidad vertical de una vía férrea mediante el método de elementos finitos. *AIT* (15).

Lopez, A. (1983). La heterogeneidad resistente de una vía y su incidencia en la evolución de la nivelación longitudinal: una aproximación al problema. *Revista de Obras Públicas,* , 719-735.

López, A. (2001). La rigidez vertical de la vía y el deterioro de las líneas de alta velocidad. *Revista de Obras Públicas ,* 222-232.

Ministerio de fomento. (1999). *Recomendaciones para el proyecto de plataformas ferroviarias.* Madrid: Servicio de publicaciones del Ministerio de Fomento.

Mira, P., Férnández, J., Pastor, M., Nasarre, J., & Carrillo, J. (2000). Aplicaciones del método de elementos finitos a la ingeniería ferroviaria. *Revista de Ingeniería Civi* (118), 71-82.

Oliver, X., & Arlet, C. (2000). *Mecánica de medios continuos para ingenieros.* Barcelona: Ediciones UPC.

Profillidis, V. (1983). *La voie et sa fondation modelisation mathematique.* París: Doctoral Thesis, École Nationale des Ponts et Chaussées.

Prud`Homme, M. A. (1970). La voie. *Revue Générales des Chemins de Fer* .

Sahu, J., Rao, & Yudhbir. (1999). Parametric study of resilient response of tracks with a sub-ballast layer. *Canadian Geotechnical Journal* , *36*, 1137-1150.

Sauvage, G., & Larible, G. (1982). La modélisation par éléments finis des couches d`assise de la voie ferrée. *Revue Générale des Chemins de Fer* , 475-484.

Teixeira, P. F. (2003). *Contribución a la reducción de los costes de mantenimiento de vías de alta velocidad mediante la optimización de su rigidez vertical.* Doctoral Thesis, Polytechnic University of Cataluña.

15

Influence of the Phreatic Level on the Stability of Earth Embankments

Shodolapo Oluyemi Franklin[1] and Gbenga Matthew Ayininuola[2]
[1]University of Botswana, Gaborone
[2]University of Ibadan, Ibadan
[1]Botswana
[2]Nigeria

1. Introduction

Slopes in soils and rocks are common place in nature and man-made structures partly due to the fact that they are generally less expensive than constructing walls. However slope stability problems may arise due to the construction of artificial slopes in cuttings and embankments for roads and railways, or the construction of earth dams and water retaining embankments. Other reasons may include the study of the process of large scale natural slips or the application of remedial measures when such slips have taken place (Capper & Cassie, 1971). Existing slopes that have been stable may experience significant movement due to natural or man-made conditions. Such changes can result from the occurrence of earthquake, subsidence, erosion, the progression of tension or shrinkage cracks coupled with water ingress, changes in groundwater elevation or changes in the slope's subsurface flow which induces new seepage forces (McCarthy, 1998). Further causes may include the removal of earth below the toe of a slope or increased loading close to the crest of the slope. Slips may occur suddenly or gradually, commencing with a crack at the top of an earth embankment and slight upheaval near to the bottom and subsequently developing to a complete slip. All the foregoing actions make slip surface stability analysis of earth embankments complex and very difficult.

Several notable methods of analyzing slip surface stability have been developed over the years. Among the earliest was one that had its slip circle passing through soil materials whose shear strength is based upon internal friction and effective stresses (Fellenius, 1927). In this method an area of unit thickness of the volume tending to slide is divided into vertical strips and it is assumed that for each slice the resultant of the interslice forces is zero. A more significant and certainly more widely used approach utilizing the method of slices assumed a circular failure surface and fulfilled moment equilibrium but did not fully satisfy force equilibrium (Bishop, 1955). Yet another method considered a cylindrical slip surface and assumed that the forces on the sides of the slices are parallel (Spencer, 1967). A generalized approach (Morgenstern & Price, 1965) was developed in which all boundary and equilibrium equations are satisfied and the failure surface could assume any shape. The method involved solving systems of singular simultaneous equations and was unduly long in obtaining approximate answers despite the several assumptions made. An alternative generalized procedure (Bell, 1968) has been advanced which satisfies all conditions of equilibrium and assumed any failure surface. Here a solution is obtained by assuming a distribution of normal stress along the rupture surface. An approach involving the determination of the critical earthquake acceleration required to produce a condition of limiting equilibrium has been developed (Sarma, 1979). Also stability charts have been proposed which were partly based on the work of previous investigators and are applicable to a wide range of practical conditions (Cousins, 1978).

In practically all the afore-mentioned methods, the mass of soil assumed to be associated with the slope slide is divided into vertical slices. A slice is selected, the free body diagram of the forces acting on the slice is drawn and subsequently, based on limit equilibrium methods, an expression is derived for determining the factor of safety of the slope. The stability problem is dealt with by assuming that the tangential interslice forces are equal and opposite (Bishop, 1955). An iterative method for analyzing the stability problem in non-circular slip failures (Janbu, 1973) utilized a full rigorous algorithm and assumed a known line of thrust for the interslice horizontal forces. The method is best suited for computer solution. Similarly analytical solutions have been presented (Morgenstern & Price, 1965) which incorporate all interslice forces but are dependent on several assumptions and are quite lengthy.

1.1 Role of water pressure forces

The action of water is highly significant in slope movement. In clay and shale, softening by rain may lead to slip of a whole layer of material as a mud run. In addition water percolating into fissured clay may result in progressive deterioration and weakening that eventually results in reduction of shear strength so that a rotational or translational slip occurs. Consequently, for the various methods of slices highlighted above, it is important to stress the need to consider the water pressure forces acting not only at the interslice but also at the slice base, for such neglect may produce erroneous results.

More accurate but lower factors of safety are claimed for methods which account for the variation in seepage forces acting on and in the slice (King, 1989). Nevertheless such refinements depend on good estimates of water pressure. The factors of safety obtained for a given slope using different methods of slices are sensitive to the assumptions made in deriving them (Morrison & Greenwood, 1989). Furthermore the interslice forces play an important role in the resulting factors of safety. Hence the present study focuses on the effects of omitting the interslice pore water pressure forces on the overall stability of earth embankments and also endeavours to reduce the complexity features common to the more recent methods of slices outlined earlier.

2. Methodology

In general, any proposed method to evaluate the factor of safety must satisfy several requirements such as fulfilling limit equilibrium laws, account for all forces acting on the slice, adopt few assumptions, which should be easily comprehensible, be applicable to non-homogeneous soils, account for the water pressure distribution at the base of the slice as well as at the interslice, and treat stability problems in terms of both effective and total stresses. Furthermore, it is immaterial whether the horizontal interslice forces are considered as total or effective together with the force due to water pressure. When the force due to water pressure is correctly accounted for at the base as well as on the vertical sides of the slice, then the equilibrium of the body (soil skeleton) and water (hydrostatic) as well as normal and shear forces is maintained.

In order to properly assess the effect of water pressure forces on the stability of earth embankments, two algorithms which utilize both the limit equilibrium approach and method of slices are presented. The algorithms satisfy all the conditions stated above and in addition, the stability problem is treated as a 2-dimensional one in order to arrive at the final solution more quickly. A comprehensive description of the two algorithms and their application to a number of embankments reported in the literature is given elsewhere (Ayininuola & Franklin, 2008). Only the more important features will be considered here. The basic assumptions adopted for the present purpose are as follows: (a) The line of thrust on a slice and the pore water pressure forces are within the slice, preferably at one-third distance from the base or mid-point of the slice vertical sides (b) Factor of safety is defined in terms of the average shear stress developed along the potential failure surface and the average shear strength within the soil (c) Failure occurs simultaneously throughout the soil mass within the assumed rupture boundary.

2.1 Formulation A

Consider an elemental slice in Fig. 1. Due to the many unknown forces acting on the nth slice, the free body diagram in Fig. 2(a) is further divided into two separate bodies (Figs. 2(b) and 2(c)) under the basic requirements for the analysis of earth embankment stability. Since Fig. 2(a) is in equilibrium with all the forces acting on it, consequently Figs. 2(b) and 2(c) are in equilibrium as well.

On examining Fig. 2(a) critically, it is observed that of the twelve forces acting on the slice, only four have known magnitudes. This makes the analysis statically indeterminate of order eight. For the remaining eight forces to be determined there is need to establish logical relationships between the forces. With reference to Fig. 2(c) and letting $\Delta P_{w(n)}$ be the elemental increment of water pressure forces $P_{w(n+1)}$ and $P_{w(n)}$ across the slice, then

$$P_{w(n+1)} - P_{w(n)} = \Delta P_{w(n)} \tag{1}$$

The hydrostatic force U_n is obtained by measuring the free standing height of water in an installed piezometric tube at the slice base. If the piezometric height at the base is $H_{w(n)}$ then

$$U_n = \gamma_w . H_{w(n)} \cdot g \tag{2}$$

where g is the acceleration due to gravity. Alternatively, since the forces acting in Fig. 2(a) are in equilibrium, it implies that the forces in Figs. 2(b) and 2(c) are also in equilibrium. Hence the hydrostatic force U_n at the base of the slice can be determined from the weight of water in the slice. From force equilibrium in the x and y directions the following expressions are obtained:

$$U_n = W_{w(n)} \sec \alpha_n \tag{3}$$

$$\Delta P_{w(n)} = -U_n \sin \alpha_n \tag{4}$$

where $W_{w(n)}$ is the weight of water in the nth slice and α_n is angle at the base of the slice. Substituting the value of U_n in equation (3) into equation (4) yields

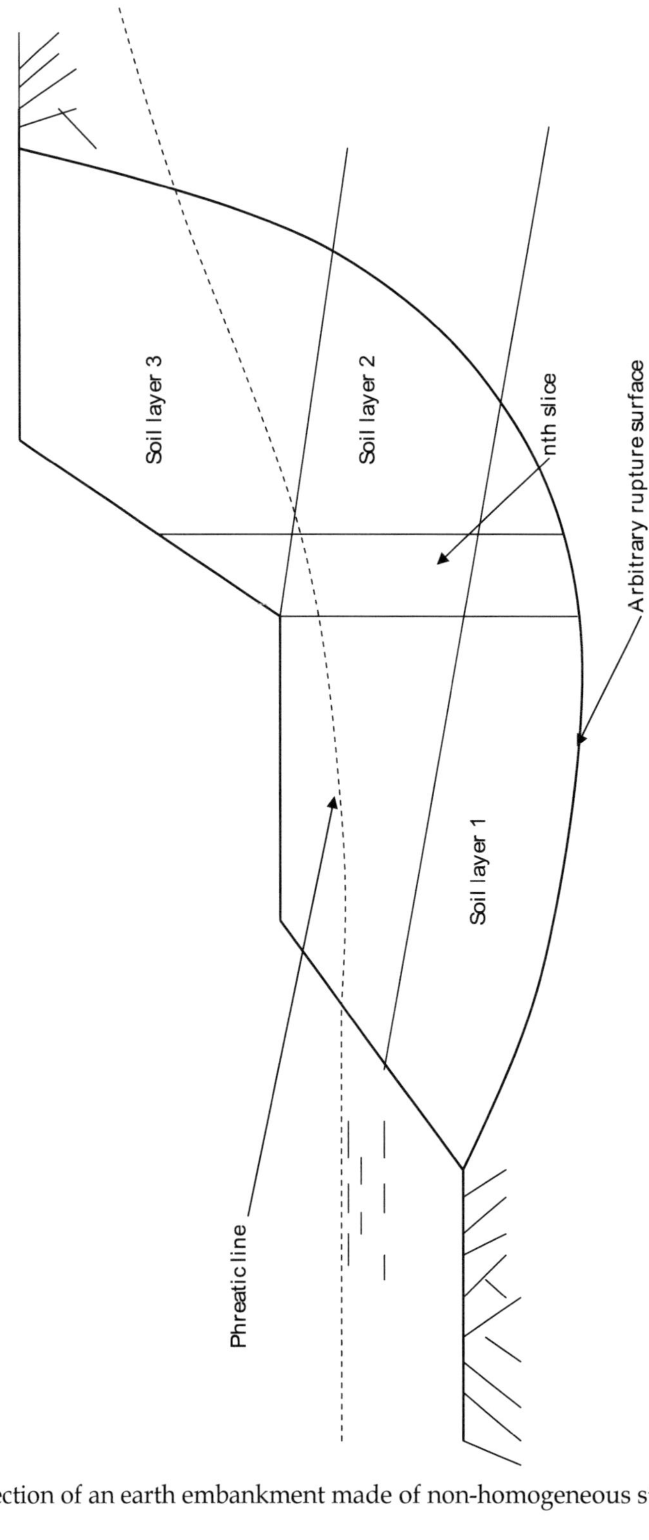

Fig. 1. Cross-section of an earth embankment made of non-homogeneous strata or soil layers

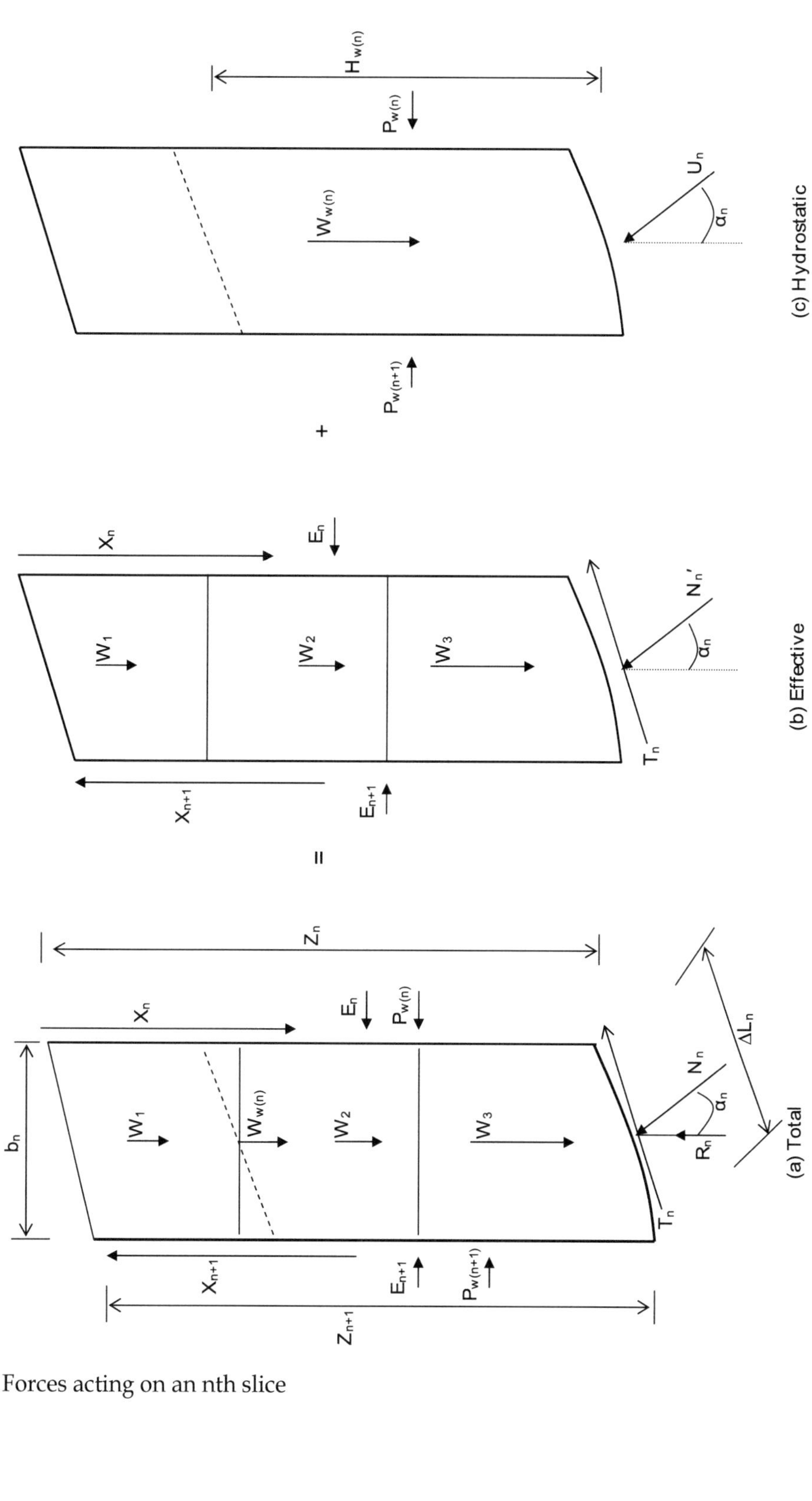

Fig. 2. Forces acting on an nth slice

$$\Delta P_{w(n)} = -W_{w(n)} \tan \alpha_n \tag{5}$$

From Fig. 2(b) let ΔE_n and ΔX_n be the elemental increments of horizontal interslice lateral thrusts E_n, E_{n+1} and vertical interslice shear forces X_n, X_{n+1} respectively across the slice, that is

$$\Delta E_n = E_n - E_{n+1}\,, \quad \Delta X_n = X_n - X_{n+1} \tag{6}$$

In Fig. 2(a) since the total normal force N_n and shear force T_n acting at the base of the slice are orthogonal, they have the same resultant R_n. In order to reduce the number of unknowns at the base of the slice the forces T_n and N_n are expressed in terms of R_n as follows:

$$T_n = R_n \sin \alpha_n\,, \quad N_n = R_n \cos \alpha_n \tag{7}$$

It should be noted at this stage that the total and effective normal forces, N_n and N_n' respectively, are related to U_n as follows:

$$N_n' = N_n - U_n \tag{8}$$

Limit equilibrium laws can be applied to the slice in Fig. 2(a). Firstly resolving the forces acting on the slice in the N_n-direction yields

$$N_n - W_n \cos \alpha_n - \Delta X_n \cos \alpha_n + \Delta E_n \sin \alpha_n + \Delta P_{w(n)} \sin \alpha_n - W_{w(n)} \cos \alpha_n = 0 \tag{9}$$

Substituting the values of $\Delta P_{w(n)}$ and N_n from equations (5) and (7) respectively into equation (9) and re-arranging yields

$$R_n \cos \alpha_n + \Delta E_n \sin \alpha_n - \Delta X_n \cos \alpha_n = W_{w(n)} \left(\cos \alpha_n + \tan \alpha_n \sin \alpha_n \right) + W_n \cos \alpha_n \tag{10}$$

Also, resolving the forces acting on the slice in the T_n-direction yields

$$T_n - W_n \sin \alpha_n - \Delta X_n \sin \alpha_n - \Delta E_n \cos \alpha_n - W_{w(n)} \sin \alpha_n - \Delta P_{w(n)} \cos \alpha_n = 0 \tag{11}$$

Substituting the values of $\Delta P_{w(n)}$ and T_n from equations (5) and (7) into equation (11) and simplifying and re-arranging results in

$$R_n \sin \alpha_n - \Delta X_n \sin \alpha_n - \Delta E_n \cos \alpha_n = W_n \sin \alpha_n \tag{12}$$

Examination of equations (10) and (12) reveals that there are three unknowns in the two equations which render them indeterminate; in order to solve for these unknowns it is necessary at this stage to introduce an equilibrium equation based on moments. From Fig. 2(b), taking moments of the resultants of the interslice forces and other forces about the mid-point of the slice base width gives

$$\Delta X_n . b_n/2 - \Delta E_n \left\{ Z/3 + \left(b_n/2 \right) \tan \alpha_n \right\} - \Delta P_{w(n)} \left(H_{w(n)}/3 \right) = 0 \tag{13}$$

where Z is the elevation of one side of the slice, or more correctly, the greater of Z_n and Z_{n+1}, and $H_{w(n)}$ is the height of water in the nth slice. Substituting for the value of $\Delta P_{w(n)}$ from equation (5) into equation (13) and re-arranging results in

$$-\Delta E_n\left\{Z/3+\left(b_n/2\right)\tan\alpha_n\right\}+\Delta X_n.b_n/2=-W_{w(n)}\tan\alpha_n\left(H_{w(n)}/3\right) \tag{14}$$

Equations (10), (12) and (14) can now be assembled together as a set of simultaneous equations in the following form

$$\left.\begin{aligned} R_n\cos\alpha_n+\Delta E_n\sin\alpha_n-\Delta X_n\cos\alpha_n &= W_{w(n)}\left(\cos\alpha_n+\tan\alpha_n\sin\alpha_n\right)+W_n\cos\alpha_n \\ R_n\sin\alpha_n-\Delta X_n\sin\alpha_n-\Delta E_n\cos\alpha_n &= W_n\sin\alpha_n \\ -\Delta E_n\left\{Z/3+\left(b_n/2\right)\tan\alpha_n\right\}+\Delta X_n.b_n/2 &= -W_{w(n)}\tan\alpha_n\left(H_{w(n)}/3\right) \end{aligned}\right] \tag{15}$$

In matrix format the simultaneous equation (15) becomes

$$\begin{bmatrix} \cos\alpha_n & \sin\alpha_n & -\cos\alpha_n \\ \sin\alpha_n & -\cos\alpha_n & -\sin\alpha_n \\ 0 & -\{Z/3+(b_n/2)\tan\alpha_n\} & b_n/2 \end{bmatrix}\begin{bmatrix} R_n \\ \Delta E_n \\ \Delta X_n \end{bmatrix} = \begin{bmatrix} W_{w(n)}(\cos\alpha_n+\tan\alpha_n\sin\alpha_n)+W_n\cos\alpha_n \\ W_n\sin\alpha_n \\ -W_{w(n)}\tan\alpha_n(H_{w(n)}/3) \end{bmatrix} \tag{16}$$

In compact form this can be written as

$$K.D=F \tag{17}$$

where $K=\begin{bmatrix} \cos\alpha_n & \sin\alpha_n & -\cos\alpha_n \\ \sin\alpha_n & -\cos\alpha_n & -\sin\alpha_n \\ 0 & -\left\{Z/3+\left(b_n/2\right)\tan\alpha_n\right\} & b_n/2 \end{bmatrix}$, $\quad D=\left[R_n \quad \Delta E_n \quad \Delta X_n\right]^{-1}$

and $F=\left[W_{w(n)}\left(\cos\alpha_n+\tan\alpha_n\sin\alpha_n\right)+W_n\cos\alpha_n \quad W_n\sin\alpha_n \quad -W_{w(n)}\tan\alpha_n\left(H_{w(n)}/3\right)\right]^{-1}$

In the above expressions the matrices D and F represent the nodal unknown forces and the nodal applied forces respectively.

Equation (17) above is for the nth slice, and when such sets of equations are assembled for the whole rupture mass this yields

$$K_g.D_g=F_g \tag{18}$$

where K_g is the global stiffness matrix, D_g is the global forces matrix and F_g is the global applied forces matrix. The simultaneous equation (18) can be solved for various values of R_n, ΔE_n and ΔX_n using the Gaussian elimination, Jacobi or Gauss-Siedel iterative techniques. The Gaussian elimination method is, relatively speaking, the simplest and easiest of the three procedures to implement. Once the values of R_n are obtained, the values of N_n and T_n can also be found using equations (7) and (8).

The factor of safety can be defined in terms of the shear strength of the soil and the shear stress developed along the potential failure surface based on the Coulomb-Mohr failure criteria in terms of effective stress as follows:

$$T_n = \left(c_n'\Delta L_n + N_n'\tan\phi_n'\right)/F_s \tag{19}$$

where F_s is the factor of safety and ϕ_n' is the angle of shearing resistance with respect to effective stress. Substituting for the values of T_n and N_n' from equations (7) and (8) into equation (19) yields

$$R_n\sin\alpha_n = \left[c_n'\Delta L_n + (R_n\cos\alpha_n - U_n)\tan\phi_n'\right]/F_s \tag{20}$$

Also substituting the value of U_n from equation (3) into equation (20) and then considering the whole of the rupture mass consisting of the set of slices will give

$$\sum R_n\sin\alpha_n = \sum\left[c_n'\Delta L_n + \left(R_n\cos\alpha_n - W_{w(n)}\sec\alpha_n\right)\tan\phi_n'\right]/F_s \tag{21}$$

Making F_s the subject of the expression in equation (21) will yield

$$F_s = \frac{\sum\left[c_n'\Delta L_n + \left(R_n\cos\alpha_n - W_{w(n)}\sec\alpha_n\right)\tan\phi_n'\right]}{\sum(R_n\sin\alpha_n)} \tag{22}$$

Equation (22) can be use to analyze stability problems involving both homogeneous and non-homogeneous soils types.

Irrespective of whether the earth embankment is partially or wholly drained, the equation can be applied because during its formulation both states of stress were taken into account.

2.2 Formulation B

The present study seeks to investigate the effect of hydrostatic pore water pressure forces on the overall stability of earth embankments and as such, in order to establish a basis of comparison with the earlier algorithm presented, an alternative approach is developed. This treats stability problems in terms of effective stresses and assumes that the influence of water pressure forces acting at the interslice can be neglected. The lines of action of $P_{w(n+1)}$ and $P_{w(n)}$ are taken to be coincident and also $\Delta P_{w(n)} = 0$. Proceeding along the same lines as the previous formulation, the following set of simultaneous equations can be arrived at:

$$\left.\begin{array}{c} R_n\cos\alpha_n + \Delta E_n\sin\alpha_n - \Delta X_n\cos\alpha_n = \left(W_{w(n)} + W_n\right)\cos\alpha_n \\ R_n\sin\alpha_n - \Delta X_n\sin\alpha_n - \Delta E_n\cos\alpha_n = \left(W_{w(n)} + W_n\right)\sin\alpha_n \\ -\Delta E_n\left\{Z/3 + (b_n/2)\tan\alpha_n\right\} + \Delta X_n.b_n/2 = 0 \end{array}\right] \tag{23}$$

For the typical nth slice the above equation in matrix format becomes

$$\begin{bmatrix} \cos\alpha_n & \sin\alpha_n & -\cos\alpha_n \\ \sin\alpha_n & -\cos\alpha_n & -\sin\alpha_n \\ 0 & -\{Z/3+(b_n/2)\tan\alpha_n\} & b_n/2 \end{bmatrix} \begin{bmatrix} R_n \\ \Delta E_n \\ \Delta X_n \end{bmatrix} = \begin{bmatrix} (W_{w(n)}+W_n)\cos\alpha_n \\ (W_{w(n)}+W_n)\sin\alpha_n \\ 0 \end{bmatrix} \tag{24}$$

Again proceeding along the same lines as the previous formulation, an expression very similar to equation (22) can be obtained as follows:

$$F_s = \frac{\sum\left[c_n'\Delta L_n + \left(R_n\cos\alpha_n - W_{w(n)}\sec\alpha_n\right)\tan\phi_n'\right]}{\sum\left(R_n\sin\alpha_n\right)} \tag{25}$$

Although equations (25) and (22) are very similar, the procedures for evaluating the values of R_n in both equations are certainly not the same. Consequently different values of factors of safety will be obtained using both approaches. The methods developed can be used for slip surface stability analysis either manually or with a programmable calculator. However while this may be true for fairly homogeneous slopes, for real or non-homogeneous soils the computation work is quite daunting for practical design. This is on account of the number of rupture surfaces that may need to be analyzed in order to obtain the most critical rupture surface for design purposes as well as the fact that the global stiffness matrix K_g mentioned earlier may be of the order 60 x 60 or more, depending on the number of slices within the rupture mass. Hence comprehensive computer software was developed involving two minimization computer programmes which can handle problems of up to three soil strata; some details of the programmes are given elsewhere (Ayininuola, 1999).

3. Results

In order to assess the effect of the pore water pressure forces, the procedures developed in the present study have been applied to a number of earth embankments some of which are reported in the literature. Firstly, the Lodalen Landslide (Sevaldson, 1956) is examined and then, the case of a non-homogeneous earth dam (Sherard et al, 1978) is investigated. Finally the effect of altering the phreatic level on the Okuku dam in South-Western Nigeria (Ayininuola & Franklin, 2008) is studied.

3.1 Stability analysis of the Lodalen Landslide (Sevaldson, 1956)

Fig. 3 shows a sectional view of the Lodalen Landslide. A stability analysis of the earth embankment prior to the occurrence of the slide will be carried out. Towards this end the initial rupture surface has been divided into 13 slices. The necessary data have been taken from the initial rupture surface and fed into the computer programme mentioned earlier. A total of 100 rupture surfaces have been considered in the analysis. Details of the computer output are not presented here, but a summary of the main findings are shown in Table 1 and these results are discussed at a later stage.

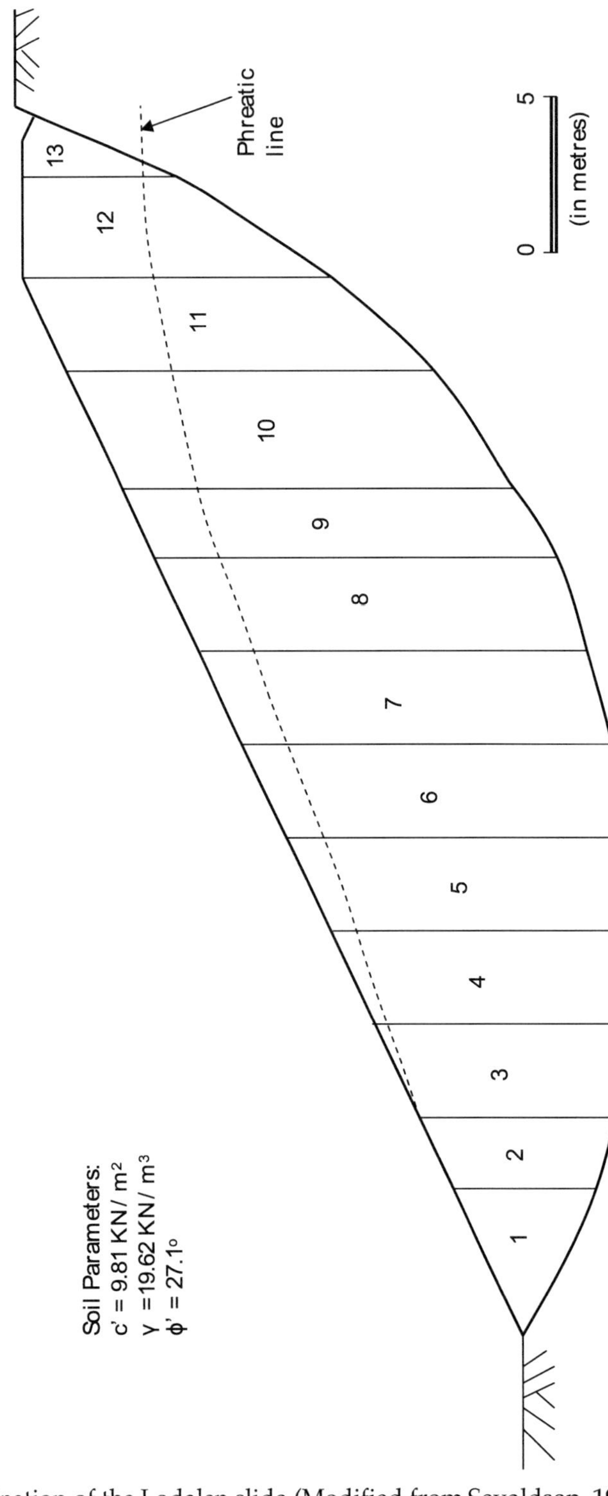

Fig. 3. Re-examination of the Lodalen slide (Modified from Sevaldson, 1956)

3.2 Stability analysis of a non-homogeneous earth dam (Sherard et al, 1978)

A non-homogeneous earth dam is shown in section in Figs. 4 and 5 and it is required to carry out a stability analysis of both the upstream and downstream sides of the dam. The assumed initial rupture surface on both sides of the dam has been divided into 12 slices each and the necessary data taken from the rupture surfaces have been fed into the computer programme referred to earlier. A total of 200 rupture surfaces have been considered at both the upstream and downstream sections. Again details of the computer output are not given in the present study, however a summary of the main findings are shown in Table1.

Earth Embankment	Method	Factor of Safety
Lodalen slide (Sevaldson, 1956)	Authors' Formulation A Authors' Formulation B Bishop's Simplified	0.80 0.90 0.90
Downstream section of non-homogeneous earth dam (Sherard et al, 1978)	Authors' Formulation A Authors' Formulation B Bishop's Simplified	1.49 1.83 1.83
Upstream section of non-homogeneous earth dam (Sherard et al, 1978)	Authors' Formulation A Authors' Formulation B Bishop's Simplified	0.66 0.78 0.78

Table 1. Results of stability analysis of Lodalen slide and a non-homogeneous earth dam

3.3 Stability analysis of Okuku earth dam, Nigeria

On account of the accessibility to data, the Okuku dam has been utilized as a case study in order to investigate and understand the response of the proposed formulations to changes in the phreatic levels in the earth embankment due to variation in water levels in the storage reservoir. The dam was constructed in 1995 at Okuku town located on the 8° 02'N and 4° 40'E coordinates and approximately 40 km North-East of Osogbo in Osun State, South-Western Nigeria. The dam axis located across River Anle, a seasonal stream, is about 1.5 km South-East of Okuku town. The dam is a homogeneous earth dam built with poorly graded sand clay mixtures which possess the following soil characteristics, namely, cohesion $c' = 45$ KN/m^2, angle of shearing resistance $\phi' = 12^{o}$ and additionally, average dry density of dam construction materials, $\gamma = 19.63$ KN/m^3. The height of the crest above the base of the dam is 10 metres and the upstream and downstream sections are sloped at ratios 1:3 and 1:2.5 respectively. In Figs. 6 and 7, diagrams of the dam embankment for both the upstream and downstream sections at different levels of water in the storage reservoir are shown. Additional details in respect of the dam design may be found elsewhere (Ayininuola, 1999). The factors of safety at different phreatic levels for both the upstream and downstream sections of the dam have been estimated using the two formulations developed earlier. A total of 500 rupture surfaces have been examined for each section. A summary of the results is presented in Table 2 and Figs. 8 and 9.

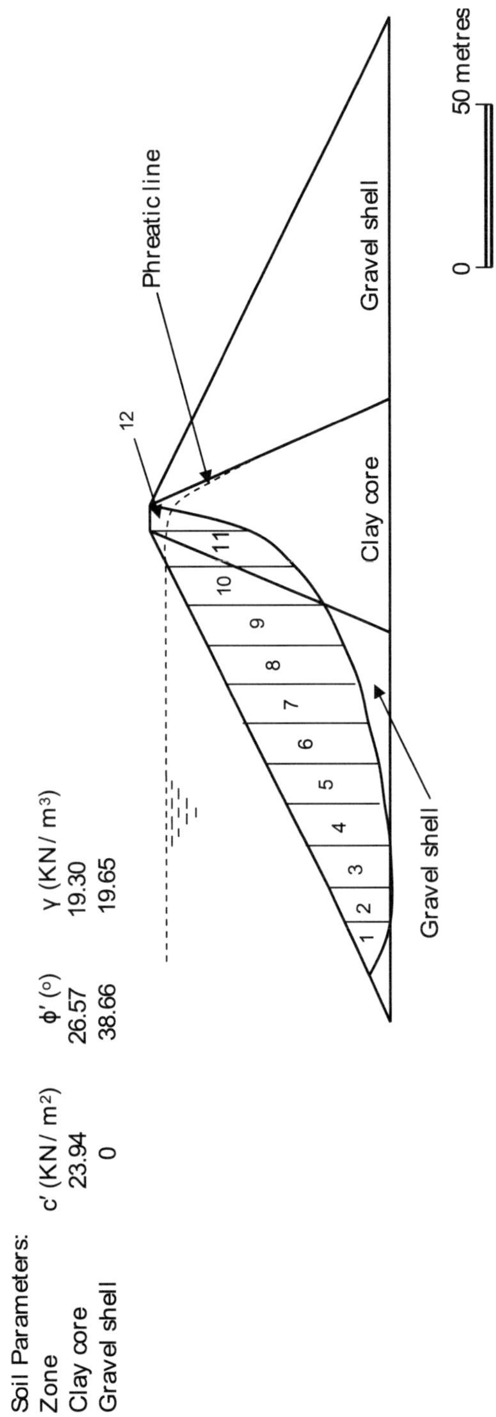

Fig. 4. Upstream section of an earth dam embankment (Modified from Sherard et al, 1978)

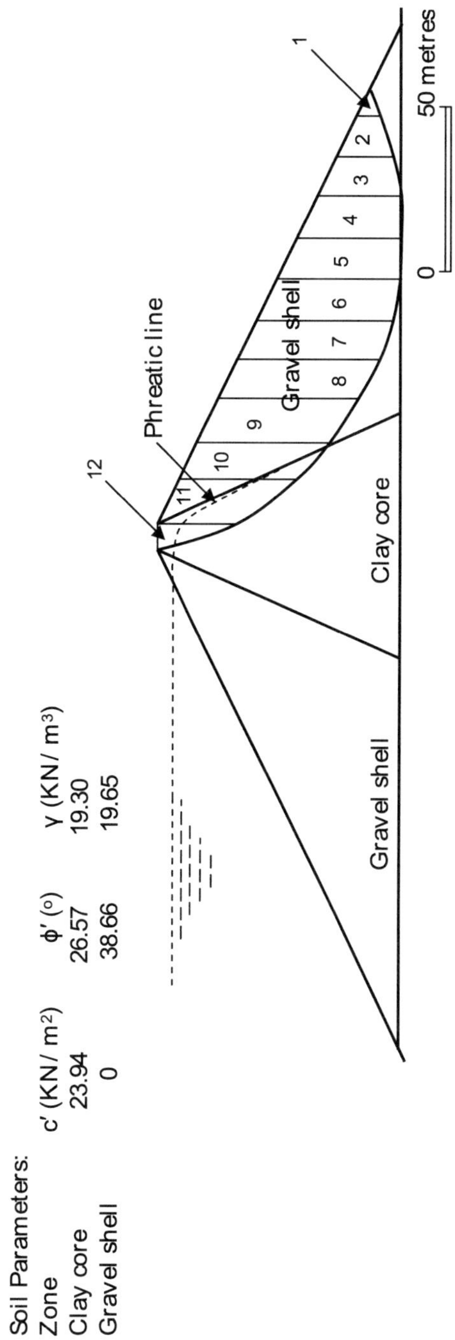

Fig. 5. Downstream section of an earth dam embankment (Modified from Sherard et al, 1978)

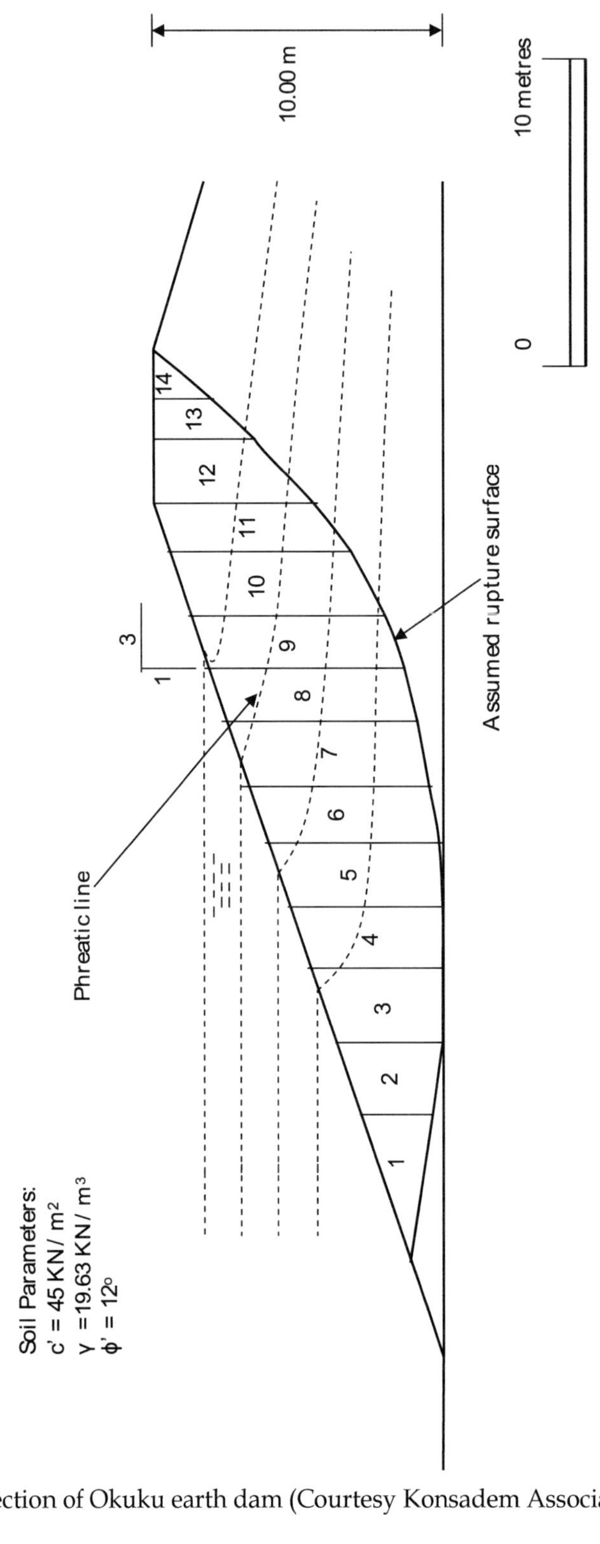

Fig. 6. Upstream section of Okuku earth dam (Courtesy Konsadem Associates Ltd., Nigeria)

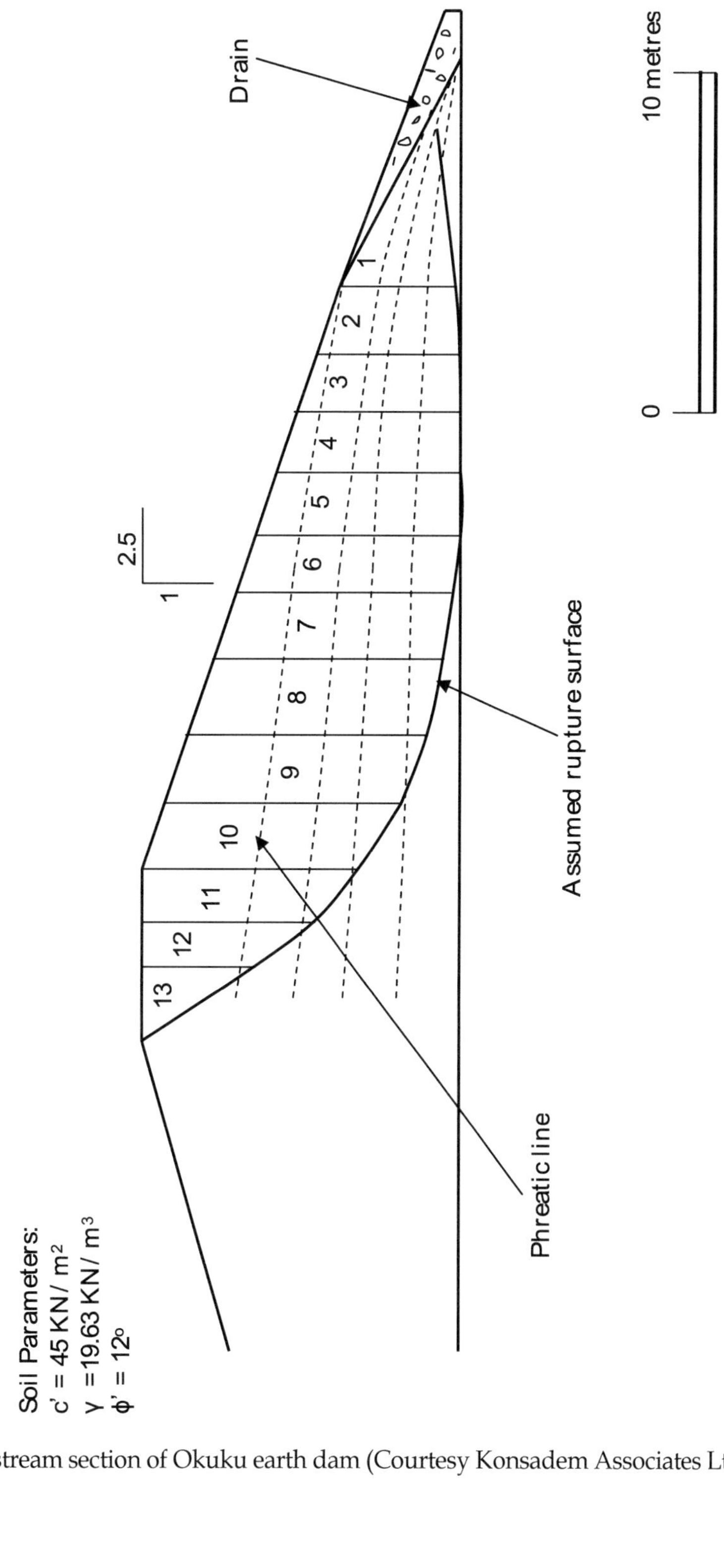

Fig. 7. Downstream section of Okuku earth dam (Courtesy Konsadem Associates Ltd., Nigeria)

Section of dam under consideration	Water level (metres)	Stability values for Formulation A	Stability values using formulation B	Difference (%)
Upstream	9.00*	2.43	2.78	14.40
	6.55*	2.60	2.85	9.62
	5.25*	2.81	2.93	4.27
	3.70*	3.01	3.03	0.66
Downstream	9.00**	2.25	2.67	18.67
	6.55**	2.53	2.76	9.09
	5.25**	2.74	2.83	3.28
	3.70**	2.89	2.90	0.35

Table 2. Results of stability analysis of Okuku Dam using Formulation A and B

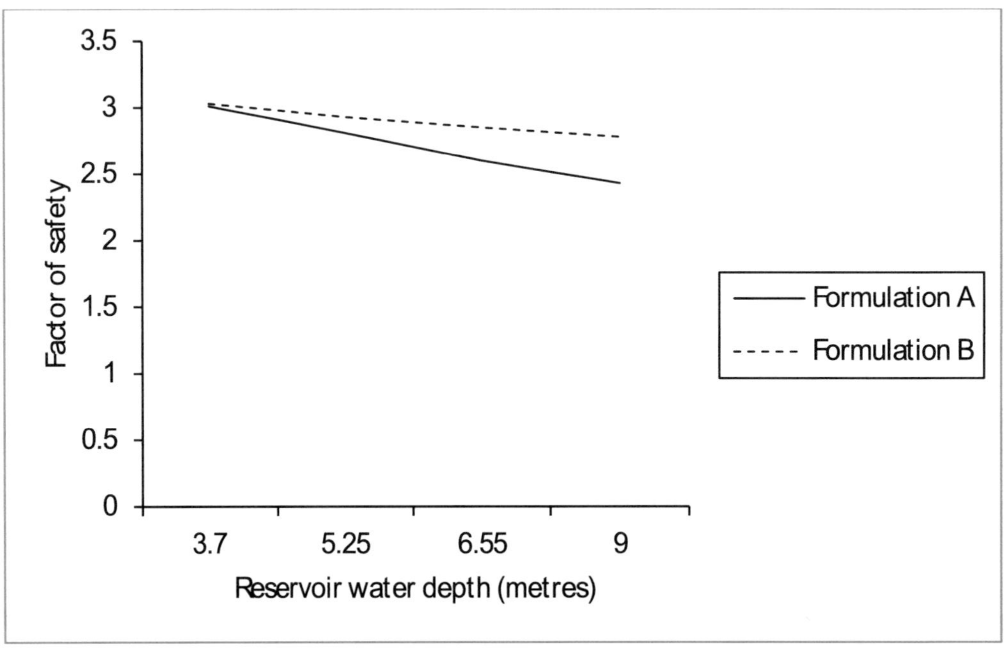

Fig. 8. Variation of factor of safety of dam embankment with reservoir water depth (upstream section)

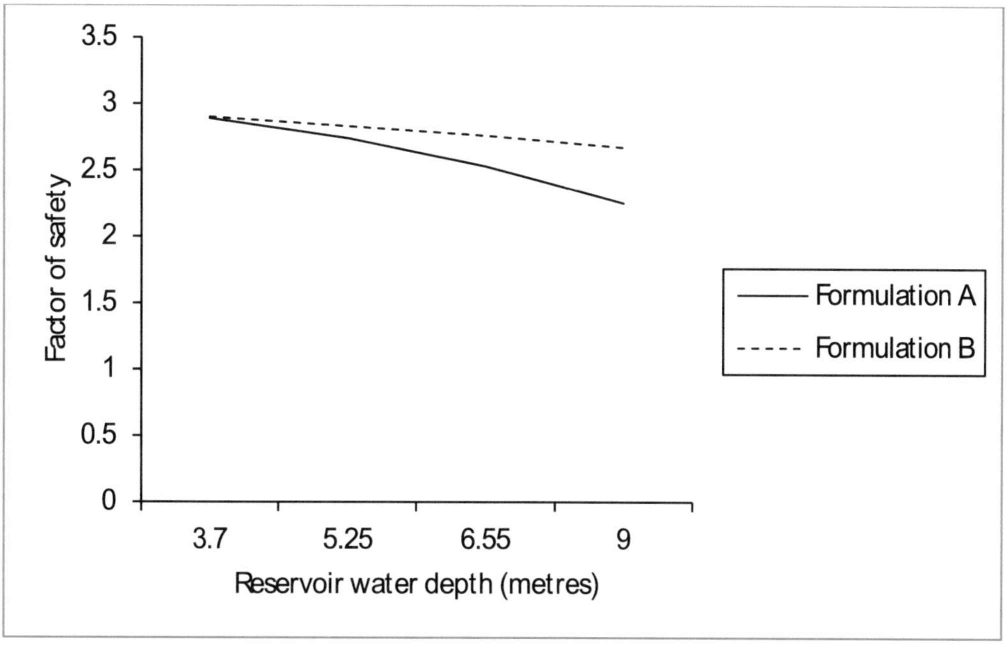

Fig. 9. Variation of factor of safety of dam embankment with reservoir water depth (downstream section)

4. Discussion

The comments outlined here are based primarily on the results presented in Tables 1 and 2, as well as in Figs. 8 and 9. As noted in the preceding section, details of the computer output in respect of the stability analysis carried out are not presented here, but may be obtained elsewhere (Ayininuola & Franklin, 2008; Ayininuola, 1999).

4.1 Comparison between formulations A and B and Bishop's simplified method

The procedures developed in the present study as well as Bishop's have been applied to the Lodalen landslide as well as the non-homogeneous earth dam described in the preceding section. Using Table 1 and the results of the computer generated output for the Lodalen landslide as guide, several points may be noted. Firstly the water pressure forces acting at the interslice have great influence on the elemental horizontal thrusts generated at the interslice. They also have a direct influence on the elemental shear forces at the interslice. In this region when the elemental water pressure forces are assumed to be zero, the values of the elemental horizontal thrusts and shear forces that develop are much smaller than those obtained when water pressure forces are taken into account.

In addition to the above, water pressure forces, elemental horizontal thrusts and elemental shear forces are directly affected by the slice inclination angles. When for example the inclination angle is zero, the effect of all the forces mentioned is practically negligible. At the interslice when the piezometric height $H_{w(n)}$ is zero, the values of elemental horizontal thrust and elemental shear force are zero. Furthermore the factors of safety obtained when the net effect of water pressure forces at the interslice is taken to be zero are higher than the corresponding values when the afore-mentioned forces are considered by the order of 8% - 24% depending on the phreatic level. Also the results given by Formulation B, which ignores the water pressure forces effect at the interslices, are practically identical to those obtained using Bishop's method.

4.2 Influence of the phreatic level on earth dam stability

With reference to Table 2 and Figs. 8 and 9, the results of stability analysis of Okuku earth dam reveal that the higher the phreatic level in the storage reservoir of the earth dam, the greater the variation between the stability values obtained using Formulations A and B. When the water pressure forces are ignored, higher factors of safety of the order of 0.35% - 18.67% are obtained. With regards to the curves in Figs. 8 and 9, and also from a study of the computer generated output, it is observed that at low phreatic levels in the storage reservoir of between 20% and 30% of the dam height, both approaches yield similar stability values.

5. Conclusions

An in-depth study of the effect of the pore water pressure forces acting at the interslice on the stability of earth dams has been carried out. This has been achieved by developing two procedures and applying the formulations to a number of practical cases. Based on the results of the investigation, a number of conclusions can be drawn: Firstly the magnitudes of effective horizontal thrusts and shear forces generated at the interslice when pore water pressure forces induced are taken into consideration are higher than those obtained when these forces are ignored. This demonstrates that the pore water pressures developed have an influence on the values of other interslice forces. Secondly the inclusion of net water pressure forces in the stability analysis of the earth embankments studied clearly show that the action of the water pressure forces serves to promote instability, as would be expected. Thirdly the popular practice amongst geotechnical engineers of resolving the water pressure within a given slice in a direction of normal at the slice base in order to estimate its value, whilst the horizontal effect of the slice base water pressure is taken as zero, constitutes a grave error. This action is not in line with limit equilibrium procedures and yields erroneous results. Fourthly at low phreatic levels the proposed approaches give practically similar factors of safety. This implies that the effect of water pressure forces acting on the interslice can only be neglected when the phreatic line in an embankment is at its lowest stage, a considerable period after drawdown, or preferably between 20% and 30% of the overall height of the embankment. Finally the factors of safety found using Bishop's simplified method and that based on Formulation B, which ignores the pore water pressure forces effect, are in very close agreement. This simply implies that the inclusion of only the effective horizontal thrusts and shear forces acting at the interslice has little influence on the resulting factors of safety.

6. Acknowledgments

The present investigators would wish to thank the management of Konsadem Associates Limited, Ibadan, Nigeria for making readily available the data and other aspects relating to the design of the Okuku earth dam embankment. In addition the assistance received from staff of Ete-Aro and Partners Limited, Ibadan, is gratefully acknowledged. The authors would also wish to place on record the encouragement of staff of the Department of Civil Engineering, University of Ibadan, Nigeria as well as staff of the Department of Civil Engineering, University of Botswana, Gaborone, Botswana.

7. References

Capper, P. & Cassie, W. (1971). *The Mechanics of Engineering Soils,* 5th Edition, E. & F.N. Spon, ISBN 419-10700-2, London, United Kingdom

McCarthy, D. (1998). *Essentials of Soil Mechanics and Foundations - Basic Geotechnics,* 5th Edition, Prentice-Hall, ISBN 0-13-506932-7, New Jersey, USA

Fellenius, W. (1927). *Erdstatische Berechnungen mit Reibung und Kohasion Adhasion und unter Annahme Kreiszylindrischer Gleitflachen,* W. Ernst, Berlin

Bishop, A. (1955). The Use of the Slip Circle in the Stability Analysis of Slopes, *Geotechnique,* Vol.5, No.1, (March 1955), pp 7-17, ISSN 0016-8505

Spencer, E. (1967). A Method of Analysis of the Stability of Embankments Assuming Parallel Interslice Forces, *Geotechnique,* Vol.17, No.1, (March 1967), pp 11-26, ISSN 0016-8505

Morgenstern, N. & Price, V. (1965). The Analysis of the Stability of General Slip Surfaces, *Geotechnique,* Vol.15, No.1, (March 1965), pp 79-93, ISSN 0016-8505

Bell, J. (1968). General Slope Stability Analysis, *J. Soil Mech. & Found. Div., ASCE,* Vol.94, No.6, (June 1968), pp. 253-270, ISSN 04447994

Sarma, S. (1973). Stability Analysis of Embankments and Slopes, *Geotechnique,* Vol.23, No.2, (June 1973), pp. 423-433, ISSN 0016-8505

Cousins, B. (1978). Stability Charts for Simple Earth Slopes, *J. Geotechnical. Eng. Div., ASCE,* Vol.104, No. GT2, (February 1978), pp. 267-279, ISSN 00936405

Janbu, N. (1973). Slope Stability Computations, In: *Embankment Dam Engineering- Casagrande Memorial Volume,* R.C. Hirschfield & S.J. Poulos (Ed.), 47-86, John Wiley & Sons, ISBN 0-471-40050-5, New York, USA

King, G. (1989). Revision of Effective Stress Method of Slices, *Geotechnique,* Vol.39, No.3, (September 1989), pp. 497-502, ISSN 0016-8505

Morrison, I. & Greenwood, J. (1989). Assumptions in Simplified Slope Analysis by the Method of Slices, *Geotechnique,* Vol.39, No.3, (September 1989), pp. 503-509, ISSN 0016-8505

Ayininuola, G. & Franklin, S. (2008). Water Pressure Forces Effect on Earth Embankments Stability, *Global Journal of Engineering and Technology,* Global Research Publication, Vol.1, No.2, (June 2008), pp. 169-188

Ayininuola, G. {1999). *The Effect of Hydrostatic Pore Water Pressure Forces on the Stability of Earth Embankments,* Unpublished M.Sc. Dissertation, University of Ibadan, Nigeria.

Sevaldson, R. (1956). The Slide in Lodalen, Oct. 6th 1954, *Geotechnique,* Vol.6, No.4, (December 1956), pp 167-191, ISSN 0016-8505

Sherard, J., Woodward, R., Gizienski, S. & Clevenger, W. (1978). *Earth-Rock Dams: Engineering Problems of Design and Construction*, J. Wiley and Sons, ISBN 0-471-78547-4, New York, USA

16

Evolutionary Algorithms in Embedded Intelligent Devices Using Satellite Navigation for Railway Transport

Anatoly Levchenkov, Mikhail Gorobetz and Andrew Mor-Yaroslavtsev
Riga Technical University
Latvia

1. Introduction

Nowadays, the most widely spread type of a computer is an embedded system. The embedded systems consist of the following hardware (i.e. nano-electronic components) - programmable microcontrollers or microprocessors; transmitters, including the receivers of global positioning information systems, which demonstrate the state and measuring parameters of a controlled object, and which relay that to the programmable microcontroller; actuators, which receive a signal from the programmed microcontroller and relay it to an antenna, a display or an electro-drive device; and communication devices, including wireless communication with other devices and the software with algorithms of artificial intelligence (Russel, Norvig, 2006).

Railway traffic flow is limited by safety criteria. Therefore, routing and scheduling task is actual for a railway transport. As well an optimal braking control and safety of braking process are very important (Luo, Zeng, 2009). The analysis of human behavior and simulation of train's braking are investigated (Hasegawa et al., 2009). An intelligent transport control system gives a possibility to make traffic control safer and more cost-effective (Gorobetz, 2008). It may find an optimal solution to a conflict faster than a human as a decision support system (Levchenkov et al., 2009). In addition in case of emergency it may prevent crashes and accidents without human intervention.

Authors propose the intelligent braking control device, which warns the driver about the necessity of starting the working braking, taking into account the signal of the traffic light. If working braking has not been started the controller activates emergency brakes with a purpose to stop before the beginning of the next block-section if it is possible, or to choose a free way with enough free distance to stop without a crash. The primary task of the diagnostics device is to separate dangerous situations in braking system by critical values of sensor measurements from the regular states of the system, to detect and to warn about changes in the system and to prevent emergencies immediately. The system allows stopping the train timely before the problem has occurred.

2. Problem formulation

Railway safety is an actual and important task. Nowadays a human factor is the main reason for 74% of railway accidents and crashes. This problem is actual as in Latvia as all over the world. Various crashes like in Riga (Latvia) in February 2005, in Ventspils (Latvia) in December 2008, in Aegvidu (Estonia) in December 2010, in Brussels (Belgium) in February 2010, in Magdeburg (Germany) in January 2011 prove the necessity of finding the problem solution.

The main reason of railway accidents is a human factor, when

- train driver does not stop a train on the restrictive signal
- incorrect decisions were made by signalmen (station-on-duty) in extraordinary situations

Therefore, auxiliary embedded electronic devices are needed

- to help a train driver and station signalmen to make the best decision faster
- to prevent accidents if a human does not react

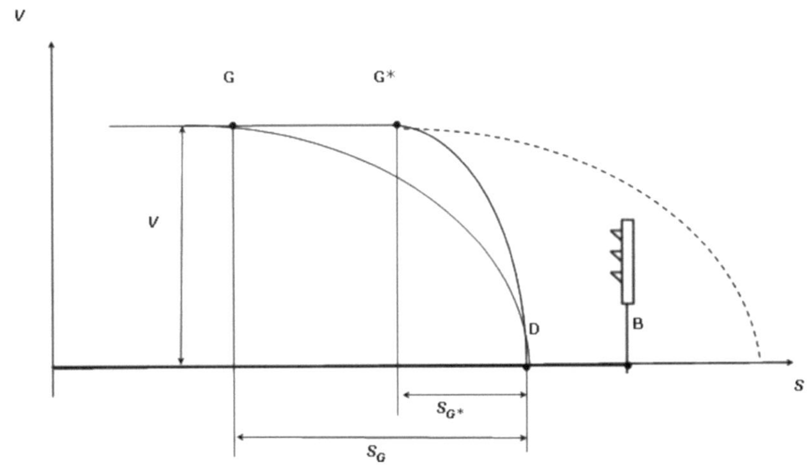

Fig. 1. Graphical representation of the task.

The primary task is to stop the train before the restrictive signal. For this task a warning point G should be detected, where regular braking still may be applied (Fig. 1.). The point G* is a marginal point and also must be defined, where only emergency braking may be applied to stop the train before the signal. Using of regular braking in G* follows the passing of the restrictive signal.

The purpose of the research is to develop a prototype of a new control command and a signalling track-side and on-board locomotive's devices to improve train movement safety.

The following functions of the on-board locomotive's embedded device are required:

- receiving location and other necessary data from satellite navigation system
- reception of the data from the server about railway infrastructure (tracks, points) and control points (signals, section points) in the location of the train;
- selecting the necessary signal on the way;
- establishing the wireless connection with the selected signal (Fig. 2.)

- receiving the data from a track-side signal
- checking if the train has reached the route control point;
- detecting the acceleration of the train
- detecting the regular braking distance;
- detecting the emergency braking distance;
- detecting the speed limitation;
- detecting the starting point of regular braking;
- starting emergency braking process
- a module for checking the reaching of the starting point of regular braking;
- warning about the necessity to start regular braking;

The following functions of the track-side embedded device are required:

- establishing the wireless connection with the train
- transmitting signal value to the locomotive's device
- detection of the safest state of the station points using evolutionary algorithms to avoid the collision (Fig. 3.).

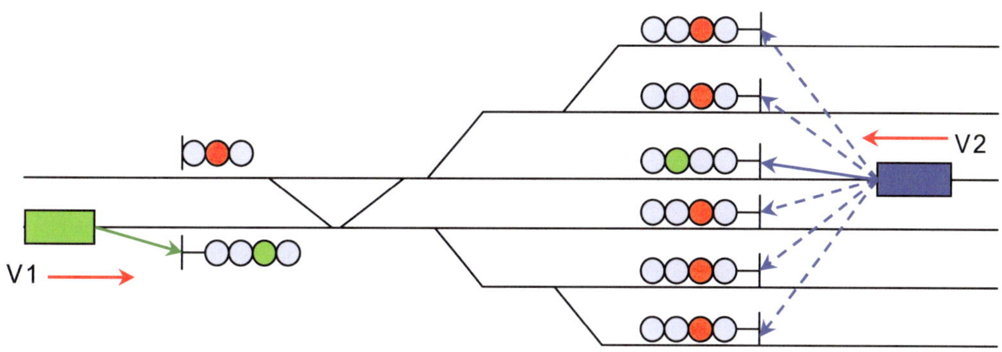

Fig. 2. Selection of the necessary signal on the way.

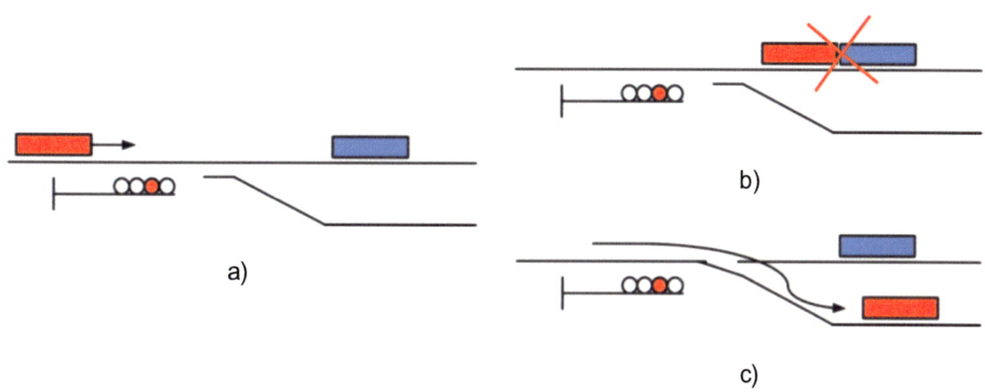

Fig. 3. Detection of the safest state of the station points to avoid the collision.
a) Initial state, b) Collision is possible, c) Collision is avoided.

There are many embedded transport control systems on the market which are designed to provide safety for a vehicle, its passengers or cargo, and other traffic participants.

In the commercial railway transport segment an example of such a safety system is KLUB-U, currently used on Russian Railways. It is installed in the locomotives and by interacting with existing signaling systems and its own modules provides information about the train's and its closest neighbors' coordinates, diagnostics of the brakes, current railway segment profile and maximum allowed speed, and controls the vigilance of the locomotive driver. Still, despite the wide array of features, it lacks automation and many decisions require manual operation.

A significant component of the whole safety system is the circuits, engines and brakes diagnostics complex. While the most complete diagnostics can be performed only in the technical service environment, most failures can be detected during its operation using circuit integrity indicators and different sensors designed to uncover electrical mechanical damage.

All kinds of damages which could lead to failures can be combined into distinctive value sets, thus recognizing them in the stream of incoming data allows early identification of problems in the engine.

Artificial immune systems (AIS) were mentioned in some papers in mid 1980s but became a subject in its own right in 1994 in the papers on negative selection (Forrest et al., 1994, Kephart, 1994). Currently the systems are actively explored for possible use cases. For example, there are studies on a real-valued negative selection algorithm for an aircraft fault detection (Dasgupta et al., 2004).

3. Structure of proposed system for railway safety tasks

The chapter will demonstrate some issues of design and modelling of a part of a modern embedded system for a rail transport (Fig. 4.). This embedded system is intended for managing the rail transport's electrical drive and the traffic lights, and it consists of the microcontrollers, the developed software and the information system, the wireless communication possibilities and the global positioning system.

Fig. 4 illustrates a complete scheme of a structure of the rail transport's embedded system. In the figure, the brown colour shows the train's embedded devices; the green colour stands for the devices of the traffic lights' embedded system; the purple colour - the devices of the crossing's embedded system; the yellow colour - coordination embedded system devices; the light blue colour - software which provides operation of the network of the wireless embedded system, operation of the communication network of the wireless devices and which is responsible for making operational decisions.

The device receives the signal from the defined traffic light and defines its position using a wireless communication network and has data storage with route control points as well.

Artificial immune systems use evolutionary data processing paradigm based on biological immune systems. It differs from computational immunology which models biological immune systems. Immune algorithms are mainly used to solve anomaly recognition, data collection and analysis tasks. From the computational point of view the most interesting features of the immune systems are self-learning, diversity maintenance and memory.

The problem is represented as an antigen and a solution candidates as antibodies which are randomly generated from the library of available solutions or genes. The evaluation of affinity or degree of binding between the antigen and the antibody is similar to complementarity level in biological IS and it defines the fate of each individual antibody as well as the termination of the whole algorithm. Individual antibodies are replaced, cloned and hypermutated until a satisfactory level of affinity is reached. Partial replacement of the solutions' population with fresh randomly generated candidates maintains diversity which allows solving a wider set of problems. The probability of cloning or hypermutating a candidate depends on its affinity.

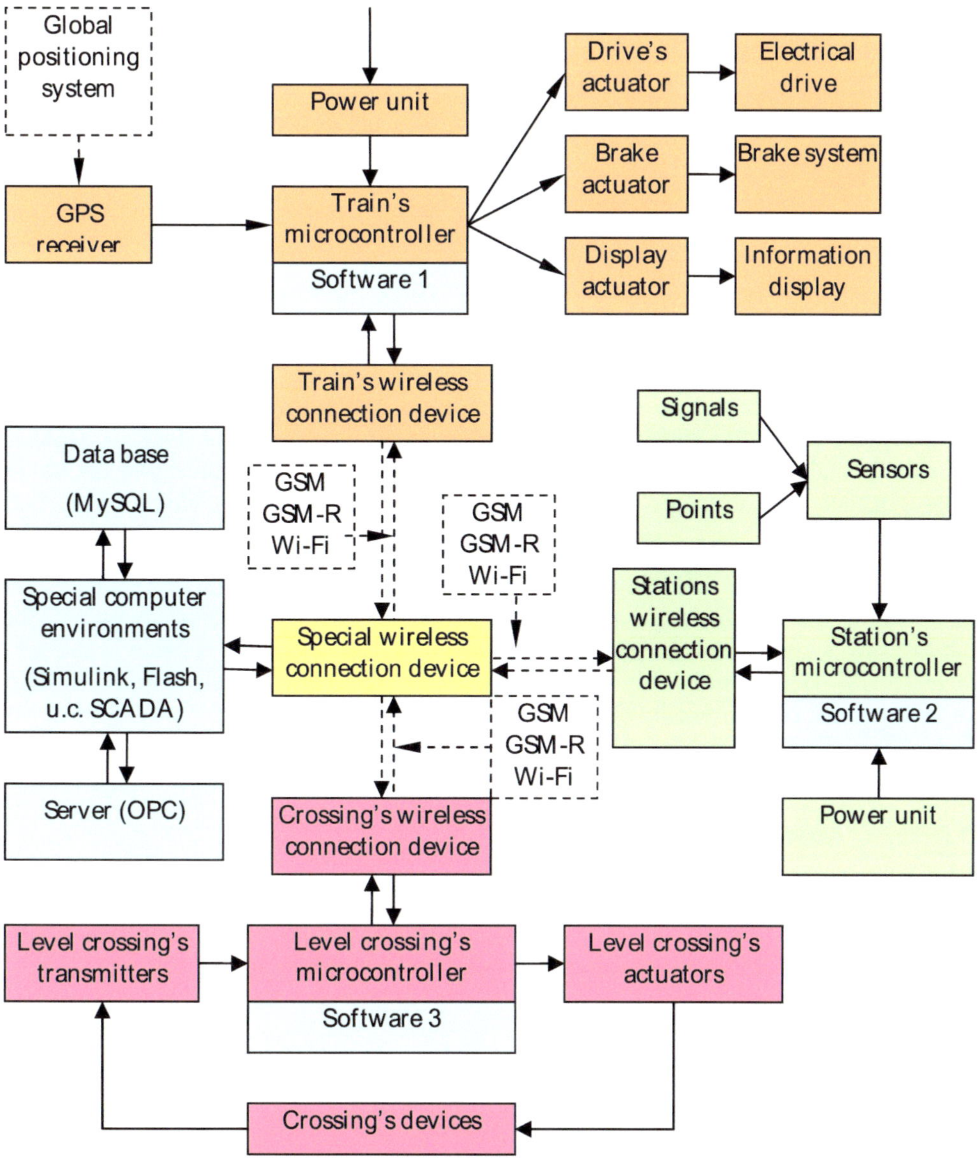

Fig. 4. Wireless network structure's scheme of embedded devices for a rail transport

In the rolling stock safety system (Fig. 5) (Mor-Yaroslavtsev, Levchenkov, 2011), the invading object *I* is picked up by sensors *S* and the data is transmitted to the nearest cell tower *CT*, which relays it to the control center *CC* and the nearest locomotives wireless modems *M*. Through the same modem the locomotive *L* receives data about the closest neighbors' rolling stock position and status, railway segment profile and maximum allowed speed.

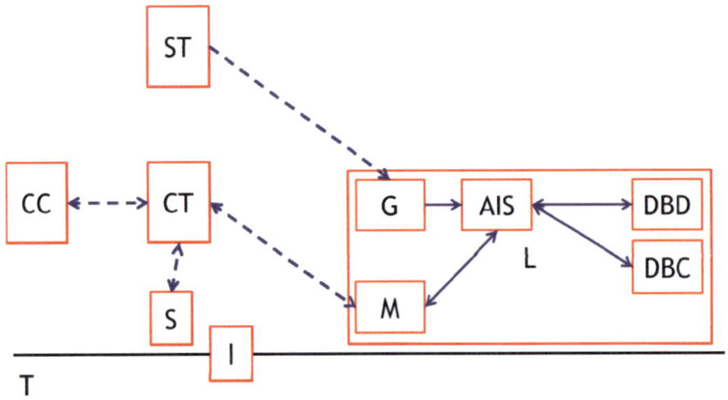

Fig. 5. The intelligent rolling stock safety system functional design

L also hosts: a positioning receiver *G* which receives data from a positioning satellite *ST*; data analysis module *AIS* which communicates to the immune detector database *DBD* and control cell database *DBC*. Depending on the results of control cell maturation the module makes a decision and executes it by sending a control signal or displaying an alert to the driver.

Analogous to the hybrid IDS (Powers, He, 2009) the most feasible way to implement such a system would be through the two phases of anomaly detection and determination of their type to draw a conclusion. In this case the incoming data from the sensors is the set of antigens. The data includes but is not limited to speed, acceleration, voltage, rotation, temperature, and presence of smoke.

4. Mathematical models for problem solution

4.1 Model of differential positioning system

Differential satellite navigation systems are used to increase precision of the positioning systems that is very significant for safety-critical systems, such as transport.

Differential satellite navigation systems (Fig. 6.) contain the following object types:

- S - satellite;
- M - base station;
- R - receiver.

Each satellite S is described by the following parameters:

- α - slope of the satellite's orbit to the equator plane;
- Ω - slope of up-going node of the satellite's orbit to the Greenwich meridian;
- ω - perigee angle from up-going node
- t - time of crossing perigee or up-going node of the satellite's orbit;

- e – eccentricity of the satellite's orbit:

$$e = \sqrt{1-(b/a)^2}\ , \tag{1}$$

where a and b are half-axes of the ellipsoid orbit;

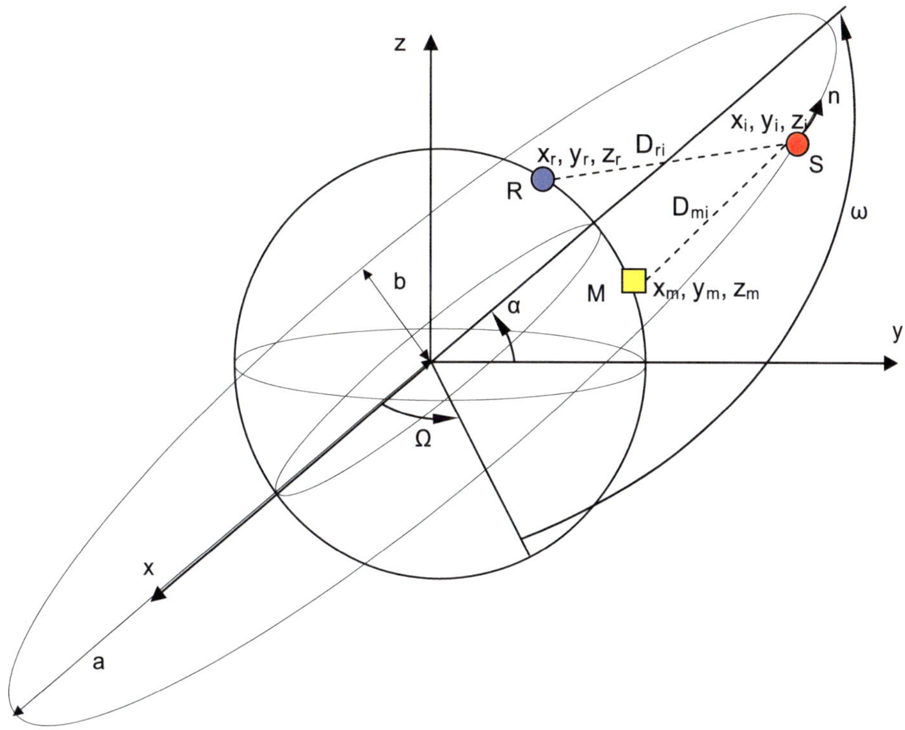

Fig. 6. Differential satellite navigation system elements

- n – angular velocity of the satellite,

$$n = \sqrt{\mu / a^3}\ , \tag{2}$$

where μ – Earth gravimetric constant.

- x, y, z – coordinates of the satellite

Base station M parameters:

- x_m, y_m, z_m – coordinates of the base station
- D_{mi} – known distance between i-th satellite and base station m:

$$D_{mi} = \sqrt{(x_i - x_m)^2 + (y_i - y_m)^2 + (z_i - z_m)^2} \tag{3}$$

- ρ_{mi}, $\Delta\rho_{mi}$ - distance measurement result and necessary correction between measured and real distance:

$$\Delta\rho_{mi} = \rho_{mi} - D_{mi} = \varepsilon_{m,sat} + \varepsilon_{m,con} + \varepsilon_{m,rec} + c \cdot \delta t_m , \tag{4}$$

where $\varepsilon_{m,sat}$ - satellite apparatus error, satellite clock error, $\varepsilon_{m,con}$ - control error, incorrect ephemerid forecast, $\varepsilon_{m,rec}$ - receiver's error, ionosphere, troposphere and other noises, δt_m - base station clock deviation from satellite clock, c - light speed.

- ρ_{ri} - corrected distance measurement between recipient and satellite:

$$\begin{aligned}\rho_{ri} &= D_{ri} + \varepsilon_{r,sat} + \varepsilon_{r,con} + \varepsilon_{r,rec} + c \cdot \delta t_m - \Delta\rho_{mi} = D_{ri} + \varepsilon_r + c \cdot \delta t_{mr} = \\ &\sqrt{(x_i - x_r)^2 + (y_i - y_r)^2 + (z_i - z_r)^2} + \varepsilon_r + c \cdot \delta t_{mr}\end{aligned} , \tag{5}$$

where ε_r – receiver's result segment error, δt_{mr} – combined clock deviation,

x_r, y_r, z_r – coordinates of the receiver.

4.2 Model of railway station

The model of the station may be described with the following sets of objects:

- V = {v1, v2, ..., vn } – set of trains;
- M = {m1,m2,…,mn} – set of train goals;
- L = {l1, l2, ..., lk } – set of signals;
- P = {p1, p2, ..., pq } – set of points;
- R = {r1, r2, ..., rw } – set of routes.

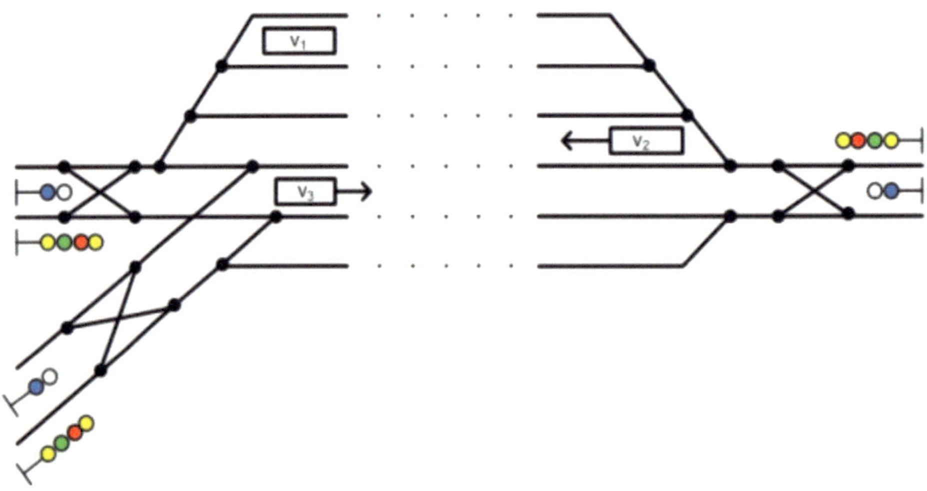

Fig. 7. Graphical interpretation of station model

4.3 Model of braking of rolling stock

The braking way consists of preparation and real segments:

$$S_T = S_P + S_D ; \tag{6}$$

where S_T is the braking distance, S_p – distance of moving during the preparation of braking system, S_D – real braking distance

The braking power of the train should be defined taking into account a real force of the braking chock influencing the train wheels. A real friction factor depends on the braking chock material.

The following factor characterises the braking chock made of cast iron:

$$\varphi_K = 0{,}6\frac{16K+100}{80K+100}\cdot\frac{v+100}{5v+100}; \qquad (7)$$

The cast iron braking chocks containing phosphorus of 1,0-1,4% are characterised with the factor

The mentioned above factor for braking chocks of composite materials can be defined with the following expression :

$$\varphi_K = 0{,}44\frac{K+20}{4K+100}\cdot\frac{v+100}{2v+100}; \qquad (8)$$

The calculations of braking force of the chocks also depend on a type of chocks.

For the standard cast iron braking chocks:

$$K_P = 2{,}22K\frac{16K+100}{80K+100}; \qquad (9)$$

The cast iron- phosphorus braking chocks:

$$K_P = 1{,}85K\frac{16K+100}{52K+100}; \qquad (10)$$

For the braking chocks of composite materials:

$$K_P = 1{,}22K\frac{K+20}{4K+20}; \qquad (11)$$

The total braking factor:

$$\vartheta_{PO} = \frac{\sum K_P}{Q+P_u}; \qquad (12)$$

The main force resistive to the motion in idle running

$$W_{Ox} = W_{O1} + N\cdot W_{Oc}; \qquad (13)$$

where N - quantity of the carriages.

Locomotive without the train:

$$W_{01} = 24 + 0{,}11v + 0{,}0035v^2; \tag{14}$$

Cargo carriages:

$$w_{0c} = 7 + \frac{a + v + 0{,}025v^2}{q_0}; \tag{15}$$

Passenger trains:

$$w_{0c} = 12 + 0{,}12v + 0{,}002v^2 . \tag{16}$$

As within the time interval Δt the braking force and the opposite self-resistive force ω_{Ox} to the motion of the train are assumed as constant values then the increasing of the speed can be calculated according to :

$$\Delta v = \frac{\xi(b_T + \omega_{Ox} + i_c)\Delta t}{3600}. \tag{17}$$

The speed of the braking force distribution is a braking wave: $v_t = \frac{L}{t_t}$; where: L is the length of the train; t_t - time from the moment when the driver turns the handle of the hoist till the pressure appears (?) in the braking cylinders; air wave : $v_v - 20\sqrt{T}$; where: $T = 273 + t^{\circ}C$ - absolute temperature of gas.

The preparation braking distance:

$$S_P = 0{,}278 v_0 t_p = \frac{v_0 t_n}{3{,}6}. \tag{18}$$

Real braking distance:

$$S_D = \sum \frac{500\left(v_N^2 - v_K^2\right)}{\zeta\left(w_{ox} + b_m + i_C\right)}; \tag{19}$$

Thus the total braking way:

$$S_T = \frac{v_0 t_n}{3{,}6} + \sum \frac{500\left(v_N^2 - v_K^2\right)}{\zeta\left(w_{ox} + b_m + i_C\right)} \tag{20}$$

4.4 Model of railway infrastructure and command and control system

Rail ways can be represented as a graph $R = \{C,S\}$, where rails are divided into sections S, and each section $s \in S$ is connected with each other by two connectors $s = < c_i, c_j >$.

Each section $s \in S$ has a constant length l_s, a curve a_s, and a speed limit v^*_s.

Each point $p \in C$ has a connecting set W of three or more sections and a set of possible states of point D_p, where $d_p^n = < s_i, s_j >$ means opened in both directions from s_i to s_j and from s_j to s_i is following for different point types:

single point: $D_p = \{< s_i, s_j >, < s_i, s_k >\}$;

dual point: $D_p = \{< s_i, s_j >, < s_i, s_k >, < s_i, s_m >\}$;

cross point: $D_p = \{< s_i, s_j >, < s_i, s_k >, < s_m, s_j >, < s_m, s_k >\}$.

Each state of point $d_p \in D_p$ has a speed limit $v^*_{d_p}$; maximal each point's $d_p \in D_p$ switching time: t_{d_p}.

Railway signal G is an object with fixed coordinates x_0, y_0 connected to the fixed position on the track.

Each signal $g \in G$ has the following states of signals $L_g \subseteq \{R, Y, YG, G, V, W\}$, where "R" - red, and rolling stock must stop before the signal; "Y" - yellow, can move and be ready to stop, the next signal is red; "YG" - yellow and green, next two sections are free; "G" - green, "V" - violet, "W" - moonlight white.

Each signal sets up speed limits for the next block-section: v_{def} - maximal predefined speed on the section, v_0 - 0 kmh, stop; v_1 - < 50 kmh, movement on turnouts 1/9 and 1/11 types; v_2 - < 80 kmh for movement on turnout 1/18 type; v_3 - < 120kmh for movement on turnout 1/22 type.

4.5 Assessment functions

Multi-criteria target function for braking:

$$\begin{cases} F^{br}(DL, CL, EL, \Delta) \to \min \\ DL = \Delta S \to 0 \\ CL = \varepsilon(t) \to \varepsilon^* \\ EL = \dfrac{d\varepsilon}{dt} = const \\ \Delta \to \{\vec{0}, C_1, C_2, \ldots\} \\ \Delta = |\Phi(t) - \mathrm{E}(t)| \\ \mathrm{E}(t) = \{\vartheta_p, S_B, \varepsilon_i, t_m, \tau, I_{dcp}, Q_v\} \\ \Phi(t) = \{\vartheta^f, S_B^f, \varepsilon_i^f, t_m^f, \tau^f, I_{dcp}^f, Q_v^f\} \end{cases} \tag{21}$$

- ΔS – distance between the closed section and rolling stock – danger level criteria (DL)
- $\varepsilon(t)$ – deceleration of rolling stock
- ε^* – optimal deceleration for passengers – comfort level criteria (CL)
- $d\varepsilon / dt$ – changes of deceleration and braking torque – optimal energy consumption criteria (EL)

- Fbr – function for braking process optimization
- Δ – difference/similarity vector, which defines the difference between the estimated and the actual values.
- Ō – Zero vector, which means a compliance with the normal situation, where the estimates and the actual value of the difference is zero;
- Ci – situation of danger classes, according to differences between estimated and the actual values;
- E(t) – vector of estimated values;
- Φ (t) – vector of actual values;
- f – index, which represents the actual values.

The routing task for accident prevention consists of a generation of a new route and schedule for rolling stocks V moving on points P.

The target function for scheduling and routing is to arrange points for each train to reach a destination and assigning of time moments t to each train and each point.

- Train's schedule: $\sigma_v : P \rightarrow \{t_{v1}, t_{v2}, ..., t_{vs}\} \subset \Re$
- Point's schedule: $\sigma_{p^1} : V \rightarrow \{t_{p1}, t_{p2}, ..., t_{pm}\} \subset \Re$

The target function for an optimal point state on the station is the following:

$$T_{\Sigma} = f(t_1, t_2, ..., t_n; \quad x_{11}, x_{12}, ..., x_{1q}; \quad ... \quad x_{n1}, x_{n2}, ..., x_{nq}) \rightarrow \min \tag{22}$$

t_i – the i-th time moment of switching points
x_{ij} – state of the j-th point in the i-th time moment

5. Evolutionary algorithms for problem solution

5.1 Fitness function for genetic algorithm

Step 0. Initialization

$T_{i\,sum} = 0$ – for each i-th train summary time

$G_i = 0$, for each i-th train goal achievement

i = 1 - selected train number

Step 1. Moving the time calculation of the i-th train on the j-th railway section, $t_j^i = S_j / v_i$ where

S_j – length of the j-th section,

v_i – current i-th train speed

Step 2. Check if the j-th section ends with signal.

Step 2.1. Check the occupancy of all tracks to the next signal.

Step 2.2. If any of sections in Step 2.1 is busy and the train is moving, then recalculate time with braking conditions and Goto 4.

$$S_b^i = \frac{-v_i^2}{2\varepsilon};\quad S_r = S_j - S_b;\quad t_r = \frac{S_r}{v_i};\quad t_b = \frac{v}{\varepsilon};\quad t_j^i = t_r + t_b;\quad v_i = 0$$

Step 2.3. If train is already stopped then check If all sections in Step 2.1 are free, then recalculate time with acceleration conditions

Step 2.4. j = j + 1, select next section of the route

else if the j-th section is not a signal then j = j + 1, select the next section of the route

Step 3. Check if the train reached the goal, then Gi = 1

Step 3.1. Check if all trains reached the goal then $T_\Sigma = \max(T_{SUM}^1, T_{SUM}^2, ..., T_{SUM}^n)$

Step 3.2. Function STOPS

Step 4. Update summary time of i-th train $T_{SUM}^i = T_{SUM}^i + t_j^i$

Step 5. Check if the summary time is less than the next point switching time $T_{SUM}^i \le t_p$?

Step 5.1. If $T_{SUM}^i \ge t_p$ then $T_{SUM}^k = \min(T_{SUM}^1, T_{SUM}^2, ..., T_{SUM}^n)$

Step 5.2 If *k* = *i* then Goto 6, else *i* = *k*, Goto 1

Step 6. Check occupancy of tracks in switching moment

Step 6.1. Recalculate position of the train $S_{beg}^i = v_i \cdot (T_{SUM}^i - t_j^i - t_p)$

Step 6.2. Find "tail" of the train $S_{end}^i = S_{beg} - L_i$

Step 6.3. Check the occupancy of all tracks.

If switching points are busy, then $T_\Sigma = \infty$, algorithm ENDS; else Goto 1

5.2 Genetic algorithm

A genetic algorithm for the task solution may be described with the following steps.

1 step: Initialize random set of possible solutions: $S^{(0)} = \{s_1^{(0)}, s_2^{(0)}, ..., s_{s_{\max}}^{(0)}\}$;

2 step: Evaluate each solution with a target function: $V^S = \{F(s_1), F(s_2), ..., F(s_p)\}$;

3 step: Arrange solutions by evaluation: $\overline{S} = \{\overline{s}_1, \overline{s}_2, ..., \overline{s}_p\}$, $F(\overline{s}_1) = opt(V^S)$;

4 step: Duplicate the best solutions in the elite set: $S_E \subset \overline{S}$;

5 step: **Selection.** Select from the set of solution pairs according to the defined selection: $S_C = \overline{S}$;

6 step: **Crossover:** Generate a new population from the set of the solution pair according to the defined crossover algorithm:

$$\bar{s}_i \Pi \bar{s}_j \rightarrow s'_i = s_{ij};\ s'_j = s_{ji},\quad i,j=\overline{1,p}; \tag{23}$$

7 step: **Mutation:** Random change of one of solution parameter that helps to find a global optimum of the function:

$$x_j^{s'_i} = x_j^{s'_i} + 1,\quad s'_i \in S',\quad j = rand(\overline{1,k}),\quad i = rand(\overline{1,p}); \tag{24}$$

8 step: Evaluate the new population using the target function:

$$V^{S'} = \{F(s'_1), F(s'_2),\dots, F(s'_p)\}; \tag{25}$$

9 step: Arrange the new population by the evaluation values:

$$\bar{S}' = \{\bar{s}'_1, \bar{s}'_2,\dots,\bar{s}'_p\},\quad F(\bar{s}'_1) = opt(V^{S'}); \tag{26}$$

10 step: Add the new population of solution to the elite set: $S = (S_E + \bar{S}')$;

11 step: Delete the last solutions from the population S if its size exceeds predefined population size p: $S = S / \{s_{p+1}, s_{p+2|},\dots\}$;

12 step: Algorithm stops by time, generation, convergence or by another predefined criteria. If stop criteria is false then repeat the algorithm from step 4. If true then the result of the algorithm is solution s_1.

5.3 Algorithm for an artificial immune system

5.3.1 Shape-space concept

AIS are modeled after biological IS and carry the terms of antigens and antibodies. They can be modeled using the shape-space concept (see Fig. 8.) (Musilek et al., 2009). The shape-space S allows defining antigens, receptors and their interactions in a quantitative way.

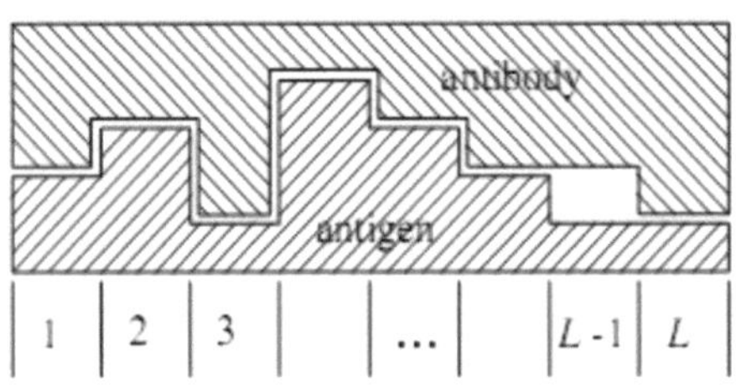

Fig. 8. A shape-space model of an antigen and an antibody.

Like chromosomes in the in the evolutionary algorithms, depending on a problem being solved it also could be a set of integers or binary numbers - $m \in Z^L$ or $m \in \{0,1\}^L$.

The affinity of an antigen-antibody pair is related to their $\in$ distance in the shape-space S and can be estimated using any distance measure between the two attribute strings. The distance between an antigen, *Ag*, and an antibody, *Ab*, can be defined, for example, using a general class of Minkowski distance measures:

$$D_M(Ag, Ab) = \sqrt[p]{\sum_{i=1}^{L} | Ag_i - Ab_i |^p}$$

By varying the value of the parameter p a suitable measure of distance can be obtained.

5.3.2 The negative selection algorithm

Negative selection is the paradigm describing the evolution of the T-lymphocytes where they are randomly generated and learn to recognize all except the self structures, specific to the host. Negative selection algorithms need training samples only from one class (self, normal), thus, they are especially suited for the tasks such as novelty, anomaly or change detection including those in engines and other devices.

The key advantage of anomaly detection systems is their ability to detect novel attack patterns for which no signature exists, while their most notable disadvantage is a larger false positive rate.

The algorithm:

Step 1. Define a set S which needs to be monitored and the set P of the know self $m = \langle m_1, m_2, \ldots, m_L \rangle \in S^L$ elements in a feature space U. The set U corresponds to all the possible system states, P - normal states and S - the current state which changes in time.

Step 2. Generate a set of candidate detectors $C = \{c_1, c_2, \ldots, c_n\}$.

Step 3. Compare each candidate c_i to the set of known good elements P.

Step 4. If a match occurs, discard the individual c_i, otherwise store it in the mature detector set D. Or, to maximize the nonself space coverage with a minimum number of detectors, move the matched candidate away from the closest element p_j, then store it in D.

Step 5. Monitor S for changes by continually matching it against the detectors in D. If any detector matches, the change which has occurred most likely is dangerous, as D is designed not to match any normal system state.

This algorithm produces a set of the detectors capable to recognize non-self patterns. The action following the recognition varies according to the problem under consideration. In the case of transport safety control system it could be an alarm or issue of an immediate stop signal depending on the detected fault.

The detectors and the caught fault conditions are stored in an immune memory for further processing and to provide further information about the consequences of the attack and possible future actions instead of simply reporting the incidents.

6. Computer experiments

6.1 Computer experiment of genetic algorithm

The task of the experiment is to minimize idle time of trains on the station and to minimize the risk of their collision.

The station (Fig. 9.) with 4 points p_1, p_2, p_3, p_4 is given and two trains V1 and V2 are approaching. Railway tracks of the station are split into the sections, where start and end of each section is a point or a signal. The length of the trains L_{v1} = 500 m and L_{v2} = 300 m is given.

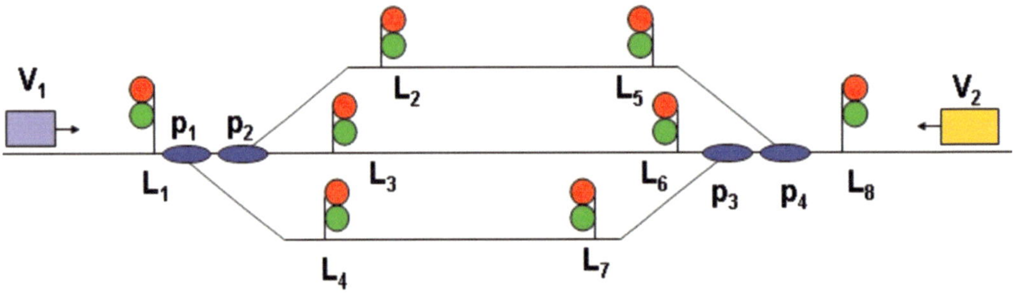

Fig. 9. Structure of the station for the computer experiment

Generation 1

0) 001100001001001000010100010 0

f(x) = 610.47947316824; T1 = 379, T2 = 564 Points states: 0 1 0 0 | 0 1 0 0

1) 00110000100100100001010000 10

f(x) = 610.47947316824; T1 = 379, T2 = 564 Points states: 0 1 0 0 | 0 0 1 0

2) 011001000000110100010100010 0

f(x) = 639.47947316824; T1 = 781, T2 = 408 Points states: 0 1 0 0 | 0 1 0 0

3) 001110100101110100010100111 1

f(x) = 686.47947316824; T1 = 455, T2 = 908 Points states: 0 1 0 0 | 1 1 1 1

4) 110011110111100011010100010 0

f(x) = 850.47947316824; T1 = 619, T2 = 775 Points states: 0 1 0 0 | 0 1 0 0

5) 010111010111101010010100010 0

f(x) = 960.47947316824; T1 = 729, T2 = 830 Points states: 0 1 0 0 | 0 1 0 0

6) 110110101100000111000101110 1

f(x) = 100000; T1 = 709, T2 = 55 Points states: 0 1 0 1 | 1 1 0 1

7) 011011010100001101101010000 0

f(x) = 100000; T1 = 854, T2 = 105 Points states: 1 0 1 0 | 0 0 0 0

8) 110011000010101011000110010 1

f(x) = 100000; T1 = 594, T2 = 336 Points states: 0 1 1 0 | 0 1 0 1

a)

Generation 12

0) 1000110001001101000001000100

f(x) = 270; T1 = 96, T2 = 406 Points states: 0 1 0 0 | 0 1 0 0

1) 1000110000000101001001000100

f(x) = 270; T1 = 94, T2 = 160 Points states: 0 1 0 0 | 0 1 0 0

2) 1000111000000101000001001111

f(x) = 270; T1 = 109, T2 = 156 Points states: 0 1 0 0 | 1 1 1 1

3) 1000110010000101010101000110

f(x) = 270; T1 = 98, T2 = 166 Points states: 0 1 0 0 | 0 1 1 0

4) 1000110000000101010101001000100

f(x) = 270; T1 = 94, T2 = 332 Points states: 0 1 0 0 | 0 1 0 0

5) 0001000010011010000101000110

f(x) = 270; T1 = 129, T2 = 814 Points states: 0 1 0 0 | 0 1 1 0

6) 1000110010000101010101000010

f(x) = 270; T1 = 98, T2 = 166 Points states: 0 1 0 0 | 0 0 1 0

7) 1000110001001101001001000100

f(x) = 270; T1 = 96, T2 = 410 Points states: 0 1 0 0 | 0 1 0 0

8) 1000110111000101010001000100

f(x) = 270; T1 = 107, T2 = 164 Points states: 0 1 0 0 | 0 1 0 0

b)

Fig. 10. Results of Genetic Algorithm - a) first iteration, b) last iteration

The fitness function and the algorithm are realised in the program and the following parameters for genetic algorithm are used:

- Crossover rate - 0.8;
- Mutation rate - 0.01;
- Population size - 50
- Random parent selection
- Single point crossover

Table 1. shows the dynamics of genetic algorithm. The algorithm is performed in 2 seconds and the algorithm converges completely in the 12th generation, where an average value of the population is equal with the best value.

Generation	Average	Best
1	88087.15753678	610.47947316824
2	82128.38630517	413.47947316824
3	70197.754252487	270
4	50320.950968194	270
5	32391.38814628	270
6	26405.504351969	270
7	14391.055583579	270
8	2366.4284573351	270
9	2332.525484097	270
10	329.57671570692	270
11	307.08794731682	270
12	270	270

Table 1. Average and best values of fitness function on each generation

6.2 Computer experiment for an artificial immune system

6.2.1 Collecting the location data

One data set for the experiment was taken from the two PLCs in the field attached to a vehicle and a level crossing. The data collection scheme is presented in Fig. 11.

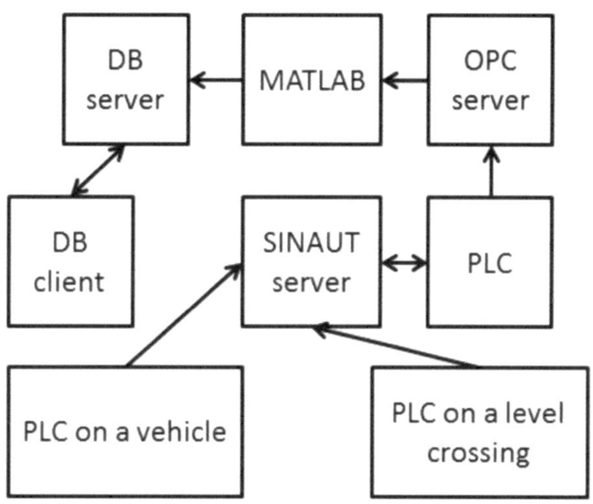

Fig. 11. The location data collection scheme.

The communication between the PLCs is facilitated by GPRS modules and a server running on a PC. Through the chain of software tools the data is piped from the PLCs to DB tables.

The data was collected into two tables for records related to a vehicle and a level crossing. The report (Fig. 4) contains this data cross-matched using the date and time – as all the data was simultaneously recorded with discrete steps of 1 s, the matches are 1:1.

This provides a set of data for test runs of the algorithms.

6.2.2 The real-value negative selection algorithm

The RNS detector generation starts with a population of candidate detectors, which are then matured through an iterative process. In particular, the center of each detector is chosen at random and the radius is a variable parameter which determines the size of the detector in m-dimensional space. The basic algorithmic steps of the generation algorithm are given in 5.3.2.

The whole detector generation process terminates when a set of mature (minimum overlapping) detectors are evolved which can provide significant coverage of the non-self space.

A detector is defined as $d = (c, r_d)$, where $c = (c_1, c_2, \ldots, c_m)$ is an m-dimensional point that corresponds to the center of a hypersphere with r_d as its radius. The following parameters are used (Fig. 12):

- r_s: threshold variation of a self point;
- a: variable movement of a detector away from a self sample or existing detectors;
- ξ: maximum allowable overlap among the detectors, allowing some overlap can reduce holes in the non-self coverage.

Settings:

- Maximum self-element variation: 0.2
- Maximum detector overlap: 0.1
- Dimensions (sensors): 4
- Maximum detector population: 10
- Number of tests: 20
- Next generation after 5 tests
- Number of top detectors to clone: 2

Problem space

Dimension	Minimum	Maximum
1. Train speed, km/h	0	100
2. Crossing car speed, km/h	0	150
3. Distance from the train to the rendezvous, m	-100	1000
4. Distance from the car to the rendezvous, m	-10	1000

Self elements

- [0, 5, 1000, 0]
- [50, 30, 900, 10]
- [20, 30, 600, 2]

Fig. 12. A screenshot from the computer program running a real-valued negative selection algorithm showing the initial settings for training the detector set.

During the straightforward detection process the matured detectors are continually compared to new test data samples. The distance D between a sample pattern $p = (c_p, r_s)$ and a detector $d = (c_d, r_d)$ is computed in the same way as in the detector generation phase. If $D < (r_s + r_d)$ then the detector d gets activated indicating possible fault.

#	Centre	Radius	Overlap	Over max	Score
1	[45.66, 40.56, 810.52, 663.2]	659.19916560457	11.712630436924	yes	4
2	[3.5377912542611, 122.88340464607, 832.85177432791, 102.2142227637]	154.21666004677	6.3919845832792	yes	0
3	[2.5695947048491, 55.544483105892, 626.76977100429, 818.4035867151]	816.98189370543	6.3486551611765	yes	5
4	[27.187694470002, 7.383584927742, 834.12094957612, 165.16710031129]	171.35523483706	1.6998396082232	yes	0
5	[2.4970453904292, 107.56920206545, 147.79921302734, 600.96580139848]	754.27876623534	1.4457514612364	yes	9
6	[-333.39525695325, -33.332929255992, -551.13755538503, 1176.2567277667]	636.717573631	1.0569522747208	yes	8
7	[41.257027357444, 43.532006533725, 972.66947431812, 415.34266049214]	412.06540262312	12.106804100201	yes	2
8	[-246.31283775086, 22.874262963601, -500.71681310562, 933.80196469498]	665.99707917402	0.63332636294142	yes	5
9	[89.731672862793, 90.455627923196, 650.94499861178, 713.53201403252]	719.01647151605	0.8099069742284	yes	9
10	[94.182276086748, 244.90184263729, -312.34914808896, -15.984607496821]	496.19048469929	6.9579249445827	yes	4

Fig. 13. The first generation of detectors with unscaled values.

#	Centre	Radius	Overlap	Over max	Score
1	[-333.39525695325, -33.332929255992, -551.13755538503, 1176.2567277667]	754.27876623534	0.16176065722673	yes	3
2	[-246.31283775086, 22.874262963601, -500.71681310562, 933.80196469498]	496.19048469929	1.8652038367886	yes	2
3	[94.182276086748, 244.90184263729, -312.34914808896, -15.984607496821]	496.19048469929	6.9579249445827	yes	4
10	[-246.31283775086, 22.874262963601, -500.71681310562, 933.80196469498]	754.27876623534	0	no	0
9	[-333.39525695325, -33.332929255992, -551.13755538503, 1176.2567277667]	496.19048469929	0	no	0

Fig. 14. The fourth generation of detectors after running several suppressions.

#	Antigen	Result	Generation #
1	[60.06, 126.02, 978.89, 427.11]	Alarm!	1
2	[94.81, 125.18, 859.17, 743.68]	Alarm!	1
3	[52.6, 107.84, 421.43, 204.17]	Alarm!	1
4	[42.88, 9.89, 700.14, 28.52]	Alarm!	1
5	[10.83, 146.65, 80.78, 34.18]	Alarm!	1
6	[69.71, 141.42, 182.66, 172.82]	Alarm!	1
7	[17.43, 5.43, 377.09, 72.29]	Alarm!	2
8	[49.87, 143.89, 156.74, 168.58]	Alarm!	2
9	[8.02, 122, 819.52, 693.07]	Alarm!	2
10	[75.09, 60.43, 344.95, 944.64]	Alarm!	2
11	[1.85, 41.12, 91.18, 882.67]	Alarm!	2
12	[25.76, 6.93, -52.54, 313.08]	Alarm!	3
13	[82.26, 83.93, -34.98, 594.2]	Alarm!	3
14	[32.93, 40.11, 100.63, 897.77]	Alarm!	3
15	[82.7, 35.2, 952.59, 465.22]	OK	3
16	[86.71, 112.21, 250.6, 758.03]	OK	3
17	[5.75, 65.84, 106.16, 465.41]	OK	4
18	[18.47, 73.92, 604.71, 450.33]	OK	4
19	[78.26, 3.03, 679.72, 340.8]	OK	4
20	[90.77, 148.39, 265.3, 284.15]	OK	4

Fig. 15. Test runs with a sample of antigens on each detector generation with detection results.

The testing of the algorithms on a 2-dimentional space proves that the detectors show good coverage of the non-self space and a stable detection of non-self antigens. Fig. 13. shows the coordinates, radii, overlap and detection score of the first detector generation. The population should stay the same but after 3 generations the detector population decreased (Fig. 14.) but still detected the pathogens (Fig. 15). The chosen actions did not differ much probably because of the implementation which needs further research and improvement.

6.3 Computer simulation of the railway station

For the experiments the program for programmable controller was implemented. The controller performs all the calculations and controls the electric drive and traffic lights on the functional prototype.

The computer model is created to show the results of controller's operations to perform an emergency stop before the red signal of the traffic lights.

The specific environment is developed by the authors for the modelling of railway system for safety improving algorithms (Fig. 16).

The data from the specific memory addresses of the controller is read by the server (Fig. 17.) and transferred to the model.

Fig. 16. Simulation environment

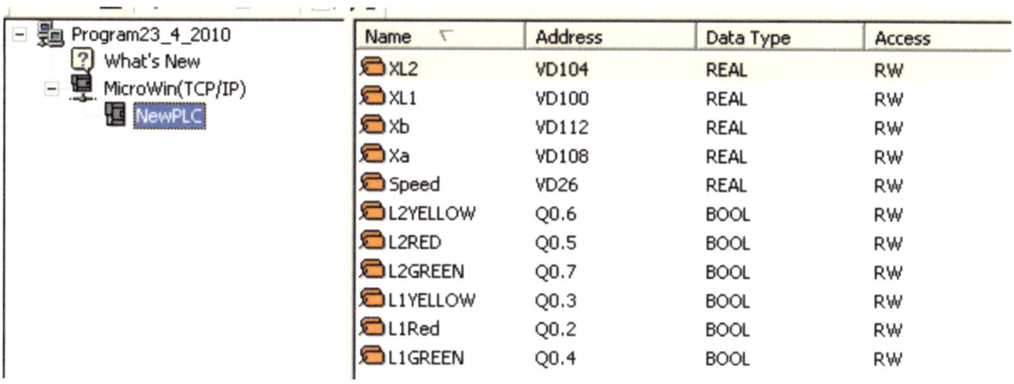

Fig. 17. Input data from PLC to computer model

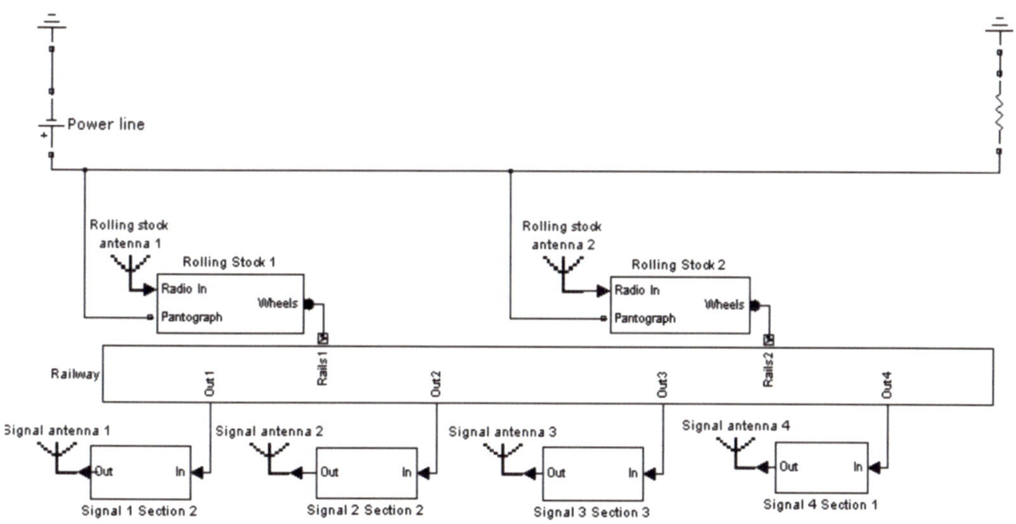

Fig. 18. Model of the railway system

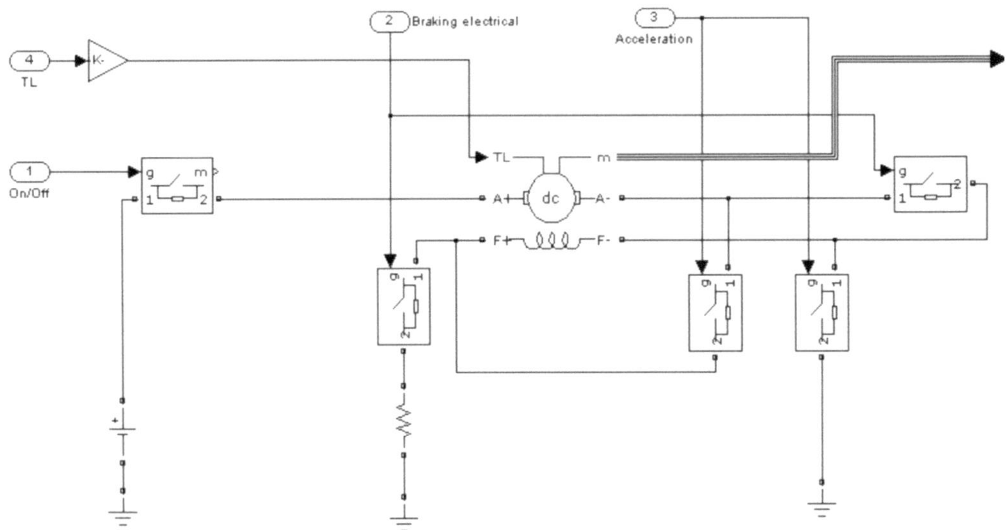

Fig. 19. Fragment of the electrical part of the computer model of the rolling stock

The current experiment is proposed for modelling of crash prevention of two trains moving towards each other. The model consists of 3 series block-sections; 2 rolling stocks; 4 railway signals (Fig. 18.).

Each rolling stock and signal is equipped with receiving and transmitting devices that give a possibility in a multi-agent system.

Electrical part of the model (Fig. 19.) consists of a DC drive with characteristics of 8 DC motors, 1 switch to connect or disconnect the electric drive from the electric contact network, and 2 pairs of switches for acceleration and for braking that changes direction of field current I_f flow. A braking branch contains braking resistance. An output of a DC drive is an electrical torque which handles the mechanical part of a rolling stock.

7. Experiments of prototype in real conditions

The result of this work is a train emergency braking device. The invented device is proposed to increase safety on railway transport. It gives possibility to stop rolling stock automatically before a closed signal timely.

In contrast to the known devices that actuate brake only after the passing of a closed signal, the invented device provides a train emergency braking and stopping before a closed section, even if it is not equipped with automatic locomotive signalling. The device also provides a distance control and an emergency braking way calculation.

The detector of the regular braking distance determines an emergency braking distance, but the detector of the starting point of regular braking defines the point on the route when braking should be started. The module for checking the reaching the starting point of regular braking evaluates location of the train, defines starting point of regular braking and operates braking signalling device in the cabin, the module for controlling starting of regular braking evaluates whether the regular braking is started. The device warns the driver about the necessity of starting process of the emergency braking taking into account the signal of the traffic light and speed limitation and allows to perform an automatic operation of the emergency braking in time and stops the train preventing trains collision at any sector of a railway.

Fig. 20. presents the demonstration of this device that can be installed on the train. Two traffic lights; the electric motor; sensors and wireless communication equipment are installed on the demonstrator.

According to the traffic light signal the controller selects the appropriate engine speed. When the red light is on, the control system automatically stops the engine. In response to the light sensor, the control unit in addition to the fan is turned on and switches to another mode of operation. Remote monitoring and control of the processes is possible using wireless communication. In a real system, it could be dispatching control centres, from which it is possible to switch both signals and also take over control of the train speed.

Taking into account the pieces of advice and the recommendations from the State joint stock company "Latvian Railways" (Latvijas Dzelzceļš/LDz) specialists, the prototypes of the locomotive and the signal devices have been created. Both inventions were issued Latvian and International Patents No. LV13978 B, LV14156 B, LV14187 B, WO 2011/115466 A2, PCT/EP2011/067474.

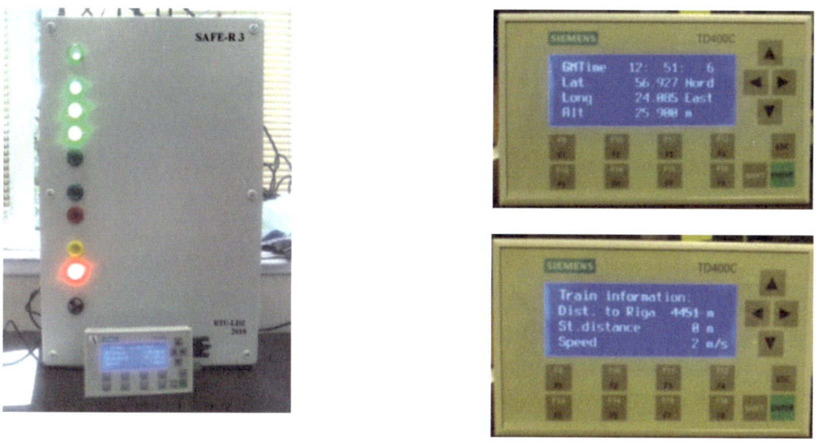

Fig. 20. Functional prototype and information screen with satellite navigation data

The authors and the LDz staff had tested the prototypes of the devices in real service conditions. A non-busy section of the railway was chosen to play the role of a proving ground. During the experiment all the devices were working steadily and without troubles, thus the experiment proved that the ideas adopted in the devices can be implemented into practice.

The task of the locomotive's embedded SAFE-R 3 device and the traffic lights' embedded SAFE-R 4 device, which was designed by the RTU and LDz, is to stop the train automatically at the restrictive signal of the traffic lights, in those cases when a driver does not react to this restrictive signal. It is provided that these devices will also work in unencoded railway sections, where the automatic locomotive signalling did not work.

8. Conclusions

Advantages of the proposed device are the following: The device is not using rail circuits and works independently of automatic locomotive signalization system. The proposed device is an alternative or auxiliary to existing safety systems. As opposed to existing systems the new device uses wireless communication network and may work in railway sections without an automatic interlocking system. The possibility to prevent a dangerous situation and a crash corresponding to the condition of the braking system of rolling stock allows stopping the train before dangerous failure time point; possibility of using of already existing measurement devices and sensors together with the new sensors.

The results of the experiment show the possibility to use the proposed system as an auxiliary safety device to prevent breaches of red signal and crashes on the railway.

The most relevant features of immune algorithms are self-learning, diversity maintenance, memory about the past decisions, and detection of previously unknown but related elements, noise rejection and classifying ability.

An intelligent rolling stock safety control system could benefit from using a combination of both an immune negative selection algorithm and a clonal selection algorithm. A fault detection system for railway electric transport could benefit from using an immune negative selection algorithm.

The most feasible way to implement a railway electric transport safety control system would be through the two phases of anomaly detection and determination of their type to draw a conclusion about further action.

Single string data encoding is better suited for use on PLC. The PLC program needs a data buffer to eliminate the risk of data loss due to unstable radio signal.

The authors need to assess the possibility to run the data analysis using these algorithms in real time. The detector maturation and control cell selection processes need improvement.

9. References

Russel S. J., Norvig P.. Artificial Intelligence. A Modern Approach, 2 edition. Prentice Hall, 2006, 1408p.

Luo R., Zeng J. Computer simulation of railway train braking and anti-sliding control. //In Proceedings of 21st International Symposium on Dynamics of Vehicles on Roads and Tracks (IAVSD'09), 2009 - 189 p.

Hasegawa Y., Tsunashima H., Marumo Y., Kojima T. Detection of unusual braking behavior of train driver //In Proceedings of 21st International Symposium on Dynamics of Vehicles on Roads and Tracks (IAVSD'09), 2009 - 166 p.

Gorobetz M. Research of Genetic Algorithms for Optimal Control of Electric Transport. Promotional thesis. Riga, 2008 - 189 p.

Levchenkov A., Gorobetz M., Ribickis L., Balckars P. „Generating of Multi-Criteria Alternatives for Decision-Making in Electric Light Rail Control" //In China-USA Business review, December 2009, pp. 49-55.

Forrest S., Perelson A.S., Allen L., Cherukuri R. Self-nonself discrimination in a computer. In proceedings of the 1994 IEEE Symposium on Research in Security and Privacy. Los Alamitos, CA.

Kephart J.O.. A biologically inspired immune system for computers. In proceedings of Artificial Life IV: The Fourth International Workshop on the Synthesis and Simulation of Living Systems, 1994. MIT Press.

Dasgupta D., Krishna Kumar K., Wong D., Berry M.. Negative Selection Algorithm for Aircraft Fault Detection. In proceedings of ICARIS 2004.

Mor-Yaroslavtsev A., Levchenkov A. Rolling Stock Location Data Analysis Using an Immune Algorithm on an Intelligent Embedded Device. // In proceedings of 19th Telecommunications Forum TELFOR 2011.

Powers S. T., He J.. A hybrid artificial immune system and Self Organising Map for network intrusion detection. Elsevier IS 179, 2009.

Musilek P., Lau A., Reformat M., Wyard-Scott L.. Immune Programming. Elsevier IS 179, 2009.

Gorobetz M., Alps I., Levchenkov A.. Mathematical Formulation of Public Electric Transport Scheduling Task for Artificial Immune Systems. Proceedings of ITELMS '2009, Kaunas.

Masutti, T. A. S. Castro L. N. de. A self-organizing neural network using ideas from the immune system to solve the traveling salesman problem. Elsevier IS 179, 2009.

Tavakkoli-Moghaddam R., Rahimi-Vahed A., Mirzaei A. H.. A hybrid multi-objective immune algorithm for a flow shop scheduling problem with bi-objectives: Weighted mean completion time and weighted mean tardiness. Elsevier IS 179, 2009.

PCT/EP2011/067474. Device for Safe Passing of Motor Vehicle over Level Crossings Using Satellite Navigation Systems. A.Ļevčenkovs, M.Gorobecs, I.Raņķis, L.Ribickis, P.Balckars, A.Potapovs, I.Alps, I.Korago, V.Vinokurovs, 6.10.2011. (26.07.2011.)

WO 2011/115466 A2, (PCT/LV2011/000004) Controlling Device of Railway Track Sections. A.Ļevčenkovs, M.Gorobecs, J.Greivulis, P.Balckars, L.Ribickis, I.Korago, A.Bobeško. 22.09.2011. (17.03.2010.)

Patent application Nr. P-11-102. Device for Safe Passing of Motor Vehicle over Level Crossings Using Satellite Navigation Systems. A.Ļevčenkovs, M.Gorobecs, I.Raņķis, L.Ribickis, P.Balckars, A.Potapovs, I.Alps, I.Korago, V.Vinokurovs, 26.07.2011.

Patent application Nr. P-11-76. Train anticollision device with satellite navigation. A.Ļevčenkovs, M.Gorobecs, I.Raņķis, L.Ribickis, P.Balckars, A.Potapovs, 23.05.2011.

Patent Nr. LV13978 B. Train Emergency Braking Device. Gorobecs M., Greivulis J., Ļevčenkovs A., Balckars P., Ribickis L. 14.05.2009.

Patent Nr. LV 14156 B. Controlling Device of Railway Track Sections. A.Ļevčenkovs, M.Gorobecs, J.Greivulis, P.Balckars, L.Ribickis, I.Korago, A.Bobeško. 17.03.2010.

Patent Nr. LV 14187 B. Train's Braking Way Control Device. A.Ļevčenkovs, M.Gorobecs, J.Greivulis, I.Uteševs, P.Balckars, L.Ribickis, V.Stupins, S.Holodovs, I.Korago

17

Study and Design of an Electro Technical Device for Safety on Railway Network

Clavel Edith, Meunier Gérard, Bellon Marc and Frugier Didier
G2Elab Grenoble Electrical Engineering Laboratory, Saint Martin d'Hères, BP46, Saint Martin d'Hères, Cedex, France

1. Introduction

The security on the railway network is a real important challenge since today, the number of trains is growing and the saturation of the network is close to be reached.

In this chapter, an electrical system is presented in order to improve the electrical detection of trains on the network and correctly manage the lights. That is how security is ensured.

In the next section, the context of the study is presented. The management of lights on network is explained to emphasize the possible trouble which may occur.

The third part deals with the electrical system which is experimented by the French National Railway Company to overcome this problem. The way it works and its main characteristics will be detailed.

In order to base the further developments of this solution not only on experiments, a modeling process is presented in the following part. For other industrial fields, it has proved to be efficient and its use in the case of a railway system seems possible.

In order to validate the modeling approach, an experimental set is developed since it is very difficult to make measurements in situ. It will be presented in the fifth part of this chapter.

Finally in a last section, the results of the modeling process are successfully compared to the measures.

The outlook of improving the studied system is important since the impact of some dimensional parameters on its performances is analyzed whether being geometrical or physical parameters.

2. Context of the study

The distance between trains on a railway is controlled by signals similar to road lights. The railroad line is divided into several sections from 1500m to 20km. Every section is protected by a signal. When a train enters a section ①, an electrical device detects it and makes the light becomes red ②. When the train penetrates into the following section, its signal turns to a red light, while the signal of the first section becomes yellow③. When the train penetrates into the third section, its signal indicates a red light. The signal of the second section gives the indication yellow, and the signal of the first section gives the indication that the way is clear by a green light④. So if a train is stopped in a section, the following train will meet a yellow light announcing to the driver that the following light is red. He will have to reduce the speed of the train to be able to stop if necessary. This is illustrated of Fig. 1.

As said before, an electrical system is used to guarantee this security working. It is constituted by:

- A generator connected to one of the extremities of the isolated section and which imposes a difference of potential between the two rails,
- A receptor connected to the other extremity of the isolated section which measures the difference of potential between the two rails,
- A transmission line which is electrically limited to the extremities of the considered section.

The generator sends a coded signal through the electric circuit constituted by the rails and the receptor. When the zone included between the generator and the receptor is free, the receptor is able to detect the coded signal: the way is then considered free. When a train enters this section, a great part of current is derived by wheels and axles (what we call "shuntage"); the receptor does not receive this coded signal coming from the generator anymore: the way is then considered busy. Such a device is thus able to detect the presence of trains on a section by the change of the impedance value of the circuit between the rails. When a train is on a section, the rails are short-circuited by the train and the group wheels, axles, rails 1 and 2 have lower impedance. So the measure of the voltage drop at the receptor implies a busy section. This change of electrical circuit is illustrated on Fig. 2.

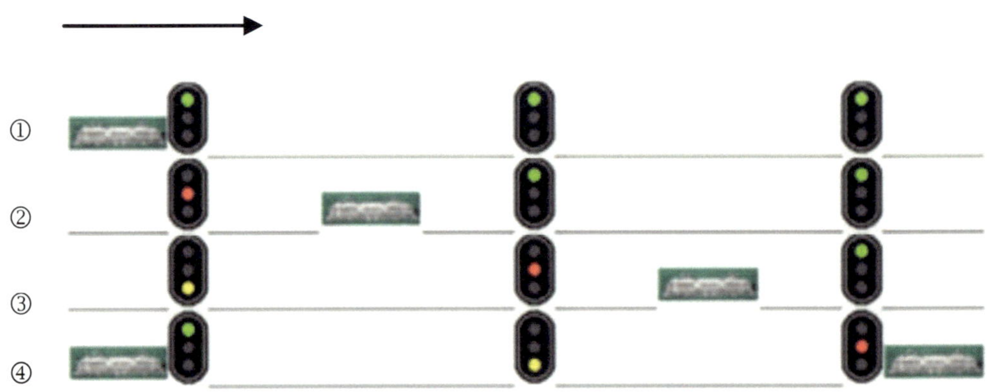

Fig. 1. The principle of railway lights

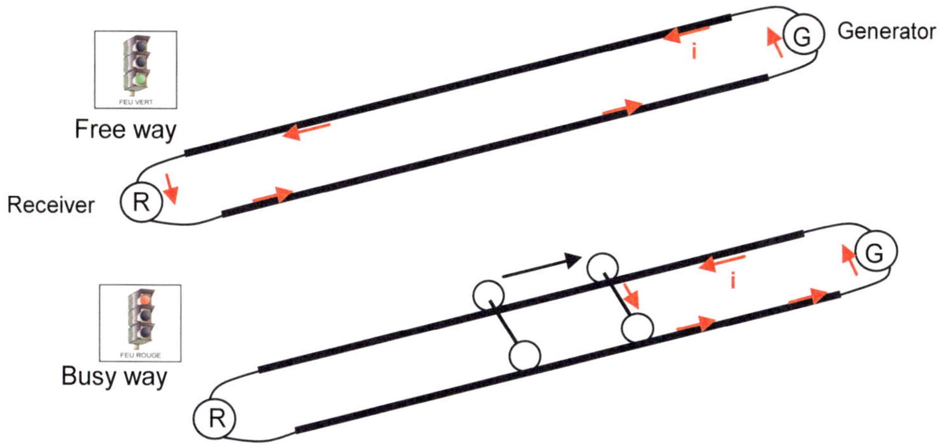

Fig. 2. The principle of the detecting electrical system

In the case of the electrified lines, which represent 90 % of the French network, rails are very often used to ensure the return of drive current towards the substations. Drive currents are about 1000A. And they have to coexist with currents of 1A from the detecting electrical system.

In order to be able to easily separate these two currents, the kind of current inside the detecting system depends on the electrified lines. In the case of DC lines, alternative currents are used for the detecting electrical system and in the case of alternative power supply different frequencies are used for the detecting circuit. In that case, the insulation between two detecting systems is carried out using inductive connections which allow the return of the drive current.

Current of the detecting circuits can be modulated with various frequencies, which can be detected by equipments on board the trains to give to the drivers all the indications in the cabin. This principle of the transmission between the way and the machine is used in the famous French High Speed Train (TGV).

Sometimes, the short circuit between wheels and rails is not of good quality. This is the case of weak machines, parts of the railway network rarely used, bad wheel/rail contact due to insulating body. So errors of detection by the system can lead to dangerous situations. The light is green even if there is a train in the section. If a more rapid train arrives, an accident could occur.

As a consequence, in order to be sure that the signal is well interrupted by the train on a section, the French National Railway Company (SNCF) has added an electrical device which will be described in the following part.

3. The proposed device

To avoid this problem of detection, several solutions have been explored:

- to replace all the detecting electrical systems,
- to improve the contact between the wheels and the rails.

This last solution has been generally adopted. This can be made by cleaning the rails. A special accessory has been designed to scrape the rails.

But another way to improve the contact is to help the short circuit created by the wheels and the rails to be efficient by adding an electrical device.

This last option is the topic of this paper.

3.1 Working principle of the inductive loop

The principle of the proposed device is to create a voltage drop of sufficient level in order to make possible a current to flow. This is done by an inductive loop able to induce in the wheels/axles/rails circuit a current of sufficient value.

The proposed device is presented on Fig. 3. It is constituted by:

- a high frequency current controlled generator,
- a parallel LC circuit made of capacitors located on the bogies and the inductance of the inductive loop.

This loop is in fact the primary circuit of a transformer. The secondary part is constituted by the axles and the rails. The current induced by the loop flows through the axles and the rails and must be high enough to guarantee the electrical contact between the wheels and the rails. During a bad contact rail / wheel, the secondary circuit of the transformer is open and an induced voltage appears between the wheel and the rail and establishes again the electrical contact insuring the "shuntage".

The chariot near the inductive loop constitutes a third circuit of the transformer which can reduce the value of the current inside the axles - rails circuit.

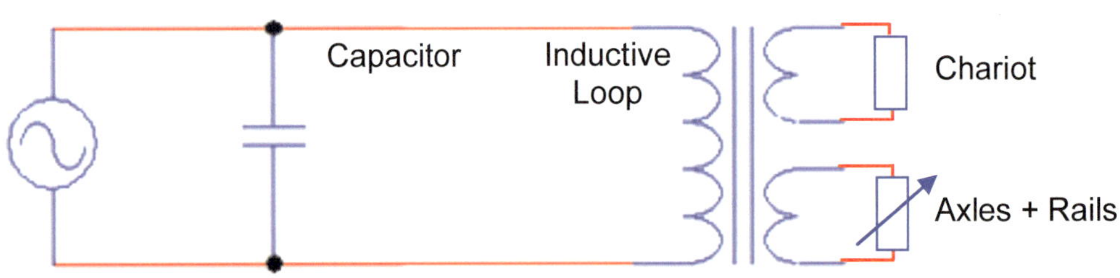

Fig. 3. The principle of the detecting electrical system

3.2 Electrical and mechanical requirements

To be efficient, this system has to create a minimum voltage drop between a wheel and the rail. The working frequency of this system is imposed by the French National Railway Company. The design data for the loop are:

- minimum voltage drop : 3V,
- frequency : 147kHz,
- range of frequencies : 140kHz to 152kHz (adjusted using capacitors),
- control of the current : if the current remains under a certain limit during 10s, a default signal is sent to the driver. An emergency braking is used and special rules are adopted by the driver,
- mechanical constraints (under the train, easy to access, easy to change and repair,...),
- material for the loop : copper is used and several other cables are added.

3.3 Design of the loop

The performances of the loop are directly linked to its geometrical size.

But the degrees of freedom are limited because a lot of mechanical constraints exist:

- the surface of the loop is limited by the space under the train,
- the height of the loop is limited by the distance between the train and the rails,
- leakage currents are flowing inside the metallic structure.

In practice, the environment of the loop changes (shunt, ballast, metallic parts ...) and the value of the inductance of the loop changes too. So the resonance frequency is modified. In order to avoid this, it is necessary to continuously adjust the frequency so that the current is sufficient inside the loop.

In order to perform this, a Phase-locked loop (PLL) is used.

Thanks to three electronic cards, every change of frequency is detected and automatically compensated:

- a ALIM card allows the galvanic insulation and adapts the voltage which supplies the system (24V),
- a BIAS card controls the value of 147kHz,
- a TFER card is a controlled amplifier for the BIAS card.

In order to reach an optimal design of the loop, experiments are not sufficient.

A modeling process has to be run in order to take into account all the requirements and desired performances.

This is the aim of the following part: to present a modeling approach.

4. Modeling method

As said before, the aim of the modeling process is to be able to evaluate the voltage and current in the studied structure. This implies to solve the Maxwell's equations and to establish an electrical equivalent circuit on which the circuit equations could be solved. In the field of electromagnetic approach, two families of modeling methods are facing:

- the finite element method (FEM),
- the integral method.

4.1 Choice of the modeling method

Each family of modeling method has advantages and drawbacks.

Concerning the FEM, it is well known and gives good results on the evaluation of the electromagnetic fields everywhere in the space. But it requires the meshing of all the space, i.e. the studied structure but also the air around. Moreover good results imply a good use of the formulations and assumptions at the limits of the studied domain. The results are principally the electromagnetic fields. Then to obtain the electrical equivalent circuit, post-processing evaluations have to be done.

Concerning the integral methods, the only meshing of the conductive parts makes them very attractive. The number of unknowns is limited. Moreover, an electrical equivalent circuit can directly be deduced from the solving.

The most famous integral methods are the method of moment (MoM) and the PEEC method (Partial Element Equivalent Circuit).

Since the studied structure could be very large and the amount of air around significant, the use of FEM could lead to solve problems with too high a number of unknowns.

So an integral method is chosen to model the studied structure and more particularly the PEEC method which is detailed in the following paragraph.

4.2 Principle of the PEEC method

The PEEC method was firstly introduced by A. Ruehli (Ruehli, 1974). Based on low frequency exact analytical formulae, it consists in extracting the electric parameters from the geometries of conductors. This allows taking into account the electrical parasitic effects of interconnections while evaluating the electromagnetic behavior of an electronic or electrical system.

Full PEEC method takes into account resistive, inductive and capacitive parts.

4.2.1 Assumptions

According the frequency range of the study, the electrical equivalent model could be more or less complicated. Indeed for not so high frequencies, only resistive and inductive effects of cabling are involved in the electromagnetic behavior of the studied system. That is why it is possible to partly use the PEEC method which allows reducing the size of the equivalent model. For the present studied application, frequency is low enough (around some kHz) to limit the evaluation to the resistive and inductive parts.

Nevertheless the capacitive aspect is detailed in (Ardon et al., 2009) to complete the study.

The mains equations will be detailed in the following paragraph to obtain the equation (13) which is the base to establish the electrical equivalent circuit.

In order to easily solve the equation (13) for each considered part, the most important assumption is that current density is uniform. But according the frequency and geometry of the studied structure, skin and proximity effects have to be taken into account during the modeling. So in a first step, all the conductors of the studied structure have to be meshed so that the current density is uniform inside each mesh.

Moreover, as presented in (13), relative permeability μ_r is assumed to be equal 1. Otherwise further developments of PEEC method are presented and detailed in (Aimé et al., 2009b) and (Kéradec et al., 2005) in order to take into account magnetic material influence on current distribution.

Finally no propagation aspect is considered for this first presentation. Otherwise, another modeling method has to be applied such as rPEEC or transmission lines approach (Antonini et al., 2007) and (Clavel et al., 2007).

4.2.2 Equations and associated model

The problem can be better formalized by considering the integral form of the Maxwell's equations and by assuming:

- quasi-static conditions;
- only surface location for the free-charges ρ;
- uniform and constant currents I in each volume element of conductors;
- constant material conductivity σ,
- vacuum permeability μ_0 surrounding the objects;
- a homogeneous medium of permittivity $\varepsilon = \varepsilon_r \varepsilon_0$;
- negligible losses in dielectric materials.

In such conditions, the following equations can be written:

$$\mathbf{rotH} = \mathbf{J} \tag{1}$$

$$\mathbf{rotE} = -j\omega\mathbf{B} \tag{2}$$

$$div\mathbf{B} = 0 \tag{3}$$

$$\mathbf{B} = \mu_0.\mathbf{H} \tag{4}$$

$$\mathbf{J} = \sigma\mathbf{E} \tag{5}$$

From (3), it comes:

$$\mathbf{B} = \mathbf{rotA} \tag{6}$$

$$\mathbf{E} = -j\omega\mathbf{A} - \mathbf{grad}V \tag{7}$$

$$\Delta\mathbf{A} = -\mu_0\mathbf{J} \tag{8}$$

(7) is the Faraday's law and (8) the Poisson's equation.

Considering a conducting volume Ωc in an external electrical field (V_{ext}), the total electrical field E_{total} at any point P in the conductor is given by (9):

$$\mathbf{E}_{total}(P) = -j\omega\mathbf{A}(P) - \mathbf{grad}\,V_{charge}(P) - \mathbf{grad}\,V_{ext}(P) \tag{9}$$

V_{charge} is the potential due to the electrical charges in the conductor.

If, in a first approximation, the capacitive effects can be neglected, this term is null.

From (8), without propagation effects and magnetic material, it comes (10):

$$\mathbf{A}(P) = \frac{\mu_0}{4\pi}\int_{\Omega c}\frac{\mathbf{J}}{r}d\Omega \tag{10}$$

where **J** is current density and r the distance between the integration point and P.

Taking (5) and (10) in (9) gives (11):

$$\frac{\mathbf{J}(P)}{\sigma} + j\omega\frac{\mu_0}{4\pi}\int_{\Omega c}\frac{\mathbf{J}}{r}d\Omega = -\mathbf{grad}\,V_{ext} \tag{11}$$

In order to reach the desired model, it is necessary to suppose a uniform current density. For that, the volume Ωc is divided into m elementary conductors. On each elementary conductor, (12) is written:

$$\mathbf{J}_k = \mathbf{J}_{0k}.I_k \tag{12}$$

J_{0k} is the electrokinetic solution for a 1A current.

Multiplying (11) by J_{0k}, it comes (13):

$$\int_{\Omega c_k}\mathbf{J}_{0k}\cdot\frac{\mathbf{J}_k}{\sigma}d\Omega c_k + j\omega\frac{\mu_0}{4\pi}\int_{\Omega c_k}\mathbf{J}_{0k}\left(\sum_{i=1}^{m}\int_{\Omega c_i}\frac{\mathbf{J}_{0i}.I_i}{r}d\Omega c_i\right)d\Omega c_k = -\int_{\Omega c_k}\mathbf{J}_{0k}\mathbf{grad}\,V_{ext}.d\Omega c_k \tag{13}$$

From (13), the electrical equivalent circuit of a conductor can be deduced (Fig. 4) (Ruehli & Cangellaris, 2001).

(12) inside the first term of (13) gives (14):

$$\int_{\Omega c_k}\mathbf{J}_{0k}\cdot\frac{\mathbf{J}_k}{\sigma}d\Omega c_k = \frac{1}{\sigma}I_k\int_{\Omega c_k}\mathbf{J}_{0k}^2.d\Omega c_k \tag{14}$$

Knowing (15), (16) can be deduced.

$$\frac{1}{\sigma}\int_{\Omega c_k}\mathbf{J}_{0k}^2.d\Omega c_k = R_k \tag{15}$$

$$\int_{\Omega c_k}\mathbf{J}_{0k}\cdot\frac{\mathbf{J}_k}{\sigma}d\Omega c_k = R_k.I_k \tag{16}$$

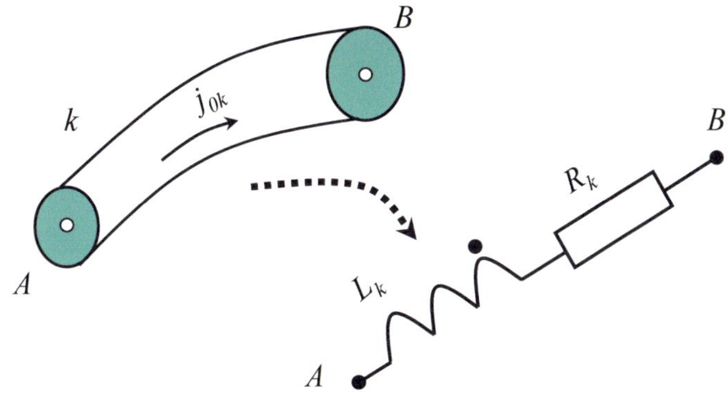

Fig. 4. k[th] element of the conductor

Introducing the mutual inductance between two elementary conductors k and i (17), the second term of (13) leads to (18).

$$L_{ki} = \frac{\mu_0}{4\pi} \int_{\Omega c_k} \int_{\Omega c_i} \frac{\mathbf{J}_{0k} . \mathbf{J}_{0i}}{r} d\Omega c_i d\Omega c_k \tag{17}$$

$$j\omega \frac{\mu_0}{4\pi} \int_{\Omega c_k} \mathbf{J}_{0k} \left(\sum_{i=1}^{m} \int_{\Omega c_i} \frac{\mathbf{J}_{0l} . I_i}{r} d\Omega c_i \right) d\Omega c_k = j\omega \sum_{l=1}^{m} L_{kl} . I_l \tag{18}$$

Concerning the term on the right of (13) the following equations can be established assuming the electrokinetic solution gives $div\mathbf{J}_0 = 0$:

$$div(V.\mathbf{J}_{0k}) = \mathbf{J}_{0k} . \mathbf{grad} V + V.div\mathbf{J}_0 \tag{19}$$

$$\int_{\Omega c_k} \mathbf{J}_{0k} \mathbf{grad} V_{ext} . d\Omega_k = \int_{\Omega c_k} div(V_{ext} . \mathbf{J}_{0k}) d\Omega_k = \oint_{\Gamma_k} V_{ext} . \mathbf{J}_{0k} . \mathbf{n}_k . d\Gamma_k \tag{20}$$

$$\mathbf{J}_{0k} . \mathbf{n}_k = 0 \tag{21}$$

$$\oint_{\Gamma_k} V_{ext} . \mathbf{J}_{0k} . \mathbf{n}_k . d\Gamma_k = V_a \int_{\Gamma_a} \mathbf{J}_{0k} . \mathbf{n}_a . d\Gamma_k + V_b \int_{\Gamma_b} \mathbf{J}_{0k} . \mathbf{n}_b . d\Gamma_k = -V_a + V_b = -U_k \tag{22}$$

with Γ_k is the edge of volume Ωc_k; $\mathbf{n}_k$ is the normal vector oriented towards the exterior of the surface, Γ_a and Γ_b are the current outputs surfaces.

Hence, the electrical equation is deduced (23) and (24):

$$-\int_{\Omega c_k} \mathbf{J}_{0k} \mathbf{grad} V_{ext} . d\Omega c_k = U_k \tag{23}$$

$$R_k . I_k + j\omega \sum_{i=1}^{m} L_{ki} . I_i = U_k \tag{24}$$

So each term of (13) leads to an electrical characteristic of conductor which is only function of its geometry.

To compute the parasitic resistance R_i in each volume element V_i of length ℓ_i, section S_i, and resistivity ρ, the following analytical formula is used:

$$R_i = \rho \frac{l_i}{S_i} \tag{25}$$

Because of the parallelepiped shape of the elements the double integral in (17) can be expressed in an analytical form and easily computed. For the case of parallel elements, the mutual inductances M_{ij} are computed thanks to an analytical expression. And for the general case presented on Fig. 5, the expressions (26) can be used either to evaluate M_{ij} but also partial inductance L_i if a=d, b=c, l1=L2 and E=p=l3=0 (Hoer & Love, 1965). If elements are not parallel an analytical/numerical integration technique is used (an analytical expression for the first integral is used, the second one being computed thanks to an adaptive gauss point integration ensuring a good accuracy). All values of L_i and M_{ij} can then be organized in a dense and square matrix [L-M] whose size is equal to the number of mesh elements (Aimé et al., 2007).

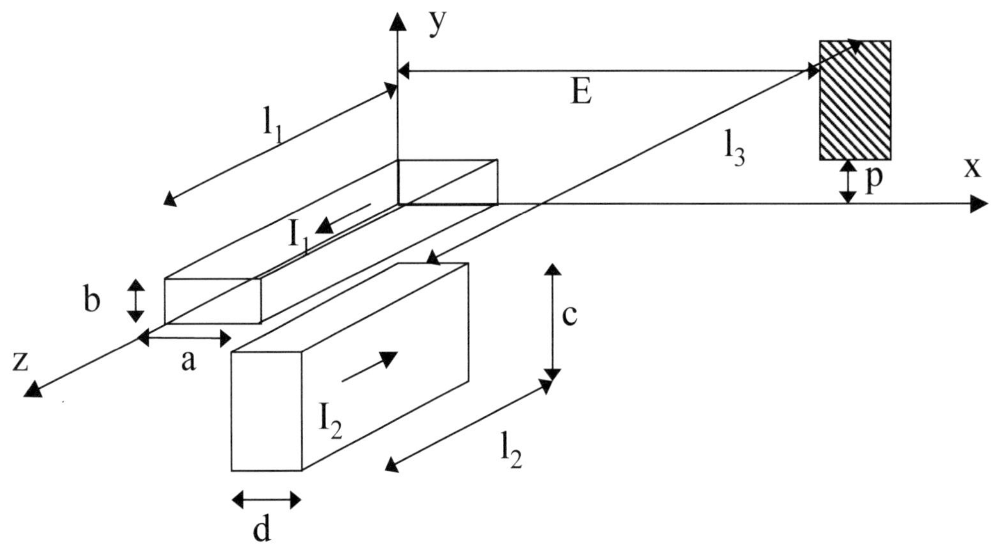

Fig. 5. General case of parallel elements for the evaluations of M_{ij}

$$M_{ij} = \left[\left[\left[f(x,y,z)\right]_{E+d-a,\,E}^{E+d,\,E-a}(x)\right]_{p+c-b,\,p}^{p+c,\,p-b}(y)\right]_{l3+l2-l1,\,l3}^{l3+l2,\,l3-l1}(z) \tag{26}$$

$$\left[\left[\left[f(x,y,z)\right]_{x_2,x_4}^{x_1,x_3}(x)\right]_{y_2,y_4}^{y_1,y_3}(y)\right]_{z_2,z_4}^{z_1,z_3}(z) = \sum_{i=1}^{4}\sum_{j=1}^{4}\sum_{k=1}^{4}(-1)^{i+j+k+1} f\left(x_i,y_j,z_k\right)$$

$$f(x,y,z)=\frac{\mu_0}{4.\pi}\frac{1}{a.b.c.d}\left(\begin{array}{l}\left(\frac{y^2z^2}{4}-\frac{y^4}{24}-\frac{z^4}{24}\right)x\ln\left(x+\sqrt{x^2+y^2+z^2}\right)+\\ \left(\frac{x^2z^2}{4}-\frac{x^4}{24}-\frac{z^4}{24}\right)y\ln\left(y+\sqrt{x^2+y^2+z^2}\right)+\\ \left(\frac{y^2x^2}{4}-\frac{y^4}{24}-\frac{x^4}{24}\right)z\ln\left(z+\sqrt{x^2+y^2+z^2}\right)+\\ \frac{1}{60}\left(x^4+y^4+z^4-3y^2x^2-3y^2z^2-3x^2z^2\right)\sqrt{x^2+y^2+z^2}-\\ \frac{xyz^3}{6}\mathrm{Arctan}\frac{xy}{z\sqrt{x^2+y^2+z^2}}-\frac{xy^3z}{6}\mathrm{Arctan}\frac{xz}{y\sqrt{x^2+y^2+z^2}}-\\ \frac{x^3yz}{6}\mathrm{Arctan}\frac{zy}{x\sqrt{x^2+y^2+z^2}}\end{array}\right)$$

The evaluation of the electrical equivalent circuit of the meshed structure has been implemented into the dedicated software InCa3D® which offers a robust and fast solver combined to a user friendly and efficient graphical interface.

For simple shapes of elements, analytical formulations (25) and (26) are used. But if the geometrical configuration is more complex, a numerical technique is used to find the values of the equivalent circuit.

4.2.3 Meshing techniques

In order to properly describe the current distribution inside conductors, an adapted meshing technique has to be applied. Indeed the shape of conductors often allows the assumption of 1D or 2D current flowing inside them.

For massive bars, or cables the 1D current leads to only mesh the cross section of conductor. Since no propagation effect is described, the length of conductors has not to be subdivided (Fig. 6). For this kind of conductor, the skin effect can be taken into account concentrating the meshes on the edges of the conductors so that the number of subdivisions is not too big.

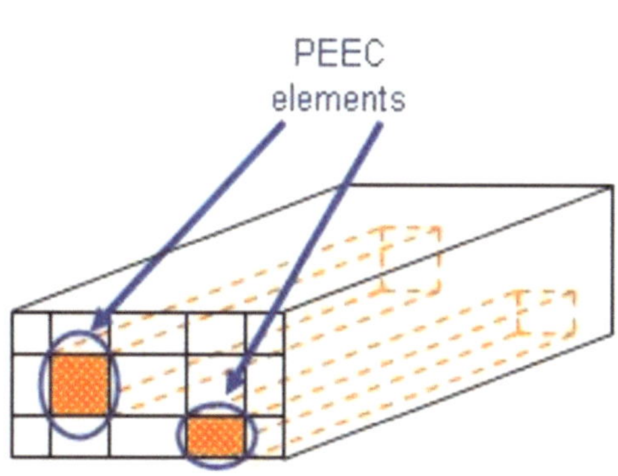

Fig. 6. The 1D meshing of conductors

In the case of very thin and large conductors such as sheet of copper, ground plane, the 1D current assumption is no more valid. Indeed current is generally flowing in a plane so that a 2D approximation can be sufficient in order to properly describe the physical phenomenon. Two quadrate directions for current inside the conductor are defined. So the developed 2D meshing technique consists in dividing the plane as presented on Fig. 7.

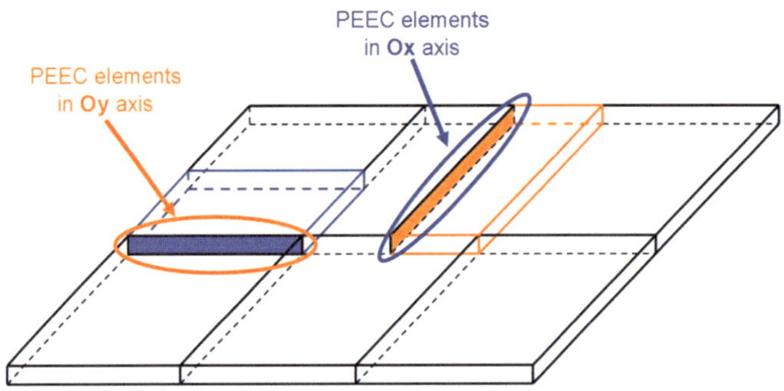

Fig. 7. The 2D meshing of conductors

For both cases, according to the shape of the cross section or conductor, a refinement meshing technique is applied so that the description is close to the real structure (Fig. 8).

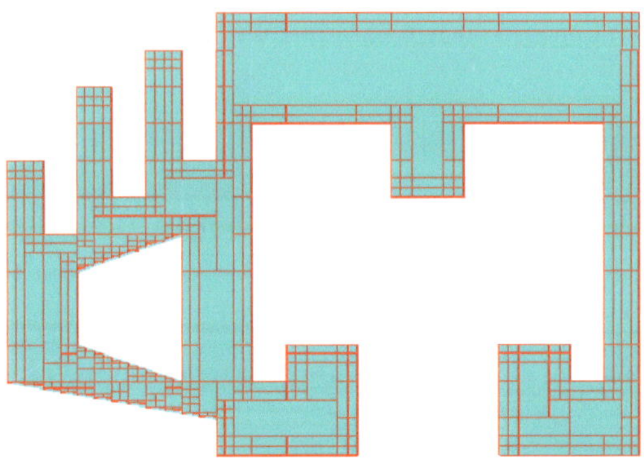

Fig. 8. Examples of meshing for complex cross section for 1D assumption

The associated electrical equivalent models for the 1D and 2D elements are summarized on Fig. 9.

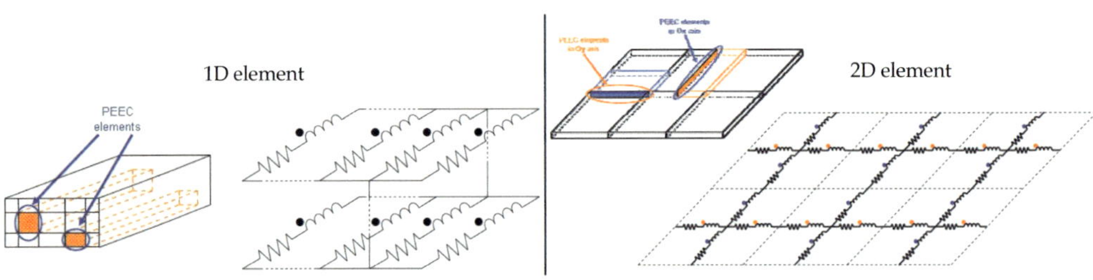

Fig. 9. R-L-M electrical equivalent circuit for 1D and 2D elements (mutual inductances are not represented on the figure)

4.3 Modeling process

On Fig. 10, the modeling process of every kind of electrical structure is detailed.

It consists of four steps:

- Geometry description from designers' data or directly imported from CAD tools,
- Meshing according the appropriated assumption for the current,
- Solving the equations to obtain the electrical equivalent circuit,
- Circuit solving.

For the studied case, the unknowns are the currents inside the equivalent circuit. For that purpose it is necessary to describe the electrical environment of the problem in order to solve the right circuit equations. This last step can be achieved in a circuit solver like SPICE® or Portunus® exporting automatically the equivalent circuit inside these tools. But the size of the equivalent circuit is linked to the number of meshes and can be too big for these tools. Moreover it is not necessary to keep the information of local current inside each mesh. What is interesting is the global current inside the conductors. So a reduced equivalent circuit is better appropriated for this goal. To that aim, the user has to clearly identify the outputs of each conductor and then using parallel and series associations, the equivalent impedance between these points can be evaluated for each frequency. This reduced circuit is afterwards more practical in order to evaluate all necessary currents and voltages. Even if it is frequency dependent, a time simulation can be efficiently done. Indeed, according the frequency range, this dependence can be negligible and if not, numerical techniques to find a non dependent circuit with more components exist (Tan & He, 2007).

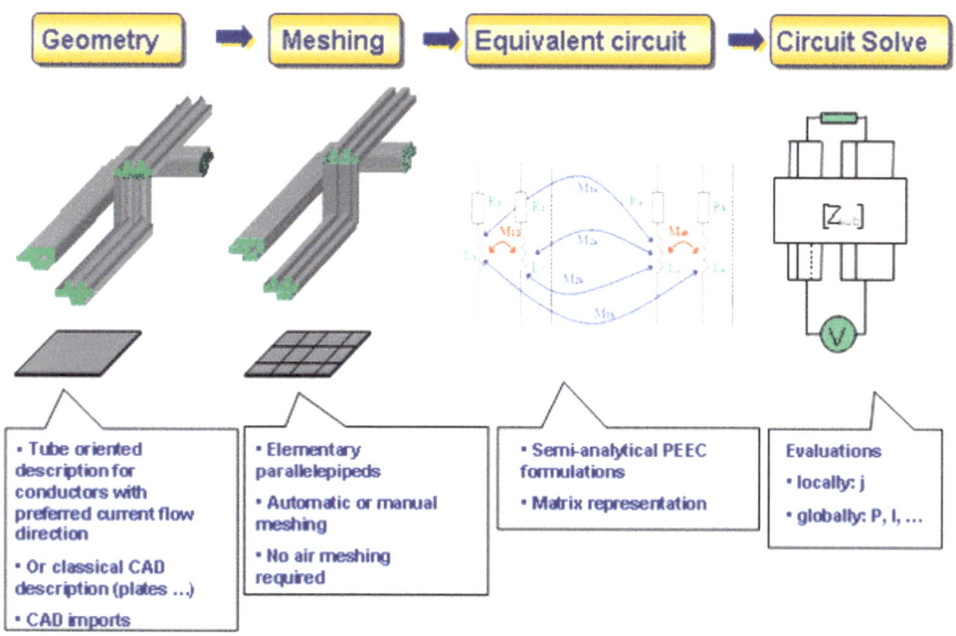

Fig. 10. Modeling process

This proposed modeling process has proved to be very efficient not only for high current electrical systems (Gonnet et al., 2004) but also power electronics devices and structures (Aimé et al., 2009a), electronic card (Clavel et al., 2007a), PCB application (Vialardi et al., 2010) and aircraft structures (Jazzar et al., 2011).

5. The experimental structure

In order to validate the modeling process, experiments have been undertaken.

Unfortunately, measures on real structures are very tricky. A simplified system has thus to be defined. It is presented on Fig. 11. The chariot has been replaced by copper tubes (diameter 42mm); the rails and axles have been replaced by cables (35 mm^2). Sizes have been chosen so that the experimental set is close to a real system (Fig. 12).

The characteristics of the voltage source are the same as described in the requirements paragraph.

Fig. 11. The experimental system

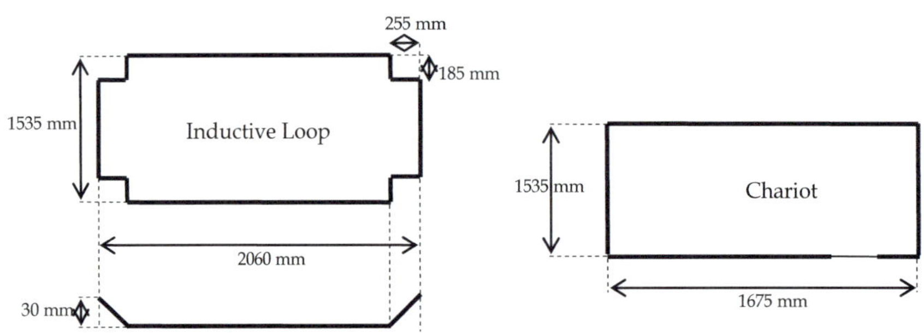

Fig. 12. Sizes of the experimental system

The measurement procedure consists in measuring (Fig. 13-14):

- current inside the chariot (so the cross section has been locally reduced in order to insert a current probe);
- current inside the inductive loop;
- voltage drop between the wheel and the rails (Urw).

Different configurations have been tested and for each of them maximum data have been measured:

- the inductive loop alone;
- the inductive loop + rails + axles;
- the inductive loop + rails + axles + chariots;
- the inductive loop + rails + axles + chariots + 4 shunts for the wheels;
- the inductive loop + rails + axles + chariots + 4 shunts for the wheels with 4 resistors to represent the non perfect electrical contact between the wheel and the rail.

For each of them open circuit and short circuit measurements have been made.

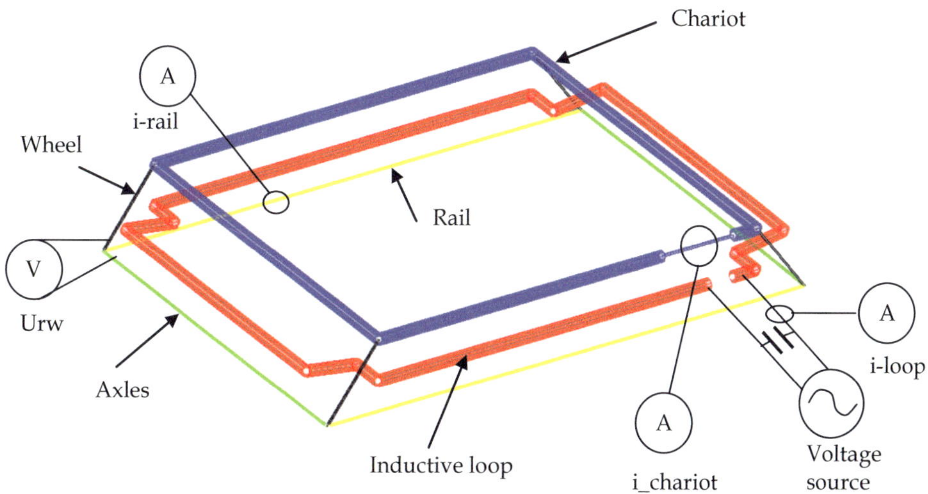

Fig. 13. Measurements procedure

Current measure Voltage drop measure

Fig. 14. Used probes (Fluke 867B, current probe PR30)

Knowing the current I in the loop, the voltage source V, their phase φ and the frequency (f, ω=2πf) an equivalent circuit for the measured system can be calculated using (27) (Fig. 15).

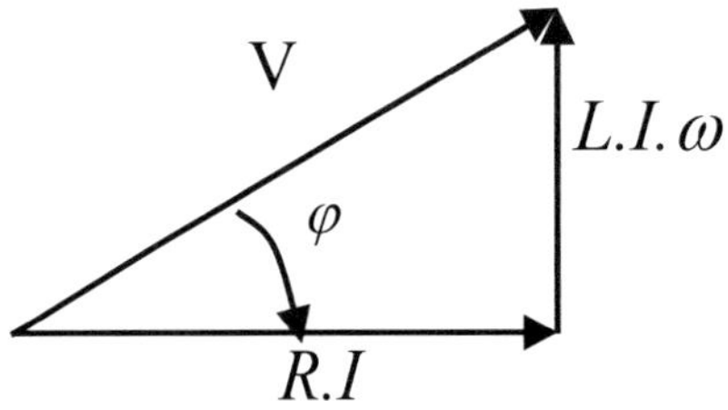

Fig. 15. Fresnel's diagram for the evaluation of equivalent circuit

$$\begin{cases} R = \dfrac{V\cos(\varphi)}{I} \\ L = \dfrac{V.\sin(\varphi)}{2\pi f.I} \end{cases} \tag{27}$$

The second configuration is very close to the classical case of two coupled inductances like a simple transformer. The first one is supplied with an alternative source and an induced current (short-circuit situation) or voltage (open circuit situation) is created on the second one (Fig. 16).

The theoretical study is briefly reminded in the following equations (28) with only one turn for our case (n1=n2=1).

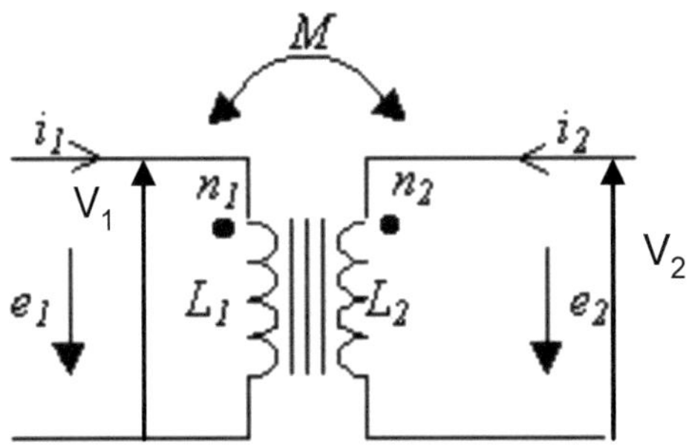

Fig. 16. The simple transformer

$$\begin{aligned} V_1 &= R_1.I_1 + L_1\frac{dI_1}{dt} + M\frac{dI_2}{dt} \\ V_2 &= R_2.I_2 + L_2\frac{dI_2}{dt} + M\frac{dI_1}{dt} \end{aligned} \tag{28}$$

The open circuit configuration gives (29). The voltage drop is directly the Urw voltage characterizing the working of the loop.

$$\begin{aligned} V_{10} &= R_1.I_{10} + L_1\frac{dI_{10}}{dt} = R_1.I_{10} + L_1.I_{10}.\omega \\ V_{20} &= M\frac{dI_{10}}{dt} = M.I_{10}.\omega \end{aligned} \tag{29}$$

6. Results

In this part the electrical equivalent circuit given by the modeling process is presented as well as comparisons between the circuit simulation and measurements.

6.1 Modeling of the real system

The CAD complete studied structure is presented on Fig. 17 on which the wheels, chariot, inductive loop clearly appear.

Indeed in order to correctly solve the problem, some geometrical simplifications have been done. The wheels, the rails and the axles have been replaced by straight massive conductors; four shunts have been added to connect the chariot to the axles in order to represent a realistic situation.

Using InCa3D, the result is presented on Fig. 18.

The geometry has been meshed in order to take into account the proximity and frequency effects.

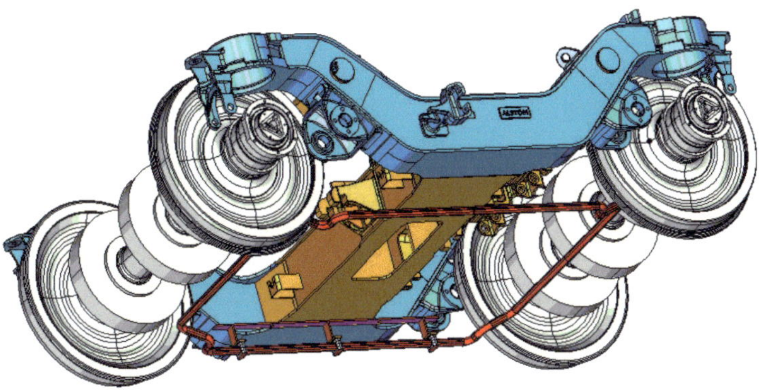

Fig. 17. The complete structure

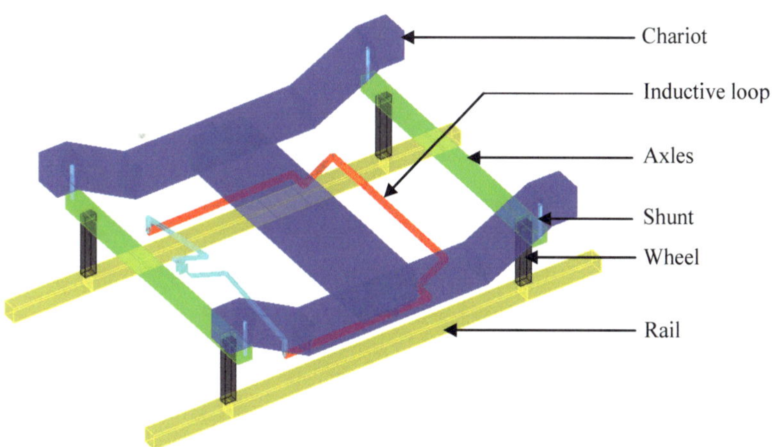

Fig. 18. InCa3D description of the studied structure

After the PEEC solving, the equivalent circuit has been reduced in order to obtain the simple SPICE-like circuit drawn on Fig. 19 where each part of the system is well identified by an L-R series equivalent circuit. On this circuit all the inductances are coupled with mutual coefficients but they have been cut off to make it clearer.

6.2 Comparison between measures and simulation on the experimental set

The experimental case has been modeled (Fig. 11) using the same process and for each configuration simulation results with the same operating conditions (value of the supply voltage source, frequency) have been compared to the measurements.

The results are presented in Table 1 for the closed circuit and Table 2 for an open one.

A good agreement between simulations and measurements can be observed.

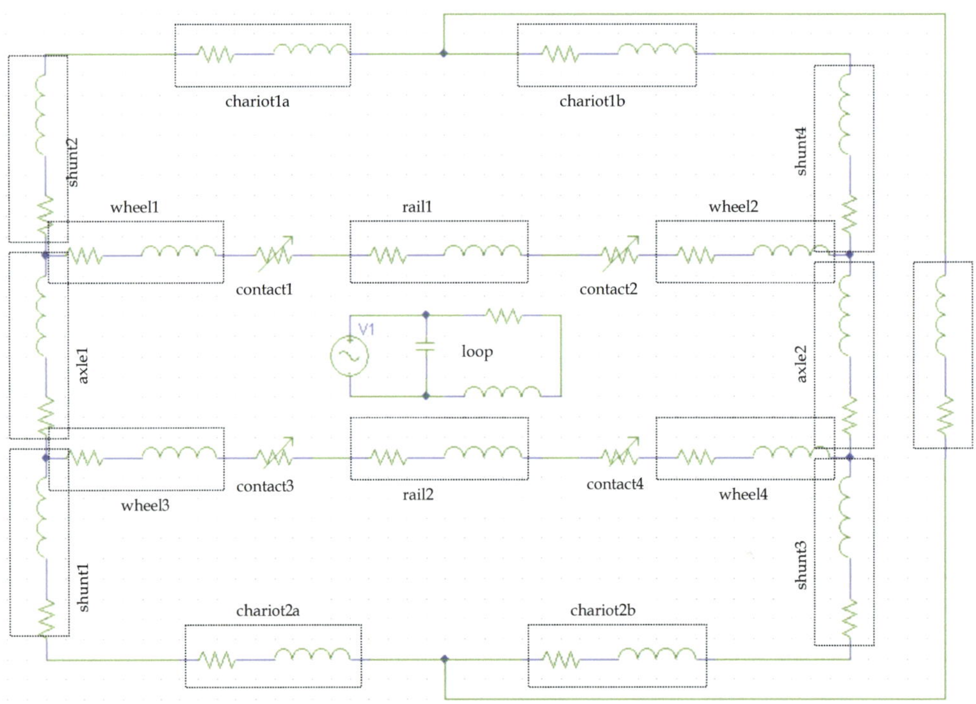

Fig. 19. Reduced electrical equivalent circuit for the studied structure (mutual inductances have been removed)

Conditions		Closed circuit I_loop (A)	Φ (°)	L (μH)	R (Ω)
Only the loop	Measure	10.07	-98	5.15	0.25
	Simulation	10.72	-89.9	4.84	0.01
Loop + Rails + Axles	Measure	11.13	-96	4.5	0.44
	Simulation	12.84	-89.8	3.98	0.01
Loop + Rails + Axles + Chariot	Measure	12.6	-95.8	4.05	0.38
	Simulation	14.65	-89.8	3.5	0.01
Loop + Rails + Axles + Chariot + 4 shunts	Measure	12.5	-95	4.16	0.34
	Simulation	14.95	-89.8	3.49	0.01

Table 1. Current in the loop and electrical characteristics for a closed circuit - simulation and measurements

Conditions		Open circuit I_loop (A)	Φ (°)	L (μH)	R (Ω)	Urw (V)	Δurw
Loop + Rails + Axles	Measure	10.8	-95.9	5.21	0.46	24.1	4.66 %
	Simulation	11.68	-89.8	4.84	0.01	25.22	
Loop + Rails + Axles + Chariot	Measure	12.04	-95.5	4.59	0.38	21.67	5.45 %
	Simulation	13.8	-89.8	4.02	0.01	22.85	
Loop + Rails + Axles + Chariot + 4 shunts	Measure	12.3	-95.3	4.35	0.37	8.4	1.9%
	Simulation	14.7	-89.8	3.65	0.01	8.56	

Table 2. Rail-Wheel voltage and electrical characteristics for an open circuit - simulation and measurements

6.3 Parametric analysis

Once the modeling process is established with satisfactory results, it is possible to make some changes to analyze the influence of some parameters on the performances of the system.

Indeed, using InCa3D, it is possible to define geometrical and physical parameters and make them varying to improve the design of the inductive loop.

For the studied structure, the following characteristics can be defined as parameters:

- height of the inductive loop;
- height of the chariot;
- size of the loop;
- position of the loop under the train;
- number of shunts;
- diameter of the loop;
- number of turns of the loop;
- resistivity of the material;
- voltage source;
- frequency;
- resistance of the wheel/rail contact.

The main performance is the rail/wheel voltage Urw.

6.3.1 Variation of the height

The distance between the inductive loop and the rails is varying in this study.

On Fig. 20, results show that the lower this distance is, the higher the Urw is.

This result is quite logical since according to (30) the induced voltage is linked to the magnetic flux. And when the distance between the two loops is low this flux is maximal.

$$e = -\frac{d\varphi}{dt} \tag{30}$$

Nevertheless, the range of variation of this parameter is quite limited because mechanical constraints on the train and required standards for some sizes. There is a minimum value to respect.

By way of contrast, if the distance between the loop and the chariot decreases, the value of Urw decreases too. This is due to the fact that, as said before, the chariot creates a supplementary winding (Fig. 3) in which an induced current can be created. The inductive coupling in this case is higher than between the loop and the rails.

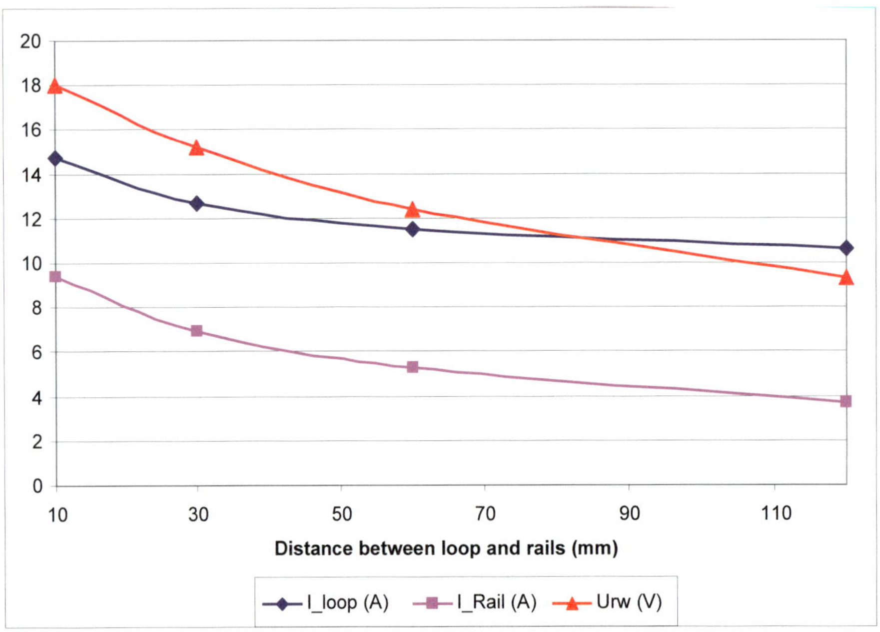

Fig. 20. Urw voltage vs. distance between the loop and the rails

6.3.2 Variation of the sizes and the position of the loop

On Fig. 21, three configurations have been modeled in order to evaluate the impact of the sizes and the position of the loop.

Along the simulations, it has been observed that it is essential that the loop is parallel to the rails and axles unless it performance drastically decreases.

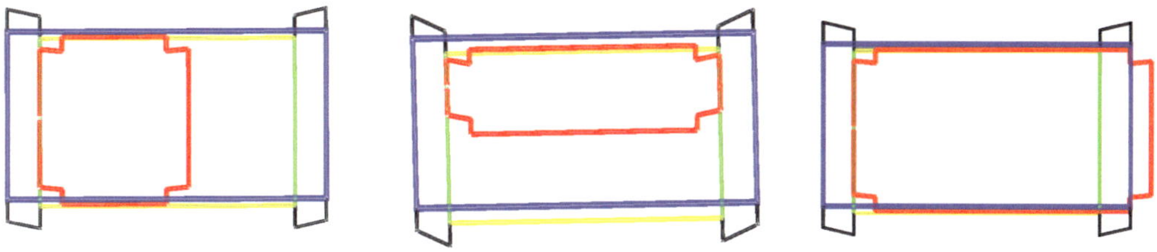

Fig. 21. Three configurations for the sizes and the position of the loop

For example, by reducing by half the length of the loop, the Urw decreases by 32.8%. And by reducing by half the width of the loop, the Urw decreases between 26% and 52% (depending on the place where it is measured -close to the loop or not). If, on the contrary, the length of the loop is increased of 25%, Urw decreases by 11.6%.

The results obtained during these simulations confirm that the coupling between the primary constituted by the loop, and the secondary constituted by the circuit rails-axles, depends on the geometry of the loop. The purpose is obviously that the secondary circuit shares as many lines of magnetic field as possible with the primary one. Since there is no magnetic material which can guide the magnetic flux, contrary to a classic transformer, it is necessary to move as close as possible from the inductive loop of the secondary conductors to increase the coupling.

Keeping the same loop and moving it under the train so that it is no more under the chariot, the distance between the primary part and the secondary one will increase and the performances will decrease. A secondary part constituted by both rails and axles of two different chariots can also be imagined. Yet the distance between the axles of two consecutive chariots is approximately 15 meters. The length of the loop is then considerably increased, and thus its inductance (ratio 3.7). The voltage source of the loop would then be oversized to obtain an Urw voltage equivalent to that of the current loop. Moreover the cost of the loop would be drastically increased and additional mechanical constraints to fix the loop would appear.

6.3.3 Variation of the diameter of the loop

Keeping the same external sizes for the loop, the diameter of the copper tube has been increased in this part of the study. As presented on Fig. 22, this implies a reduction of the internal surface of the loop.

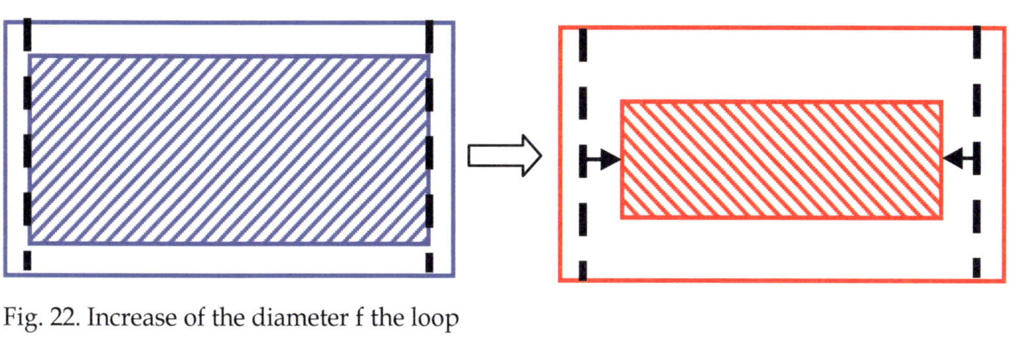

Fig. 22. Increase of the diameter f the loop

The simulations show that for a constant supply voltage the magnetic flux remains constant too. Since Φ=BS, when the diameter of the tube increases the internal surface of the loop decreases. So the induction B increases. As the current is directly proportional to the induction and as Φ=LI, the increase of the induction leads to an increase of the current and thus a decrease of the inductance of the system.

By dividing the diameter of the tube by 2 (21mm), the surface increases by 7.5% and the current in the loop decreases by 13.6%. The inductance increases by 16.7%. The result of this variation of diameter is the decrease of Urw voltage by 8%.

On the contrary, if the diameter of the tube is doubled (84mm), the surface decreases by 13.8%, the current in the loop increases by 20%. The Urw voltage increases by 8%.

But it is not possible to increase this parameter too much because the increase of the current would imply more Joule losses and a higher temperature.

At this step, a question could be to determine the maximum value for the current in the loop regarding the losses but also the EMC performances of the whole system.

6.3.4 Variation of the number of turns for the loop

Keeping the same external shape as the current loop, a second turn has been added using a copper tube of 20mm of diameter.

This additional turn considerably increases the inductance (ratio 3.7). Theoretically, the inductance is proportional to the square of the number of turn (4 in this case).

So with a constant voltage source, the current inside the loop thus decreases with the same ratio and the created magnetic field is also weaker. The Urw voltage is then decreasing by 46%. So it is necessary to increase the voltage source to keep the same performances.

6.3.5 Variation of the number of the shunts

On Fig. 19, it is clear that the four shunts, which connect the chariot to the axles to ensure a return path for the current and protect the persons of an electric risk, introduce new loops in which current can flow.

Using simulations it is possible to connect -or not- these shunts in order to evaluate their impact on the Urw voltage.

Indeed, considering only two shunts Urw voltage increases by 110%. And with only one shunt, no induced current can flow since the circuit is open and the Urw voltage increases then by 210%.But since these shunts ensure the electric safety of the persons, it is not possible to eliminate them.

6.3.6 Variation of the resistivity of the material

Considering the results presented in Table 1, the impedance of the loop is mainly inductive , all the more because the resistive part becomes negligible since the frequency is high.

So by changing the resistivity of material, aluminum instead of copper (ratio 1.6), no impact on the Urw voltage can be underlined.

This aspect is very interesting in an economic point of view because the loop could be made of a material cheaper than copper.

6.3.7 Variation of the electrical contact

To model the electrical contact between the wheel and the rail a resistor has been added because this is not a perfect short-circuit (Fig. 19).

By making the value of this resistor vary, it is possible to deteriorate the contact.

Results are presented on Table 3 for a resistance varying from 1Ω to 1MΩ.

For values higher than 5Ω, the Urw voltage remains constant.

And for a 42V voltage source, the Urw is about 3.7V. This characteristic has been drawn on Fig. 23 and is close to that of the voltage at the secondary winding of a transformer.

This study shows that even if the four wheels of the chariot are not in electric contact with the rails, as with a resistance of contact of 1MΩ, the inductive loop always allows to obtain a sufficient Urw voltage which corresponds in this case to the open circuit voltage of a transformer.

The bad quality of contacts which is one of the causes of non detection of trains inside a section is then swept away.

R	I loop	I Rail	Urw
0 Ω	12.9 A	2.55 A	4.2 V
1 Ω	12.57 V	1.50 A	3.96 V
5 Ω	12.42 V	0.367 A	3.73 V
10 Ω	12.42 A	0.185 A	3.68 V
100 Ω	12.41 V	1.85 10^{-2} A	3.68 V
1 kΩ	12.41 V	1.85 10^{-3} A	3.68 V
10 kΩ	12.41 V	1.85 10^{-4} A	3.68 V
50 kΩ	12.41 V	3.71 10^{-5} A	3.68 V
100 kΩ	12.41 V	1.85 10^{-5} A	3.68 V
500 kΩ	12.41 V	3.71 10^{-6} A	3.68 V
1 MΩ	12.41 V	1.85 10^{-6} A	3.68 V

Table 3. Comparison between simulation and measurements for an open circuit

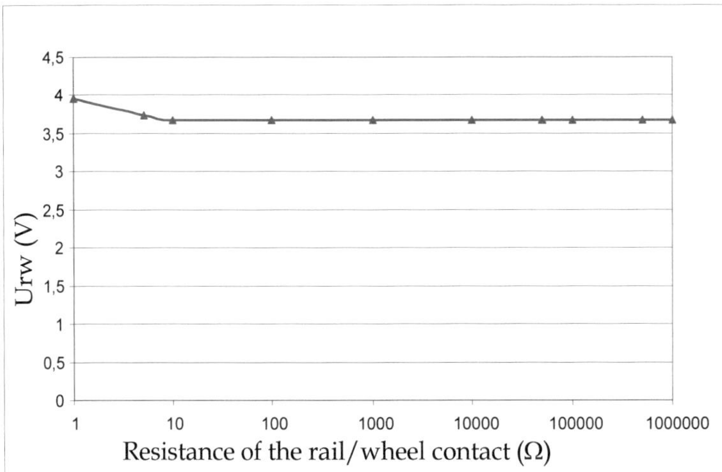

Fig. 23. Urw vs. resistance of the rail/wheel contact

6.3.8 Variation of the frequency of the source

The supply source of the inductive loop is a 147kHz sinusoidal voltage which is obtained from the 72V battery embedded on the train and electronic cards. This is a specific source and no particular constraints are linked to the other electric systems embedded on the train. So it is possible to imagine a variation of the level of voltage as well as the frequency.

It is clear that, the Urw voltage is directly linked to the voltage value (proportionality). This is logical since no magnetic material has been taken into account in this study.

If there is a magnetic material, with the increase of the voltage, saturation will appear and this proportionality relation will be wrong.

Concerning the frequency, using simulations, it has been changed from 50Hz to 1MHz.

On Fig. 24 the frequency evolution of the Urw voltage is represented.

With the obtained results, since the resistive part of the loop is very low, the famous relation $U = L\omega I$ even for low frequencies is valid. So the current into the loop could be very high for low frequency and but could decrease with increasing frequency.

And to ensure a sufficient value for the Urw voltage it must be higher than some 10A.

So compromise has to be reached to improve the inductive loop, between the values of the voltage source, its frequency, losses, possible saturation if magnetic materials are used.

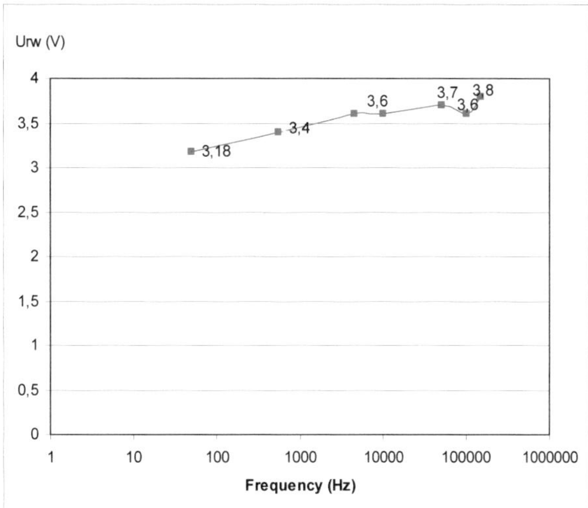

Fig. 24. Urw (V) vs. frequency (Hz) of the voltage source

7. Conclusions

The number of trains on the national network is increasing. In order to ensure a maximum security, it is necessary to localize all the trains all the time.

For that purpose, the French National Railway Company (SNCF) uses an electronic detection based on the fact that the set wheel/axles/chariot short circuits the rails.

In case of a bad shuntage, an additional device is used; its working, as well as its main parameters, is studied in this article.

A modeling process is applied with the support of the PEEC method to generate a complete electrical equivalent circuit of the device. Thanks to measurements, the accuracy of the modeling approach has been validated. The influence of geometrical and physical parameters on the performances of the studied device has been analyzed in order to find the main parameters and to optimize the structure.

Future works concerning the evaluation of the supplementary losses, the modeling of magnetic material as well as the modeling of the contact wheel/rail which is not fixed but flowing have to be achieved.

8. Acknowledgment

Authors want to thank the French National Railway Company for its financial support for this study and its help for the experimental sets.

9. References

Hoer, C. & Love, C. (1965). Exact Inductance Equations for Rectangular Conductors with Applications to More Complicated Geometries. *Journal of Research C. Engineering and Instrumentation*, Vol. 69C, No.2, (April-June 1965), pp. 127-137

Gonnet, J-P.; Clavel, E.; Mazauric; V. & Roudet, J. (2004). PEEC Method dedicated to the design of electrical systems. *Proceedings of PIERS 2004 Progress In Electromagnetic Research Symposium*, Pise, Italie, March 28-31 2004

Aimé, J.; Ardon, V.; Clavel; E., Roudet, J. & Loizelet, Ph. (2009a). EMC behavior of static converters thanks to radiated field modeling using an equivalent electrical circuit. *IEEE-IECON 35th International Annual Conference of the IEEE Industrial Electronics Society*, Porto, Portugal, November 2-5 2009

Ruehli, A.E. (1974). Equivalent circuit models for three dimensional multiconductor systems, *IEEE transaction on microwave theory and techniques*, Vol. MTT 22, No.3, (March 1974), pp. 216-221

Ardon, V.; Aimé, J.; Chadebec, O.; Clavel, E. & Vialardi, E. (2009). MoM and PEEC Method to Reach a Complete Equivalent Circuit of a Static Converter. IEEE - EMC'09, Zurich, Switzerland, January 12-16 2009

Clavel, E. & Prémont, Ch. (2007a). Function transfer sensitivity of an electronic filter versus capacitors location on a printed circuit board. *2EMC07* Rouen, France, October 18-19 2007

Jazzar, A.; Clavel, E.; Meunier, G.; Vincent, B.; Goleanu, A. & Vialardi, E. (2011). Modeling and simulating the lightning phenomenon: aeronautic materials comparison in conducted and radiated modes. *IEEE - ISIE11 20th International Symposium on Industrial Electronics*, Gdansk, Poland, June 27-30 2011

Ruehli, A. E. & Cangellaris, A. C. (2001). Progress in the methodologies for the electrical modeling of interconnects and electronic packages. *Proc. IEEE*, Vol. 89, no. 5, May 2001.

Kéradec, J-P.; Clavel, E.; Gonnet, J-P. & Mazauric, V. (2005). Introducing Linear Magnetic Materials in PEEC Simulations. Principles, Academic and Industrial Applications. *IEEE - IAS Industrial Applications Society*, Hong Kong, October 2-6 2005

Aimé, J.; Tran, S-T.; Clavel, E. & Meunier, G. (2009b). Far field extrapolation from near field interactions and shielding influence investigations based on a FE-PEEC coupling method. *IEEE Energy Conversion Congress and Exposition ECCE*, San José, Californie, USA, September 20-24 2009

Aimé, J.; Roudet, J.; Clavel, E.; Aouine, O.; Labarre, C.; Costa, F. & Ecrabey, J. (2007). Prediction and measurement of the magnetic near field of a static converter. *IEEE-ISIE 16th International Symposium on Industrial Electronics*, Vigo, Spain, June 4-7 2007, pp. 2550-2555.

Antonini, G.; Deschrijver, D. & Dhaene, T. (2007). Broadband Macromodels for Retarded Partial Element Equivalent Circuit (rPEEC) Method. *IEEE Transactions on Electromagnetic Compatibility*, Vol 49 , Issue: 1, 2007, pp. 35-48

Clavel, E.; Roudet, J.; Chevalier, Th. & Postariu, D. (2007b). Modelling connections taking into account return plane: application to EMI modelling for railway. . *IEEE-ISIE 16th International Symposium on Industrial Electronics*, Vigo, Spain, June 4-7 2007, pp. 2526-2531.

Vialardi, E.; Clavel, E.; Chadebec, O.; Guichon, J-M. & Lionet, M. (2010). Electromagnetic Simulation of Power Modules via Adapted Modelling Tools. *EPE-PEMC 14th International Power Electronics and Motion Control Conference*, Ohrid, Republic of Macedonia, September 6-8 2010.

Tan, S. X.-D.; He L. (2007) Advanced Model Order Reduction Techniques in VLSI Design, *Cambridge University Press*, New York, USA

18

General Principles Regarding the Rehabilitation of Existing Railway Bridges

Petzek Edward and Radu Băncilă
"Politehnica" University of Timişoara & SSF-RO Ltd
Romania

1. Introduction

Rehabilitation and maintenance of existing steel constructions, especially steel bridges is one of the most important actual problems [1],[2],[3],[8],[14]. The infrastructure in Romania and in other East - European countries has an average age of about seventy to ninety years. Many of these structures are still in operation after damages, several phases of repair and strengthening. Replacement with new structures raises financial, technical and political problems. The budget of the administration gets smaller. The present tendency to raise the speed on the main lines to a level of v ≤ 160 km / h must be emphasized (Figure 1).

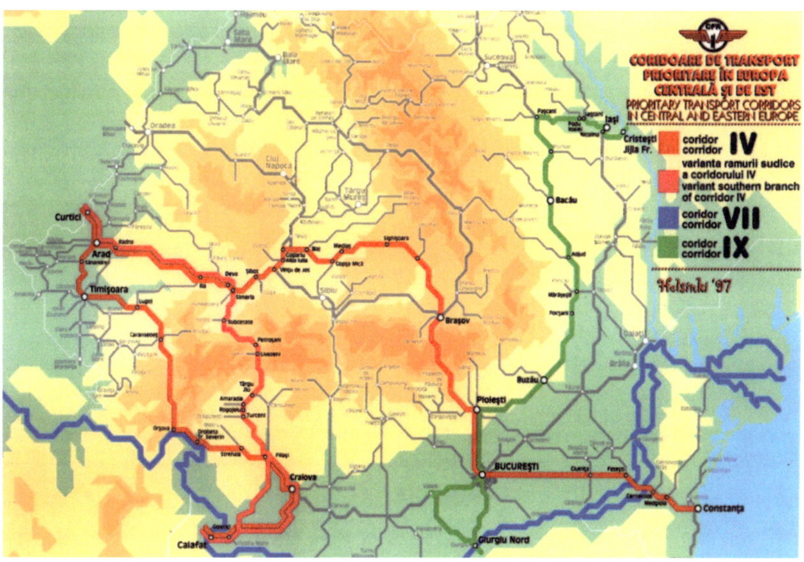

Fig. 1. The European corridors crossing Romania's territory: IV, VII and IX

During service, bridges are subject to wear. In the last decades the initial volume of traffic has increased. Therefore many bridges require a detailed investigation and control. The examination should consider the age of the bridge and all repairs, the extent and location of any defects etc (Figure 2).

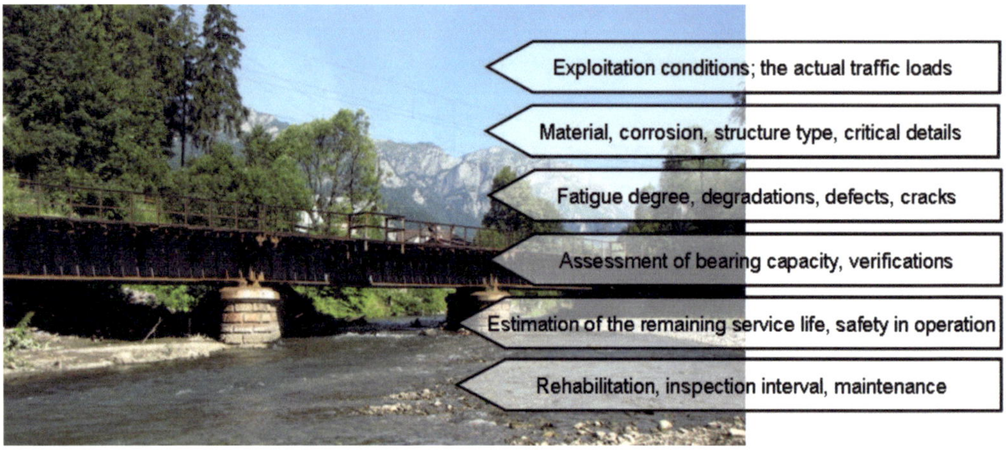

Fig. 2. Assessment and control of existing steel bridges

Carefully inspection of the structure is the most important aspect in evaluating the safety of the bridge. On the accuracy of the in situ inspection depends the level of evaluation.

The check of existing structures should be based on the complete bridge documentation (drawings with accuracy details, dimensions and cross sections of all structural elements, information about structural steel, stress history. However, in many cases these documentations are incomplete or missing. But these informations can be recovered due to the carefully investigations and inspections of the structures, experimental determination of the material characteristics and stresses in structural elements, full scale in situ tests (static and dynamic), calibration of structure and spatial static analysis.

Today, the budget of the administration and the owners (i.e. the railways and highway companies) get smaller. In consequence it is necessary to invest the available money where there will be the greatest benefit. Therefore, those responsible for the decisions need information about the safety of the structure, the remaining life, the costs for maintenance etc. Nobody will take the responsibility for failure of a structure as a result of budget restrictions.

Bridge life is generally given by fatigue; difficult is the estimation of the loading history. For bridges where the stress history is known the fatigue life may be calculated using the Miners's rule and an appropriate S-N curve; also the assumption of the same spectrum for bridge life (or for certain periods) must be made. During the process of assessment the fatigue life of old riveted bridges, is important to establish the proportion of the whole fatigue life that has been already got through. For stringers and cross girders of existing railway bridges the number of 10^7 cycles is exceeded. It might be affirmed that for such bridges if no fatigue cracks can be detected, no fatigue damage has occurred! Subsequently, if the loading spectrum remains the same in the future as in the past, fatigue cracking might not take place! In almost all the cases the loadings increased; in this situation survival for 100 years (or more) without cracking would not justify the assumption that no damage has yet occurred! Minor cracking is difficult to detect during usual inspections.

2. Technical condition of existing bridges

Rehabilitation and maintenance of existing steel bridges is one of the most important actual problems.

A continuous maintenance, which generally must increase in time, is important in order to assure the safety in operation of the existing structures.

The evaluation of the current technical condition of an existing steel bridge structure with the help of in situ evaluation depends, in a high percentage, on the engineer's qualification. In the case of in situ inspection it is recommended to insist on the appeared fatigue defects, of riveted or welded connections (the expert must to insist on the connections between the stringers with cross girders and cross girders with main girders), critical details (which are included in standards or catalogues), corrosion level, the structure deformations due to traffic, bridge bearings. The expert can use non-destructive techniques to determine the integrity of base material or structural components. The non destructive testing of the inspection of in service damage in bridge elements are the following:

- *visual inspection* - the most common method which includes microscopes, mirrors, portable video cameras, robotic crawlers; this method is very useful in case of surface cracks;
- *magnetic particle inspection* - this method is also very simple and does not need high qualification personnel, but can be applied just in case of ferromagnetic materials (not for austenitic steels). The method consists in the magnetization of the high stress elements or critical details and indicates directly the surface discontinuity through forming a distorted magnetic field, which can be detected under proper lighting conditions;
- *liquid penetration inspection* - is a simple method including the qualification of the personnel; it uses penetrate liquids with fluorescent pigment and UV - light in order to indicate the surface defections;
- *radiographic inspection* - the method is applied for hidden defects and it uses Gamma or Roentgen radiation. The inspected element is placed between the radiation and the film. The interpretation of the radiographic images should be done by experts, otherwise defects could be ignored;
- *ultrasonic inspection* - this testing is used for flaws and cracks in the material thickness, on the surface or hidden defects; highly qualified personnel is needed. High frequency sound waves are introduced into a material and they are reflected back from surfaces or flaws. This process is recorded by an oscilloscope. This method cannot be used for elements made of multiple plates (riveted sections).
- *Eddy Current testing* - this method can detect surface defects but can also be used for thickness inspection.

The usual simplified analysis methods do not always give the lowest resistance values for the structure, but usually the more refined assessment methods which give greater resistance are expensive. According to the experience of the expert a progressive analysis can be applied. In a first step simple classical methods can be applied [2]. If they fail, more sophisticated methods can be used, until either it is shown that the bridge is adequate, or it is concluded that strengthening is needed. An engineer with experience can jump over some time consuming steps which do not give any benefit (an interesting proposal is that the engineer will be paid on a percentage on the saving basis [4]). Actual loads are lower than those used for design purposes. Fatigue tests on elements taken from demolished structures gives - generally - greater fatigue life than the values according the codes [5].

The applied stress range, the geometry of the detail and the number of stress cycles has a decisive effect on the remaining fatigue life of the structures.

By differences of more than 5 % of the cross section - due to corrosion, the actual values must be introduced.

However, from the overall examination of a large number of bridges many defects can be pointed out. The defects are widespread, having a heterogeneous character from the point of view of location, development and development tendency; their amplification was also due to the climate and polluting factors that caused the reduction of the cross section due to corrosion. Statistically, in 283 from among 1088 welded bridges, and in 356 from among 3201 steel riveted bridges cracks were detected and repaired. It is not allowed to weld cracks! Old bridges can have welds executed in the early years; a special attention must be paid to these parts. Generally the riveted connections have a good behavior in time due to the initial pre-stressing force which can reach 70 - 80 N/mm^2. In Table 1 some typical defects in stringers, cross girders, main girders wind bracings and orthotropic deck and their repair are presented [6],[7].

In figure 3 there is presented a crack of a joint plate from a wind bracing and in figure 4 also cracks in the lower flange joint of a double T girder, near to the bearing.

Fig. 3. Crack in the joint plate

Fig. 4. Crack in the lower flange near to the bearing

<table>
<tr><th>Element</th><th>CRACK</th><th>REPAIR</th></tr>
<tr><td>0</td><td>1</td><td>2</td></tr>
<tr><td>Main girder</td><td>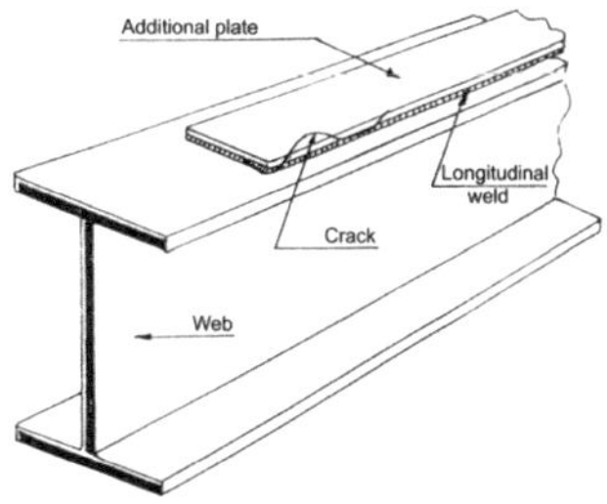
</td><td>⇨ Grinding
⇨ New additional plate
⇨ Observation</td></tr>
<tr><td>Main girder</td><td>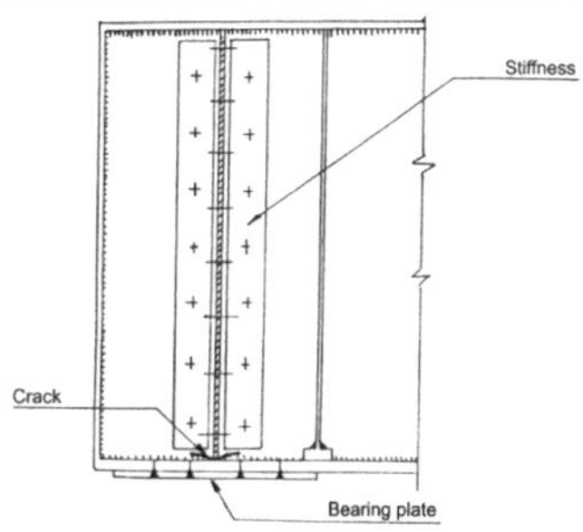
</td><td>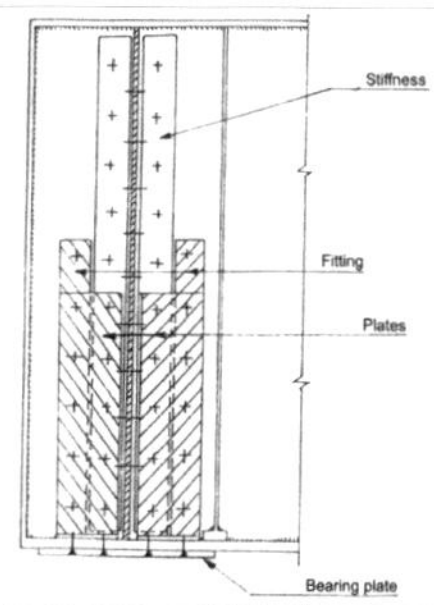
</td></tr>
<tr><td>Main girder</td><td>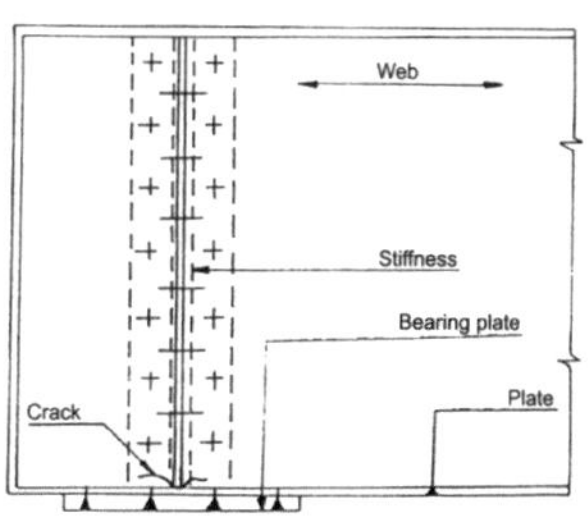
</td><td>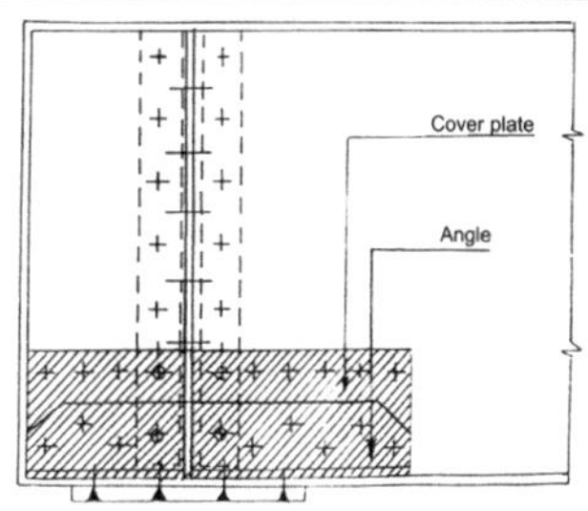
</td></tr>
<tr><td>Stringer</td><td>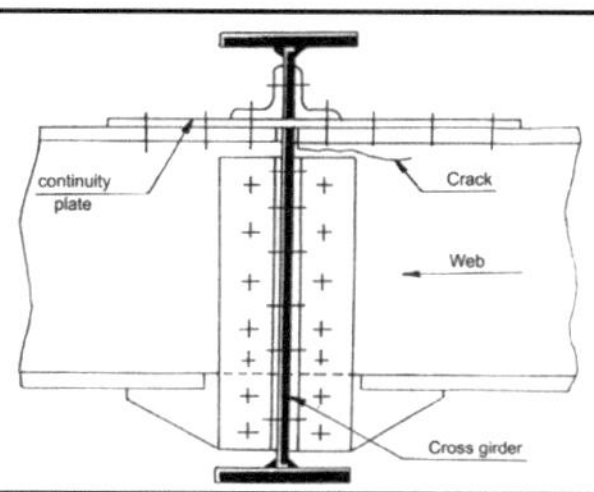
</td><td>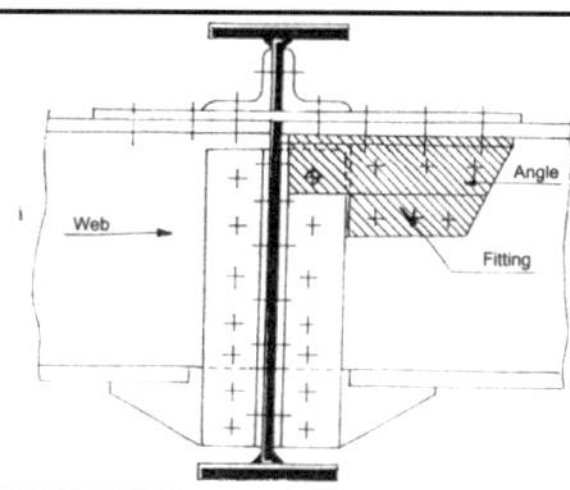
</td></tr>
<tr><td>Stringer</td><td>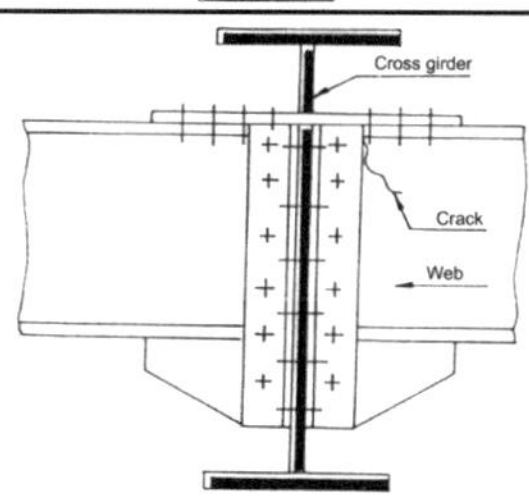
</td><td>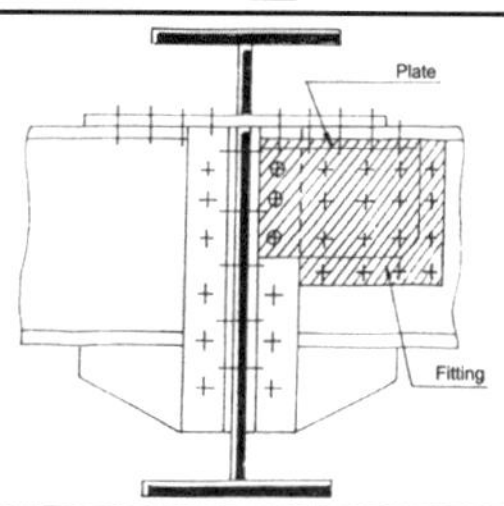
</td></tr>
</table>

Table 1. Typical defects in welded and riveted steel bridges

<table>
<tr><th>Element</th><th>CRACK</th><th>REPAIR</th></tr>
<tr><td>Stringer</td><td>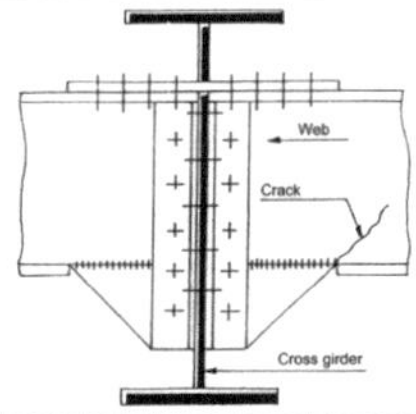
</td><td>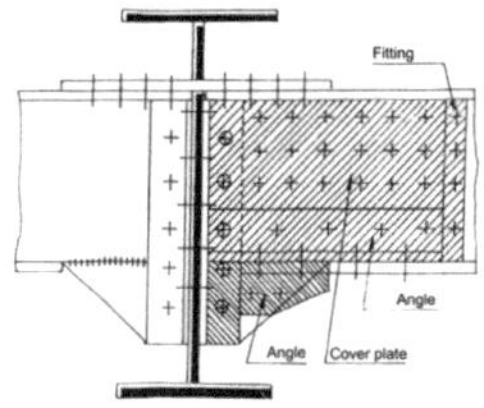
</td></tr>
<tr><td>Stringer</td><td>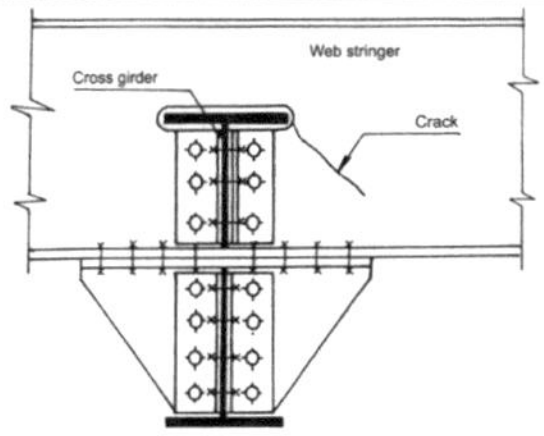
</td><td>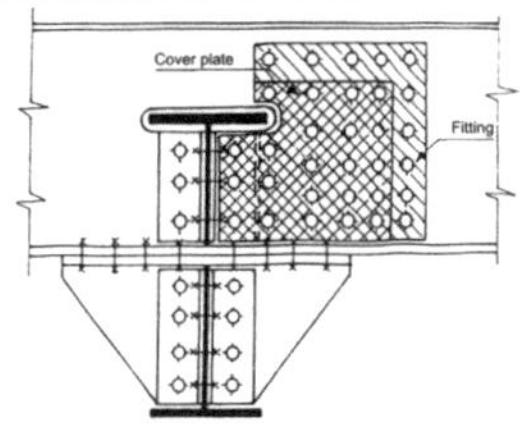
</td></tr>
<tr><td>Cross girder</td><td>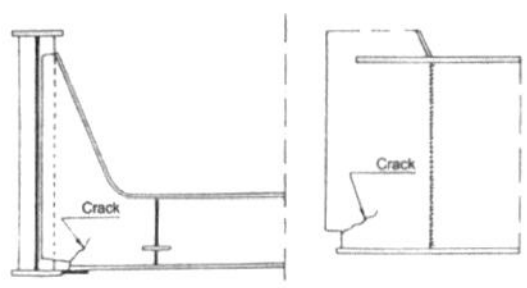
</td><td>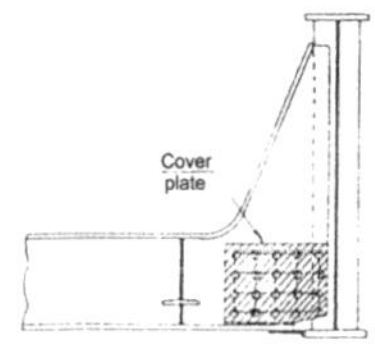
</td></tr>
<tr><td>Cross girder</td><td>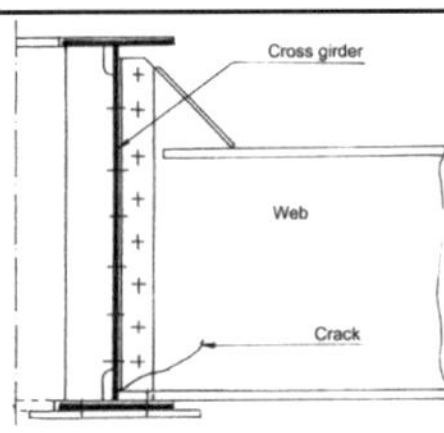
</td><td>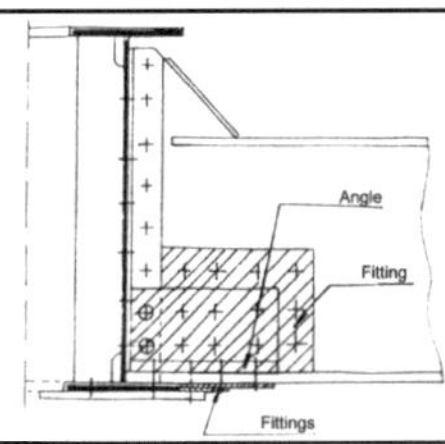
</td></tr>
<tr><td>Orthotropic deck</td><td>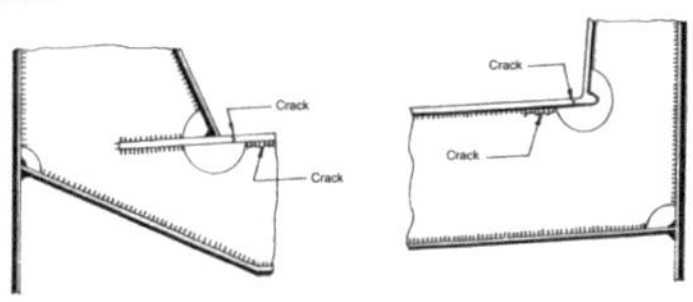
</td><td>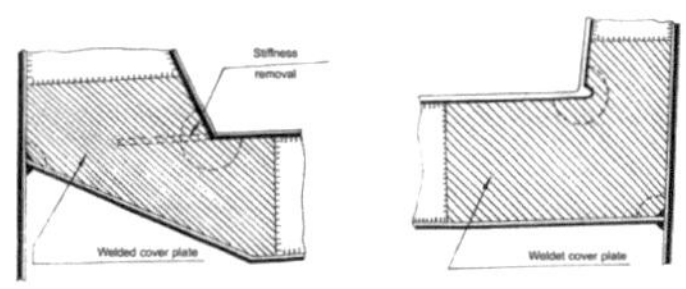
</td></tr>
<tr><td>Orthotropic deck</td><td>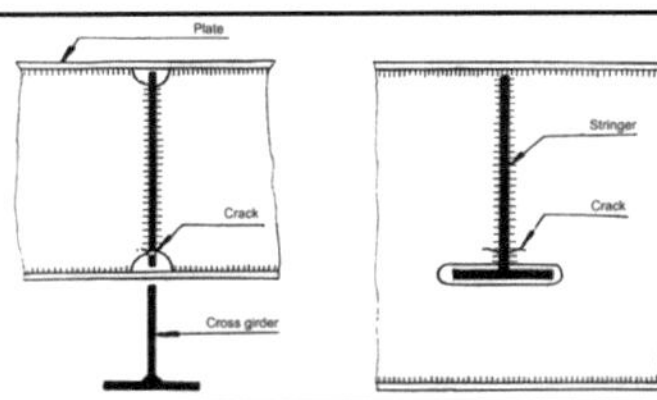
</td><td>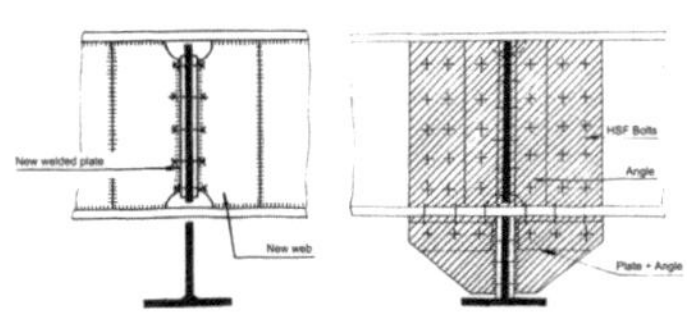
</td></tr>
<tr><td>Stringer</td><td>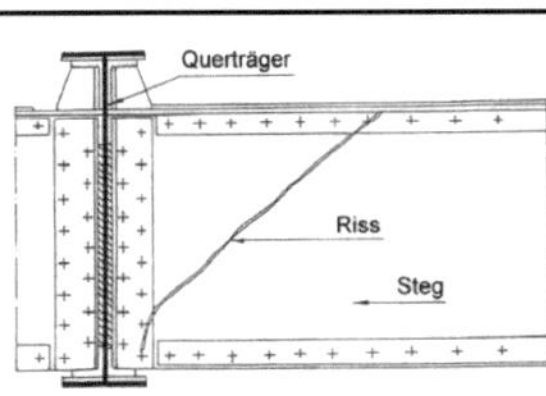
</td><td>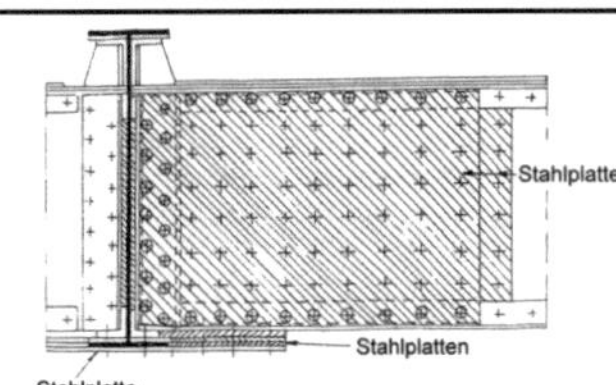
</td></tr>
</table>

Table 1. Typical defects in welded and riveted steel bridges - continuation

<table>
<tr><th>Element</th><th>CRACK</th><th>REPAIR</th></tr>
<tr><td>Stringer</td><td>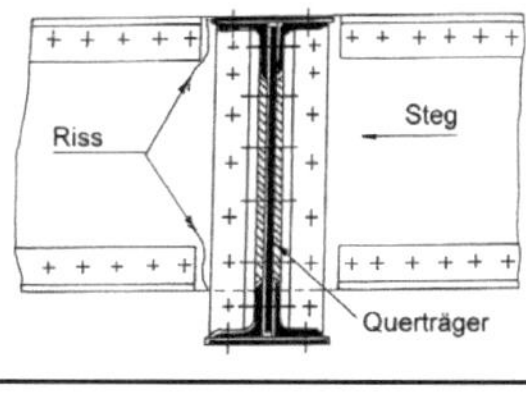
</td><td>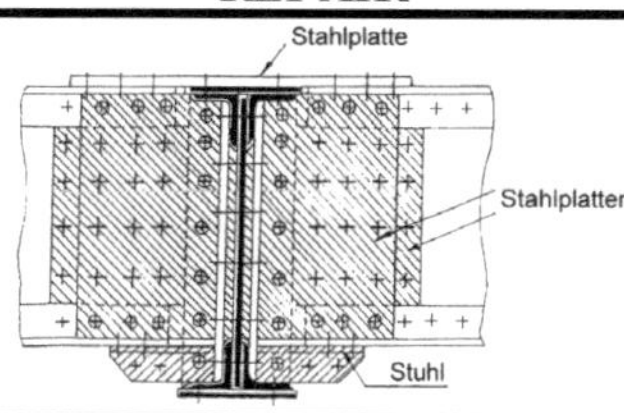
</td></tr>
<tr><td>Stringer</td><td>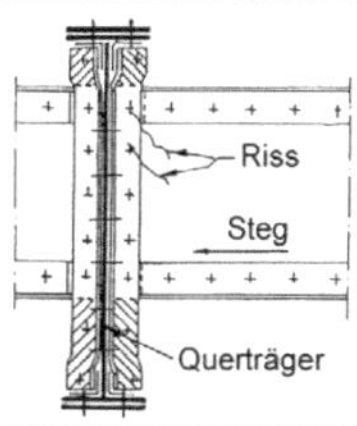
</td><td>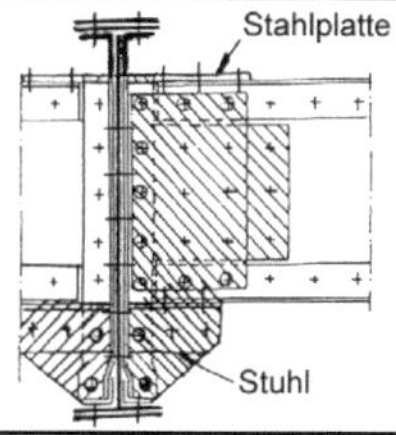
</td></tr>
<tr><td>Stringer</td><td>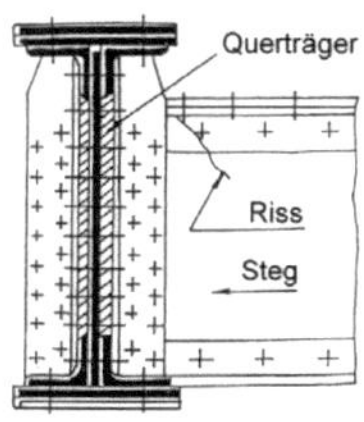
</td><td>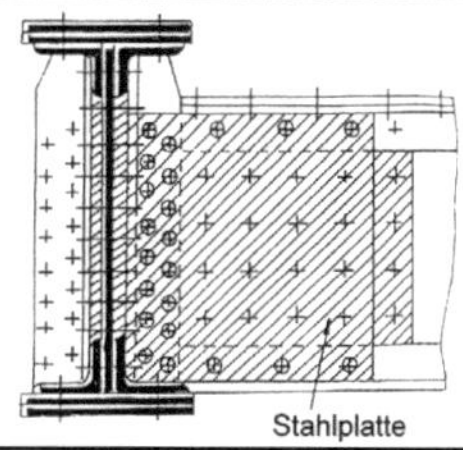
</td></tr>
<tr><td>Stringer</td><td>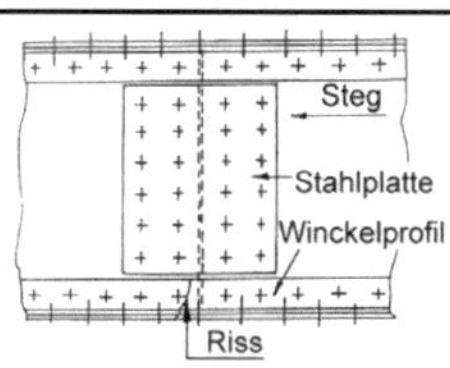
</td><td>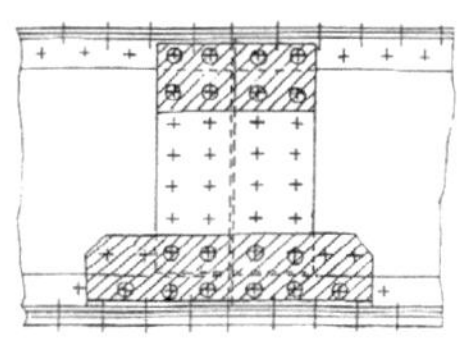</td></tr>
<tr><td>Twin girders</td><td>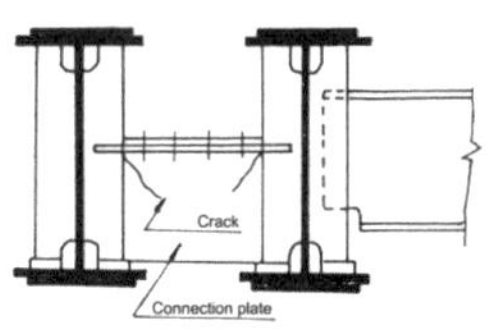
</td><td>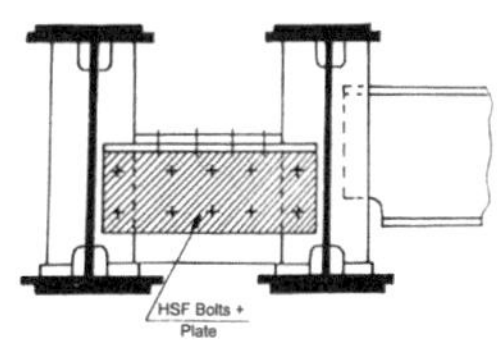
</td></tr>
<tr><td>Twin girders</td><td>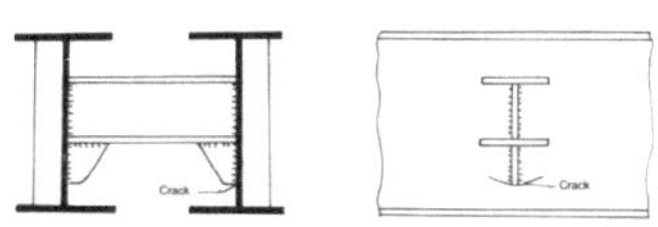
</td><td>⇨ Hole at the end of the crack
⇨ Replacement</td></tr>
<tr><td>Wind bracing</td><td>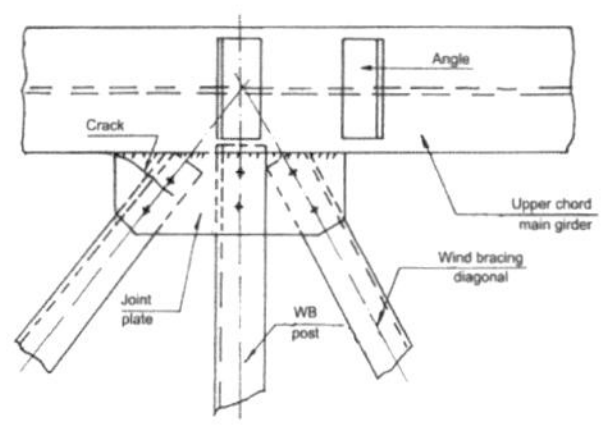
</td><td>⇨ Replacement of the gusset</td></tr>
</table>

Table 1. Typical defects in welded and riveted steel bridges - continuation

Signs of cracks and defects are rust traces, which occur by friction between jointed plates of the elements. These are of relevance for hidden constructive elements.

Defects of the bearings are frequent as well. In figure 5 there are presented two examples.

Fig. 5. Defects on bearings

Due to deficient maintenance the riveted structures are strongly corroded, especially in the lower zones (figure 6). These corroded surfaces represent also a critical detail for fatigue. Fatigue under corrosion factor is an aspect treated only qualitatively and not quantitatively. The interaction between the two aspects is obvious. There can also found inefficient joints, like weakened rivets (figure 7).

Fig. 6. Strong corrosion at the lower flange - railway bridge on a main line

Fig. 7. Inefficient joints - railway bridge on a main line

3. Characteristics of materials

The following facts show that a material analysis for old riveted bridges is very useful:

- Old bridges are in many cases erected using material with very poor welding qualities and basing on railway administration data and specialized literature it is known that cast iron was used to build bridges;
- The specialized literature doesn't offer enough information about this structural steel;
- The structural material comes from several producers (for South Eastern Europe mostly from Reschitz - Romania and Györ - Hungary).

Span	Year of construction	Position	Type of structures
101,76 m	**1912** (St.E.G. Reschitz)	**Mehadia**	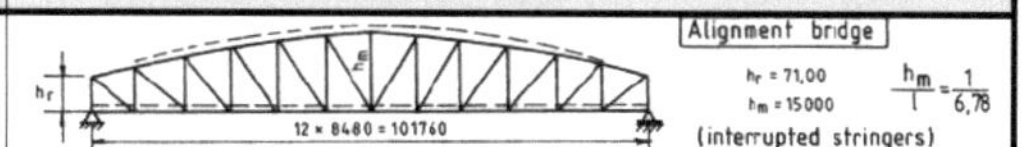
81,6 m	**1911** (St.E.G. Reschitz)	**Valea Cernei**	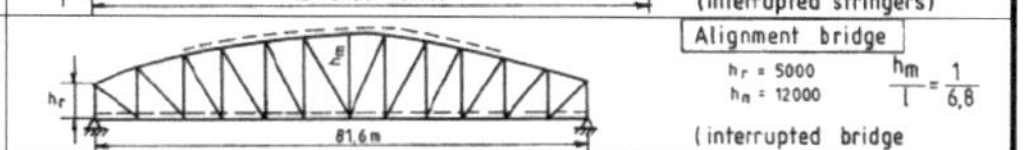
71,76 m	**1913** (St.E.G. Reschitz)	**Balta Sărată**	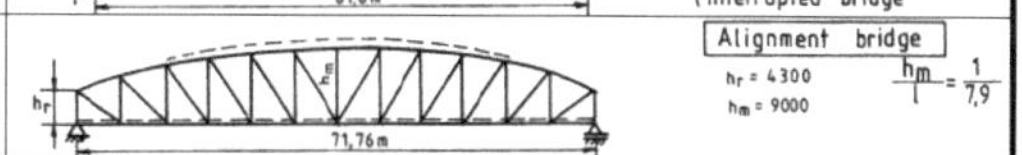
50,65 + 66,20 + 36,70 + 2x30,40 m	**1912** (St.E.G. Reschitz)	**Şag** replaced 1997	
2x76,80 + 3x51,42 m	**1912** (Györ)	**Arad** replaced 2000	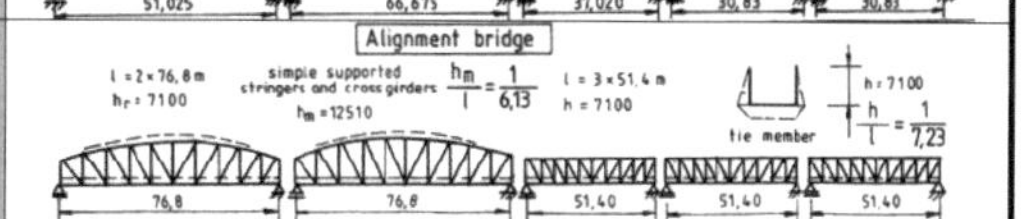
56,2 + 2x31,0 + 4x 32,7 m	56,2 m 1927 (Reschitz) 2x31,0 + 4x32,7 m **1907**	**Brănişca**	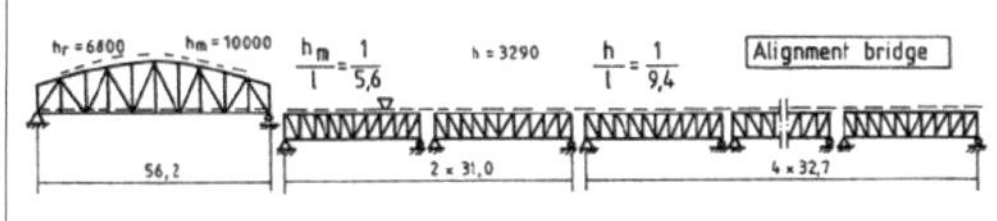
			* Possible construction year

Table 2. The bridges on which the present material study basis

The presented study's results can be extended to Middle and South Eastern Europe when the history of communication ways and the state of old railway and highway steel bridges in this region is regarded. In this context the following event can be mentioned: on the 1st of January 1855 the "Kaiserliche und Königliche Privilegierte Österreichische Staatseisenbahngesellschaft" (St.E.G.) took over all steel producers in Banat. The investments in Reschitz (in present Resita) turn the steel mill into an important bridges' factory. The production of steel bridges reached 3960 tonnes in 1910. Bridges made by St.E.G. Reschitz are still in use in Romania, Austria and Hungary. Between 1911 - 1913, 1620 tonnes of bridge structures made of cast iron were replaced in the western part of Romania (Banat), namely on the railway segment Timişoara - Orşova. In this sense the material study took into account bridges from this region, built around 1911. Following material analysis were performed in order to determine the characteristics of the material: chemical analysis, metallographic analysis, tensile tests, Brinell tests, Charpy "V Notch" tests. The samples were taken from secondary elements, but also in some cases (Bridges in Arad and Şag which were replaced) from main elements: stringers, cross girders, main girders [7]). The results of the chemical analysis are presented in the following table.

No.	Bridge / Specimen number & position	Chemical composition						
		C %	S %	Mn %	P %	Si %	Ni %	N %
1.	Valea Cernei Bridge / S1-VCB / Secondary elements	0,12	0,034	0,39	0,021	0,023	-	-
2.	Valea Cernei Bridge / S2-VCB / Sec. Secondary elements	0,11	0,020	0,32	0,024	0,010	-	-
3.	Mehadia Bridge / S1-MB / Secondary elements	0,11	0,014	0,50	0,028	0,050	-	-
4.	Balta Sărată Bridge / S1-BSB / Secondary elements	0,07	0,072	0,38	0,016	0,010	-	-
5.	Şag Timiş Bridge / S1-STB / Secondary elements – Span I	0,16	0,058	0,46	0,035	0,035	-	-
6.	Şag Timiş Bridge / S2-STB / Secondary elements – Span II	0,14	0,066	0,63	0,030	0,061	-	-
7.	Şag Timiş Bridge / S3-STB / Secondary elements – Span III	0,13	0,060	0,50	0,051	0,112	-	-
8.	Şag Timiş Bridge / S4-STB / Secondary elements – Span IV	0,18	0,054	0,48	0,066	0,056	-	-
9.	Şag Timiş Bridge / S5-STB / Secondary elements – Span V	0,18	0,017	0,44	0,059	0,010	-	-
10.	Brănişca Bridge / S1-BB / Secondary elements	0,12	0,070	0,39	0,013	0,060	-	-
11.	Brănişca Bridge / S2-BB / Secondary elements	0,11	0,055	0,38	0,014	0,060	-	-
12.	Arad Bridge / S1-AB / Secondary elements – Span I	0,14	0,061	0,53	0,038	0,069	-	-
13.	Arad Bridge / S2-AB / Secondary elements – Span II	0,19	0,033	0,64	0,051	0,018	-	-
14.	Arad Bridge / S3-AB / Main elements – Stringers Span III	0,089	0,032	0,531	0,009	0,018	0,067	-
15.	Arad Bridge / S4-AB / Main elements – Cross girders Span III	0,058	0,059	0,485	0,017	0,018	0,037	-
16.	Arad Bridge / S5-AB / Main elements – Main girder Span III	0,056	0,032	0,493	0,001	-	0,031	-
17.	Arad Bridge / S6-AB / Secondary elements – Span IV	0,18	0,035	0,46	0,063	0,078	-	-
18.	Arad Bridge / S7-AB / Secondary elements – Span V	0,1	0,047	0,43	0,020	0,030	-	-
19.	St 37 (STAS 500/2-80)	0,25	0,065	0,85	0,065	0,07	0,30	0,015
20.	St 34 (STAS 500/2-80)	0,17	0,055	0,60	0,055	-	-	-
21.	INCERTRANS Bucharest Research for cast iron	0,04... 0,11	0,014...0,043	0,15... 0,48	0,121...0,32	0,07... 0,31	-	-
22.	German Research for cast iron*	0,16	0,056	0,100	0,470	0,100	0,007	-

* German research – Stahlbau 05.1985 (Prof.Dr.Ing. D. Kosteas, Ing. W. Stier, Ing. W. Grap)

Table 3. Chemical analysis results

The statistical interpretation of the tensile tests results shows a minimal value for the yield stress of 230 N/mm^2.

The impact tests on Charpy V Notch specimens lead to conclusion that the transition temperature is situated in many cases in the range from -10°C to 0°C (figure 8).

By analyzing the laboratory results it can be concluded that the steel is a mild one, that could be associated to the present steel types St 34 or St 37.1 (S 235 according to the Eurocodes). Also, on the two dismantled bridges - Arad and Şag - fracture mechanics tests were made [7] in order to establish the integral value J_c according to ASTM E813-89 (figure 9), the CTOD and to determine the fatigue crack growth rate and the material constants C and m according to ASTM E647-93. For these tests compact specimens CT (thickness 8 mm) as well as bending specimens for CTOD, have been used.

They have been obtained from the stringers, cross girders and main girder - lower chord. The minimal value of material toughness in term of J-Integral for these old riveted steel bridges is J_{crit} = 10 ... 20 N/mm for a temperature of -20°C.

The method of fatigue assessment for structural elements with defects, was developed basing on the possibility of crack propagation modeling under fatigue loads and with the help of known laws.

The method is founded on the recommendations of the BS 7910:1999.

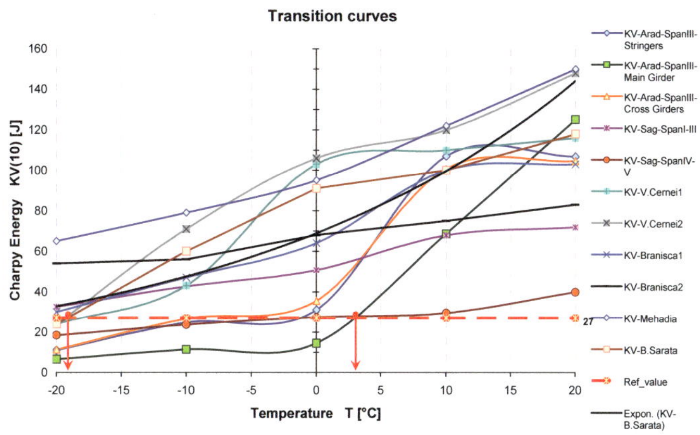

Fig. 8. Transition curves for the analyzed bridge structures

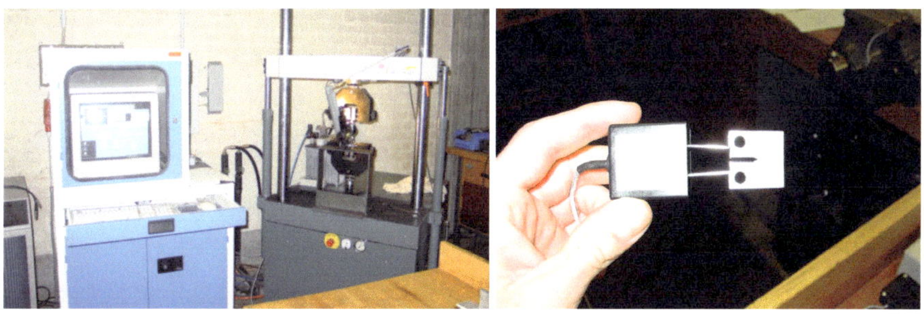

Fig. 9. Experimental tests for determining of J values

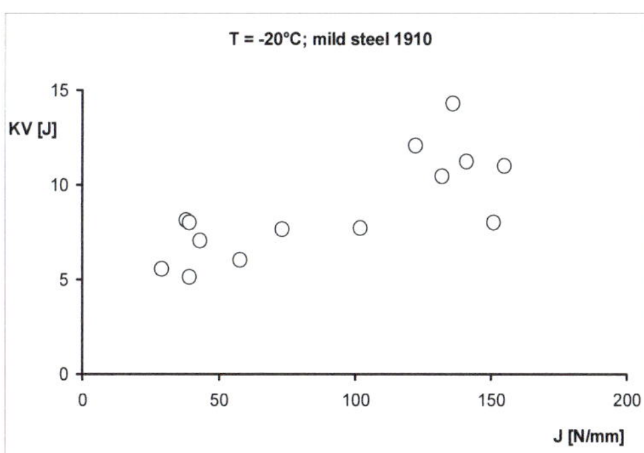

Fig. 10. Charpy V Notch Energy vs. material toughness J_c (Arad Bridge 1912)

In the present state of knowledge it is generally accepted that the fatigue failure of materials is a process containing three distinct steps: (1) initiation of defect (crack), (2) crack propagation in material, (3) separation through complete failure of the material in two or more pieces. Practically, the safety service life of an element under fatigue conditions can be expressed as follows (figure 11):

$$N_f = N_i + N_p \tag{1}$$

N_i = number of cycles necessary for the initiation of the defect (crack)
N_p = number of cycles necessary for the propagation of the defect until the occurrence of failure.

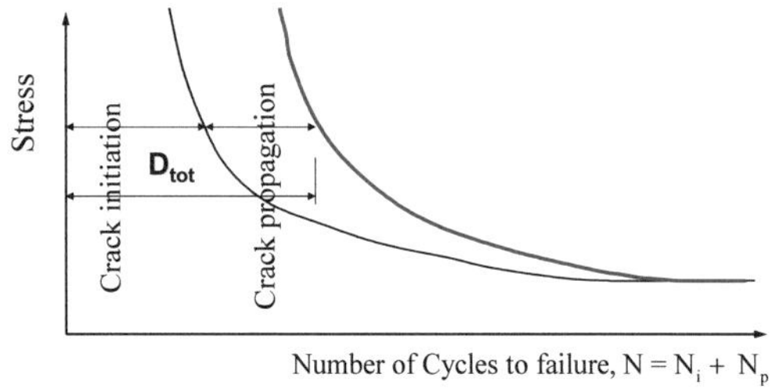

Fig. 11. Fatigue life of structural elements

The evaluation of crack propagation conditions can be accomplished with the help of characteristically values, which are founded on fracture mechanics concepts: material toughness express by the stress intensity factor K or J integral value and the crack growth rate da/dN (crack growth for each load cycle). A relation of the following type can express the crack growth rate (Figure 12):

$$\frac{da}{dN} = f(\Delta K, R, H) \tag{2}$$

da/dN - crack extension for one load cycle; ΔK- stress intensity range, established basing on the stress range Δσ; R- stress ratio; H- indicates the stress history dependence.

The crack growth rate da/dN, defined as a crack extension - da obtained through a load cycle dN (it can also be defined as da/dt, in which case the crack extension is related to a time interval), represents a value characteristic of the initiation phases respectively the stable crack propagation. It has been experimentally observed that the connection between the crack growth rate and stress intensity factor variation represents a suitable solution for the description of the behavior of a metallic material containing a crack, as in the case of steel. In a logarithmic graphical representation of the crack growth rate da/dN versus the stress intensity range ΔK a curve as the one in the following figure (Fig. 12), is obtained.

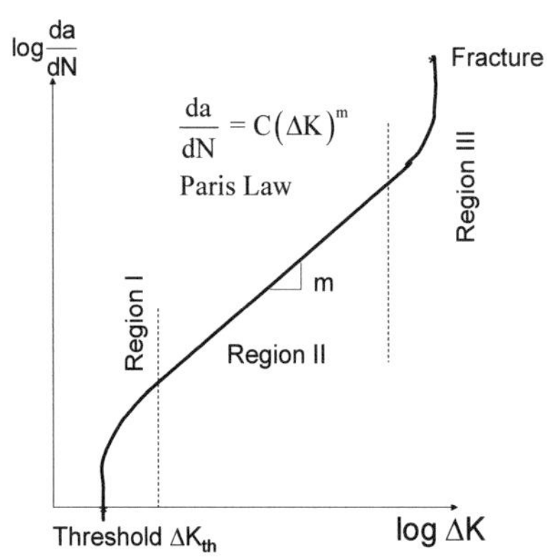

Fig. 12. Logarithmic representation of the fatigue crack growth in steel

The calculation of the structural elements remaining service life can be done basing on the Paris law, more precisely by integration of this law:

$$N = \int_0^N dN = \int_{a_0}^{a_{crit}} \frac{da}{C \cdot \Delta K^m} \tag{3}$$

N - number of stress cycles necessary in order that the crack extends from its initial dimension a_0 to the critical value a_{crit}, where failure occurs;
a - crack length;
C, m - material constants from the crack propagation law;
ΔK - stress intensity factor range.

This integral can be numerically calculated by taking into account a critical detail knowing the crack values (initial and critical), basing on the following relation:

$$N = \int_{a_0}^{a_{crit}} \frac{da}{C \cdot \Delta\sigma^m \cdot Y^m \cdot (\pi a)^{-m/2}} \tag{4}$$

The number of cycles Ni obtained with the help of relation (4) represents the remaining service life of the detail, by regarding the initial length a_0 up to the critical length a_{crit}, by admitting stable crack propagation (Figure 13).

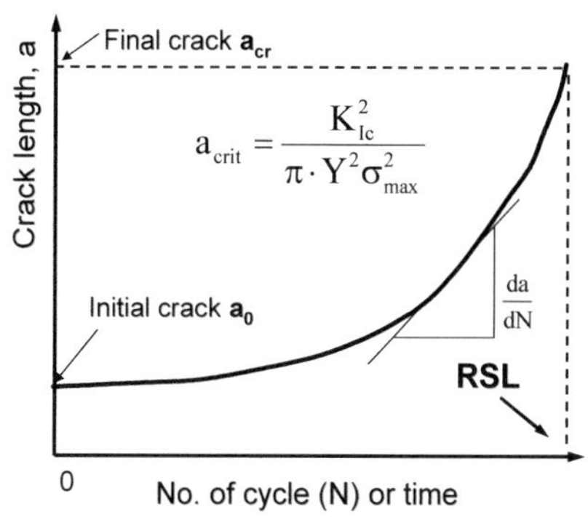

Fig. 13. Principles for determining the remaining service life

The critical crack value can be calculated basing on the K criterion respectively on the J or CTOD criterions or with the help of the failure assessment diagram.

The C and m material constants from the Paris law are experimentally determined by fracture mechanical tests. In this sense in most cases compact specimens C(T), three point bended specimens SEN(B) and middle central panels M(T) are used. Such a standard which describes the test methods for the determination of the crack growth rate, is the American Standard ASTM E 647 (Standard Test Method for Measurement of Fatigue Crack Rates).

Basing on the determined values da/dN and ΔK the program presented below, also automatically determines the C and m material constants by the Paris relation:

$$\ln\left(\frac{da}{dN}\right) = \ln C + m \cdot \ln \Delta K \tag{5}$$

The experimental tests on 26 CT specimens from two old bridges have shown that for the oldest mild steels the values of the material constants from the Paris relation are in the following intervals: m = 2,05 … 5,65 and C = 2,2 x 10^{-11} … 10^{-18}.

Relatively large value of m corresponding to very small values of C, for example for m > 4 → C ≅ 10^{-15} … 10^{-18}.

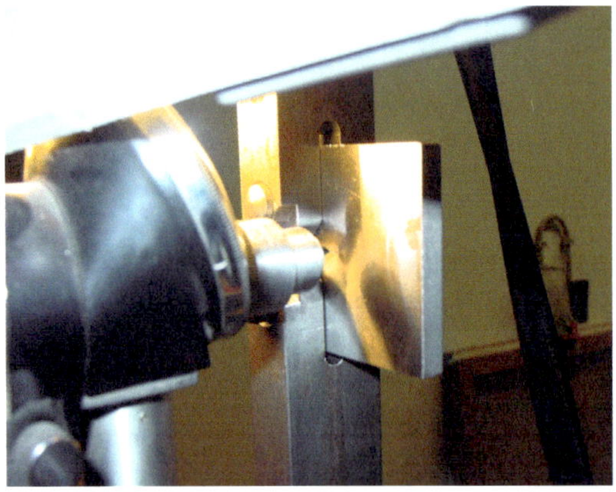

Fig. 14. Experimental tests in order to establish the material constants C and m

4. Present verification concept

During service bridges are subjected to repeated loadings causing fatigue. Therefore many bridges require an inspection. The examination should consider the age of the bridge and all repairs, the extent and location of any defects etc. [8]. A continuous maintenance, which generally must increase in time, is important in order to assure the safety in operation of the existing structures. The present methodology includes the following stages [9]:

Step 1. estimation of the loading capacity of the structure based on a detailed inspection; analysis of drawings, inspection reports, repairs, reinforcements, analysis of the general behavior of the bridge (displacements, vibrations, corrosion and cracks). In this phase the stresses in the structure can be calculated with the usual simplified hypothesis;

Step 2. the accurate determination of the stresses in the structure and of the remaining safety of the elements. This phase includes: tests on materials, computer aided analysis of the space structure, remaining safety calculated on the base of the real time - stress history;

Step 3. in situ static and dynamic tests.

This methodology adopted by the Romanian standard is illustrated in Figure 15.

The calculation of remaining fatigue life is normally carried out by a damage accumulation calculation. The cumulative damage caused by stress cycles will be calculated; failure criteria will be reached.

$$D = \sum \frac{n_i}{N_i} \leq 1 \tag{6}$$

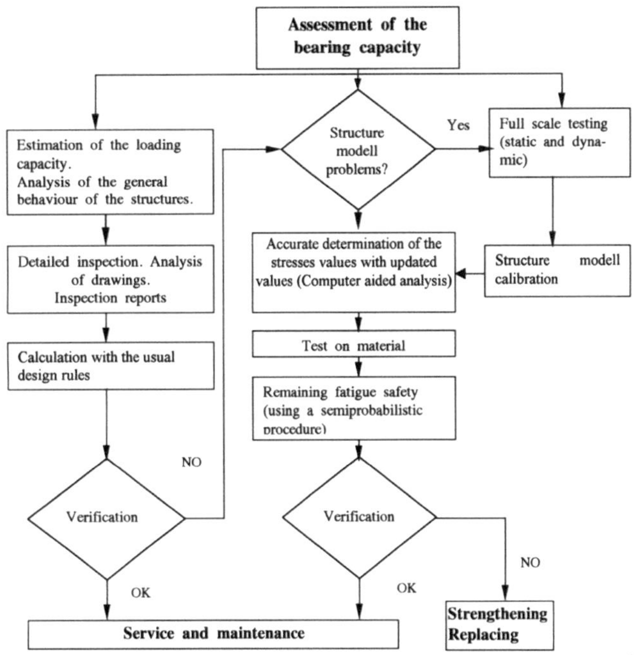

Fig. 15. Methodology of the Romanian standard SR 1911-98 [3]

The classical fatigue concept is based on the assumption that a constructive element has no defects or cracks. However, discontinuities and cracks in the components of structures are unavoidable, basically because of the material fabrication and the erection of structures. It is very clear that the kind of fatigue cracks, which are initiated by structural non-homogeneity (possible non-metallic inclusions or other impurities), surface defects (including corrosion) and the stress factor, are present in the old riveted structures.

The presence of cracks in structural elements modifies essentially their fracture behavior. Fracture, assimilated in this case as crack dimensions growth process under external loadings, will be strongly influenced by the deformation capacity of material. The FM approach has acceleration in damage increase; with increasing damage a smaller stress range contribute to the damage increase. The authors proposed [9], a complementary method based on the fracture mechanics basic concept

$$J_I \le J_{Ic} \tag{7}$$

in order to calculate the remaining fatigue life.

In practice two situations can be distinguished:

- D < 0,8 the probability to detect cracks is very low. The inspection intervals (generally between 3 - 6 years) can be established on criteria independent of fatigue. Nevertheless, a special attention must be paid to critical details.
- D ≥ 0,8 cracks are probable and possible. An in situ inspection and the analysis of critical details are strongly necessary. Also a fracture mechanics approach is recommended.

Generally, the establishing of the maintenance program, the determination of inspection intervals, the inspection priorities of structural elements and finally the calculation with high accuracy of the remaining service life of old riveted bridges takes into account the following main data:

- type of structure and exploitation conditions (traffic events);
- information about structural steel (mechanical properties - yield strength, tensile strength, hardness, transition curve ductile - brittle and transition temperatures, chemical composition, metallographic analysis);
- determination of critical members and details;
- crack detection and inspection techniques for evaluation of the initial crack size - a_0 and crack configuration;
- recording of the stress spectrum for the critical members under the actual traffic loads;
- evaluation of the critical crack size - a_{crit} based on failure assessment diagrams;
- fracture mechanics parameter - K_{crit}, δ_{crit}, J_{crit} (fracture toughness);
- simulation of the fatigue crack growth;
- temperature, environment conditions.

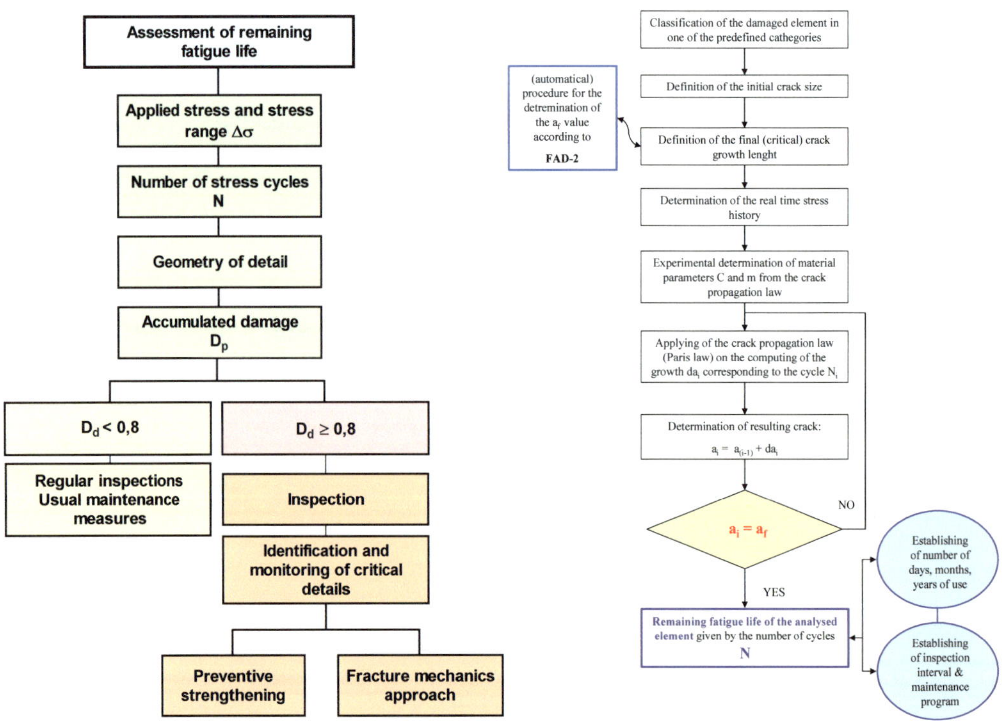

Fig. 16. Assessment of the remaining fatigue life and the crack growth procedure

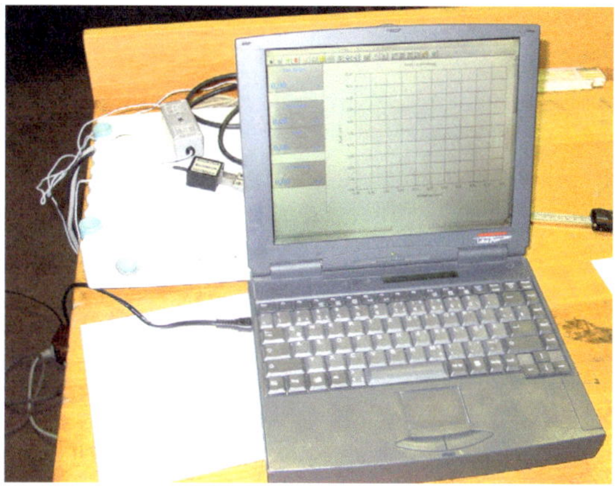

Fig. 17. Soft for life prediction analysis

The methodology is conceived as an advanced, complete analysis of structural elements containing fatigue defects, being founded on fracture mechanics principles and containing two steps: namely one of determination of defects' acceptability with the help of Failure Assessment Diagrams (level 2) [10] and of determination of final acceptable values of defect dimensions; this is followed by a second step which in fact represents a fatigue evaluation of the analyzed structural elements basing on the present stress history recorded on the structure, on the initial and final defect dimensions and the FM parameters, namely the material characteristics C and m from the Paris relation (crack growth under real traffic stress) and further on the exact determination of the number of cycles N needed in order that a fracture take place, respectively the determination of the remaining service life of the structural elements (years, months, days).

In order to determine the remaining service life it is important to know how long it will take the crack to grow from the minimum detectable size to the critical value. In this situation the safe inspection intervals can be calculated with the following relation:

$$\Delta T_{insp} = N_{RFL} / N_{\Delta a}, \qquad (8)$$

where

N_{RFL} = the remaining fatigue life calculated for the structural element
$N_{\Delta a}$ = the number of cycles computed for a crack extension rate of 5 mm during two successive inspections.

The life prediction computing was performed with the help of a soft developed by one of the authors [7].

5. Case studies

Three European corridors cross Romania's territory: IV, VII and IX. Of a special interest for many European countries is the Pan-European Corridor IV „Berlin - Nürnberg - Prague - Budapest - Constanta /Istanbul/ Thessaloniki". On Romania's territory the railway component of Pan - European Corridor IV has the following route: Curtici - Arad - Sighisoara - Brasov – Predeal – Campina - Bucharest - Constanta.

Due to the fact that the Campina - Predeal railway route crosses the sub-Carpathians area and the southern part of the Southern Carpathians it was necessary to adapt it to the difficult geographical conditions; actually this section is the most complicated part from the whole Romanian route. The railway line was built step by step beginning with 1879. The doubling of the 43 km long railway line Campina - Predeal was accomplished between 1939 - 1942. The line electrification was completed during 1961 - 1965.

Generally, the line is oriented from the south to the north. It follows the valley of the Prahova River crossing this river in 17 points.

This section includes 41 bridges. From this total number 22 are plate girder bridges, 6 are truss structures, 12 concrete bridges and 1 are conceived in the filler beams deck solution.

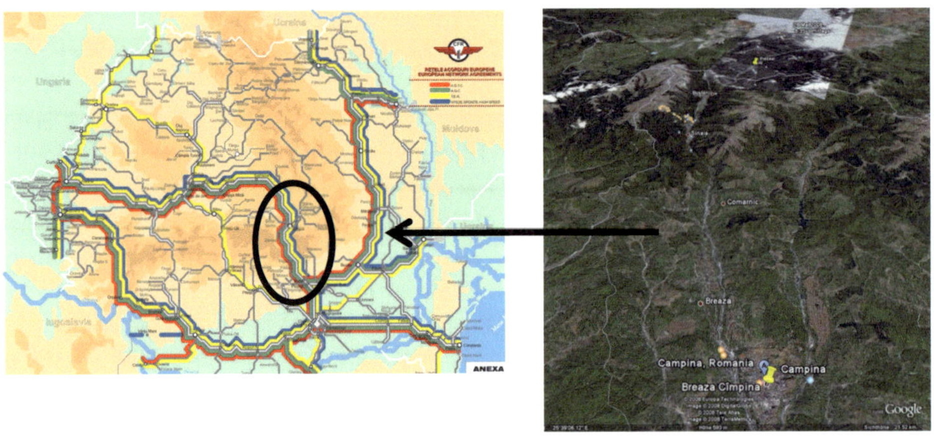

Fig. 18. Emplacement plan

All these structures were verified. In the following some aspects regarding the verification of the steel bridges are presented. The methodology which was adopted is showed in the figure 19.

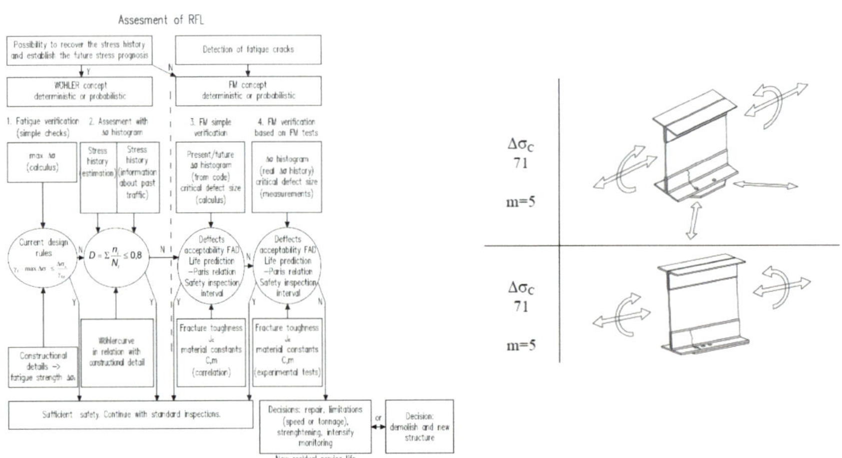

Fig. 19. The methodology and the constructive details for the evaluation of the RFL

After the stress analysis (Figure 20) and the fatigue verifications based on the Wöhler concept which were made in relation with the prescriptions of the Romanian standard SR 1911-1998, Swiss code 161 & SBB Richtlinie 2002 and the German code DS 805-2002, a life prediction analysis based on fracture mechanics principles was performed in order to evaluate the remaining fatigue life for these structures for the new traffic UIC conditions.

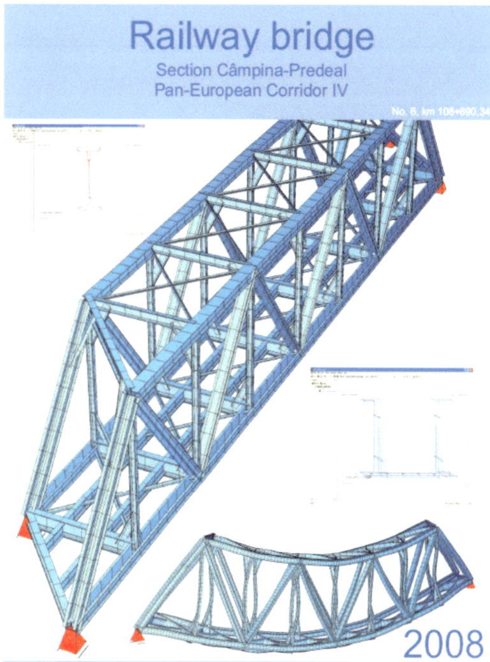

Fig. 20. The static model of the structure no. 6, km 108+690,34

For example, a plate girder bridge (no. 28, at km 125+323.25) is presented, which is a riveted structures crossing the Prahova River. The structure has four spans, one of 13.90 m and three of 14.10 m each and is skew (skew to the right - 48°37') and its superstructure is made out of independent plate girder decks.

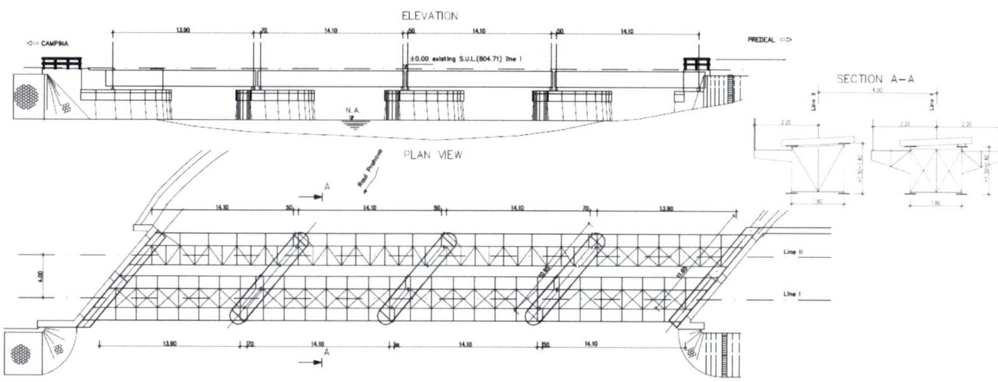

Fig. 21. The general disposition of the bridge no. 28

Calculation elements:

- Construction year – 1940.
- Due to lack of any documentation, the dimensions of the main girders elements were measured in the emplacement (Figure 22).

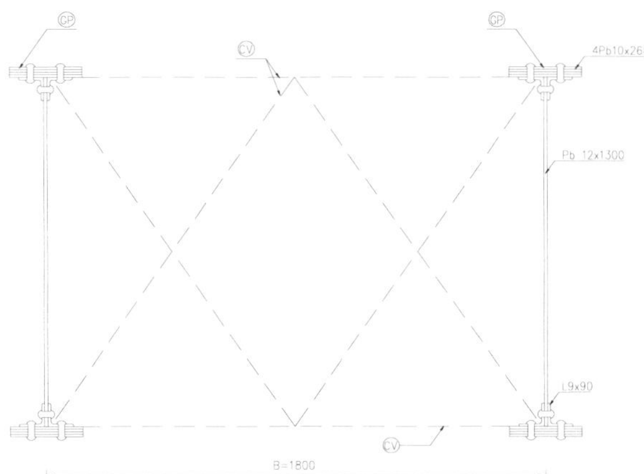

Fig. 22. Dimensions of the steel elements – Main girders

- The basic material to be considered is mild steel similar with St 37 (S235).
- The static scheme - simply supported girders.

The traffic on one main railway line was considered:

- the reference traffic 20-24x10^6 t/ line/ year
- the Calculation Centre of SNCF CFR S.A. for the year 2004: 13,7 t/ line / year =>12-16 t / line/ year.

The traffic expressed in pairs of trains per day in the peak month that took place on the line Câmpina – Predeal in the year 2004, was:

- 28 pairs of passenger trains / day;
- 17 pairs of freight trains / day;

Total – 48 assimilated pairs of trains / day

For the main girders of the steel decks, all the fatigue checks were performed in the section from the middle of the span.

According to the stress analysis, the maximal stress range for UIC 71 convoy is:

$$\max \Delta\sigma_{UIC} = 97.13\ N / mm^2$$

The resulted damage accumulated value (Miner rule) is D = 0.98.

Also, the complementary method of fracture mechanics was applied. For the material characteristics followings values were considered: the material is mild steel similar to the former steel St 34 - 37. n (Romanian standard - STAS 500/2 – 80); yield stress is σ_y = 230 N/mm^2; tensile stress σ_{ult} = 360 N/mm^2.

For the material toughness in terms of J_{crit} a minimal value of 20 N/mm at a temperature of -20°C was chosen. For the life prediction procedure in the case of the material constants following values have been chosen: m = 3 and C = 3 x 10^{-12} (see also [11]).

A stress history was established using the following mixed traffic from EC 1. This is actually the future traffic which will be characteristic for the new rehabilitated railway corridor. All these trains were moved on the structure in order to establish the multi-block stress history.

Train type	No. of trains / day	Train weight [t]	Traffic volume [Mil. t/year]
1	12	663	2,90
2	12	530	2,30
3	5	940	1,72
4	5	510	0,93
5	7	2160	5,52
6	12	1431	6,27
7	8	1035	3,02
8	6	1035	2,27
	67		24,95

Table 4. Mixed traffic from EC 1

All these trains were moved on the structure in order to establish the multi-block stress history.

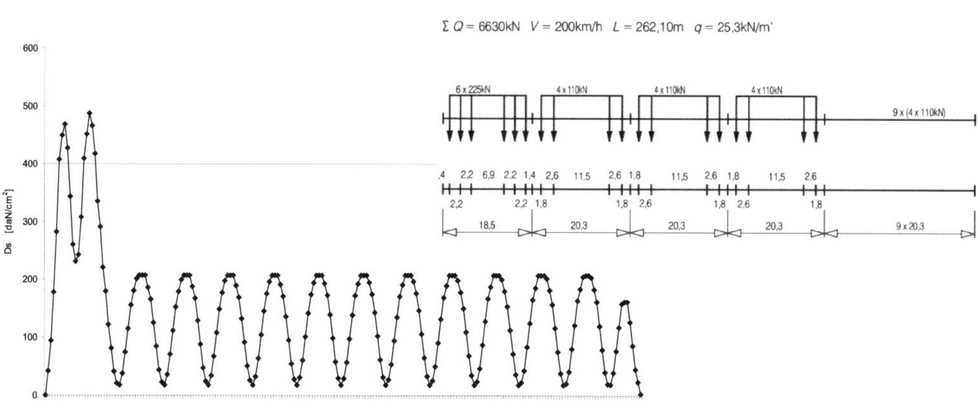

Fig. 23. The stress range diagram Δσ for train type 1

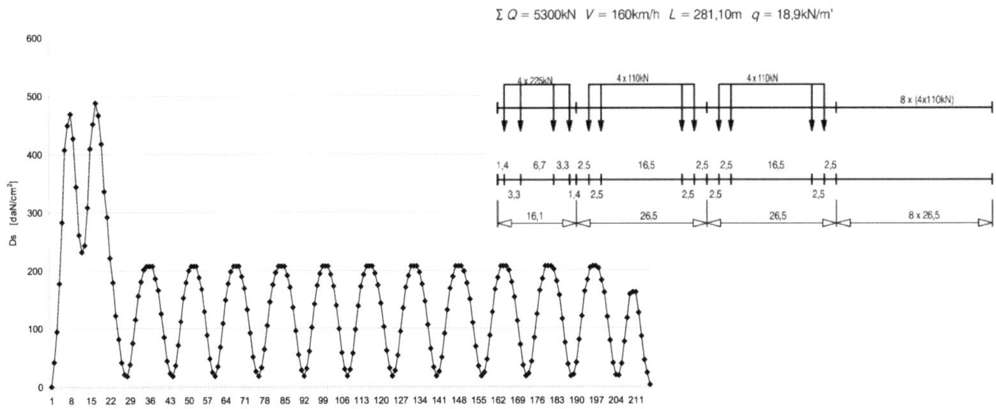

Fig. 24. The stress range diagram Δσ for train type 2

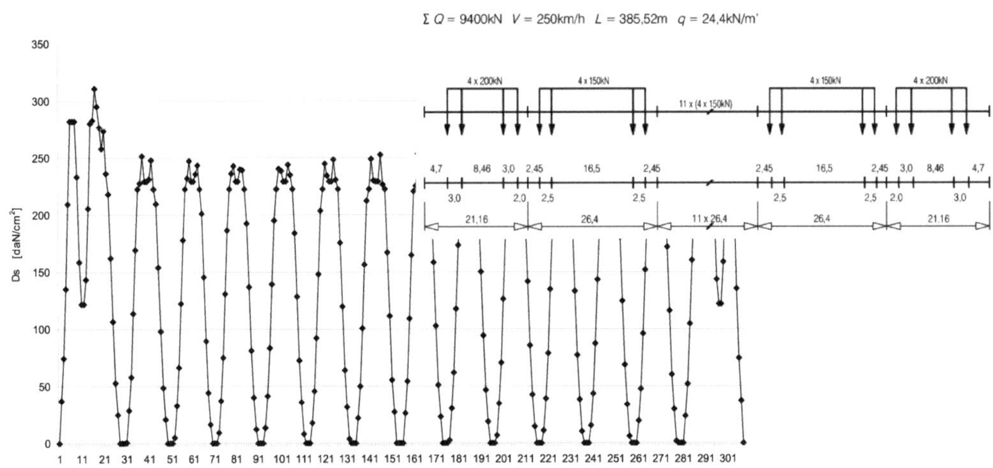

Fig. 25. The stress range diagram Δσ for train type 3

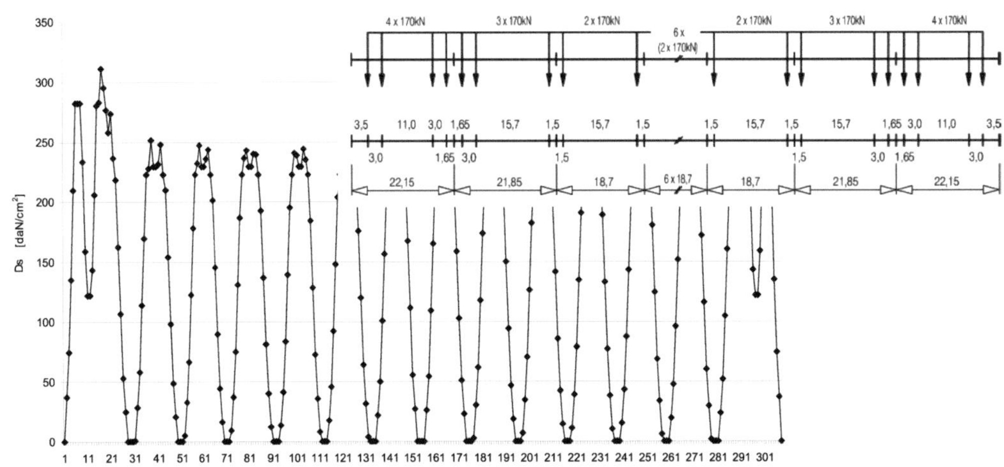

Fig. 26. The stress range diagram Δσ for train type 4

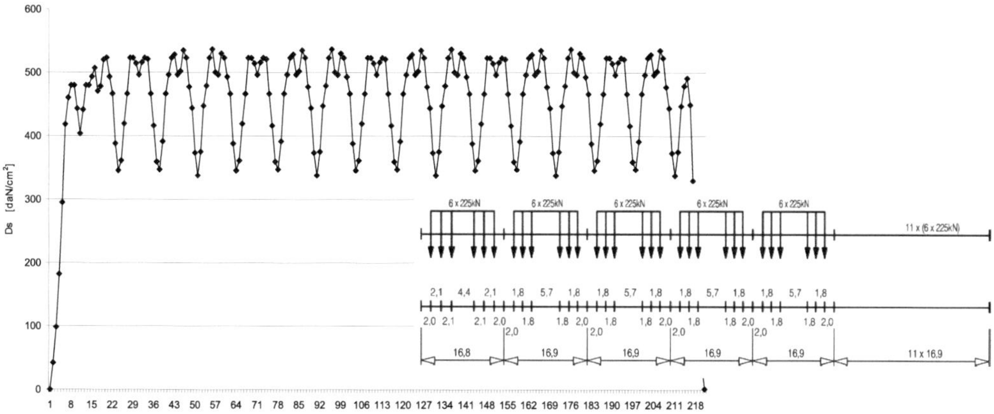

Fig. 27. The stress range diagram Δσ for train type 5

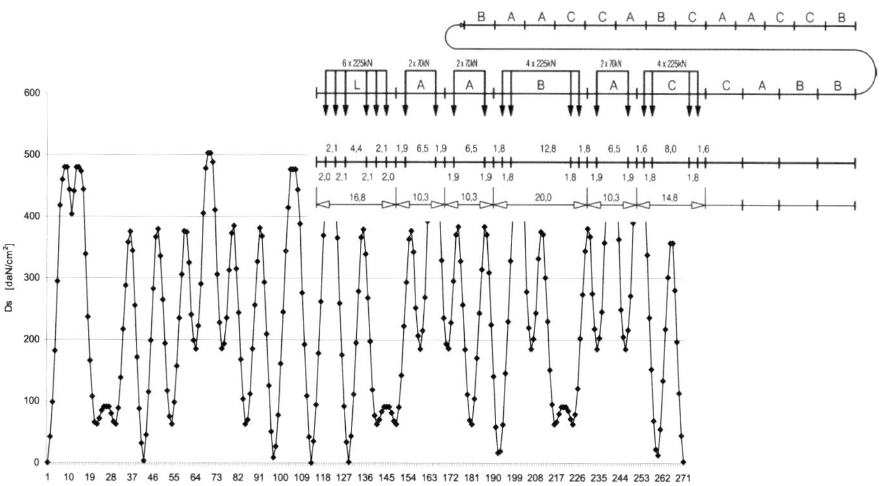

Fig. 28. The stress range diagram Δσ for train type 6

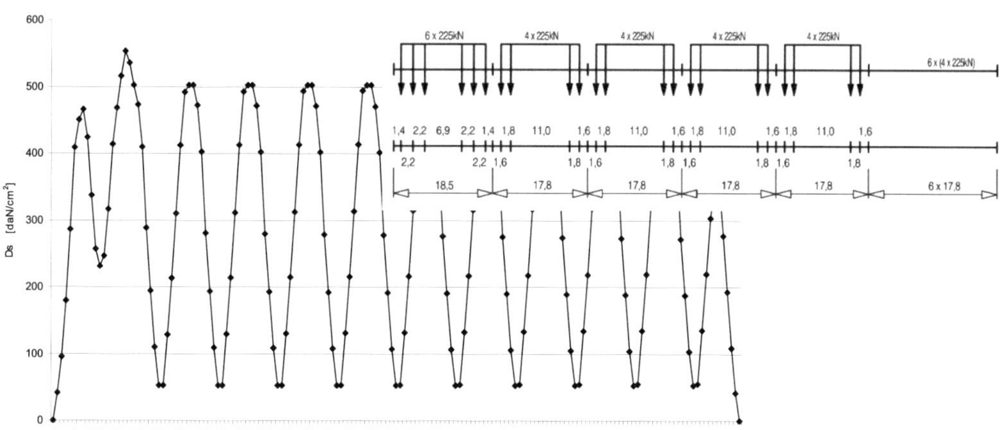

Fig. 29. The stress range diagram Δσ for train type 7

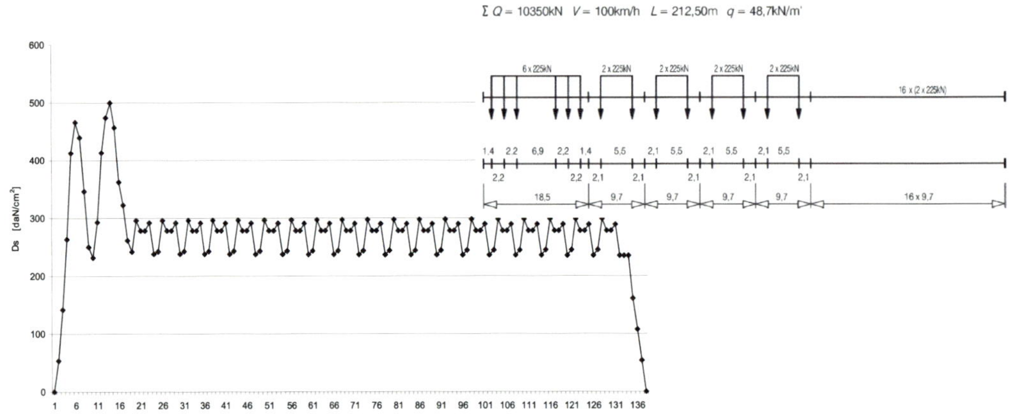

Fig. 30. The stress range diagram Δσ for train type 8

Δσ	MAIN GIRDER (L/2) mix								
[daN/cm²]	T1	T2	T3	T4	T5	T6	T7	T8	**Total**
50	0	0	0	0	105	36	0	108	**249**
100	0	0	0	0	7	12	0	0	**19**
150	12	12	10	0	0	0	0	0	**34**
175	0	108	0	0	0	0	0	0	**108**
200	132	0	0	10	105	36	8	0	**291**
250	12	0	65	60	0	0	0	12	**149**
275	0	0	10	10	0	0	0	0	**20**
300	12	12	10	0	0	60	0	6	**100**
350	0	12	0	0	0	60	8	0	**80**
400	0	0	0	0	0	36	0	0	**36**
450	0	0	0	0	0	0	88	6	**94**
475	12	0	0	0	0	0	0	0	**12**
500	0	0	0	0	14	72	8	0	**94**
$\Delta\sigma_e$ [daN/cm²] =						**381.36**		Total/day	**1037**

Table 5. Classification of the stress range intervals

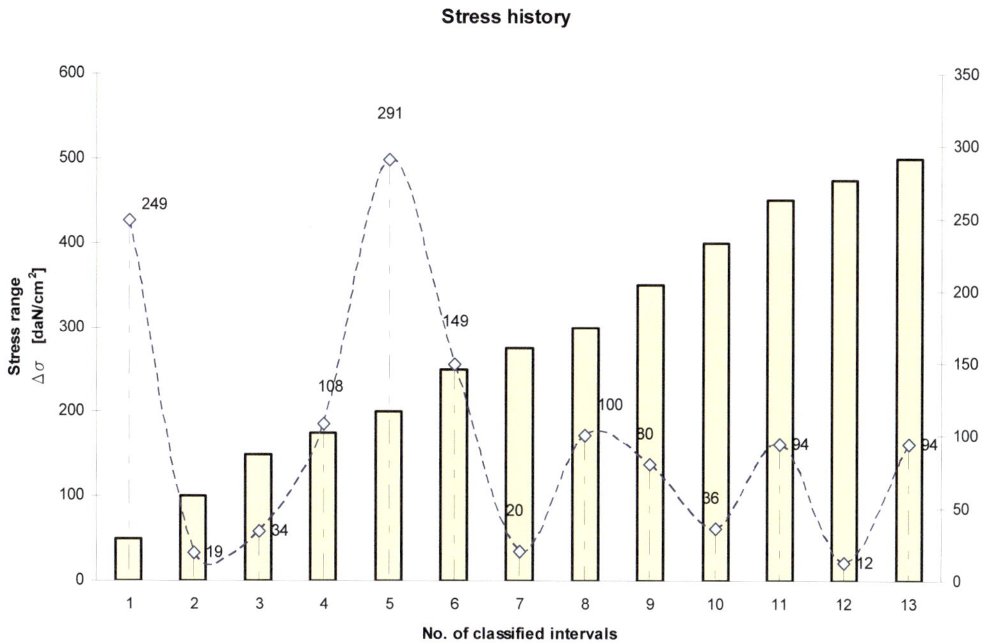

Fig. 31. Stress history

The crack case assumed for these structures (riveted bridges) is illustrated in figures 32; it is a trough thickness defect.

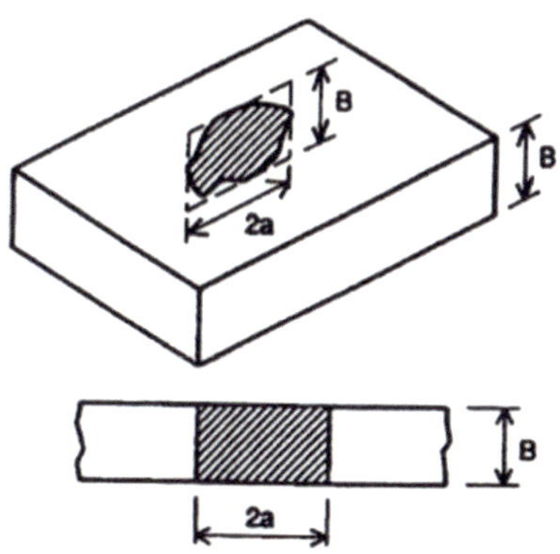

Fig. 32. Theoretical crack models

In order to determine the remaining service life it is important to know how long it will take the crack to grow from the minimum detectable size to the critical value.

Two cases were studied:

- through thickness flaw with initial size a_0 = 2.0 mm which is undetectable because it is situated under rivet head in the web steel plate, and
- through thickness flaw with initial size a_0 = 2.0 mm which is also undetectable during a visual inspection appearing in the corner (lower flange).

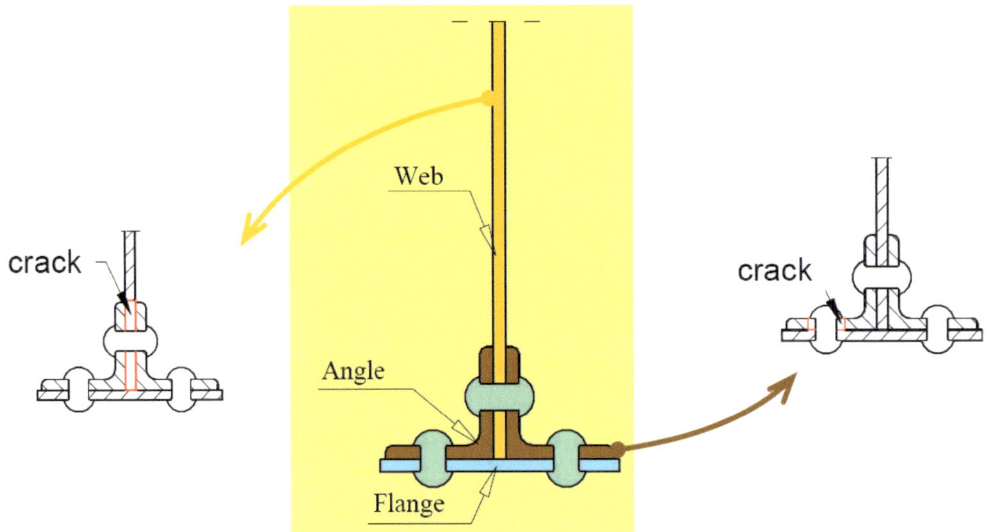

Fig. 33. The two studied defect cases

In the first case the remaining fatigue life is estimated at a value of 8.31 years (corresponding to a number of cycles of 1 566 000). In the second case the administrator must be acquainted with the fact, that propagation of the crack to the critical value will occur in 8.89 years (N=1 676 000 cycles).

In this case the safe inspection intervals, calculated with formula 8, are situated between 1.2 and 2.0 years.

Finally due to the high corrosion level (Figure 34) and also based on the fatigue assessment, the superstructure was proposed for the replacement.

Fig. 34. Corrosion attack details

Also all the other plate girder bridges were verified. They are presented in the next table.

Bridge nr.	Constr. year	Span	Cross section type / main elements	Photo
4	1940	6,60 m	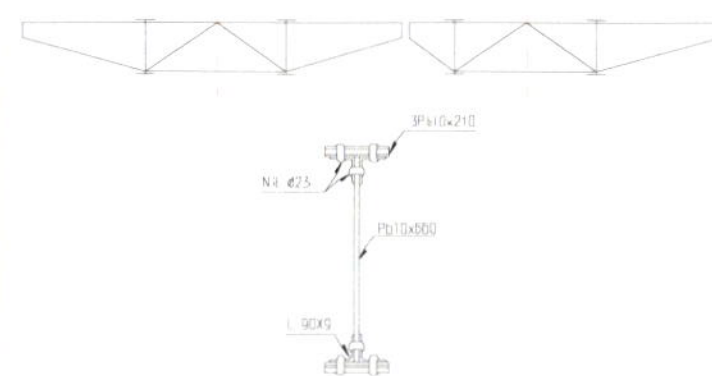	
5	1942	2 x 11,8 m	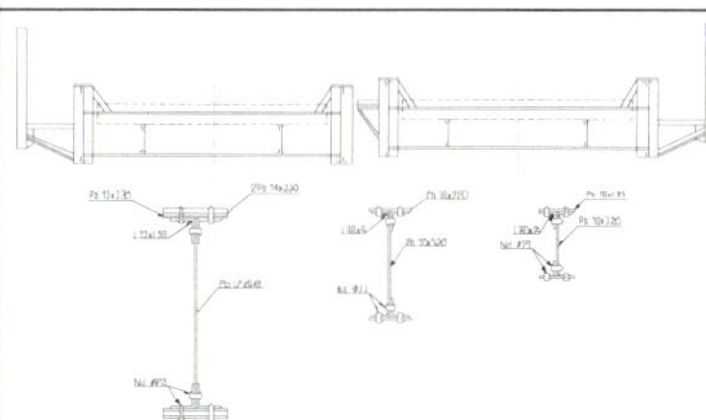	
8	1942	2 x 11,8 m	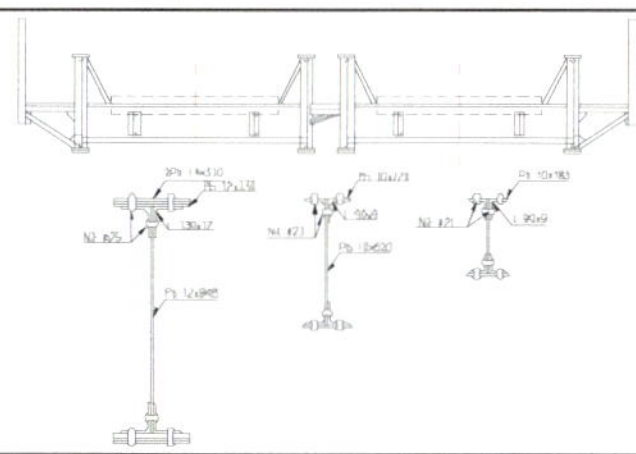	
9	1942	8,60 m	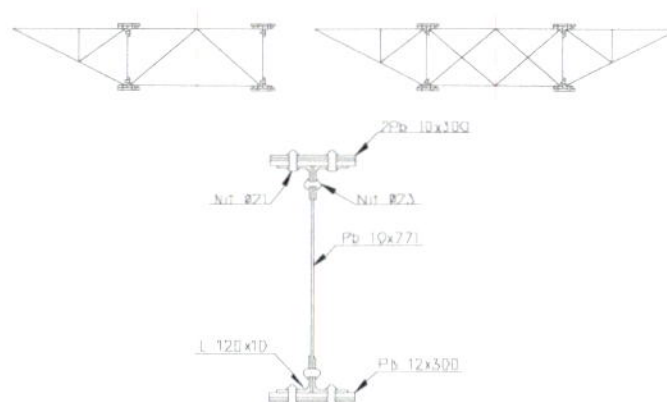	
13	1942	9,80 m + 26,90 m	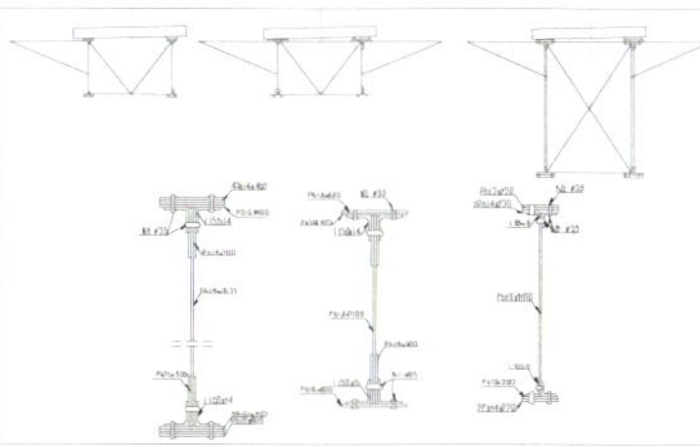	
15	1942	19,00 m + 18,80 m	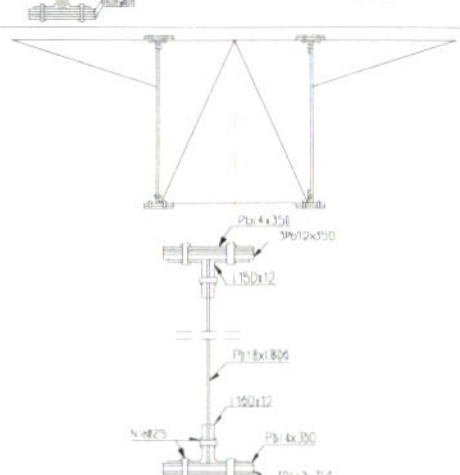	

Table 6. Analyzed plate girder bridges

16	1942	26,90 m	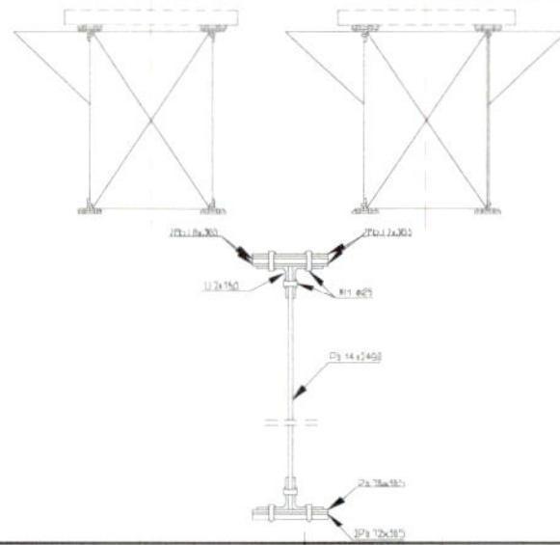	
17	1880/ 1942	8,40 m	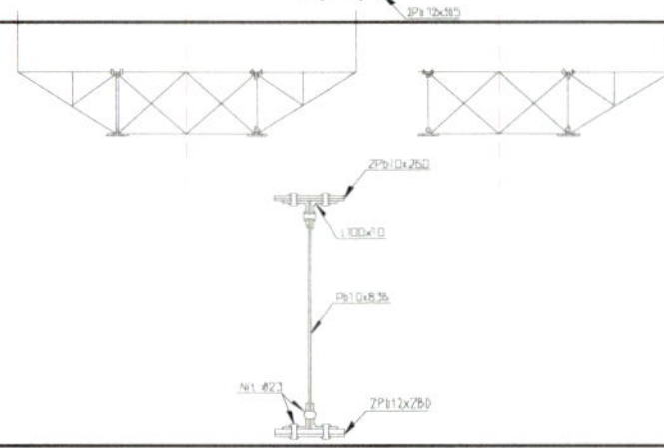	
20	1942	37,20 m	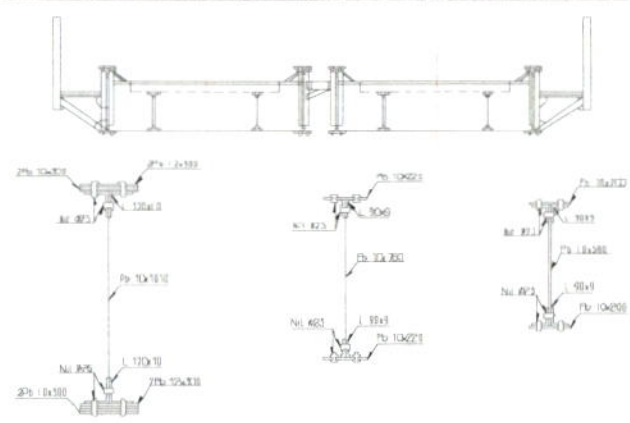	
22	1942	14,00 m		
24	1944	53,60 m	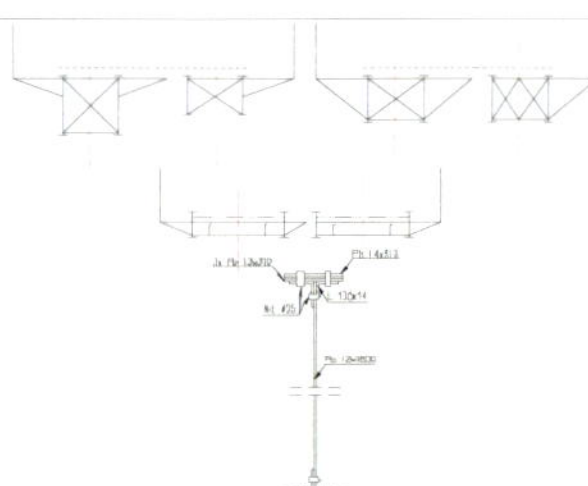	
25	1940-1944	58,25 m	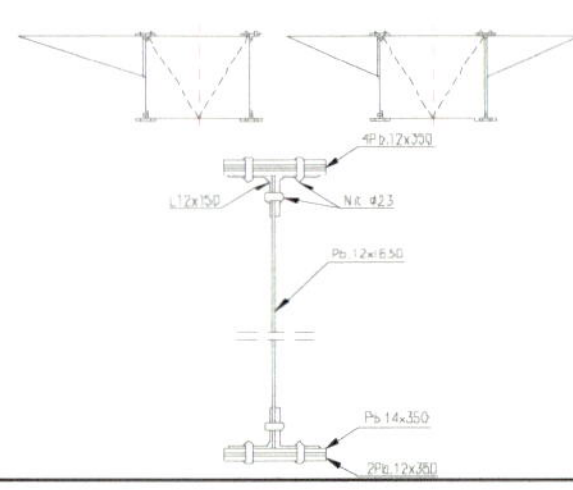	

Table 6. Analyzed plate girder bridges - continuation

28	1940	56,20 m	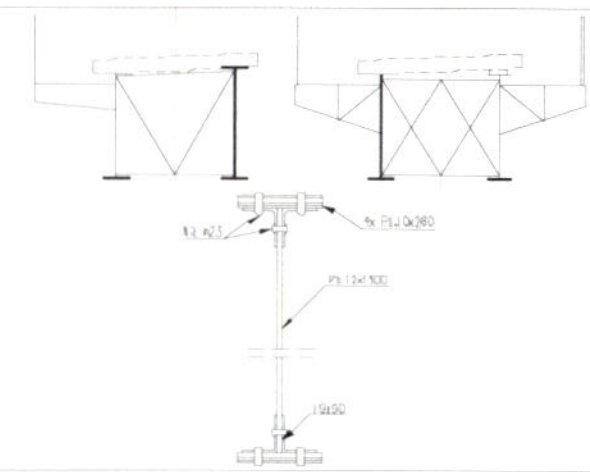	
29	1940	9,00 m	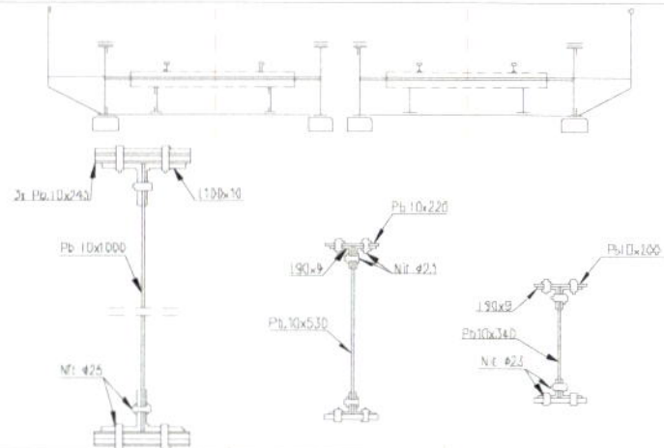	
30	1979	16,30 m	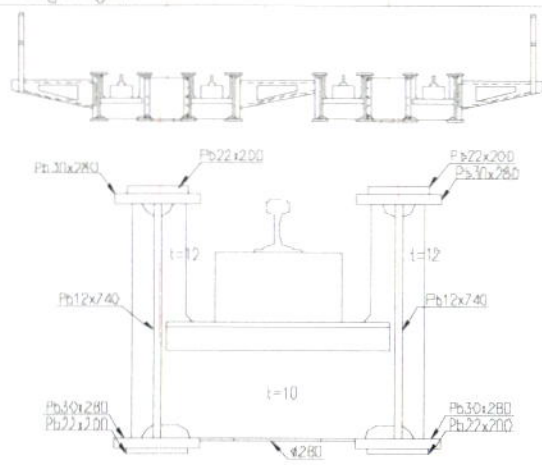	
31	1942	8,30 m	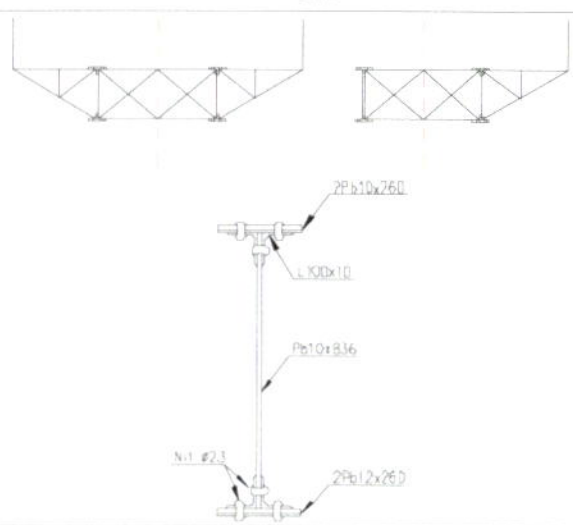	
32	1940	53,40 m	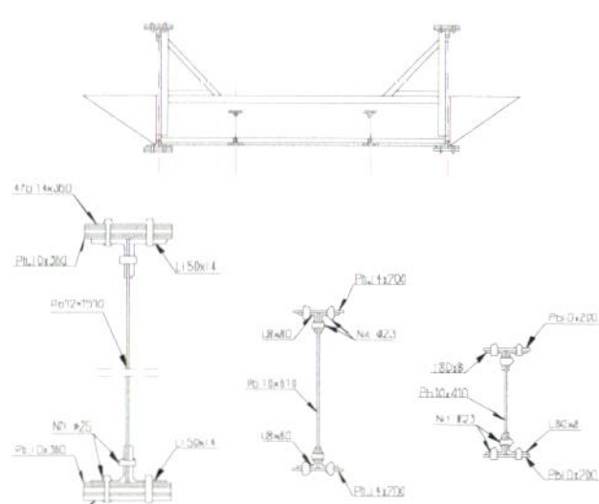	
33	1942	10,60 m	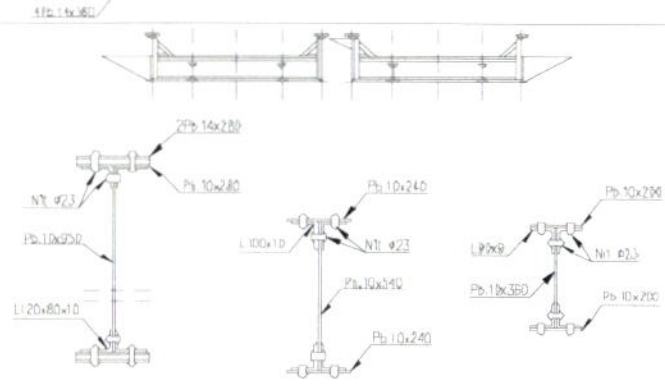	

Table 6. Analyzed plate girder bridges - continuation

34	1966	14,20 m	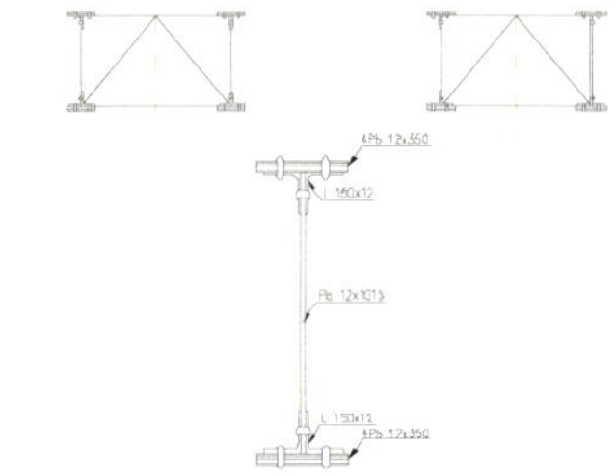	
37	1960	23,00 m	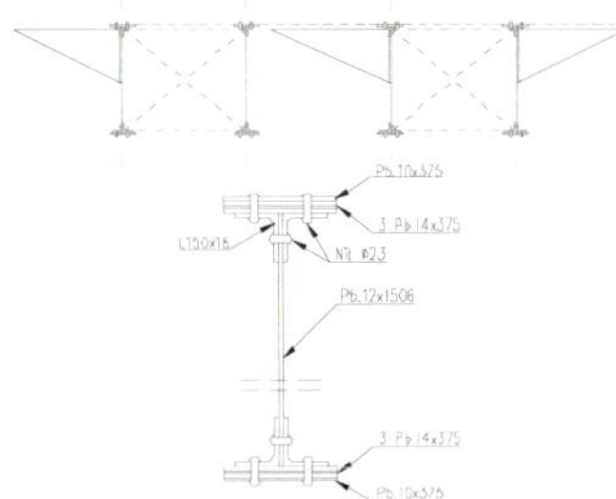	
40	1942-1944	31,20 m	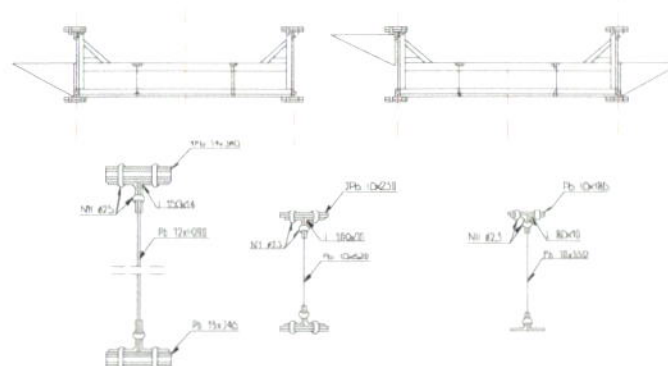	
41	1880/1942	16,00 m	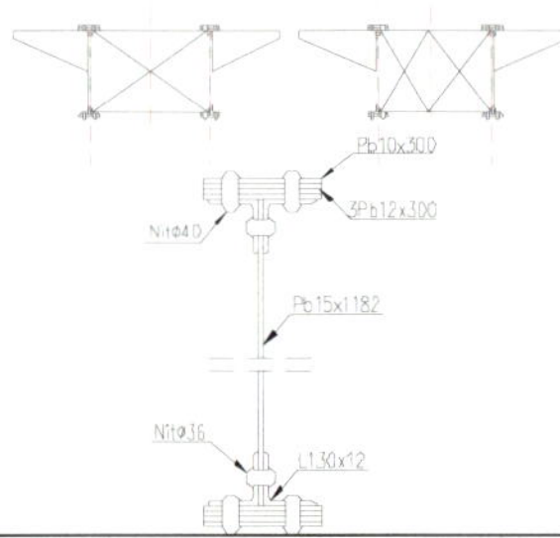	

Table 6. Analyzed plate girder bridges - continuation

For the replacement different solutions can be proposed. A modern economical and robust structure is the Schmitt Stumpf Frühauf solution (VFT-WIB®), [12] - which has a high degree of prefabrication and goes along with a significant improvement of working conditions, as weather-independent working; also improved environment conditions for the workers while erecting formwork, placing re-bars and casting concrete is guaranteed. Furthermore a smaller amount of man-hours outdoors at the construction site is needed.

The constructive depth was optimized by a computer specialized program (Figure 35).

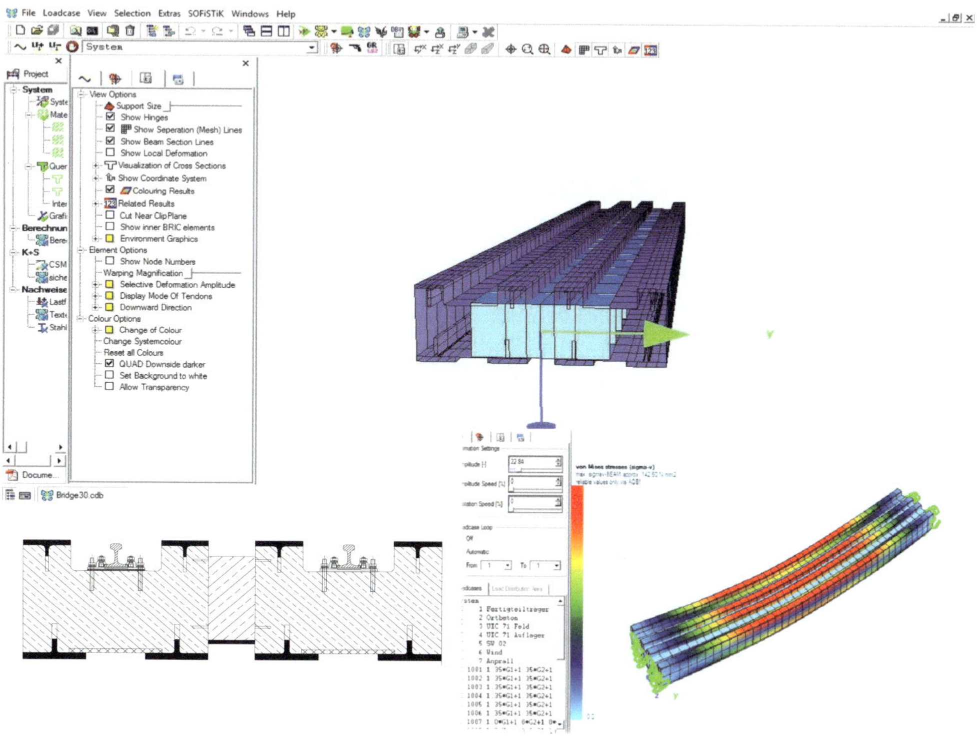

Fig. 35. Cross section of the new structure and calculation model

6. Conclusion

The progress recorded in the last decades allows on one part an accurate evaluation of the remaining safety of the structures and on the others part the proposal of new efficient, economical solutions with the high level sustainability.

7. References

[1] * * * Code UIC 778-2R; Recommandations pour la détermination de la capacité portante des structures métalliques existantes; Union Internationale des Chemins de fer, Paris, 1986.

[2] **, DS 805 „Bestehende Eisenbahnbrücken. Bewertung der Tragsicherheit und konstruktive Hinweise, August 2002.

[3] **, SR 1911-98, „Poduri metalice de cale ferată. Prescripții de proiectare", Institutul Român de Standardizare, Bucuresti, 1998.

[4] Jackson, P. "Is bridge assessment losing its credibility?" The Structural Engineer, Volume 79 / No 9, May 2001.

[5] Xie, M., Bessant G.T., Chapman, J.C., Hobbs, R.E., "Fatigue of riveted bridge girders" The Structural Engineer, Volume 79 / No 9, May 2001.

[6] * * *, "Maintenance of steel bridges", Romanian Pre-standard, Bucharest - 2000.

[7] Petzek, E., „Safety in Operation and Rehabilitation of Steel Bridges", Doctoral Thesis, UP Timişoara, 2004.

[8] ** I-AM 08/2002. Richtlinie für die Beurteilung von genieteten Eisenbahnbrücken, SBB CFF FFS.

[9] Petzek E., Kosteas D., Bancila R., 2005. Sicherheitsbestimmung bestehender Stahlbrücken in Rumänien", Stahlbau Nr. 8, 9, ISSN 0038-9145, Ernst & Sohn.

[10] ***, BS 7910:1999, „Guide on the Methods for Assessing the Acceptability of Flaws in Metallic Structures", British Standards Institution, London, 1999.

[11] Eriksson, K., Toughness requirements for old structural steel, IABSE Report Congress, 2000.

[12] SCHMITT, V., et alt.: VFT-Bauweise, Entwicklung von Verbundfertigteilträgern im Brückenbau, Beton- und Stahlbetonbau 96, 2001, Heft 4.

[13] SEIDL, G. et al. (2006), "Prefabricated Enduring Composite Beams based on Innovative Shear Transmission - Proposal RFSR-CT-2006-00030 ".

19

Special Tunnel Blasting Techniques for Railway Projects

More Ramulu
Central Institute of Mining & Fuel Research, Nagpur, India

1. Introduction

Excavations of tunnels are common features in mining and civil engineering projects. In absence of initial free face, solid blasting method is employed for excavation of tunnels, drifts and mine roadways, which have many similarities in configurations and in different cycles of operation followed during excavation. A greater proportion of world's annual tunnel advance is still achieved by drilling and blasting. In spite of inherent disadvantages of damaging the rock mass, drilling and blasting has an unmatched degree of flexibility and can overcome the limitations of machine excavations by Tunnel Boring Machine (TBM) or road headers. In spite of no major technical breakthrough, the advantages like low investment, availability of cheap chemical energy in the form of explosives, easy acceptability to the practicing engineers, the least depreciation and wide versatility have collectively made the drilling and blasting technique prevail so far over the mechanical excavation methods.

Since tunnels of different sizes and shapes are excavated in various rock mass conditions, appropriate blast design including drilling pattern, quantity and type of explosive, initiation sequence is essential to achieve a good advance rate causing minimal damage to the surrounding rock mass. The cost and time benefit of the excavation are mostly decided by the rate of advance and undesired damage.

Excavation of tunnels, except in geologically disturbed rock mass conditions, is preferred with full face blasting. It is common to excavate large tunnels of 80-90 m^2 cross-section in sound rock masses by full face in a single round. However, tunnels larger than 50m^2 cross-sectional area driven through incompetent ground condition are generally excavated in smaller parts.

Introduction of electro-hydraulic jumbo drills with multiple booms, non-electric initiation system, small diameter explosives for contour blasting and fracture control blasting are some of the recent developments in tunnel blasting. Prediction and monitoring the blast damage, application of computers in drilling, numerical modelling for advanced blast design, use of rock engineering systems for optimization and scheduling of activities have been the areas of intense research in today's competitive and high-tech tunnelling world

In tunnel blasting, explosives are required to perform in a difficult condition, as single free face (in the form of tunnel face) is available in contrast to bench blasting where at least two free faces exist. Hence, more drilling and explosives are required per unit volume of rock to be fragmented in the case of tunnel blasting. A second free face, called 'cut', is created initially during the blasting process and the efficiency of tunnel blast performance largely depends on the proper development of the cut. The factors influencing the development of the cut and the overall blast results are dependent on a host of factors involving rock mass type, blast pattern and the tunnel configurations.

2. Blasting mechanics

The tunnel blasting mechanics can be conceptualised in two stages. Initially, a few holes called cut holes are blasted to develop a free face or void or cut along the tunnel axis. This represents a solid blasting condition where no initial free face is available. Once the cut is created, the remaining holes are blasted towards the cut. This stage of blasting is similar to bench blasting but with larger confinement. The results of tunnel blasting depend primarily on the efficiency of the cut hole blasting. The first charge fired in cut resembles crater blasting. Livingston's spherical charge crater theory (Livingston, 1956) suggests that the blast induced fracturing is dominated by explosion gas pressure which is supported by Liu and Katsabanis (1998). Duvall and Atchison (1957), Wilson (1987) and others believe that the stress wave induced radial fracturing is the dominating cause of blast fragmentation and gas pressure is responsible only for extension of the fractures developed by the stress wave.

The natures of influence of the two pressures i.e. of stress and gas are different in the jointed rock mass where the stress waves is useful in fragmentation as the joints restrict the stress wave propagation. The gases, on the other hand, penetrate the joint planes and try to separate the rock blocks. The fragments' size and shape in jointed formations are dominated by the gas pressure and the joint characteristics. The roles of the stress wave and the gas pressures are no different in the second stage of tunnel blasting. But with the availability of free face, the utilisation of stress wave is increased. The rock breakages by rupturing and by reflected tensile stress are more active in the second stage because of cut formation in the first stage.

3. Parameters influencing tunnel blast results

The parameters influencing the tunnel blast results may be classified in three groups:

i.	Non controllable	Rock mass properties,
ii.	Semi-controllable	(a) Tunnel geometry & (b) Operating factors, and
iii.	Controllable	Blast design parameters including the explosive properties.

4. Models for prediction of tunnel blast results

Specific Charge is one of the important paprameter of prediction of tunnel blast results. Pokrovsky (1980) suggested the following empirical relation to determine the specific charge (q) in tunnels (Eq. 1):

$$q = q1.\ s_t.\ f.\ s_{wr}.\ d_{ef},\ kg/m^3 \tag{1}$$

where,

q1=specific charge for breaking of rock against a free face in kg/m^3,
s_t = factor for structure and texture of rock,

$$f = \text{rock confinement} = 6.5 / \sqrt{A}, \tag{1a}$$

A= area of tunnel (m^2),
s_{wr} = relative weight strength of explosive (ANFO = 1), and
d_{ef} = factor for diameter of explosive cartridge,

According to Langefors and Kihlstrom (1973), the specific charge (q) is related to the cross-sectional area of the tunnel (A, m^2) as given below:

$$q = (14/A) + 0.8\ kg/m^3 \tag{2}$$

The specific charge in the cut holes remain maximum and it can be upto 7 kg/m^3 in a parallel cut.

5. Rock mass damage

The aspects of blast induced rock mass damage around a tunnel opening and its assessment have been the subjects of in-depth research for quite a long time. The type of damage can be grouped into three categories: (i) fabric damage due to fracturing, (ii) structural damage exploiting discontinuities and shears, and (iii) lithological damage causing parting between two different rock units or lithological boundaries between similar rock types.

Chakraborty et al. (1996a) observed in the tunnels of Koyna Hydro-electric Project, Stage-IV poor pull and small overbreak in volcanic breccia having low Q value, P-wave velocity and modulus of elasticity. On the other hand, large overbreak on the sides due to vertical and sub-vertical joints and satisfactory pull were found in the compact basalts having comparatively much higher Q value, P-wave velocity and modulus of elasticity. The fact is attributed to the presence of well defined joints in compact basalts which is absent in volcanic breccia.

The effects of joint orientations on overbreak/underbreak and pull in heading and benching operations during tunnel excavations are explained by Johansen (1998). The work of Johansen (1998) describes that joints normal to tunnel direction are favorable for good pull and parallel to the tunnel advance direction yield poor pull. advance direction. The obtuse angle between joints and tunnel direction results in more damage and breakage towards the wall of that angle.

The dip direction of the blasted strata on pull could be well experienced while blasting in the development faces of Saoner coal mine where the pull was increased by 11 per cent in the rise galleries compared to that in the dip galleries (Chakraborty, 2002). Longer rounds in tunnels can be pulled when the dominant joint sets are normal to the tunnel axis. Whereas, better pull can be obtained in shaft sinking if the discontinuities are parallel to the line joining the apex of the Vs in a V-Cut Hagan (1984).

Chakraborty (2002) observed the following influences of joint directions on pull and overbreak (Table 1).

Joint Orientation		Face Advance	Roof Overbreak
Dip	Strike with respect to tunnel axis		
Steep	Parallel	Very poor	Very small
Steep	Across	Very good	Very large
Gentle	Across	Fair	Large
Moderate	Across/oblique	Good	Small

Table 1. Influence of joint direction on overbreak (Chakraborty, 2002)

The gentle, moderate and steeply dipping joint planes signify the dip angles as 0°-30°, 30°-60° and 60°-90° respectively. Similarly, strikes with respect to tunnel axis are mentioned as parallel, oblique and across to indicate that the joint strike intersection angle with the tunnel axis as 0°-30°, 30°-60° and 60°-90° respectively.

If the geo-mechanical properties of the constituting formations of a tunnel are quite different, the stress energy utilisation and resulting fragmentation are adversely affected. Chakraborty et al. (1996b) suggested an increase of specific charge by a per cent equal to ten times the number of contact surfaces.

Engineers International Inc. modified Basic RMR (MBR) considering blast-induced-damage adjustments, as shown in Table 2, were suggested for planning of caving mine drift supports (Bieniawski,1984). Chapter 4 in the present publication defines basic RMR.

Method of Excavation	Damage Level	Blast Damage Adjustment Factor	Per cent Reduction
1. Machine boring	No damage	1.0	0
2. Controlled blasting	Slight	0.94-0.97	3-6
3. Good conventional blasting	Moderate	0.9-0.94	6-10
4. Poor conventional blasting	Severe damage	0.9-0.8	10-20

Table 2. Blast damage adjustments in MBR (after Bieniawski, 1984)

Ouchterlony et al. (1991) observed that the damage zone could be to the extent of 0.5 m with cautious tunnel blasting. McKenzie (1994) related the threshold peak particle velocity PPV (v_{max}) for incipient fracture with uniaxial tensile strength (q_t), Young's modulus and P-wave velocity (V_p, m/s) as shown below:

$$v_{max} = \frac{q_t \times V_p \times 10^{-3}}{E}, \text{ m/s} \quad (3)$$

where

q_t = uniaxial tensile strength, MPa,
V_p = P-wave velocity, m/s, and
E = Young's modulus, GPa.

Pusch and Stanfors (1992) and others observed that the minimum disturbance by blasting is reported when the tunnel orientation was within 15° with the strike of the joint sets.

Yu and Vongpaisal (1996) concluded that the damage is a function of blast induced stress and rock mass resistance to damage. They proposed Blast Damage Index (D_{ib}) to estimate the type of damage due to blasting. It is the ratio of the blast induced stress to the resistance offered against damage.

Ramulu et al (2009) categorised blast induced damage as,

- near-field damage due to high frequency and critical vibrations
- far-field damage due to repeated low frequency and sub-critical vibrations. Ramulu and Sitharam (2011) assessed the near-filed damage by using vibration attenuation model of charge weight scaling law and dynamic tensile failure criteria instead of conventional Holmberg-Persson model (1979) and static tensile failure criteria. Ramulu (2010) correlated the far-filed damage with shear wave velocity of rock mass and found the following equation with reasonably good correlation coefficient (R^2=0.76).

$$D_{max}=322.5(V_s)^{-0.61}\ \text{m} \tag{4}$$

where,

D_{max} – Maximum extent of rock mass damage due to repeated vibrations, m
V_s – S-wave velocity, m/s

6. Contour blasting

Contour blasting in tunnelling is adopted to obtain a smooth tunnel profile and minimise damage to the surrounding rock mass. Despite a large amount of drilling required, it is preferred over conventional blasting because of the following advantages:

i. The shape of the opening is maintained with smooth profile.
ii. Stability of the opening and the stand-up-time of the tunnel are improved.
iii. Support requirement is reduced.
iv. Overbreak is reduced to minimise unwanted excavations and filling to bring down the cost and cycle time.
v. Ventilation improves due to smooth profile.

The performance of contour blasting is frequently measured in terms of `Half cast factor' (HCF) which is dominated by the design parameters of the contour holes, the joint orientation and the explosive energy distribution.

Generally, two types of contour blasting are used in tunnelling, i) pre-splitting and ii) smooth blasting. When two closely spaced charged holes are fired simultaneously the stress waves generated from the two holes collide at a plane in between the holes and create a secondary tensile stress front perpendicular to the hole axis and facilitates extension of radial cracks along the line joining the holes. The wedging action from the explosion gas acts in favour of extending the crack along the same line. It is, therefore, essential to contain the gas pressure till the cracks from both ends meet by adequate stemming. Further, the delay timing of the adjacent holes need to be very accurate so that the stress waves should collide at the mid-point and the arbitrariness of the breakage between the holes can be reduced.

The contour blasting performance largely depends on the nature and the orientation of joint planes. Gupta et al. (1988) found that the joint orientation adversely influences the pre-splitting results to a maximum when these are at an angle of 1-30° to the pre-split axis.

In smooth blasting, the delay intervals between the contour holes and the nearest production holes are kept high to facilitate complete movement of material in production holes before the contour holes detonate so that the gas expansion in contour holes occurs towards the opening. Sometimes, holes are drilled in between two charged blast holes and are kept uncharged. These are called dummy holes (Figure 1). The stress concentration at the farthest and the nearest points of the dummy holes become high to initiate cracks from the dummy holes extending towards the charged holes. The fracture is, thus, controlled along the desired contour.

In some cases, slashing or trimming techniques are used where the central core of excavation area is removed first to reduce the stress and then post-splitting is adopted to remove the remaining rock mass along the desired contour. The technique is generally referred to as 'slashing' or 'trimming' [Calder and Bauer (1983), Figure 2].

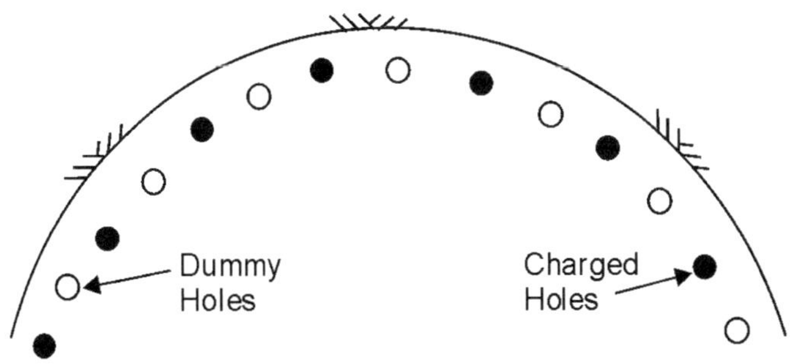

Fig. 1. Smooth blasting pattern with dummy holes

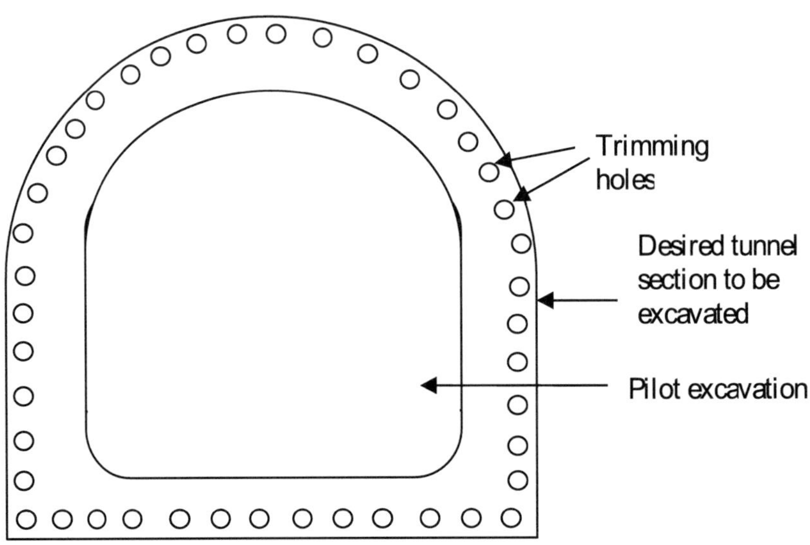

Fig. 2. Cushion blast holes for trimming of a tunnel after pilot excavation

Line drilling is adopted as an alternative technique where a number of uncharged holes are drilled along the contour with a spacing of 2-4 times the hole diameter (Du Pont, 1977). The distance of the row of empty holes from the final row of charged holes is kept as 0.5-0.75 times the normal burden. The empty holes are joined during the main blasting round and a separation is created along the contour.

According to Holmberg and Persson (1978), the spacing of pre-split holes should be 8-12 times the blast hole diameter. The following design parameters for contour hole spacing, burden to spacing ratio of contour holes and linear charge concentration in smooth blasting are suggested by Holmberg (1982) :

$$S_{dc} = 16 \times d_b, \text{ m} \quad (5)$$

$$m_{dc} = 1.25 \quad (6)$$

$$q_{lcc} = 90 \times (d_b)^2, \text{ kg/m} \quad (7)$$

where

S_{dc} = spacing of contour holes while drilling, m,
m_{dc} = burden to spacing ratio of contour holes while drilling,
q_{lcc} = linear charge concentration in the contour holes, kg/m, and
d_b = diameter of blast holes, m.

Controlled blast design details recommended by Olofsson (1988) are presented in Table 3.

Type of Blasting	Blast Hole Diameter (mm)	Spacing of Blast Holes (m)	Burden (m)	Linear Charge Concentration (kg/m)
Smooth blasting	25-32	0.25-0.35	0.3-0.5	0.11
	25-48	0.5-0.7	0.7-0.9	0.23
	51-64	0.8-0.9	1.0-1.2	0.42-0.45
Pre-splitting	38-44	0.3-0.45	-	0.12-0.37

Table 3. Recommended blast design for contour blasting (Olofsson, 1988)

7. Special tunnel blasting techniques

Some special blasting techniques were practised in tunnels and underground coal mine to attain greater advance and better safety in some critical working sites under the recommendations and supervision of Central Mining Research Institute, Regional Centre, Nagpur. Those cases are discussed in brief in the following paragraphs.

7.1 Long hole raise driving by blasting

A 123 m deep pilot shaft was excavated in 95 days time using Long Hole Raise Blasting (LHRB) method for faster and safer shaft sinking in the surge shaft, passing through various kinds of basaltic formations, in Ghatghar Hydro-electric Project of Maharashtra. The blast hole charging pattern is shown in Figure 3. Application of this techniques resulted in saving of 75% time 60% cost of excavation in comparison to the conventional shaft sinking method.

A ventilation shaft of 40m depth was also excavated by using the same technique at diversion tunnel of Latur-Osmanabad Railway tunneling project of Central Railways in 20 days. This techniques yielded in saving of time by 80% and cost of excavation by 60% in comparison to the conventional shaft sinking method, which mainly suffer from weather effects, confined working space and low cycle time.

Similarly, a pilot surge shaft of 3.0m diameter130m depth was excavated by long hole raise driving technique at a lift irrigation scheme of Koilsagar project. This swift and cost effective shaft excavation technique was completed in just 60 days with cost savings of 70% and time saving by 95% in contrast to conventional shaft sinking method. The profile of excavated pilot surge shaft at Koilsagar project is shown in Figure 4.

7.2 Lake tap blasting

The lake taping of fist of its kind with indigenous technology was carried out in India by CMRI (now CIMFR) at granitic rock mass in South India. The Andhra Pradesh Power Generation Corporation (APGENCO), India, executed a lift irrigation scheme (SLBC) for the Government of Andhra Pradesh to install 4 Nos. of 4 x 25000 hp pumps to lift 2400 cusecs of water from the Nagarjuna Sagar reservoir for irrigation purpose. A 4 m thick rock plug, designed by CMRI, was left for lake tapping at the end of project. The area of cross section of the tunnel was 40 m^2. Considering proximity of the nearby structures a controlled blast strategy in phased manner was evolved prior to final plug blasting. Vibration and damage characteristics were ascertained to finalise the blast design of the final plug.

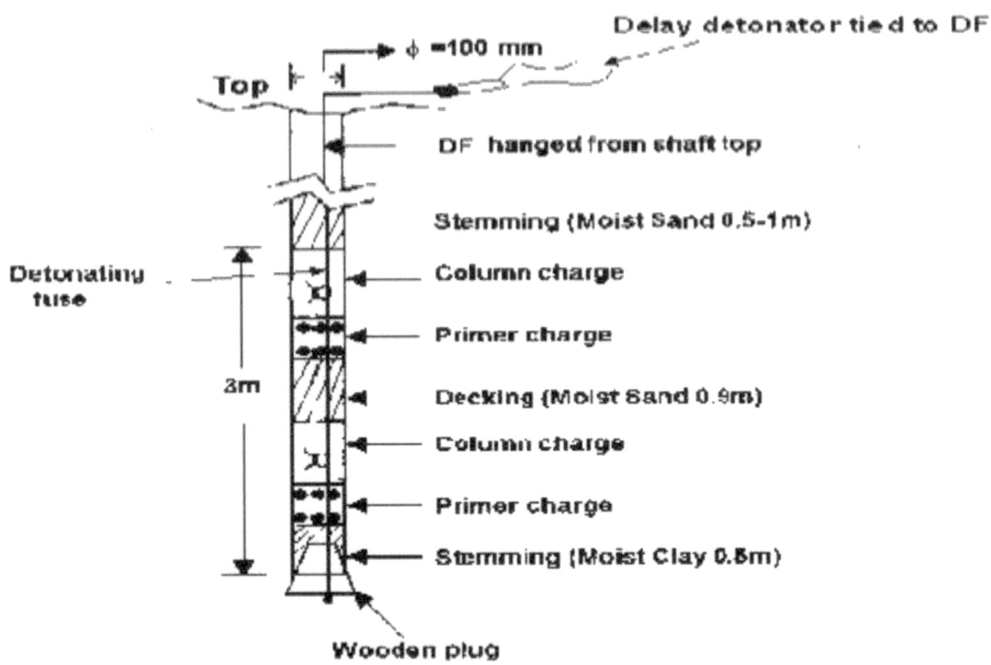

Fig. 3. Charging pattern in raise blasting

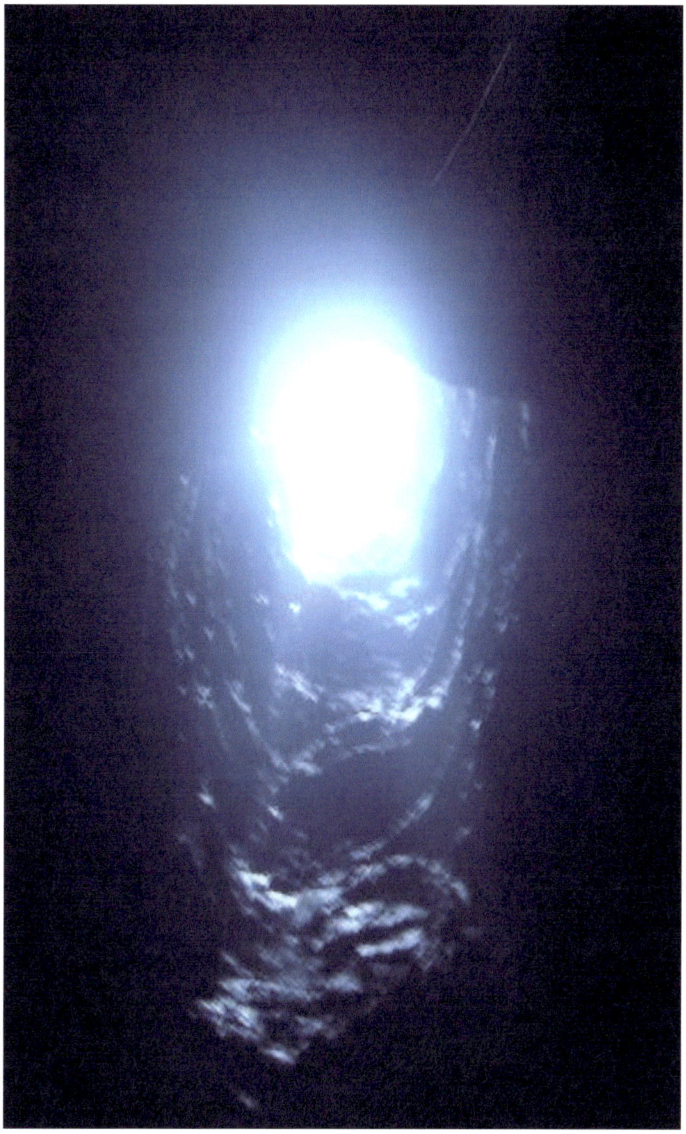

Fig. 4. Profile of pilot surge shaft excavated by long hole raise driving at Koilsagar

Based on the blast performance of the trail rock plug final plug blast design was made with the following salient features:

- Specific charge was increased from 1.25 kg/m^3 to 1.33 kg/m^3 to improve throw and fragmentation.
- Only gelatine explosive was recommended considering the water inflow from the blast holes.
- Dummy holes were made above the crown holes, at a distance of 0.3 m, to minimise rock mass damage.
- A borehole from the top was used to convey initiation to the blast holes.

The final plug-blasting pattern is shown in Figure 5. This novel technology being an indigenous one could save Crores of national exchequers.

7.3 Cautious blasting

By adopting an extremely cautious approach, all 10 reinforced concrete plugs, each of 125 m^3 volume, in 5 units were removed by controlled blasting without causing any damage to the surrounding periphery and pier nose in Srisailam left bank project of the APPGENCO while the power house was in running condition. The controlled blasting pattern is described below:

i. Line drilling holes of 1.5m depth were drilled with spacing of 0.15 m between the holes on the pier nose side and at 0.20 m inside the periphery.
ii. The periphery holes were pre-split with air-decking. The half cast factor of the periphery blasting was around 95%, which indicates low damage level. The pre-split blasting connections and the post-blast wall with half cast holes are shown in Figures. 5(a) and 5(b).
iii. A cut was created at the heading and it was widened and deepened to make a pilot hole in the plug along its axis.
iv. The balance concrete mass of the heading was slashed with less charge against the void.
v. The bottom was blasted with benching method.
vi. Mucking was done by mechanical and manual means.
vii. Continuous blast vibration monitoring was carried out during the blasts at near, intermediate and far field.
viii. Analysis of vibration data was done for subsequent blasting and to develop general predictor equation.

Pre and post blast ultrasonic measurements were taken at the exposed areas of the pier nose walls to know the change in physical property the reinforced mass due to blasting. The compressional wave velocities (P-wave) were measured by Roop telesonic ultrasonix instrument 'Ultrasonix 4600' which is shown in Figure 6. The average P-wave velocity was 2075 m/s and 2100m/s before and after blasting respectively. The values indicate that there has been no blast-induced damage to the structure under consideration.

The cautious blasting was also applied at Koldam Hydroelectric Power Project (KHEPP) to reduce overbreak and to get a smoother tunnel wall profile. The rock mass encountered in all the tunnels of KHEPP was Dolomite, which was very heterogeneous, highly weathered, metamorphosed, compact, foliated, sheared and crushed due to the effect of Chamiatar Khad fault striking N1700 E and 450 W. Joints are open, closely spaced, intersecting, which are having clay fillings due to mechanical and chemical weathering of the rocks. One main joint with angle of N 750 E/800W is running parallel to the axis of the tunnels which is very unfavourable. At some places huge wedges were formed due to the intersection of the joints, which caused excessive overbreaks in the tunnels. The Q values of most of the rock mass of tunnels range from 0.12 to 0.21, which indicates that the rock was very poor. Core samples were collected from both the monitoring locations by underground coring machine. Engineering properties like Rock Quality Designation (RQD) compressive strength, tensile strength, density and compressional wave velocity (Vp) were determined from the core samples.

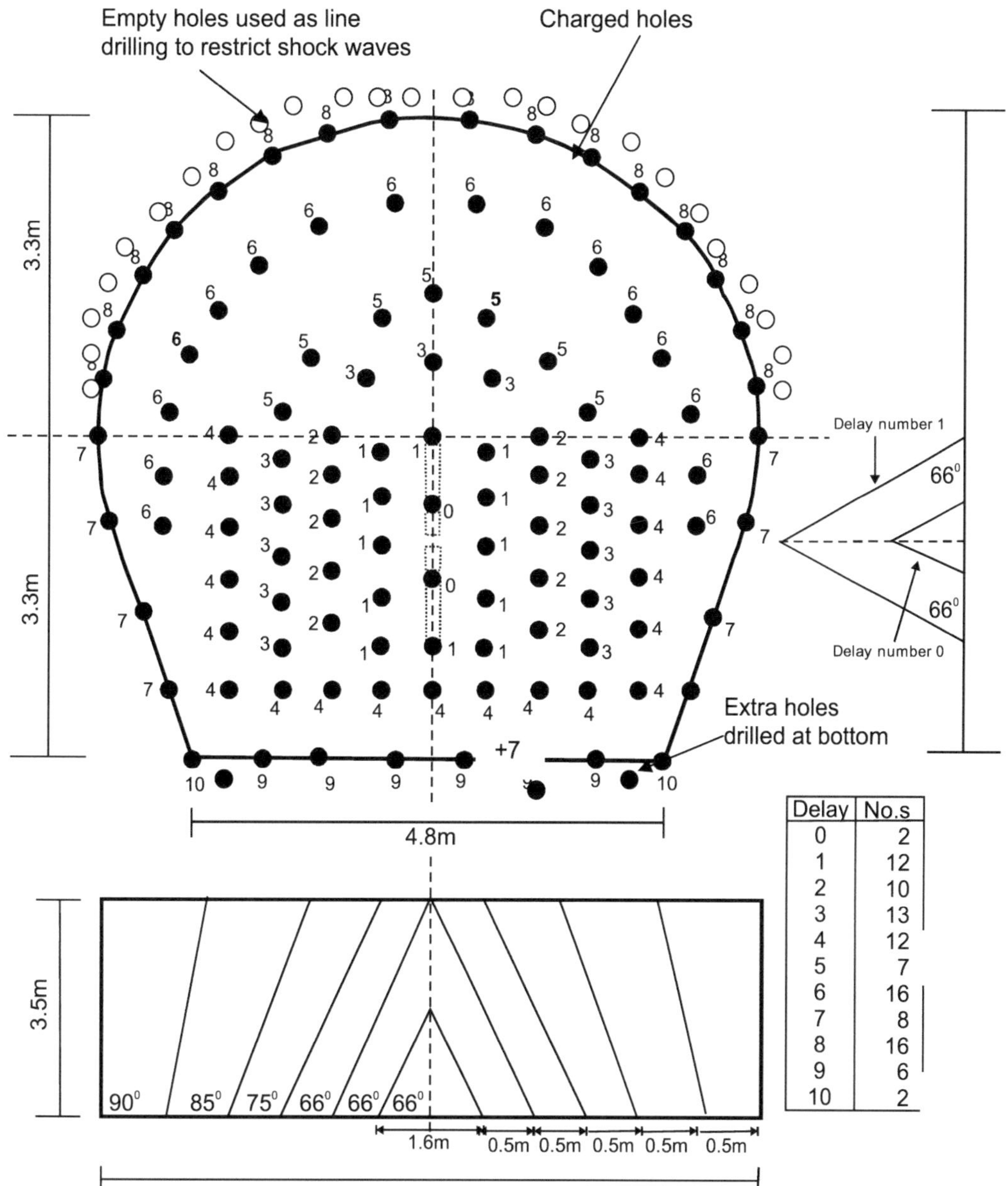

Delay	No.s
0	2
1	12
2	10
3	13
4	12
5	7
6	16
7	8
8	16
9	6
10	2

Hole Diameter = 32 mm; Total no.of blast holes = 113,
Length of blast holes = 3.5 to 4.0 m,
Specific Charge = 1.33 kg/m3

Fig. 5. Lake tap blast design

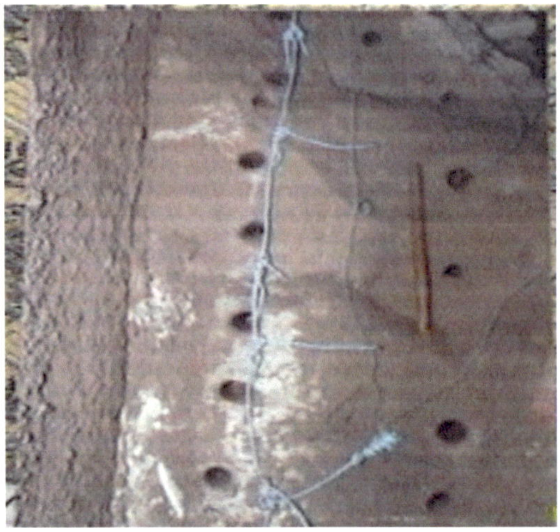

Fig. 5(a). Connections for pre-split blasting

Fig. 5(b). Pier nose wall after pre-split blasting

Fig. 6. Compressional wave velocity (P-wave) measuring device 'Ultrasonix 4600'

In-situ compressive strengths were also determined by using Schmidt hammer rebound testing. The average RQD values of Dolamite rock mass ranging from of 40-60%. Water absorption properties measured at the test site was 1.2% at both the sides. The improved blast performance of smooth blasting in the form of smooth profile is shown in Figure 7. The results were consistent for 12 trial blasts at the Dolomite tunnel. The controlled blasting restricted the overbreak to only 3%, which was 27% with the conventional tunnel blasting. The average half cast factor was calculated as 85%.

Fig. 7. Improved blast performance of smooth blasting in the form of smooth profile at KHEPP

7.4 In-hole delay blasting

Following the trend of opencast blasting, in hole delay blasting technique using delay electric detonators were used in some mines and tunnels to improve the pull per blast and reduce the ground vibration. As the confinement in the cut holes are maximum and the blast performance in tunnels depend mainly on the development of the cut portion, the in-hole delay were used in the cut holes only. The salient features of the in-hole delay pattern are:

1. The collar portion of the hole was blasted prior to the bottom. Thus, the confinement at the hole bottom was less during firing.
2. Mid-column decking between the two charges in a hole was kept at least 0.6 m to avoid sympathetic detonation. This decking provided confinement for the bottom charge.

The charging pattern is explained in Figure 8.

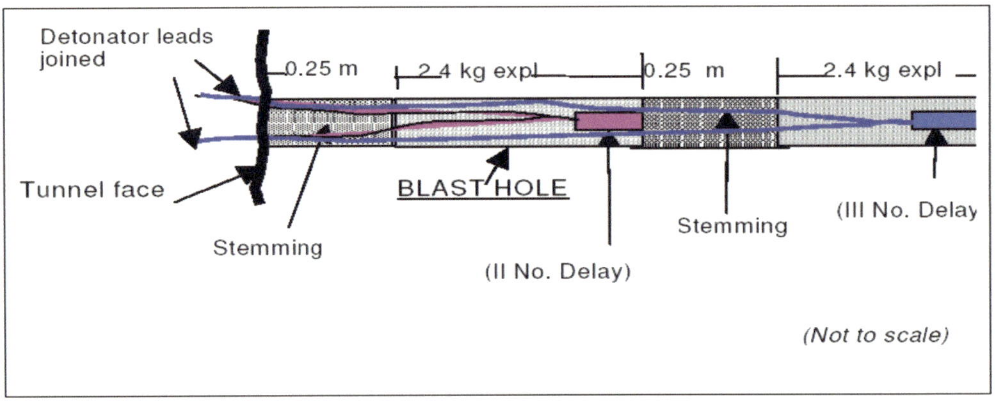

Fig. 8. Charging pattern of cut holes with in-hole delay

This technique was successfully applied at basaltic rock mass of Central railway tunnels and gneiss rock mass of Lohari Nag Pala Hydel power project.

The advantages of the in-hole delay cut blasting includes:

1. The average face pull improve by nearly 30-50%. The specific charge also reduces proportionately.
2. The blast vibration intensity reduces by 20 to 25% as the cut hole charge is distributed in two delays. This is going to reduce the overbreak proportionately.

7.5 Bottom hole decking technique

The mining industry is striving to enhance the productivity by improving fragmentation to reduce the system cost. In order to achieve this objective, development of new techniques and their application is essential. The authors at CIMFR, experimented a blasting technique called 'bottom hole decking technique' to achieve the objective of blasting productivity improvement of the mining industry. The technique consists of air decking at the bottom of the blasthole in dry holes by means of a wooden spacer or a closed PVC pipe. Although, practice of air decking is not new thing in blastholes, the concept of inserting bottom hole decking below the explosive column is relatively new. Explosives provide a very concentrated source of energy, which is often well in excess of that required to adequately fragment the surrounding rock material. Blast design, environmental requirements and production requirement limits the degree to which the explosive energy distribution within the blasthole can be significantly altered using variable loading techniques. Use of air-decks provide an increased flexibility in alteration and distribution of explosive charge in blast holes.

The bottom hole air-decking was developed to avoid the general disadvantages of middle air decking and to simplify the complex charging procedure, associated with it. The design aspects of the technique are explained in the following sections. The bottom hole decking consists of air decking at the bottom of the hole in dry holes by means of a spacer or a closed PVC pipe, covered at the upper end. The fume characteristics of the spacer are to be tested before applying in underground coal mine. If blast holes are wet, water decking will be created at the bottom by means of a spacer with a weight attached to it for sinking to the bottom. The diameter of the spacer should be preferably one third of the blasthole diameter for easy lowering and not allowing the charge to go to bottom side while loading. The reported values of air-deck length was taken as basis for optimum bottom deck length which was about 10% of the hole depth (Mead et al, 1993). The hole contains explosive and stemming column as in conventional loading but with a spacer at the bottom. The principle of bottom hole air decking in achieving optimum explosive energy interaction on rock mass is given below:

- Reduced shock energy around the blast hole due to cushioning effect of air decking, which otherwise would result in crushing
- Explosive energy-rock interaction is more at the bottom due to relative relief zone existing at that zone.
- Effective toe breakage is due to striking and reflection of shock waves at the bottom face of hole

The procedure and sequence of blast hole loading and initiation for the bottom hole decking are given below:

- Inserting the spacer in to the hole bottom by stemming rod.
- Loading the primer explosive cartridge attached by delay detonator charging the column charge conventionally
- Stemming of the hole by proper stemming material, preferably by sand mixed clay

The advantages of the bottom air decking technique in comparison to the conventional middle air decking are given below:

i. The highly confined toe is free of explosive charge but exposed to high concentration shock energy, resulting in good toe breakage and low vibration intensity.
ii. The reduced overall peak shock reduces the back break and damage.

Blast hole charge design for production blasts with bottom air-decking is Figure 9.

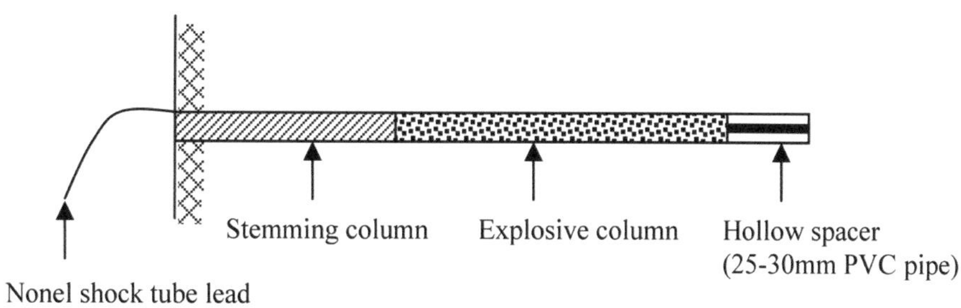

Fig. 9. Blast hole charge design for production blasts with bottom air-decking

The bottom air decking also resulted in the overall progress/pull per round of 36% with 1.5 deep rounds and 22% with 1.8 m deep rounds even with the powder factor improvement (ton/kg) upto 70%. The increase of detonator factor was very predominant in case of tests with bottom decking in comparison to tests with bottom decking technique. The technique was also resulted in reduction of ground vibrations by 20-26%. The laboratory and field experimental results prove that the bottom-hole air decking is an effective technique for improving the opencast blasting productivity as well as safety.

8. Sand stemming device for horizontal blast holes

The device of sand stemming for horizontal blast holes constitutes an assembly of a plastic pipe with proper cut and slits, a wooden block with pulley arrangement for resisting sand and an anti-static (non metallic) rope to pull out the plastic pipe from blast hole. The device essentially consists of a plastic pipe tied with an anti-static rope which is passed through a wooden resisting block to which a pulley is attached. The main objective of the device is for efficient use of sand as stemming material in horizontal blast holes. Another objective of the present device is to provide an effective and economic and fast stemming method which can find a mass application in underground blasting. The device essentially consists of a plastic pipe tied with a non metallic rope which is passed through a wooden resisting block to which a pulley is attached, an assembly of a plastic pipe cut and slit properly and a rope passed through a wooden block which can insert the sand in to blast hole and resist the sand to come out while pipe is pulled out of the blast hole. The position of stemming device while inserting the sand with plastic pipe and the position of removal of the plastic pipes are shown in the Figure 10. Actual application of the device in the field is shown in the Figure 11.

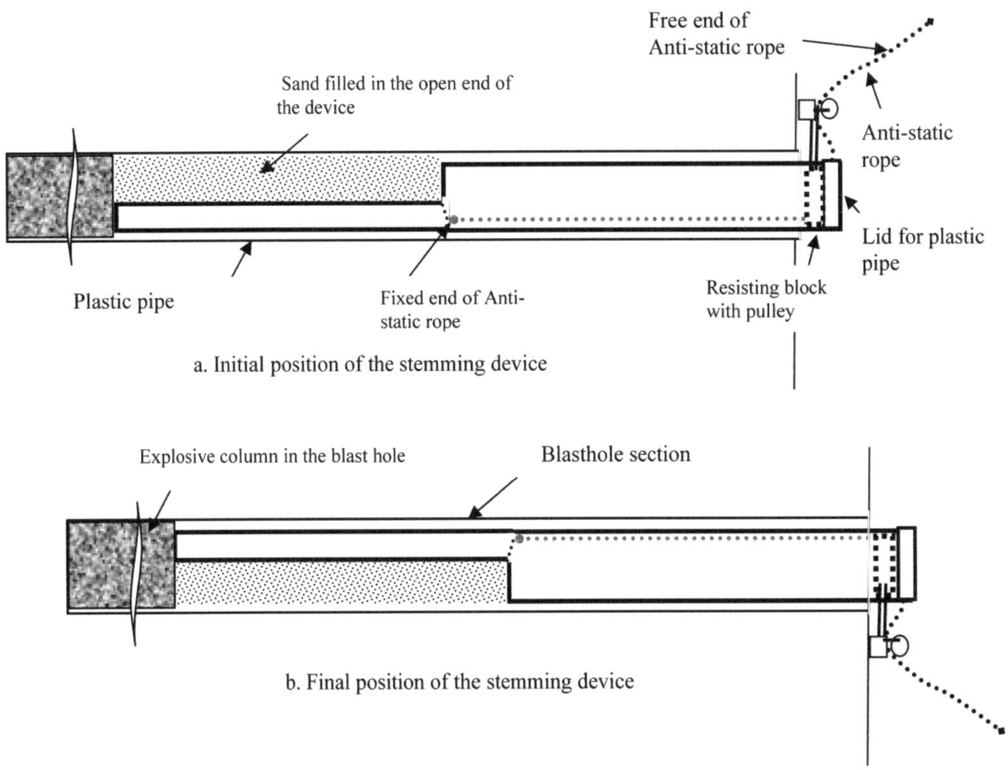

Fig. 10. Position of stemming device for loading and unloading in the blasthole.

Application of this tool in place of conventional stemming resulted in pull improvement of 5-10% in dolomite tunnels and 8-12% in gneiss tunnel. The improved blast performance was recorded consistently for 20 trial blasts at the gneiss tunnels and 25 trial blasts at the dolomite tunnels.

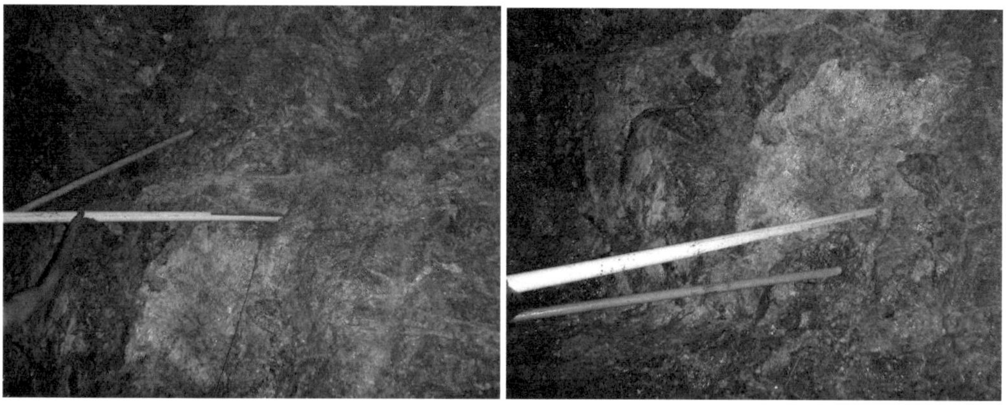

Fig. 11. Loading and unloading of sand into the stemming column of a blasthole at KHEPP

9. Computer aided blast design

Some of the software developed for blast design and optimisation are reported in Table 4. Few blasting software on tunnel blasting are commercially available and the details can be obtained through web search.

Name of Software	Purpose	Reference
OPTES	Blast optimisation in tunnels	Vierra (1984)
VOLADOR	Estimation of blast results, blast efficiency and cost analyses in tunnels	Rusilo et al. (1994)
TUNNEL BLAST	Blast design in tunnels	Chakraborty et al. (1998)
CAD	Optimum design of ring hole blasting	Myers et al. (1990)
FLAC and UDEC	Blasting effects on the near field rock mass	Pusch et al. (1993)
ABAQUS V 5.4	Mechanics of crater blasting and the effects of air decking and decoupling	Liu and Katsabanis (1996)
ALEGRA	Air-decking blasting	Jensen and Preece (1999)
PFC-2D/3D	Crack and heaving simulation	Itasca Consulting Group Inc. (2002)
Neural networking	Model free computing	Leu S. S. et al. (1998)

Table 4. Various routines for computer aided tunnel blast design

9.1 TUNNEL BLAST$^{2.0}$ software

Based on the past experience and extensive field investigations over a variety of underground structures of varying lithologies, CMRI Nagpur Centre devised a software "TUNNELBLAST" for generating blast design for Tunnels and underground workings. This software is a handy intelligent tool for the site engineers to optimise the blasting process and improve productivity without spending their valuable time on scrutinising variety of documents, books and literature available. The software is simple to operate and user friendly. The input and output parameters of the software are as under:

Input parameters:

1. Rock properties (density, compressive strength and joint spacing),
2. Tunnel (shape, width and height),
3. drilling (diameter and length of blast hole), and
4. Explosive properties (weight strength, weight and length of cartridge).

Output parameters:

1. Size of the tunnel,
2. Probable deviation of blast holes,
3. Optimum depth of round,
4. Look out angle of peripheral holes,
5. Burden, spacing and charge of holes in cut area, floor periphery and in the middle of the tunnel section,
6. Front and sectional views of the blast pattern

9.2 Field application of TUNNEL BLAST software at gneiss rock mass

The TUNNEL BLAST software was applied to design the parallel cut blast pattern at Lohari-Nag Pala Hydroelectric Power Project (LNPHPP). The LNPHPP falls in the Uttrakhand Himalayas and is located on the River Bhagirathii upstream of Uttarkashi district. The main rock type of powerhouse complex is schistose gneiss and augen gneiss with abundance of mica and geotechnically the rock mass is negotiating in "Fair Category" and it's having three prominent joint sets. The Rock Mass Quality (Q) was varying from 1-10. The main two joint sets intersecting at right angle which makes wedge continuously. Some weak zone/clay filling, altered rock, sheared rock mass and excessive flow of water at places makes the rock poor. In maximum area it is found that the regional trend of foliation is perpendicular to the tunnel alignment, another joint which is intersecting the foliation at right angle and creates wedge on roof. The strike of the foliation is going through along the tunnel alignment which is geologically not favourable because of probabilities of plane failure and wedge failure in presence of heavy joint planes.

The input geological parameters required for the blast design software are as follows:

Rock type	Metabasic (Amphibolite) & quartz vein
Joint sets	Three + Random ($045^0/35^0$, $210^0/45^0$, $130^0/80^0$)
Critical joint	$045^0/35^0$, $210^0/45^0$
Water Condition	Dry
Weathering	Highly Weathered/Fractured
Filling	Clay seam, width 10-20cm

Boundary conditions:

- Rock Type: Metabasic (Amphibolite) & quartz vein
- Av. rock density: 2.6 t/m³
- Type of explosives: Emulsion
- Blast hole diameter: 45 mm
- Explosives diameter: 40 mm & 32 mm
- Explosives strength: 80% (60% may also be required in the periphery holes
- and hence provisions may be made)
- Length of blast hole: 2 m
- Delay: Long delay (NONEL)

After feeding the input information the software process the entire data and gives the blast hole geometry and charge pattern for cut holes and other holes separately. The utput information given by TUNNELBLAST software is given in Figure 12, Figure 13 and Table 5 and Table 6.

The blast design generated by TUNNEL BLAST software was applied at intermediate adit and the blast results were satisfactory in terms of pull per round and overbreak control. The trial blast results with felid application of TUNNEL BLAST software are given in Table 7. The blast results indicate the efficacy of the TUNNEL BLAST software, as a preliminary tool for tunnel blast design for various geological conditions. The fine tuning of this design can be done for further improvements in the progress and yield of tunnel blasting.

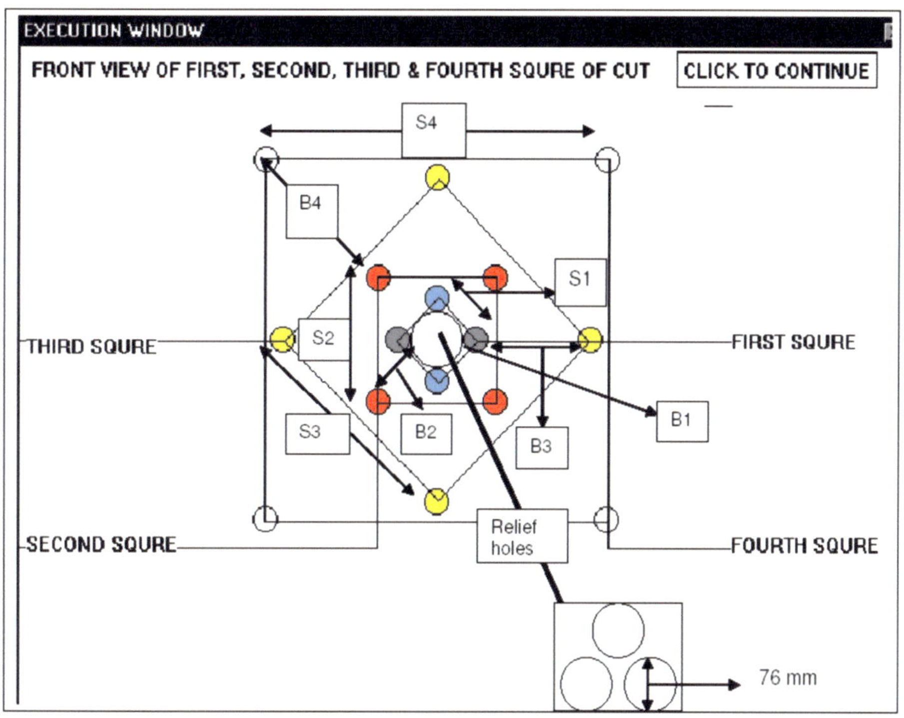

Fig. 12. Blast design output from TUNNELBLAST for cut holes of intermediate adit

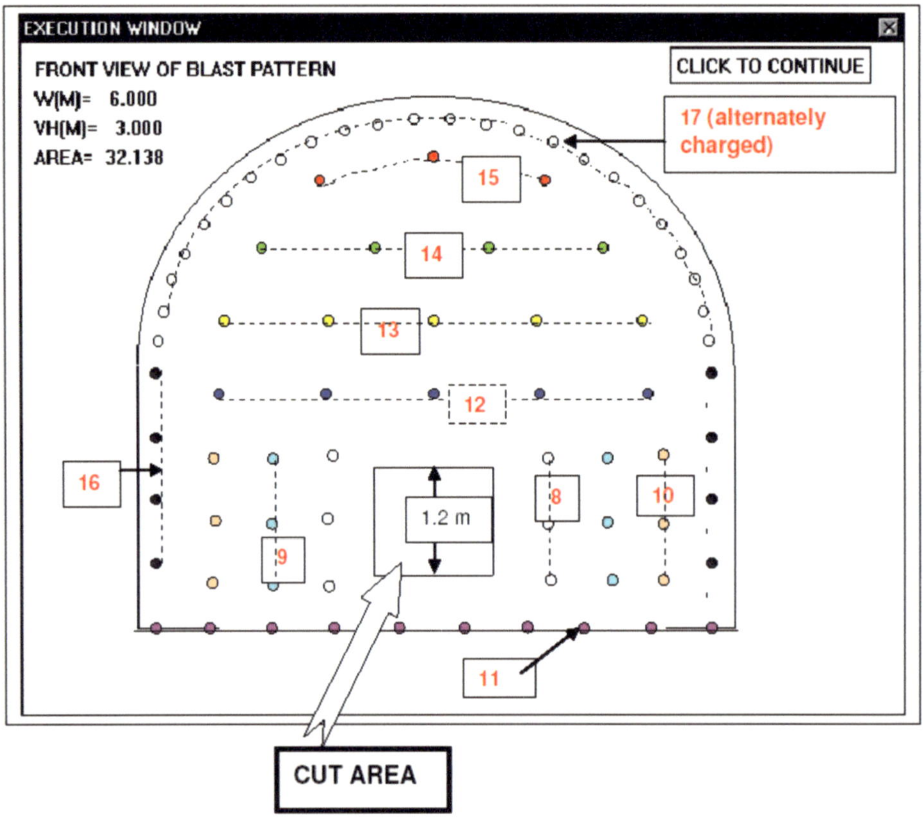

Nos. in the boxes denote the delay numbers; Total Charge per round = 97.7 kg
Total no. of holes= 3-Relief holes + 69-Charged holes+ 12-Dummy holes; Powder factor = 1.52 kg/m^3

Fig. 13. Controlled blast design output from TUNNEL BLAST for rest of the holes at intermediate adit of LNPHPP

Short Delay No. (25 ms delay)	Name of square	Burden, m	Spacing, m	No. of holes	Charge/hole, kg	Total charge, kg
1	First	0.15	0.2	4	1.2	4.8
2/3	Second	0.20	0.4	4	2.4	9.6
4/5	Third	0.35	0.75	4	2.4	9.6
6/7	Fourth	0.45	1.2	4	2.4	9.6

Table 5. Blast pattern and charge configuration of the cut holes

Description of holes	Delay No.	No. of holes	Burden	Spacing	Charge per hole (kg)	Total Charge/ delay (kg)
Easer holes	8	6	0.6	0.75	0.95	5.7
Support holes	9	6	0.6	0.75	0.95	5.7
Support holes-II	10	6	0.6	0.75	0.95	5.7
Bottom Holes	11	10	0.4	0.70	1.8	18
Crown Holes-I	12	5	0.8	1.2	1.6	8
Crown Holes-II	13	5	0.8	1.2	1.6	8
Crown Holes-III	14	4	0.8	1.2	1.6	6.4
Crown Periphery holes	15	3	0.6	0.3	0.6	1.8
Side Periphery holes	16	8	0.6	0.3	0.6	4.8

Table 6. Design and charging details of blast holes, other than cut holes

S No.	Location	Hole diameter, mm	Depth of holes, m	No. Of holes	Charge per round, kg	Specific charge, kg/m^3	Pull/round, m
1	Downside, TRT	40	3.5	91	217	2.1	1.98
2	Upsideside, TRT	40	3.5	89	225	1.85	3.1
3	Upsideside, TRT	40	3.5	89	250	1.98	3..0
3	Downside, TRT	40	3.5	91	220	2.0	1.95

Table 7. Trial blast results with felid application of TUNNEL BLAST software

10. Conclusions

The reviews on the developments in rock mass damage and contour blasting brings an important information on field application of controlled blasting and damage assessment and control The contributions of CIMFR on special tunnel blasting techniques resulted in improvement of both productivity and safety. The following conclusions can be drawn based on the various topics discussed in the paper:

i. Application of this techniques resulted in saving the time of 75-80% and cost of 60%-95% in comparison to the conventional shaft sinking method at three different projects
ii. Lake Tap Blasting of a 4 m thick 40 m2 cross sectional area was carried out as of fist of its kind with indigenous technology in India by CMRI (now CIMFR) at granitic rock mass in Andhra Pradesh Power Generation Corporation (APGENCO), which could save Crores of national exchequer.
iii. Ultra cautious blasting techniques were adopted as an extremely cautious approach, for removal of 10 reinforced concrete plugs, each of 125 m3 volume, without causing any

damage to the surrounding periphery and pier nose in Srisailam left bank project of the APPGENCO while the power house was in running condition.

iv. Successful application of in-hole delay cut blasting method at basaltic rock mass and gneiss rock mass improved average face pull improve by nearly 30-50%. Blast vibration intensity reduces by 20 to 25% which resulted in reduction of the overbreak proportionately.

v. Bottom hole decking technique resulted in the overall progress/pull per round of 36% with 1.5 deep rounds and 22% with 1.8 m deep rounds even with the powder factor improvement (ton/kg) upto 70%.

vi. Application of sand stemming device for horizontal blast holes in place of conventional stemming resulted in pull improvement of 5-10% in dolomite tunnels and 8-12% in gneiss tunnel.

11. References

Bieniawski, Z. T., (1993). Classification of rock masses for engineering: the RMR system and future trends. In J.A. Hudson (Ed.), Comprehensive rock engineering: principles, practice, and projects, Oxford: Pergamon Press. V.4, pp. 553-573

Calder, P. N. and Bauer, A. (1983). Presplit blast design for open pit and underground mines, *5th International Cong. on Rock Mechanics*, Melbourne, Vol. 2, pp. E185- E190.

Chakraborty, A. K. (2002). Development of predictive models for optimum blast design in mine roadways and tunnels under various rock mass conditions, *Ph.D. Thesis, Indian School of Mines*, Dhanbad, India, 298 p.

Chakraborty, A. K., Jethwa, J. L. and Dhar, B. B. (1996b). Predicting powder factor in mixed-face condition: development of a correlation based on investigations in a tunnel through basaltic flows, *Engineering Geology*, Elsevier Science B.V., Netherlands, No. 47, pp. 31-41.

Chakraborty, A. K., Murthy, VMSR and Jethwa, J.L. (1996a). Blasting problems in underground construction through Deccan Trap formation: Some experiences at Koyna Hydro-electric Project, Stage IV , *Tunnelling & Underground Space Technology*, Elsevier Science Ltd., Great Britain, Vol. 11, No. 3, pp. 311-324.

Chakraborty, A. K., Murthy, VMSR, Jhanwar, J. C., Raina, A. K., Ramulu, M. and Jethwa J. L. (1998). Final report on Development of Rock Mass Classification for Computer Aided Tunnel Blast Design, *Grant-in-Aid Project funded by Ministry of Water Resources*, Govt. of India, 136 pp.

du Pont, E.I., (1977). Blasters hand book, 175th Anniversary edition, E.I. du Pont de Nemours, Inc., Wilmington, Delaware. Pp.526-541.

Duvall, W. I. and Atchion, T. C. (1957). Rock breakage by explosives, USBM, RI 5356, Explosives Reference Database on CD-Rom, *International Society of Explosives Engineers*, Ohio, USA, 1997.

Gupta, R. N., Singh, R. B., Adhikari, G. R. and Singh, B. (1988). Controlled Blasting for Underground Excavation, *International Symposium on Underground Engineering*, 14-17 Apr., New Delhi, India, pp. 449-460.

Holmberg, R., and Persson, P. A. (1978). The Swedish approach to contour blasting, Proc. of Annual Conference on Explosives and Blasting Research, Explosives Reference Database on CD-Rom, *International Society of Explosives Engineers*, Ohio, USA, 1997.

Itasca Consulting Group Inc. (2202). Partcile Flow Code in 2 Diemnsions - Theory & Background, Mnneapolis, USA. p. 1.1-1.28.

Jensen, R. P. and Preece, D. S. (1999). Modelling explosive/rock interaction during pre-splitting using ALE computational method, *Proc. 6th. International Symposium for Rock Fragmentation by Blasting,* The South African Inst. of Mining and Metall., Johannesburg, Aug. 8-12, pp. 199-202.

Johansen, J. (1998). Modern trends in tunnelling and blast design, IDL Industries Ltd., Hyderabad, India, pp. 34-41.

Langefors, U. and Kihlstrom, B. (1973). *The Modern Technique of Rock Blasting,* John Willey & Sons, pp. 188-257, 299-301.

Leu S. S., Lin S. -F., Chen C. -K. And Wang S. -W. (1998). Analysis of powder factors for tunnel blasting using neural networks, The Int. Journal for Blasting and fragmentation, Balkema A. A. , Netherlands, Vol. 2., No. 4, pp. 433-448.

Liu, L. and Katsabanis, P. D. (1996). Numerical modelling of the effects of air-decking/ decoupling in production and controlled blasting, *Proc. 5th International Symposium on Rock Fragmentation by Blasting, FRAGBLAST-5,* Montreal, Quebec, 25-29.,pp. 319-330.

Livingston, C. W. (1956). Fundamentals of rock failure, *Quarterly of the Colorado School of Mines,* Vol. 51, No. 3, Jul..

Lopez Jimeno, C., Lopez Jimeno, E., Carcedo, F. J. A. and De Ramiro, Y. V. (1995). *Drilling and Blasting of Rocks,* Balkema A. A., Rotterdam, pp. 200-204 and 259-260.

McKenzie, C. J. (1994). *Blasting for Engineers,* Blastronics Pty. Ltd., Brisbane, Australia.

Olofsson, S., O. (1988). Applied Explosives Technology for Construction and Mining, Applex, Arla, Sweden, 303 pp.

Ouchterlony, F., Nyberg , Sjoberg, C., Johansson, S-E. (1991). Damage zone assessment by vibration measurements, *Aspo hard rock laboratory,* Progress Report, , No. 3, pp. 25-91.

Pokrovsky, N. M. (1980). *Driving Horizontal Workings and Tunnels,* Mir Publishers, Moscow, pp. 38-41.

Pusch, R. and Stanfors, R. (1992). The zone of disturbance around blasted tunnels at depth, *International Journal Rock Mech. Mining Sci. & Geomech. Abstr.,* Vol. 29, No. 5, pp. 447-456.

Pusch, R., Hokmark, H. and Borgesson, L. (1993). Characterisation of structure and stress state of nearfield rock with respect to the influence of blasting, *Proc. 4th International Symp. on Rock Fragmentation by Blasting, FRAGBLAST-4,* Vienna, Austria, pp.175-181.

Ramulu, M., (2009), Rock mass damage due to repeated blast vibrations in underground excavations, IISc Dept.of CE PhD Thesis- G23635, 624.15132 P09, IISc Press.

Ramulu, M., Chakraborty A. K. and Sitharam T.G., (2009), Damage assessment of basaltic rock mass due to repeated blasting in a railway tunnelling project - a case study, Tunnelling and Underground Space Technology, Vol.24, pp. 208–221.

Ramulu, M., Sitharam, T.G., (2011), Blast induced rock mass damage In underground excavations -Applications to civil engineering projects, LAMBERT Academic Publishing GmbH& Co. KG, 66123 Saarbrücken, Germany, ISBN (978-3-8433-9318-8)

Rusilo, L. C., Sansone, E. C., Hennies, W. T. and Ayres Da Silva, L. A.(1994). Design and optimization of tunnel blasting operations, *Proc. 3rd International Symposium on Mine Planning and Equipment Selection*, Istanbul, Turkey, 18-20 Oct., pp. 651-655.

Singh, S. P. (1995). Mechanism of cut blasting, Trans. Inst. Mining and Metall., Section A, Mining Industry, Vol. 104, Sept-Dec.,*The Inst. of Mining and Metall.*, U.K., pp. A134-A138

Wilson, W. H. (1987). An experimental and theoretical analysis of stress wave and gas pressure effects in bench blasting, *Ph.D. Thesis*, University of Maryland.

Yu, T. R. and Vongpaisal. S. (1996). New blast damage criteria for underground blasting, *CIM Bulletin*, No. 998, Vol. 89, pp. 139-145.

20

Susceptibility of the GSM-R Transmissions to the Railway Electromagnetic Environment

Stephen Dudoyer[1], Virginie Deniau[1],
Nedim Ben Slimen[2] and Ricardo Adriano[3]
[1]*Univ Lille Nord de France, IFSTTAR,*
[2]*Assystem,*
[3]*Federal University of Minas Gerais,*
[1,2]*France*
[3]*Brazil*

1. Introduction

The Electromagnetic radiations are non-ionising radiations; they cannot involve the ionisation of atoms or molecules. Nevertheless, they can cause various adverse effects. From a biological point of view, they provoke heating due to the occurring of induced current in the body. But, this issue will not be considered in this chapter. From a technological point of view, they can cause malfunctions, permanent damages for electronic devices or telecommunication systems. In this chapter we will focus on their impact on a telecommunication system dedicated to the European railway and potential consequences on the management of the railway network.

Today, the European railway network is undergoing significant changes, which aim at deploying a unique management system in Europe which will replace local systems. This unique management system called ERTMS (European Railway Traffic Management System), involves the deployment of a telecommunication network dedicated to railway management, the GSM-Railway network, in order to harmonize in Europe the system of communication between the trains and the infrastructures. This harmonization is intended to clear the technological boundaries between railway networks of European countries and thus to remove border for trains. GSM-R is a key element in the management system as it provides the vocal exchanges, but also the transmission of signalling data. However, as all the telecommunication systems, the GSM-R can be vulnerable to the Electromagnetic (EM) interferences and the railway environment is particularly rich in EM interferences. This chapter will then focus on this issue.

After a general background about the electromagnetic interferences and the management of the European railway network, we present the standards and approaches applied in the railway domain to control the Electromagnetic compatibility (EMC). The GSM-R and the EM disturbances which can affect it are then detailed. Finally, a methodology for testing the vulnerability of the GSM-R transmissions and the test results are presented and analysed.

2. General notions

Understanding the electromagnetic emission from the railway environment is important to prevent and control electromagnetic interference. Currently, trains are more and more often equipped with potentially sensitive systems from an electromagnetic compatibility point of view. Consequently, railway systems have to be sufficiently robust to guarantee the safety of the railway transportation. In this section the fundamental concepts related to EMC are briefly introduced. For this purpose, the following definitions given in (IEC 60050, 1990), International Electrotechnical Vocabulary (IEV), chapter 161, apply:

Electromagnetic environment: The totality of electromagnetic phenomena existing at a given location.

Immunity (to a disturbance): The ability of a device, equipment or system to perform without degradation in the presence of electromagnetic disturbance.

(Electromagnetic) Susceptibility: The inability of a device, equipment or system to perform without degradation in the presence of an electromagnetic disturbance.

Immunity level: The maximum level of a given electromagnetic disturbance incident on a particular device, equipment or system for which it remains capable of operating at a required degree of performance.

2.1 Electromagnetic disturbances and electromagnetic compatibility

A system is electromagnetically compatible with its environment if it is able to operate satisfactorily in its electromagnetic environment without introducing intolerable electromagnetic disturbances to anything in that environment. Typically, an EMC problem can be decomposed into three main parts. A source that intentionally or unintentionally produce the emission, a coupling path that transfers the emission energy to a receptor and the receptor that can be susceptible if the received energy exceeds its immunity level.

During their operation, electrical or electronic systems generally produce radiated or conducted signals, which can lead to equipment malfunctions neighbours. The "electromagnetic disturbance" term then assigns these signals that can be voltages, currents or electromagnetic fields. In general, the higher the frequency of the electromagnetic disturbance is, the more efficient the coupling path. It is important to keep in mind that the source and receiver can be classified as intend or unintended. For instance, the GSM-R system intentionally transmits and receives electromagnetic fields in some frequencies between 876 MHz and 925 MHz. Consequently, the equipment near the GSM-R antennas must be designed to operate properly under the influence of the GSM-R signals. On the other hand, the GSM-R antenna will collect all the signals generated by the railway environment at these frequencies. Depending on the coverage of the GSM-R system and the levels of the electromagnetic disturbances, the communication between rolling stocks can be affected or even interrupted.

2.2 Electromagnetic coupling

Electromagnetic disturbances produced by the emitter can be coupled to the receptor by either radiated or conductive paths. The coupling mechanism can be classified into Conductive coupling, Magnetic coupling, Electric field coupling and Electromagnetic field coupling.

Conductive coupling can be viewed as a common impedance coupling. Conductive coupling occurs when the source and the receptor circuits are physically connected with a conductor and share a common-impedance path. Magnetic coupling occurs when two objects exchange energy through their varying or oscillating magnetic fields. It can be represented by a mutual inductance between the source and the receptor. Electric field coupling (or capacitive coupling) is caused by a voltage difference between conductors. It dominates in high-impedance circuits and can be represented by mutual capacitance. Finally, Electromagnetic coupling is a combination of both electric and magnetic fields. It is the most common coupling mechanism observed. It occurs when source and receptor are separated by a large distance, (typically more than a wavelength). In this case, source and receptor act as radio antennas. The electromagnetic field radiated by the source propagates across coupling path and is picked up by the receptor.

2.3 Electromagnetic radiation, emission and immunity

Any device which drives an electric current is likely to radiate an electromagnetic field. This electromagnetic field may act in two ways on electronic and telecommunication equipment. It can either be picked up by cables of other systems (or lines of electronic circuits) on which unwanted electrical signals appear and can cause malfunctions or it can also interfere with the telecommunication signals when they reach the receivers causing losses of information. In the first situation, the interference occurs because the dimensions of the conductors in the electronic equipment are comparable with the wavelength of the electromagnetic disturbance. In this case conducting elements can act as receiving antennas.

There are numerous sources of unintentional electromagnetic radiation such as lighting, relays, electric motors and digital systems. The number of emitters is increasing rapidly. Some of these emitters employ very high power levels; others such as digital systems are using faster digital electronics and are becoming more efficient radiators of unintentional electromagnetic energy. Consequently, EMC has become a particularly important topic. In order to ensure that EMC will be not a problem, many EMC standards are used. These are often supported by EMC legislation to ensure that existing equipment conforms to the required standards. EMC standards specify a limited number of essential emission and immunity tests, as well as minimum test levels. The aim is to ensure adequate compatibility. Section 4 summarizes the major standards concerning the electromagnetic emissions in railway environment while section 5 addresses the immunity problems.

3. Management and signalling of european railway network

The management of a railway network is generally performed thanks to several key components, notably a ground-train radio which allows the vocal exchanges, a lateral signalling system including lights and traffic signs and a localization system of the trains which can also control the speed of the trains. However, these different components are not necessary ensured by similar technologies in all the European countries. This situation inhibits the carrying out of a real European railway network which would allow the different railway operators to offer their services anywhere in Europe. Today, trains crossing borders are necessarily equipped with various national systems and at the borders the trains have to change their system to be in accordance with the cross border country. This increases the costs of equipment and maintenance of the trains, the operating costs and extends the travel time.

The ERTMS (European Railway Traffic Management System) standard was then thought out in order to remove these obstacles and to optimize the use of the European railway network and to improve the reactivity, adaptability and affordability of the European railway. ERTMS would allow the interoperability of trains on the European territory (Jarašūnienė, 2005). This standard is generally presented as composed of two main components, which are the European Train Control System called ETCS, a standard for in-cab train control, and the GSM-R (Global System for Mobile communications-Railway) system, an international wireless communications standard dedicated to railway applications.

ETCS can allow automatically controlling the speed of the train if necessary. ETCS is composed of trackside and on-board modules. The trackside module transmits information to the train which enables the on-board computer, called Eurocab, to calculate the maximum permitted speed.

Nevertheless, the implementation of ETCS requires major adjustments on the European network, such as the installing of standard beacons called "Eurobalise" and GSM-R deployment. Indeed, the most complete version of ETCS relies heavily on the use of GSM-R. Three levels of deployment are then scheduled in order to progressively equip the railway network.

In the first level "ETCS level 1", the trackside equipment transmits information to the train in order that it calculates its maximum authorized speed. The information given by the trackside signalling (lights and traffic signs allowing the driver to know the permitted speed), can be forwarded to the train by the Eurobalise beacons located along the track.

The second level "ETCS level 2" includes a partial deployment of the GSM-R and information can then be forwarded to the train by the GSM-R. The position of trains is still detected by trackside systems but the trackside signalling is no longer necessary since all information is transmitted directly to the train.

Finally, the third level aims to optimise railway lines capacity and further reduce the trackside equipment. ETCS Level 3 is a major revision of the classic management system which is based on fixed intervals between the trains. In ETCS level 3, the route is thus no longer managed in fixed track sections but the intervals depend on the braking distances. The trains find their position themselves by means of positioning beacons or sensors and transmit the positioning signal to the radio block centre.

Then, this highlight the GSM-R is an essential and safety component in the management of the railway European network and it is necessary to warrant its immunity facing the railway electromagnetic environment (Midya, 2008).

4. Control of the radiated EM emissions in railway

The railway environment is a severe electromagnetic environment where railway equipment performs safety critical functions. Additionally, the railway runs very close to commercial and residential areas. For these reasons, it is important to provide guidance on EMC issues by applying specific EMC standards to railway applications. These standards fall generally into two categories: governmental standards, such as the EN50121:2006 part 1-5 (EN50121, 2006) published by European Committee for Electrotechnical Standardization

(CENELEC), describing EMC for railway applications, or railway industry standards such as Railtrack Group Standard GM/RC 1031 (GMRC1500, 1994), which provide guidance on EMC between railway infrastructure and trains.

A complete list of standards related to railway applications is presented and discussed in (Konefal et al., 2002), some of these standards are presented in the table 1 for convenience.

EN 50121 parts 1-5	Railway Applications Electromagnetic Compatibility
CISPR/C/116/CDV	Interference from overhead power lines, high voltage equipment and electric traction systems.
GM/RC 1500	Code of Practice for EMC between the Railway and its Neighbourhood
EN 55011 (CISPR 11)	ISM radio-frequency equipment - Radio disturbance characteristics - (CISPR 11) Limits and methods of measurement
UMTA-MA-06-0153-85-6	Conductive Interference in Rapid Transit Signalling Systems, Suggested Test Procedures for Conducted Emission Test Vehicle
UMTA-MA-06-0153-85-8	Inductive Interference in Rapid Transit Signalling Systems Suggested Test Procedures for Inductive Emissions of Vehicular Electrical Power Subsystem, Rail-to-Rail Voltage from 20 Hz to 20 kHz
UMTA-MA-06-0153-85-11	Radiated Interference in Rapid Transit Signalling Systems Suggested Test Procedures for Broadband Emissions of Rapid Transit Vehicles -140 kHz to 400 MHz

Table 1. List of EMC standards applied to railway domain

The standards applied in Europe in order to characterize the EM environment in the railway context are the EN 50121 while in USA, the electromagnetic emission limits are imposed by the Urban Mass Transportation Administration of the U.S. Department of Transport (UMTA standards). The EN 50121 standards notably aim to control the emission levels from the railway infrastructures to the outside world while UMTA standards aim to avoid interferences with the wayside equipment (transit signalling systems). In both cases, no method is proposed to characterize the EM environment on board trains, i.e. above, inside, and under the trains, especially where new and future sensitive systems can be located.

The standards EN 50121 indicate the methodologies and the limits to apply, relating to the EM emissions and immunity of railway equipment, vehicles and infrastructures. The emissions of the whole railway system, including vehicles and infrastructure are dealt with the section 2 of the EN 50121. The objective of the tests specified in this standard is to verify that the EM emissions produced by the railway systems do not disturb the neighbouring equipment and systems. The methodology then consists in measuring the radiated EM emissions at a distance of 10 m from the middle of the tracks and at about 1.5 m from the ground and to compare them with the maximum allowed levels. The limits are specified for the frequencies included between 9 kHz and 1 GHz. The measurement protocol is specified for four frequency bands which are 9 kHz-150 kHz, 150 kHz - 30 MHz, 30 MHz - 300 MHz and 300 MHz-1 GHz.

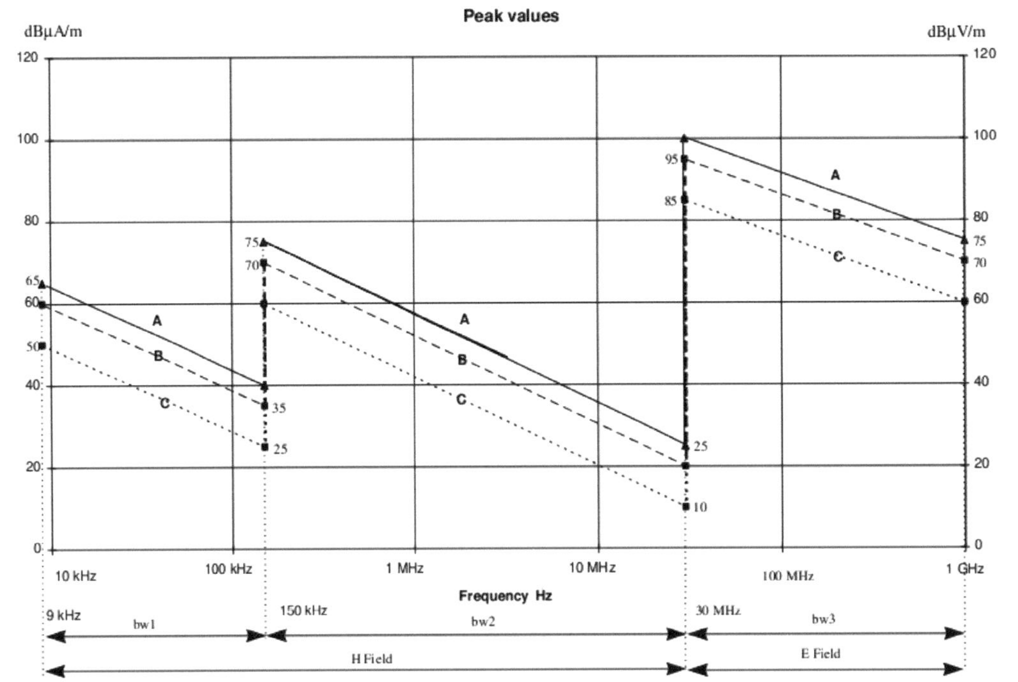

Fig. 1. Emission limits according EN 50121-2. A=25 kV ac; B = 15 kV ac, 3 kV dc or 1.5 kV dc; C = 750 V dc and bw1 = 200 Hz; bw2 = 9 kHz; bw3 = 120 kHz

The basis for the level derived in EN50121 has been the actual levels measured at a number of railways sites around Europe. While the scope of this standard covers the frequency range DC to 400 GHz, in practice limits are not set above 1 GHz. In general words, this standard does not consider the wider impact on the radio spectrum, it mostly sets the actual stage of the current levels around the railway structure.

Additionally, when comparing the EN 50121 standards to common EMC measurement standards, it is noted that there are several crucial differences in the methods of measurement. In many EMC tests, emission limits are specified in terms of a measurement with a quasi-peak detector (QP). However, the use of a quasi-peak detector in EN 50121 standards is not possible due to the highly dynamic environment. For EN 50121, a peak detector is prescribed.

5. EM immunity of the railway equipment and systems

The railway immunity levels for radiated interference are comparable with those specified by the industrial generic standard; 10V/m for trackside equipment. For rail borne equipment mounted externally to the rolling stock or within the driver's cab or passenger compartment 20V/m is specified. This level is comparable to the 24V/m specified by the Automotive Directive 95/54/EC. However, tests are different from automotive industry where full vehicle tests are performed in anechoic chamber to guarantee that all the systems can work together in the presence of electromagnetic disturbances. In the railway environment full vehicle tests are often not feasible. Due to the dimension and the speed of the trains, they cannot be tested in nominal operating condition inside an anechoic chamber. In this context, component and sub-system testing becomes very important to prevent EMC problems.

Although the immunity levels presented in EN 50121 provide an overall view of the railway electromagnetic environment, they are not suitable to perform immunity tests on the on board components, especially in the case of modern communication systems such as GSM-R. Additionally, high speed trains as other rolling stock apparatus are supplied by a catenary. In this particular context the train can be considered as a fixed equipment supplied by an electrical network. Consequently, EMC standard EN 61000-4-4 should apply. This standard aims at defining a common and reproducible basis for the evaluation of the performances of electrical and electronic equipment facing electrical fast transients on its different inputs. It is clearly adapted to test the immunity of the electronics and we would have referred to it if our objective had been to test the electronics of a GSM-R mobile.

However, as it will be shown in section 7, the test signals defined in this standard EN 61000-4-4 differ significantly from the typical transient disturbances received by GSM-R antennas. Additionally, as presented in (Knobloch, 2002), modern communication systems use digitally coded radio signals that operate with a much smaller signal-to-noise ratio (SNR) in comparison to analogical ones. The explanation lies in the fact that digital data streams are discontinuous and include redundancy to correct errors. (Knobloch, 2002) also points out that peak detector or QP detectors are not suitable to convert electromagnetic disturbance in some measure of deterioration in communication. Consequently, it is important to envisage component immunity testing solution which permits us to evaluate the telecommunication system against electromagnetic conditions representative of the railway electromagnetic environment.

6. The GSM-R communication system

The GSM-R (Global System for Mobile communication-Railways) is a wireless digital communication system, based on the public European GSM Phase 2+. This system is used to ensure the vocal exchanges and to transmit railway signalling information between trains and railway control centres. The GMS-R is currently deployed in numerous European countries in order to ensure the interoperability of trains throughout the whole European railroad network. The GSM-R system includes two parts:

- dedicated base stations, called Base Transceiver Station (BTS) installed along the railway tracks, and connected to railroad control centres, through a wired network.
- GSM-R antennas installed on the roof of train locomotives and connected through shielded cables to GSM-R mobile on board the train, as shown in Fig. 2.

The base stations are generally spaced from about 3 or 4 km and the GSM-R signal level has to be superior to -92 dBm, 95 % of the time and the space (UIC, 2003). In practise, the power of the received signal on board train varies between -20 dBm at proximity of the base station and -90 dBm at middle distance between two successive base stations (Hammi, 2009).

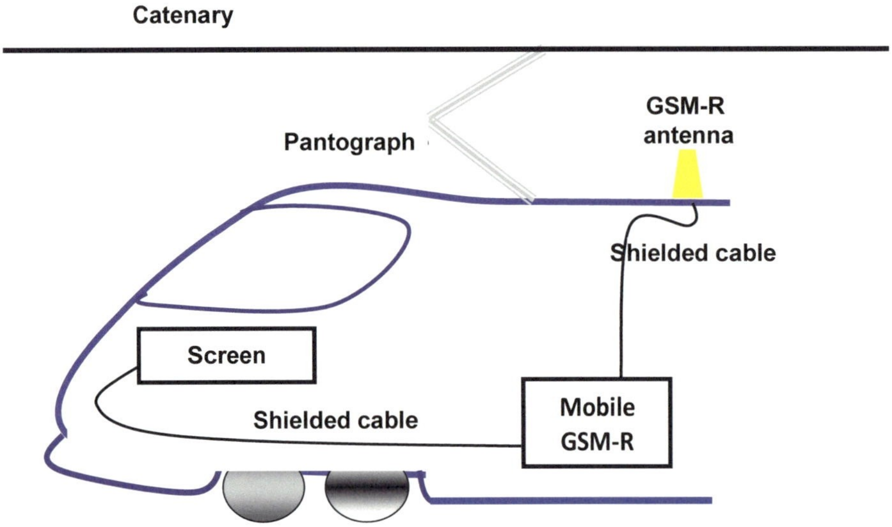

Fig. 2. Illustration of the on-board GSM-R system

The GSM-R is used in order to maintain a continuous voice and data link between the train and the control centres, and different trains located in the same neighbourhood. In the final version of ERTMS, the train sends its position through the uplink (from the train to the base stations) and receives signalling traffic information (speed limit, pass-through authorization…) through the downlink (from base stations to the train).

In Europe, the GSM-R uplink occupies the frequencies between 876 MHz and 880 MHz and the downlink between 921 MHz and 925 MHz. These frequency bands are separated by a frequency bandwidth dedicated to public and extended GSM.

Each frequency band used by the GSM-R is divided into 19 frequency channels of 200 kHz bandwidth. Only 19 channels are used by the system, in order to reduce the risk of interference with the public and extended GSM, using adjacent frequency bands.

The GSM-R is a TDMA (Time Division Multiple Access) system. The information is transmitted through each channel, according to 4.516 ms periodical TDMA frames, divided into 8 time intervals called "Time Slots" of 577 μs. During this time slot, the transmitted information is called burst, including 156 bytes, transmitted during 3.7 μs.

The data transmitted through the GSM-R system are very sensitive and the good operation of the GSM-R system is crucial to the capability and security of the European railway network. Thus, this system has been developed in order to be robust, with the capability of standing to some electromagnetic interference (EMI).

In fact, the GSM-R is included in the Euroradio protocol, which is specific to the railway and manages with altered received information, notably by resending some altered bursts until good reception. The use of such robust communication system is essentially motivated by the severity of the railway electromagnetic environment and the safety requirements.

In the next section we will focus on the different EMI that the GSM-R transmission signals can meet in the railway electromagnetic environment.

7. The EM noise sources affecting the GSM-R signals

On board a moving train and in normal operating conditions, the GSM-R system can meet different transient or permanent EMIs, with various amplitudes, time durations, repetition rates, frequency bands… Moreover, the GSM-R antennas are generally multi-band antennas and are not really selective around the frequency bands dedicated to the railway. They can thus receive GSM-R in-band and out-band EMIs (Mansson, 2008).

(Mansson, 2008) showed that out-band EMIs observed in railway environment could be a serious threat to the low noise amplifier (LNA) installed at the GSM-R receiver input. The susceptibility of this component can be reached with such EMIs and permanent damages on the system can happen.

In this effort, we will mainly focus on the in-band EMIs acting basically on the GSM-R useful signal. A description of the sources and different characteristics of these disturbances will be presented in the next part. Their impact on the GSM-R communication will also be described.

7.1 Description of the EM noise sources

From an EMC point of view, the railway infrastructure is a harsh complex EM environment where cohabitation between high power and digital communication systems with numerous eventual coupling mechanisms could be hazardous for the useful signal of the GSM-R. In this part, we will show that, on board moving trains, the GSM-R system is mainly affected by transient EM disturbances occurring between the catenary and the pantograph, in addition to the permanent disturbances coming from the public GSM base stations.

Fig. 3 synthesizes the different EMIs that could impact the GSM-R useful signals and describes the mechanism responsible of the generation of the transient disturbances on a GSM-R antenna fixed on the roof of a train. In fact, when a bad sliding contact occurs between the catenary and pantograph, a transient event could appear between these elements. This phenomenon can be observed with naked eye as a spark appearing between the catenary and the pantograph. Thus, a transient current circulates through these elements, which behave as transmission antennas, emitting EMI that the GSM-R antenna can receive.

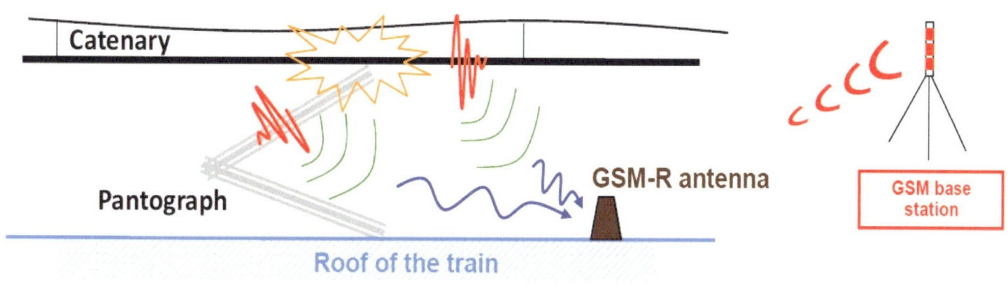

Fig. 3. EMIs received by GSM-R antenna and acting on the GSM-R useful signals

The generated wideband signal can easily cover the frequency bandwidth of the GSM-R system. However, from the train side, the GSM-R transmissions are mainly vulnerable to the EMIs covering the down-link frequency band. Indeed, on board trains, the signals emitted by the GSM-R antenna (up-link) have power levels highly superior to the power levels of the useful signals received by the antennas (down-link).

In addition, the GSM-R system uses frequency bands quietly close to the public GSM bandwidths, and when public GSM base stations use the adjacent frequency bands of the GSM-R, the risk for the GSM-R communications increases. This phenomenon is mainly observed when the train is operating in the vicinity of a city, where public GSM base stations and user numbers highly increase.

7.1.1 Transient EMI acting on the GSM-R useful signal

Measurement campaigns carried out on board moving trains (Hammi, 2009) showed that the transient events, triggered when a bad sliding contact occurs between the catenary and the pantograph, are the most penalizing events for the GSM-R useful signals. Fig. 4 (a) shows an example of a transient signal recorded by an oscilloscope connected to a GSM-R antenna. The analysis of a large number of transients collected on board trains showed that their time duration is generally inferior to 20 ns (Ben Slimen, 2009), with a typical value of 5 ns and a typical value of the rise time is 0.4 ns. Fig. 4 (b) shows the maximal EM amplitude generated by 284 successive transient events on the downlink frequency band of the GSM-R in normal operation conditions. Each point in this graph links the rank of the recorded transient and its maximal amplitude within the 921 - 925 MHz frequency band, corresponding to the down-link frequency band.

These results show that these transients generate high level EMIs that can reach - 40 dBm. Moreover, statistical analysis (Ben Slimen, 2009) of these transient disturbances highlighted that they can be very frequent, especially on high speed lines.

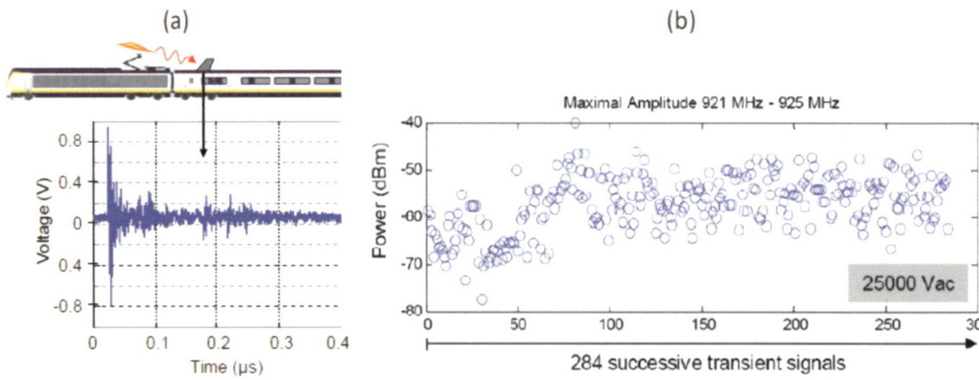

Fig. 4. (a) Example of transient disturbance in time domain and (b) maximal EM power generated by 284 successive transients in downlink GSM-R band

7.1.2 Permanent EMI acting on GSM-R antenna

Measurement campaigns have been carried out on board moving train, on railway lines equipped with the GSM-R system, in order to show the relation between the public GSM signal and the permanent interferences that can be observed in the closest downlink GSM-R frequency channel to the public GSM band (Hammi, 2009).

In Fig. 5 the top curve gives the variation of the EM power signal obtained through the last GSM-R channel along 20 km. The second one is the result obtained into an intermediary channel, supposed to be free. The last curve is obtained through the first public GSM frequency channel.

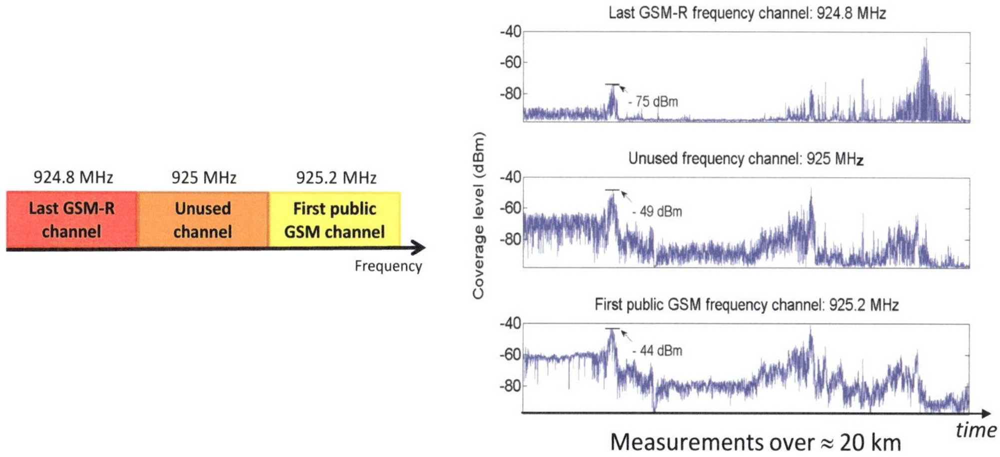

Fig. 5. EM measurement in different frequency channels used by the GSM-R and public GSM along 20 km railway line equipped with GSM-R system

Fig. 5 shows clearly that the variation of the measured amplitudes into the unused frequency channel and the last GSM-R frequency channel are partially similar to the EM noise variation observed through the first public GSM channel. This result proves that the public GSM signal sent through the first channel can disturb the GSM-R bandwidth, and confirms that the public GSM signal can be considered as a serious threat to the GSM-R useful signal. Indeed, as can be seen on Fig. 5, a signal of -44 dBm on the first public GSM channel induces an EMI of -75 dBm on the last GSM-R frequency channel. Knowing that the GSM-R signal level can decrease to a minimum value of -92 dBm (UIC, 2003), this interference level can be sufficient to severely disturb a GSM-R transmission on this channel.

7.2 Impacts of the EM noises

According to the type of the disturbance affecting the GSM-R system, the effect of the received EMI can vary. In fact, the transient events taking place between the catenary and the pantograph are wideband disturbances that can affect all the frequency channels used by the GSM-R system. It is obvious that the useful signal sent through the GSM-R system at the occurrence of the transient disturbance will be somehow disturbed. In fact, compared to the 3.7 μs time duration of one GSM-R bit, the transient duration of some ns is quietly small. So we need to investigate the real impact of such short time duration events on the interpretation of a disturbed GSM-R bit. This work will be presented in the next part of this effort.

When it comes to permanent EM disturbances coming from the public GSM base stations, the impact of these disturbances can be mainly observed when the GSM-R system is using the last frequency channel and the public GSM system is using the first frequency channel. So, even if the last frequency band of the GSM-R system is affected by the public GSM signal, the useful GSM-R signal is not necessary affected. In fact, when such scenario occurs, the GSM-R system could use different advanced protocols, mainly Euroradio protocol that can allow the system to stand to such disturbances. However, when the whole GSM-R channels are used, the system could be really affected by these disturbances.

However, among the current EMC standards, none methodology of immunity testing is adapted to the GSM-R system and to the characteristics of the EMIs that it can meet in the railway environment. The next section is then focused on a specific immunity approach to this system in order to assess the real risks for the GSM-R transmissions.

8. Immunity testing of the GSM-R signals in laboratory

8.1 Methodology

To evaluate the impact of the interferences on the quality of the GSM-R transmissions, a GSM-R mobile was employed. This mobile can be connected to one network either simulated by a specific piece of equipment or coming from a base station installed at proximity. In our case, we use a network simulator allowing controlling the network parameters such as GSM-R channel used for the communication, power of signals... and performing BER measurements. The principle of this methodology of test is to first connect the mobile to the simulated network and to establish a communication with the simulator. Then, the EM disturbance signals (permanent noise + transient signals) were generated and their impact on the quality of the GSM-R communication was evaluated thanks to criteria introduced in the following paragraph. Fig. 6 gives an illustration of this test methodology.

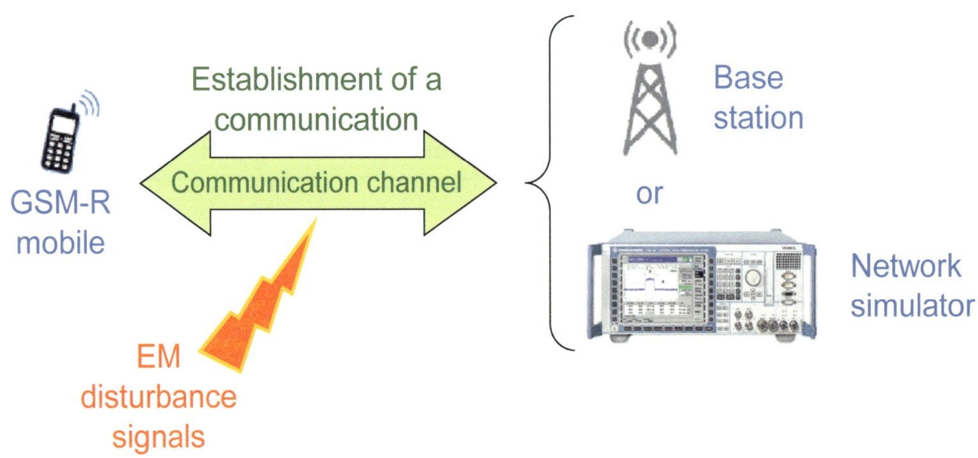

Fig. 6. Principle of the employed methodology of test

8.2 Definition of the immunity criteria

Two immunity criteria can be employed: Rxqual and Bit Error Rate (BER). The BER corresponds to the percentage of erroneous bits in a given transmission length (Breed, 2003):

$$BER = \frac{\text{Number of erroneous bits}}{\text{Total number of bits}} \times 100\% \quad (1)$$

The Rxqual is a quality parameter measured by the GSM-R mobile and it defines the quality of the received signal on a level from 0 to 7 (the lower the Rxqual is, the higher is the quality). The Rxqual is linked to the BER calculated on the learning sequences included in the GSM-R frames. The specifications defined in standards such as (ITU-T, 2003) require that the Rxqual is inferior or equal to 3 in order to ensure a good quality of communication. A relationship exists between BER and Rxqual: each value of Rxqual is associated with a range of values of BER (Lagrange et al., 1996) as can be seen in the following table.

RXQUAL	BER	
Quality level	Range of values	Typical value
0	BER < 0.2 %	0.14 %
1	0.2 % < BER < 0.4 %	0.28 %
2	0.4 % < BER < 0.8 %	0.57 %
3	0.8 % < BER < 1.6 %	1.13 %
4	1.6 % < BER < 3.2 %	2.26 %
5	3.2 % < BER < 6.4 %	4.53 %
6	6.4 % < BER < 12.8 %	9.05 %
7	12.8 % < BER	18.10 %

Table 2. Correspondence table between BER and Rxqual

8.3 Employed test bench

The test bench which was employed to perform immunity tests in laboratory is presented in Fig. 7. This test bench aims at reproducing the EM conditions that the GSM-R system is susceptible to meet on board trains. It is composed of three main parts:

- the communication system which consists in a GSM-R mobile connected to a network simulator called CMU 200 from Rohde & Schwarz.
- the EM noise generation which permits us, thanks to the two signal generators, to simulate the presence of permanent and EM transient noises simultaneously or separately.
- the area "analysis in frequency domain" is used to control the power of the exchanged signals at the input of the GSM- R mobile.

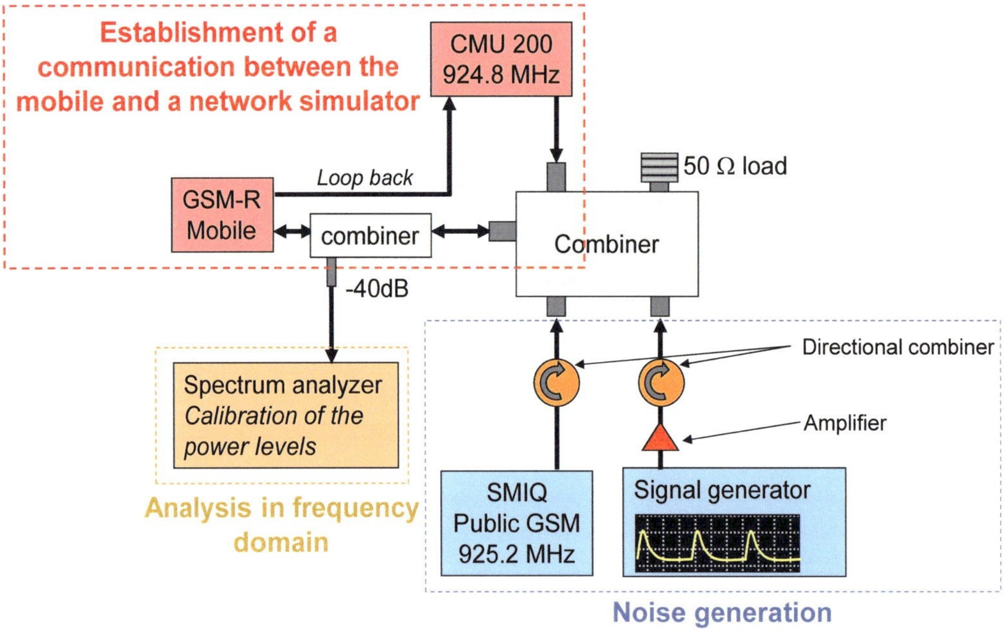

Fig. 7. Employed immunity test bench

8.4 Employed test signals

The GSM-R communication is established using the last useful GSM-R channel (924.8 MHz) from the down-link frequency band. The power level of the signals generated by the network simulator is adjusted so as to obtain a level of -70 dBm at the input of the GSM-R mobile. That corresponds to realistic operational conditions on board trains.

As for the public GSM signals, the communication channel employed is the first one (925.2 MHz) which is adjacent to the last useful GSM-R channel (924.8 MHz) used for the tests. The level of these signals is variable in order to study the effect produced on the quality of the GSM-R communication depending on the power level of the interference signals.

The signal used to simulate the presence of transient signals is a double exponential (duration=5ns, rise time=0.4 ns) modulated by a sinus at the frequency 923 MHz which corresponds to the center frequency of the GSM-R down-link frequency band. The corresponding mathematical expression is the following one:

$$\mathrm{S(t)} = \mathrm{A} \times (e^{-\frac{1}{D}t} - e^{-\frac{1}{RT}t}) \times u(t) \times \sin(2\pi F t) \qquad (2)$$

where D=5 ns, RT=0.4 ns, F=923 MHz and u is the unit step function.

The values employed for rise time (RT) and duration (D) result from a statistical analysis we performed on transients collected on board trains during one measurement campaign (Ben Slimen, 2009). Fig. 8 gives the time representation of this test signal.

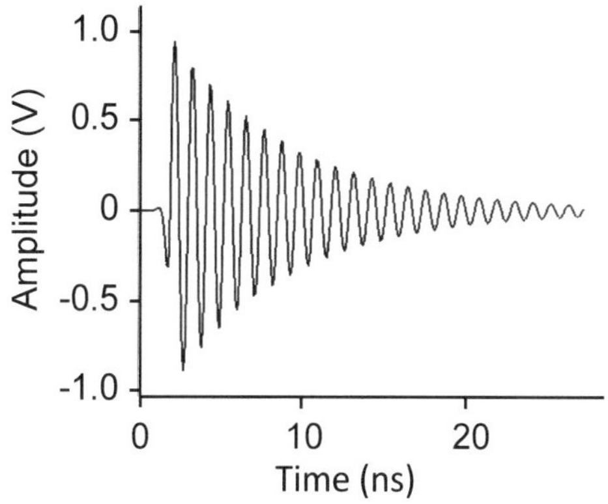

Fig. 8. Used transient test signal

Contrary to duration and rise time, it is not possible to determine a typical value for the recurrence of transients since it is very variable and depends on several operating conditions (speed of the train, one or two pantographs, state and age of the catenary and pantograph…). During the measurement campaign performed on board trains, we generally noticed that very few transients appeared at low speed whereas they could occur with a time interval of about 5 µs at about 200 km/h. As a consequence, the recurrence of transients is considered as a variable parameter for the immunity tests: for each measurement, the transient disturbances are generated with a constant time interval (TI) between two successive transients as illustrated in Fig. 9 and the immunity results are given in relation to the value of the time interval.

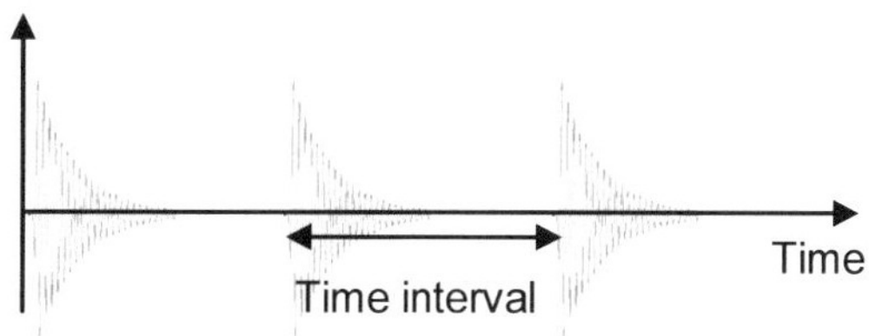

Fig. 9. Illustration of the time interval (TI) between the successive transient disturbances

9. Results of EM immunity tests on GSM-R transmissions

9.1 Configurations of test

Three different configurations, as shown on Fig. 10, are considered when studying the effect produced by the interference signals on the quality of the GSM-R transmissions:

- presence of public GSM signals only,
- presence of transient signals only,
- presence of public GSM and transient signals simultaneously.

The aim is, in a first step, to observe and quantify the impact of each type of interference separately and in different conditions of test (different power levels for permanent interferences, different time intervals for transient interferences…). In a second step, the combined effect of the two types of disturbances is assessed.

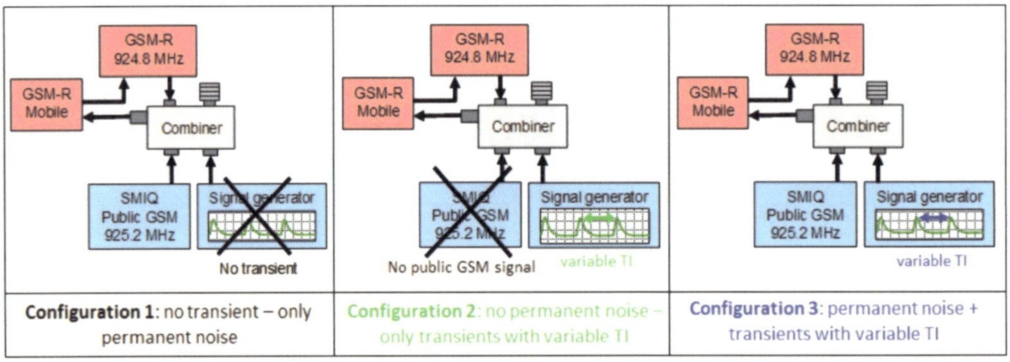

Fig. 10. The different configurations used for immunity testing

9.2 Impact of public GSM signals

The configuration of test corresponds to "configuration 1" from Fig. 10. As previously explained in paragraph 8.4, the GSM-R transmissions takes place on the channel 924.8 MHz with a power level of -70 dBm and the public GSM ones on the channel 925.2 MHz with a variable level from -72 to -12 dBm. Fig. 11 presents the results of the BER measurements as a function of the power level of public GSM signals. The vertical axis gives the value of the BER in % and the horizontal axis represents the power level of the public GSM band signal (on the channel 925.2 MHz) which induces the permanent noise on the GSM-R channel.

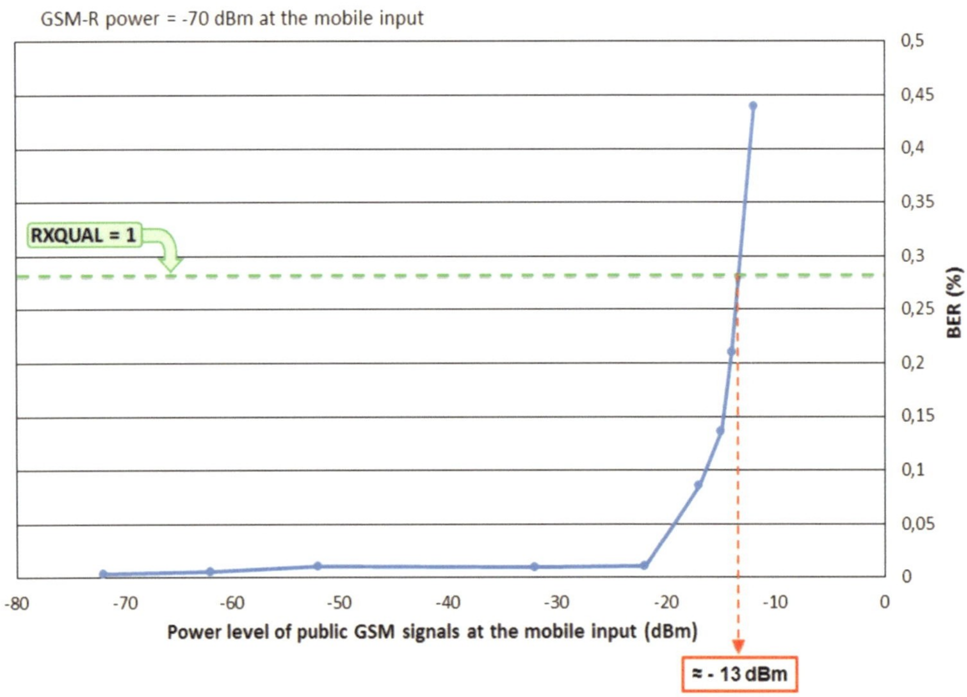

Fig. 11. Results of the BER measurement in the presence of public GSM signals

In Fig. 11, the public GSM signal has to exceed -20 dBm to start affecting the quality of the GSM-R communication (the BER starts to increase). That means that the interference signals on the 400 kHz adjacent channel have to be 50 dB higher than the wanted signal on the GSM-R communication channel to deteriorate the quality of the transmission, which well complies with the specifications (ETSI, 2000). Indeed, the standard EN 300 910 stipulates that a mobile has to tolerate a 400 kHz adjacent interference level of -50 dB.

Then, we also notice that a level of -13 dBm is necessary to induce a Rxqual equal to 1. It will be highlighted later that this level is different when transient interferences are simultaneously present.

9.3 Impact of transient EM interferences produced by catenary-pantograph sliding contact

These tests and measurements are related to "configuration 2" in Fig. 10. The GSM-R signal can be set to the desired value and the interference level produced by transient signals on the GSM-R frequency band can be controlled by using a variable attenuator in order to obtain the desired signal-to-noise ratio (SNR) at the mobile input. As for the measurements, during a test sequence we vary the time interval between two consecutive transients and one measure of BER is made for each chosen time interval. Then, the same test sequence is applied with one other signal-to-noise ratio. Three different SNR at the mobile input are tested: +5, 0 and -5 dB. The results are presented in Fig. 12 where the vertical axis of the graph corresponds to the value of the BER in % and the horizontal axis gives the time interval between two successive transients in μs.

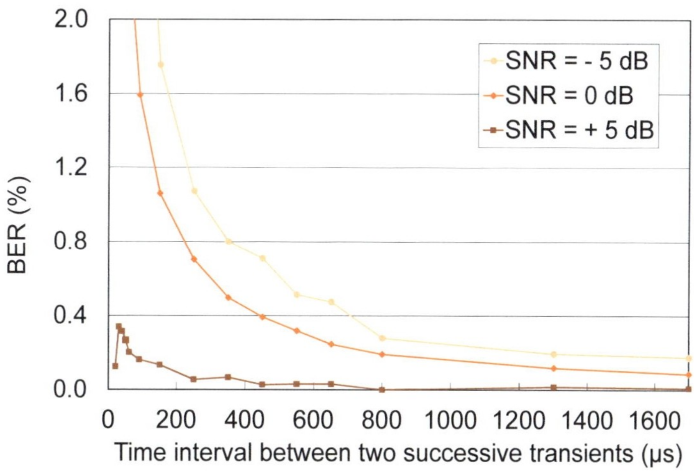

Fig. 12. Results of the BER measurement in the presence of transients for different values of the signal-to-noise ratio (SNR)

The first thing to notice is that the BER evolves with the time interval between transients: it increases with the recurrence of transients. Indeed, the BER is higher for small values of time interval whatever the value of the SNR. In (Adriano, 2008), a relation was proposed to estimate the BER from the TI between the transient interferences, under the assumption that the SNR is equal to 0 dB.

The second thing to observe is that the SNR has an impact on the BER. Indeed, if taking the curve obtained for SNR = 0 dB (at the mobile input) as a reference, we see that, when the transient level is 5 dB higher than the GSM-R signal (SNR=-5 dB), the measured BER increases. Consequently, the transmission could be more severely disturbed when the SNR decreases to -5 dB whereas, in the reverse case (SNR=+5 dB), the BER is lower (less than 0.4 %) which guarantees a good quality of communication whatever the recurrence of the transient interferences.

9.4 Tests and measurements in the presence of both types of interference signals simultaneously

In this section, we now consider "configuration 3" in Fig. 10: presence of permanent noise and transients simultaneously with two arbitrarily chosen values for the transient time interval which are TI=150 µs and TI=550 µs. The following graph, on the right of Fig. 13, shows the results of the BER measurements in this configuration of test. The first curve (black one with points) corresponds to the evolution of the BER without transient and the two others (orange with squares and blue with triangles ones) with transients for the two considered values of time interval. These values were chosen so that 3 transients can occur during the time duration of one GSM-R burst in the first case (TI=150 µs) and only one in the second case (TI=550 µs), as can be seen in the illustration on the left of Fig. 13.

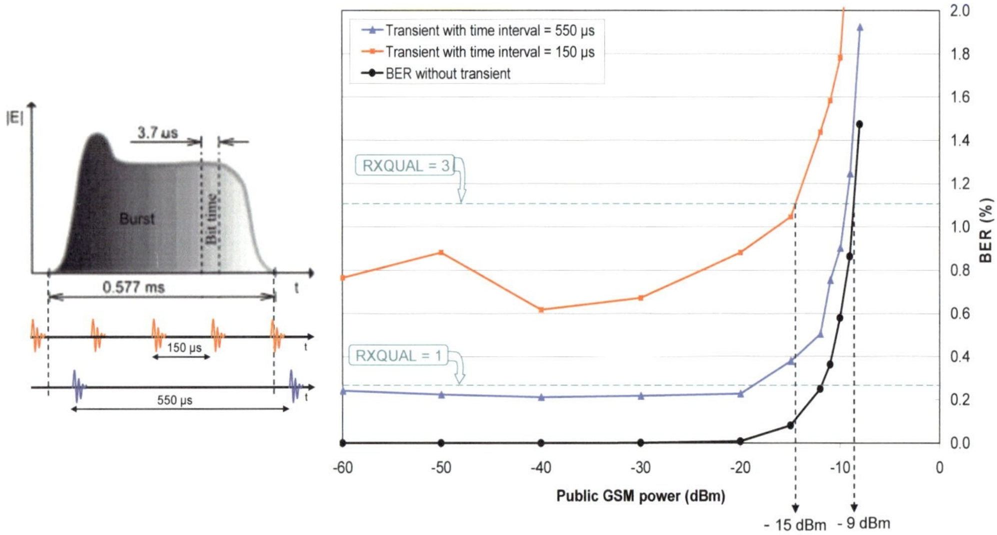

Fig. 13. Results of the BER measurements in the presence of public GSM signals and transient signals with GSM-R signal power = -70 dBm at the mobile input

In the absence of transient signals (black curve with points), the public GSM signals have to reach a power level of -9 dBm to induce a Rxqual equal to 3 whereas in the presence of transient disturbances with a time interval of 150 µs, a level of -15 dBm is sufficient. In other words, the impact on the GSM-R communication of the transient disturbances "adds" to the one of signals in the public GSM band. We thus conclude that the susceptibility of the GSM-R to permanent noise is higher in the presence of transient disturbances.

Obviously, these results are linked to the GSM-R signal power used for the test (-70 dBm) and we would obtain a better level of immunity if setting up the GSM-R signal to a higher level of power. However, we are not going to develop this point in this chapter, since further studies on the immunity of the GSM-R system can be found in (Dudoyer et al., to be published).

10. Conclusion

This chapter outlined the major developments underway on the European rail network and highlighted the electromagnetic vulnerability of the GSM-R which is a key component of the management system. Indeed, immunity testing carried out in laboratory to confront the GSM-R transmissions to EM disturbances representative to those measured on trains, have shown that the quality of the transmissions can be significantly affected. The results of the section 9 highlighted that the impact of the transient disturbances on the quality of the GSM-R transmissions is linked to two main factors: the levels of noise produced on the GSM-R down-link frequency band and the repetition rate of the transient disturbances. Moreover, their impact can also be related to the presence of permanent interferences with the GSM public. Consequently, the assess of the risks of disturbances of the GSM-R transmissions requires to monitor the spectral distribution of the EM noise over the time, and with a high temporal resolution which permits us to perform direct comparison with the transmission of the digital data.

The current European standard methodologies of measurement of the EM emissions in the railway domain (EN 50121, 2006) which only consist in spectral analysis of the radiated emissions without taking into account the time dimension are then not adapted to the control of the EM emissions in order to protect the GSM-R transmissions. This chapter which proposed a methodology to perform immunity testing of GSM-R transmissions in laboratory, has also highlighted the main features of the EM noise it is necessary to characterize on board trains to ensure that the radiated emissions will not affect the ability of the GSM-R system.

11. Acknowledgment

The authors of this chapter would like to thanks SNCB and SNCF to have given them access to their trains to perform measurements in real conditions and also ALSTOM which provided them specific railway equipment. This work was performed in the framework of the RAILCOM project supported by the PCRD 6 and CISIT projects supported by the North Region and the FEDER.

12. References

Adriano, R., Ben Slimen, N., Deniau, V., Berbineau, M. & Massy, P. (2008). Prediction of the BER on the GSM-R communications provided by the EM transient disturbances in the railway environment, *Proceeding of EMC Europe*, pp. 771-775, Hamburg, Germany, September 2008.

Ben Slimen, N., Deniau, V., Rioult, J., Dudoyer S. & Baranowski S. (2009). Statistical characterisation of the EM interferences acting on GSM-R antennas fixed above moving train. *The European Physical Journal Applied Physics,* Vol. 48, No.2, (2009), ISSN 1286-0042

Breed, G. (2003). Bit Error Rate: Fundamental Concepts and Measurement Issues. *High Frequency Electronics,* Vol.2, No.1, (January 2003), pp. 46-48

Dudoyer, S., Deniau, V., Adriano, R., Ben Slimen, M. N., Rioult, J., Meyniel, B. & Berbineau, M. (to be published). Study of the Susceptibility of the GSM-R Communications to the Electromagnetic Interferences of the Rail Environment, *accepted for publication in IEEE Transactions on electromagnetic compatibility.*

EN 50121. (2006). European standards for Railway applications: Electromagnetic Compatibility-Part 1-5

ETSI EN 300 910 V8.5.1 (2000). European Standard on Digital cellular telecommunication system (Phase 2+); Radio transmission and reception, November 2000.

GM/RC 1500, (1994). Code of Practice for EMC between the Railway and its Neighbourhood, December 1994

Hammi, T., Ben Slimen, N., Deniau, V., Rioult, J. & Dudoyer, S. (2009). Comparison between GSM-R coverage level and EM noise level in railway environment, *Proceedings of the 9th International Conference on Intelligent Transport System-Telecommunications, ITST 2009,* pp. 123-128, Lille, France, October 2009

IEC 60050-161. (1990). International Electrotechnical Vocabulary. Chapter 161: Electromagnetic compatibility, Edition: 1.0, International Electrotechnical Commission, 1990

ITU-T Recommendation K.48 (2003). EMC requirements for each telecommunication equipment - Product family Recommendation, July 2003.

Jarašūnienė, A. (2005). General Description of European Railway Traffic Management System (ERTMS) and Strategy of ERTMS implementation in Various Railway Managements, *Transport and Telecommunication,* Vol.6, No.5, (2005), pp. 21-27, ISSN 1407-6160

Knobloch, A. & Garbe, H. (2002). Critical Review of Converting Spectral Data into Prospective Bit Error Rates, *Proceeding of IEEE International Symposium on Electromagnetic Compatibility, Volume 1,* pp. 173-178, ISBN 0-7803-7264-6, Minneapolis, MN, USA, August 2002

Konefal, T., Pearce, D.A.J., Marshman, C.A. & McCormack, L.M., (2002). Potential electromagnetic interference to radio services from railways, *Final Report, University of York,* 2002, Retrieved from <http://www.yorkemc.co.uk/research/railways/>.

Lagrange, X., Godlewski, P. & Tabbane, S. (1996). GSM-DCS Networks, Hermès, Paris, 1996, pp. 207-208, ISBN 2-86601-558-4

Mansson, D., Thottappillil, R., Bäckström, M & Lundén, O. (2008). Vulnerability of European Rail Traffic Management System to Radiated Intentional EMI. *IEEE Transactions on Electromagnetic Compatibility,* Vol.50, No.1, (2008), pp. 101-109, ISSN 0018-9375

Midya, S. & Thottappillil, R. (2008). An Overview of Electromagnetic Compatibility Challenges in European Rail Traffic Management System, *Journal of Transportation Research Part C: Emerging Technologies,* Elsevier, (Ed.), Vol.16C, Issue.5, (2008), pp. 515-534, ISSN 0968090X

UIC Project EIRENE (2003). System Requirements Specification, v14.0, 21 October 2003. Source : EIRENE Project Team